Metal-Containing and Metallosupramolecular Polymers and Materials

ACS SYMPOSIUM SERIES **928**

Metal-Containing and Metallosupramolecular Polymers and Materials

Ulrich S. Schubert, Editor
Eindhoven University of Technology and Dutch Polymer Institute

George R. Newkome, Editor
The University of Akron

Ian Manners, Editor
University of Bristol

**Sponsored by the
ACS Divisions of Polymer Chemistry, Inc.,
Inorganic Chemistry, Inc., and
Polymeric Materials: Science and Engineering, Inc.**

American Chemical Society, Washington, DC

Library of Congress Cataloging-in-Publication Data

Metal-conntaining and metallosupramolecular polymers and materials / Ulrich S. Schubert, editor, George R. Newkome, editor, Ian Manners, editor ; sponsored by the ACS Divisions of Polymer Chemistry, Inc., Inorganic Chemistry , Inc., and Polymeric Materials: Science and Engineering, Inc.

    p. cm.—(ACS symposium series ; 928)

    Includes bibliographical references and indexes.

    ISBN 13: 978–0–8412–3929–6 (alk. paper)

    1. Organic conductors—Congresses. 2. Polymers—Congresses. 3. Supramolecular chemistry—Congresses

    I. Schubert, U. (Ulrich) II. Newkome, George R. (George Richard) III. Manners, Ian, 1961- IV. American Chemical Society. Division of Polymer Chemistry. V. American Chemical Society. Division of Inorganic Chemistry. VI. American Chemical Society. Division of Polymeric Materials: Science and Engineering. VII. Series.

QD382.C66M48   2006
547′.70457—dc22   2005055592

The paper used in this publication meets the minimum requirements of American National S tandard for Information Sciences—Permanence of Paper for Printed Library Materials, ANSI Z39.48–1984.

Distributed by Oxford University Press

ISBN 10: 0–8412–3929–0

PRINTED IN THE UNITED STATES OF AMERICA

Foreword

The ACS Symposium Series was first published in 1974 to provide a mechanism for publishing symposia quickly in book form. The purpose of the series is to publish timely, comprehensive books developed from ACS sponsored symposia based on current scientific research. Occasionally, books are developed from symposia sponsored by other organizations when the topic is of keen interest to the chemistry audience.

Before agreeing to publish a book, the proposed table of contents is reviewed for appropriate and comprehensive coverage and for interest to the audience. Some papers may be excluded to better focus the book; others may be added to provide comprehensiveness. When appropriate, overview or introductory chapters are added. Drafts of chapters are peer-reviewed prior to final acceptance or rejection, and manuscripts are prepared in camera-ready format.

As a rule, only original research papers and original review papers are included in the volumes. Verbatim reproductions of previously published papers are not accepted.

ACS Books Department

Contents

Metallosupramolecular Polymers

Metallodendrimers

Organometallic Polymers, Materials, and Nanoparticles: Ferrocene-Containing Systems

Organometallic Polymers, Materials, and Nanoparticles: Non-Ferrocene-Containing Systems

Preface

Introduction

The valuable physical and chemical properties and functions of many solid-state and biological materials can be attributed to the presence of metallic elements. Examples include magnetic materials used in data storage, superconductors, electrochromic materials, and catalysts including metalloenzymes. It has long been recognized that incorporation of metal atoms into synthetic polymer chains may also lead to desirable properties and thereby generate a new and versatile class of functional materials capable of enhanced processability. However, until recently, synthetic difficulties—associated with the creation of macromolecular chains possessing metal atoms as a key structural component—have been deterrents to progress in the field. During the past decade or so, these synthetic obstacles have been, in part, overcome through many creative procedures to prepare new materials. The approaches now available have led to macromolecular structures in which metals are not only incorporated via the use of traditional covalent bonds but also by potentially reversible coordination interactions. This has led to a remarkable series of "metallosupramolecular" polymers that lie conceptually at the interface of traditional polymers, coordination chemistry, and supramolecular chemistry. In addition, structures radically different from traditional linear polymer architectures, such as stars and dendrimers, have been established. Of comparable importance to these synthetic developments, these new nanomaterials have shown that they can possess a diverse range of interesting and useful properties and potential utilitarian applications that complement those of more traditional organic macromolecules.

Historical Development of Metal-Containing and Metallosupramolecular Polymers

To put the current state of the subject in context, it is both interesting and informative to briefly reflect on the historical development of the field of metal-containing and metallosupramolecular polymers. Almost 50 years ago, the first soluble metal-containing polymer, poly(vinyl ferrocene) (**1**), was prepared by radical polymerization (*1*). With the growing interest in new polymeric materials possessing novel properties, the late 1960s and early 1970s was a dynamic time in the expansion of metal-containing polymer science. However, very few new, well-characterized, soluble, high-molecular weight materials were actually reported during this period. The first well-characterized polymer of appreciable molecular weight with main-chain metal atoms was polyferrocene–siloxanes (**2**) that was prepared by Pittmann et al. in 1974 via a polycondensation strategy (*2*). In the late 1970s, noteworthy work by Neuse and Bednarik led to the well-characterized, but still rather low molecular weight, polyferrocenylenes (**3**) (*3*), after which the initial reports of the important class of rigid-rod polymetallayne polymers containing Pd and Pt (**4**) were made by Hagihara and co-workers. (*4*)

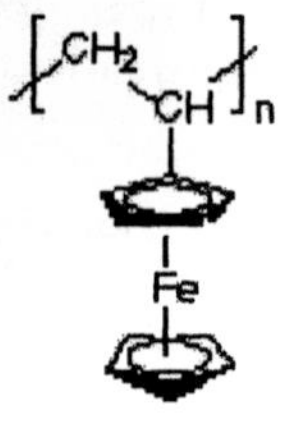

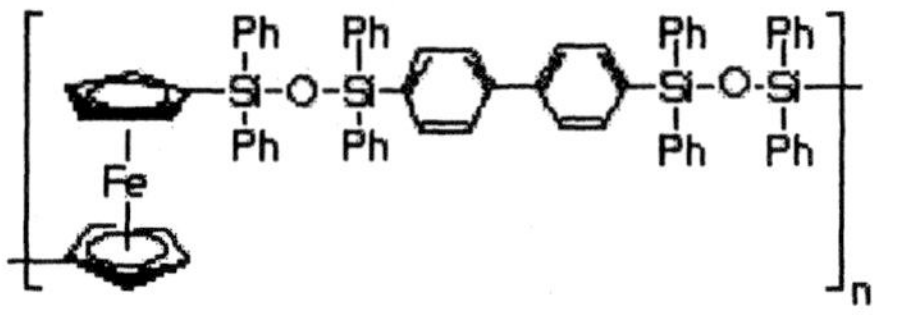

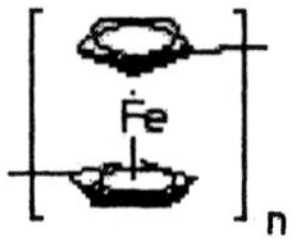

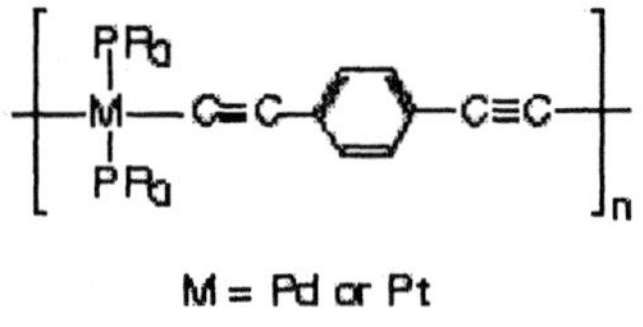

Figure 1: Selected structures of early metal-containing polymers.

In the early 1990s, many significant developments in the area of metal-containing polymers occurred as a consequence of several synthetic breakthroughs as well as of improved analytical and spectral instrumentation coupled with appropriate software to add the critical quantitative aspects, so necessary, for the advent of the nanoregime. The first examples of star and dendritic materials (e.g., **5**) containing metal atoms were described (*5*). Then, ring-opening polymerization (ROP) and ROP-related processes were developed to provide convenient access to high-molecular weight polymetallocenes, such as polyferrocenylsilanes (**6**) and analogs with disulfido spacers (**7**) (*6,7*). Homopolymers and block copolymers with metal-containing side groups (e.g., **8**) were also made available by the technique of metal-catalyzed, ring-opening metathesis polymerization (ROMP) (*8*). Transition metal-catalyzed polycondensation strategies were also utilized to prepare the first polystannanes (**9**), with main chains consisting of δ-conjugated tin atoms, and metallacyclic and cyclobutadienyl cobalt polymers (**10**) (*9–11*).

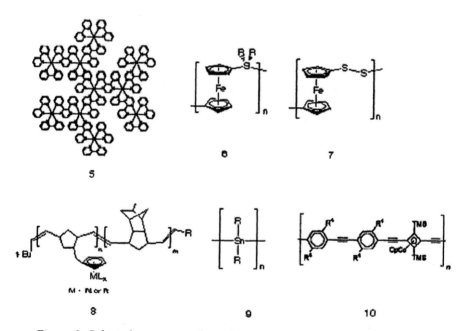

Figure 2: Selected structures of metal-containing polymers discovered after 1990.

By the mid-1990s, routes had been developed to create and characterize soluble coordination or "metallosupramolecular" polymers (e.g., **11**) that contain a variety of d-block elements or lanthanide metals bound by N- and O-donor ligands (*12,13*). The first block copolymers containing metals in the main-chain, such as polymetallocene, di- and triblock copolymers (e.g., **12**) and materials with metallosupramolecular linkers (e.g., **13**) were described (*14,15*).

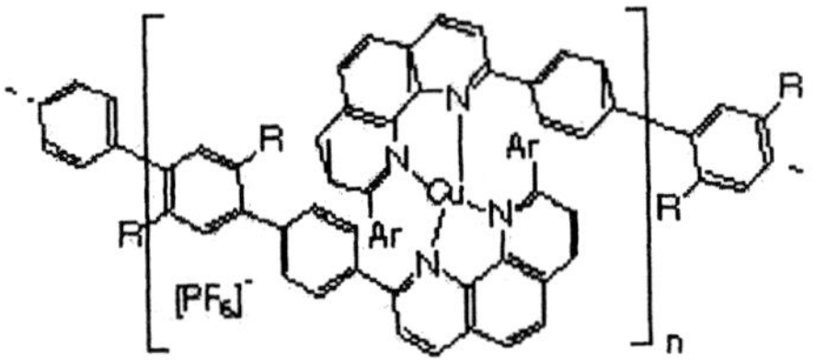

11

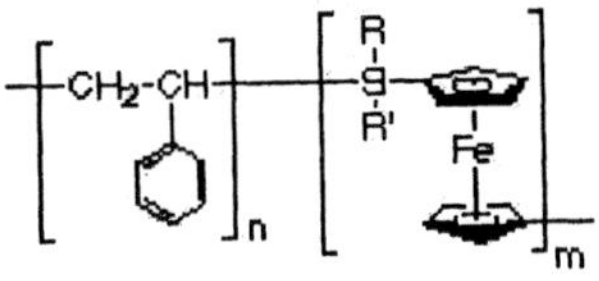

PS-b-PFS

12

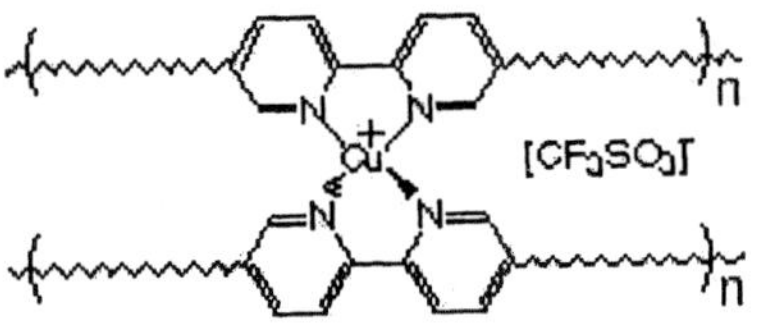

13

Figure 3: Modern examples of metal-containing macromolecules.

In the mid-to-late 1990s, an exciting development involved the creation of metalloinitiators for controlled polymerization reactions that have considerable synthetic potential for the preparation of star architectures (e.g., **14**) (*16*). Impressive routes to materials appeared (e.g., **15**) in which metal atoms were directly placed in the main chains of heteroaromatic π-conjugated polymeric frameworks (*17*). Self-assembled and hierarchical structures based on metal-containing polymers, such as liquid crystals, nanostructured self-assembled block copolymer micelles and thin films, and electrostatic superlattices are also starting to attract intense attention and are rapidly expanding into the 21st century (*18*).

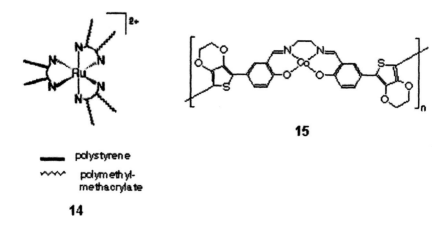

15

▬▬ polystyrene

⌇⌇⌇ polymethyl-
methacrylate

14

Figure 4: Recent metal-containing macromolecules.

Properties and Applications

The synthetic developments in the 1990s in numerous laboratories worldwide effectively reenergized the field; however, equally important to the current surge in interest and reestablished potential of this exciting area is the now growing attention from a variety of research groups who were interested in detailed studies of the physical properties and applications of these new nano-materials. This has led to applications as conductive or photoconductive materials, electrode mediators in sensors for biomolecules, phosphorescent sensors, redox-responsive materials, supramolecular gels, liquid crystalline materials, nanostructured etch resists and catalysts, as well as precursors to metal nanoparticles with useful magnetic or catalytic properties.

This Volume

The highly active state of tne metal-containing and metallosupramolecular polymer field is reflected by the numerous new books and reviews published in the past five years or so (*17–26*).

This symposium series contains contributions by authors working in most of the rapidly developing areas of the field. The volume was triggered by a highly successful American Chemical Society (ACS) symposium in Spring 2004 during the ACS National Meeting in Anaheim, California. Approximately 100 presentations clearly showed the vitality and appeal of the work in this field. We hope this selection of contributions will further stimulate the process in the field of metal-containing and metallosupramolecular polymers and materials.

We would be remiss not to thank the multitude of contributors —as well as friends, colleagues, and scientists worldwide—who have sparked the imagination and intellectual spirit to move this exciting field into this millennium. Most assuredly, the next decade will be an exciting time as supramacromolecular science and the nanoworld collide.

References

1. F. S. Arimoto, A. C. Haven, *J. Am. Chem. Soc.* **1955**, *77*, 6295.
2. W. J. Patterson, S. P. McManus, C. U. Pittman, Jr., *J. Polym. Sci. Polym. Chem.* **1974**, *12*, 837.
3. E. W. Neuse, L. Bednarik, *Macromolecules* **1979**, *12*, 187.
4. K. Sonogashira, S. Takahashi, N. Hagihara, *Macromolecules* **1977**, *10*, 879.
5. (a) G. Denti. S. Campagna, S. Serroni, M Ciano, V. Balzani *J. Am. Chem. Soc.* **1992**, *114*, 2944. (b) G. R. Newkome, F. Cuardullo, E. C. Constable, C. N. Moorefield, A. M. W. C. Thompson, *J. Chem. Soc. Chem. Commun.* **1993**, 925. (c) B. Alonso, I. Cuadrado, M. Noran, J. Losada *J. Chem. Soc. Chem. Commun.* **1994**, *2575*.
6. D. A. Foucher, B. Z. Tang, I. Manners, *J. Am. Chem. Soc.* **1992**, *114*, 6246.
7. P. F. Brandt, T. B. Rauchfuss, *J. Am. Chem. Soc.* **1992**, *114*, 1926.
8. V. Sankaran, J. Yue, R. E. Cohen, R. R. Schrock, R. J. Silbey, *Chem. Mater.* **1993**, *5*, 1133.
9. T. Imori, V. Lu, H. Cai, T. D. Tilley, *J. Am. Chem. Soc.* **1995**, *117*, 9931.
10. I. L. Rozhanskii, I. Tomita, T. Endo *Macromolecules*, **1996**, *29*, 1934.
11. M. Altmann, U. H. F. Bunz, *Angew. Chem. Int. Ed Engl.* **1995**, *43*, 569.
12. R. D. Archer, *Coord. Chem. Rev.* **1993**, *128*, 49.
13. S. Kelch, M. Rehahn *Macromolecules*, **1997**, *30*, 6185.
14. C. D. Eisenbach, A. Goldel, M. Terskan-Reinold, U. S. Schubert *Macromol. Chem. Phys.* **1995**, *196*, 1077.
15. Y. Z. Ni, R. Rulkens, I. Manners, *J. Am. Chem. Soc.* **1996**, *118*, 4102.
16. (a) X. Wu, J. E. Collins, J. E. McAlvin, R. W. Cutts, C. L. Fraser,

Macromolecules **2001**, *34*, 2812; (b) G. Hochwimmer, O. Nuyken, U. S. Schubert, *Macromol. Rapid Commun.* **1998**, *19*, 309.

17. Review: R. P. Kingsborough, T. M. Swager, *Prog. Inorg. Chem.* **1999**, 48, 123.
18. Review: I. Manners *Science* **2001**, *294*, 1664.
19. Review: G. R. Newkome, E. He, C. N. Moorefield, *Chem. Rev.* **1999**, *99*, 1689.
20. Review: P. Nguyen, P. Gómez-Elipe, I. Manners, *Chem. Rev.* **1999**, *99*, 1515.
21. Reviews: U. S. Schubert, C. D. Eschbaumer, *Angew. Chem. Int. Ed.* **2002**, *41*, 2926; H. Hofmeier, U. S. Schubert, *Chem. Soc. Rev.* **2004**, *33*, 373; P. R. Andres, U. S. Schubert, *Adv. Mater.* **2004**, *16*, 1043; H. Hofmeier, U. S. Schubert, *Chem. Commun.* **2005**, 2423.
22. Review: T. M. Swager, *Chem. Comm.* **2005**, *23*.
23. Symposium book: A. S. Abd-El-Aziz (Ed.), *Macromol. Symp.* **2003**, *196*, 1-353.
24. R. D. Archer, *Inorganic and Organometallic Polymers:* Wiley VCH: Weinheim, 2001.
25. I. Manners *Synthetic Metal-Containing Polymers* Wiley-VCH: Weinheim, 2004.
26. V. Chandrasekhar *Inorganic and Organometallic Polymers:* Springer: New York, 2005.

Metal-Containing and Metallosupramolecular Polymers and Materials

Metallosupramolecular Polymers

Chapter 1

Dendronized Copper(I)-Metallopolymers

Julia Kubasch and Matthias Rehahn[*]

Ernst-Berl-Institute for Chemical Engineering and Macromolecular Science, Darmstadt University of Technology, Petersenstrasse 22, D–64287 Darmstadt, Germany

Most metallopolymers based on kinetically labile transition-metal complexes are polyelectrolytes and dissolve in polar, coordinating solvents only. In these solutions, solvent molecules can displace the metallopolymers' original ligands and thus cause decomposition. Consequently, such systems are very difficult to characterize. Our adopted strategy to overcome this problem consists of the attachment of apolar substituents which solubilize the metal-containing polymers in non-coordinating solvents. Then, even intrinsically labile chains should behave like inert systems. Based on this strategy, we developed an efficient synthetic access to dendronized metallopolymers. The dendrons were attached very closely to the tetrahedral copper(I) complexes. Positioned in this way, they also increase the materials' solubility in non-coordinating solvents. Moreover, due to their bulkiness, the dendrons should reduce the tendency of the metal-containing polymers to aggregate in solution, and protect the kinetically labile complexes against the attack of competing ligand molecules.

Background

In recent years, a variety of fascinating supramolecular architectures were developed.[1,2] Many of them are based on transition-metal complexes. In addition to well-defined oligonuclear assemblies like helicates, catenanes, dendrimers or grids, some high-molecular-weight, linear-chain species could be created – the so-called "metallopolymers". These latter compounds bear transition-metal complexes either as lateral substituents of a classic covalent polymer mainchain ("complex polymers" **A**), or as an integral part within their polymer backbones ("coordination polymers" **B**).[3]

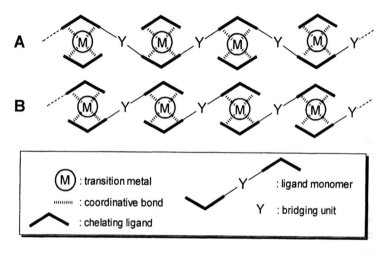

Figure 1: Schematics of (A) complex- and (B) coordination polymers

Metallopolymers require – if characterization is desired not only in the solid state but also in solution – thermodynamically very stable metal complexes. Moreover, kinetically inert complexes are advantageous because then the choice of solvent for polymer characterization is almost unlimited: ligand-exchange processes occurring between metallopolymer and a coordinating solvent can be excluded. This is the main reason why most soluble, well-defined and high-molecular-weight metallopolymers known today contain *kinetically inert* complexes. In contrast to this, most kinetically labile systems decompose simultaneously with their dissolution. Characterization is therefore possible only in the solid state, and important molecular parameters like degree of polymerization, chain conformation, and backbone flexibility etc. are not accessible. Consequently, well-defined metallopolymers from *kinetically labile* transition-metal complexes are almost unknown, and so far, only very few systems based on copper(I) and silver(I) were prepared and characterized in solution.[4-11]

Recently, we outlined a number of requirements whose fulfillment is essential for obtaining readily soluble metallopolymers from kinetically labile complexes.[6-11] The key assumption was that kinetically labile complex- and coordination polymers decompose exclusively via the displacement of their original ligand moieties by coordinating solvent molecules. Even if the latter molecules are not chelating ligands, they nevertheless compete successfully for the metal ions because they are present in a very large excess. To avoid this decomposition, steps may be taken to make the metallopolymers soluble also in strictly non-coordinating media: here, kinetically labile multinuclear complexes should behave like "true" polymers. Unfortunately, most metallopolymers are polyelectrolytes and prefer highly polar, coordinating solvents. An appropriate measure to increase solubility in less polar solvents is attaching apolar alkyl side chains to the metallopolmers. Figure 2 displays two polymers developed recently based on these considerations.[6-11]

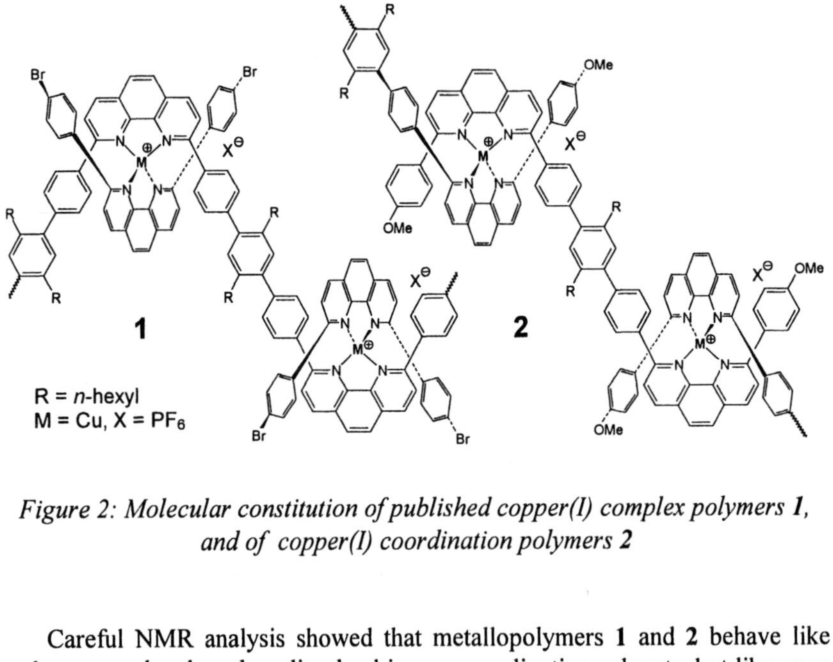

Figure 2: Molecular constitution of published copper(I) complex polymers 1, and of copper(I) coordination polymers 2

Careful NMR analysis showed that metallopolymers **1** and **2** behave like real macromolecules when dissolved in non-coordinating solvents, but like open solution aggregates in the presence of even very small amounts of coordinating molecules. Unfortunately, further characterization in non-coordinating media via, for example, viscosimetry and light scattering was still affected by aggregation. Therefore, the effect of the solubilizing substituents had to be intensified: additional substituents had to be introduced into the metallopolymers which (*i*) improve solubility even more efficiently, (*ii*) reduce the aggregation tendency of the ionic complexes, (*iii*) protect the labile complexes via steric shielding[12] but (*iv*) do not affect the complex formation process itself.

Objective

It is the objective of this contribution to present a further step towards soluble, well-defined metallopolymers based on kinetically labile copper(I) complexes. We attach sterically demanding dendrons close to the metallopolymers' coordinative centers. The synthesis of the dendrons is described as well as their introduction into the kinetically labile copper(I)-based metallopolymers **1** and **2**.

Experimental

NMR spectra were recorded on a BRUKER DRX 500 spectrometer (500 MHz for ^{1}H and 125 MHz for ^{13}C). Acetone-d_6 and 1,1,2,2-tetrachloroethane-d_2 were used as the solvents. Signal assignment was done based on gs-COSYDF, NOESY, gs-HSQC, gs-HMBC and DEPT measurements and is given according to the numbering shown in Schemes 2 and 3. Mass spectra were recorded on a VARIAN MAT 311 A and a VARIAN MAT 212 mass spectrometer. Ionization was done by field ionization (FI), field desorption (FD), fast atom bombardment (FAB) and electrospray ionization (ESI) technique, respectively. MALDI-TOF mass spectra were recorded on a KRATOS ANALYTICAL Kompact MALDI 4 mass spectrometer. Dithranol was used as the matrix, and LiCl or CuCl as salt.

All chemicals were from FLUKA, ALDRICH and ACROS. *o*-Phenanthroline derivatives **5[OH]** and **5[OCH₃]**,[13] ligand monomer **6[OCH₃]**,[6,7] precursor polymer **1** and metallopolymer **1[L₂]**[8] were prepared according to the literature. Dendrons $G_x(1{\rightarrow}2)$ were prepared in generations x = 1 – 3 according to the literature.[14] Dendrons $G_x(1{\rightarrow}3)$ were prepared according to the procedure given here for $G_1(1{\rightarrow}3)$ (Scheme 2). All reactions were carried out under nitrogen.

$G_1(1{\rightarrow}3)$COOCH₃: Compound **4** (5.00 g, 27.15 mmol), 18-crown-6 (2.87 g, 10.86 mmol), K₂CO₃ (15.01 g, 108.61 mmol) and dry acetone (500 mL) were stirred and refluxed. Benzyl bromide (11.29 mL, 16.25 g, 95.03 mmol) was added slowly, and refluxing is continued for 2 d. The mixture is cooled down, and the solvent is removed. The residue is dissolved in CH₂Cl₂, washed with water and dried (MgSO₄). After evaporation of the solvent, the residue is crystallized from acetone. The yield was 12.09 g (26.61 mmol, 98%).

$G_1(1{\rightarrow}3)$OH: Compound **$G_1(1{\rightarrow}3)$COOCH₃** (10.00 g, 22.00 mmol), LiBH₄ (0.96 g, 44.00 mmol) and dry THF (250 mL) were stirred for 1 h at 0 °C, for 1 h at room temperature, and for 2 d under reflux. At room temperature, saturated aqueous NaCl (150 mL) and *tert*-butyl methyl ether (150 mL) were added. The organic layer was separated, washed (4 x saturated NaCl, 1 x water) and dried (MgSO$_4$). The solvent was removed, and the pure product was obtained in yields of 9.10 g (97%).

G$_1$(1→3)Br: Compound **G$_1$(1→3)OH** (5.00 g, 11.72 mmol) and tetrabromomethane (11.66 g, 35.16 mmol) were dissolved in THF (40 mL). At 0 °C, triphenyl phosphine (9.22 g, 35.16 mmol) in THF (40 mL) was added. The mixture was stirred for 20 min at room temperature and filtered over a column of silica gel (toluene as the eluent). The oil obtained after removal of the solvent was again purified by column chromatography (silica gel, toluene). A crystalline product was obtained in yields of 3.50 g (61%).

5[G$_1$(1→3)]: Compound **5[OH]** (0.20 g, 0.55 mmol), **G$_1$(1→3)Br** (0.65 g, 1.32 mmol), 18-crown-6 (0.07 g, 0.28 mmol), K$_2$CO$_3$ (0.30 g, 2.20 mmol) and dry DMF (40 mL) were stirred and refluxed for 15 h. After cooling down to room temperature, saturated aqueous NaCl (100 mL) and CH$_2$Cl$_2$ (100 mL) were added. The organic layer is separated off, the aqueous one extrated with CH$_2$Cl$_2$. The combined organic layers were dried (MgSO$_4$). Removal of the solvent and recrystallization of the residue from acetone gave the pure product in yields of 0.43 g (65%).

6[G$_1$(1→3)]: Compound **6[OH]** (0.40 g, 0.43 mmol), **G$_1$(1→3)Br** (0.50 g, 1.02 mmol), 18-crown-6 (0.04 g, 0.15 mmol), K$_2$CO$_3$ (0.25 g, 1.81 mmol) and dry DMF (20 mL) were stirred and refluxed for 3 d. The work-up procedure was as described for **5[G$_1$(1→3)]**, recrystallization was done from chloroform. The yield was 0.19 g (25%).

Model complexes 7[G$_x$(1→2/3)]: 0.003 mmol of the respective *o*-phenanthroline derivative **5**, 0.0015 mmol [Cu(CH$_3$CN)$_4$]PF$_6$ and 1,1,2,2-tetrachloroethane-d$_2$ (0.6 mL) were mixed in an NMR tube. An NMR spectrum is recorded to control the 2:1-equivalence of ligand and metal species. If necessary, adequate amounts of the minor component were added. Precipitation of the pure complexes was achieved by pouring the solution into *n*-hexane.

Complex polymer 1[G$_x$(1→2/3)]: In an NMR tube, poly(2,9-*o*-phenanthroline-*alt*-2′,5′-di-*n*-hexyl-4,4′′-*p*-terphenylene) **1** (13 mg, 0.023 mol-equiv.) was dissolved in 1,1,2,2-tetrachloroethane-d$_2$ (0.5 mL). [Cu(CH$_3$CN)$_4$]PF$_6$ (8.4 mg, 0.023 mmol) was added. After shaking, a suspension of the respective phenanthroline derivative **5[G$_x$(1→2/3)]** (0.023 mmol) in 1,1,2,2-tetrachloroethane-d$_2$ (0.1 mL) was added. Isolation of the product is possible via pouring the solution into an excess of *n*-hexane.

Coordination polymer 2[G$_x$(1→2/3)]: Equimolar amounts of ligand monomer 6[G$_x$(1→2/3)] and [Cu(CH$_3$CN)$_4$]PF$_6$ (0.011 mmol for experiments in NMR tubes) were polymerized in 1,1,2,2-tetrachloroethane-d$_2$ (0.6 mL). After NMR analysis, the product is isolated via pouring the solution into *n*-hexane.

Results & Discussion

The first step towards dendronized copper(I)-metallopolymers was the preparation of appropriately functionalized bulky groups. Dendrons **G$_x$(1→2)Br** were prepared for x = 1 – 3 according to the literature (Scheme 1).[14]

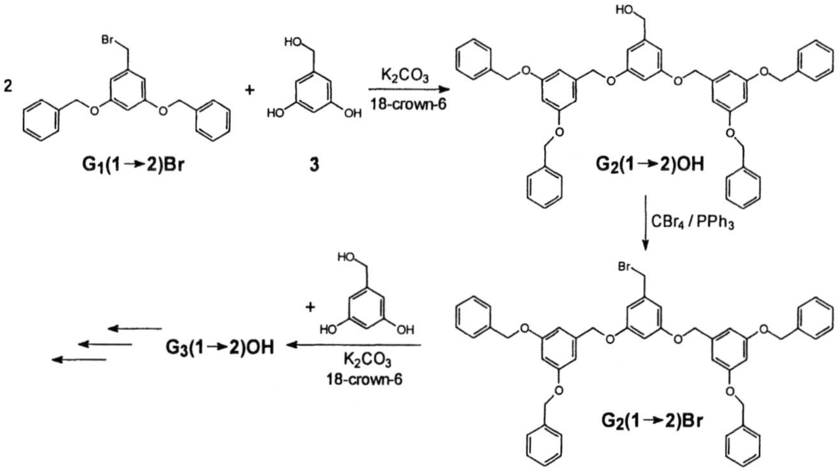

Scheme 1: Synthesis of G$_x$(1→2)Br dendrons

The synthetic protocol includes two steps. The first step is ether formation: two equivalents of dendron **G$_{x-1}$(1→2)Br** are treated with dihydroxybenzyl alcohol **3**. In the second step, the formed **G$_x$(1→2)OH** dendron is then converted into the bromo derivative **G$_x$(1→2)Br** via treatment with CBr$_4$ and PPh$_3$. The reaction cycle is then repeated.

In order to increase the dendrons' steric demand further, a fourfold branching point was introduced in some of the dendrons, leading to the **G$_x$(1→3)** series. The synthesis is shown in Scheme 2. The higher steric demand is achieved by using **4** as the core molecule, and dendrons **G$_x$(1→2)Br** of generation 1 – 3 as bromo counterparts. The conditions used for ether formation were the same as for **G$_x$(1→2)OH**. Subsequently, **G$_x$(1→3)COOCH$_3$** were converted in almost quantitative yields into $G_x(1\to3)OH$ by treatment with $LiBH_4$ in THF.[15,16] Finally, the hydroxyl derivatives were brominated using CBr_4 and PPh_3.

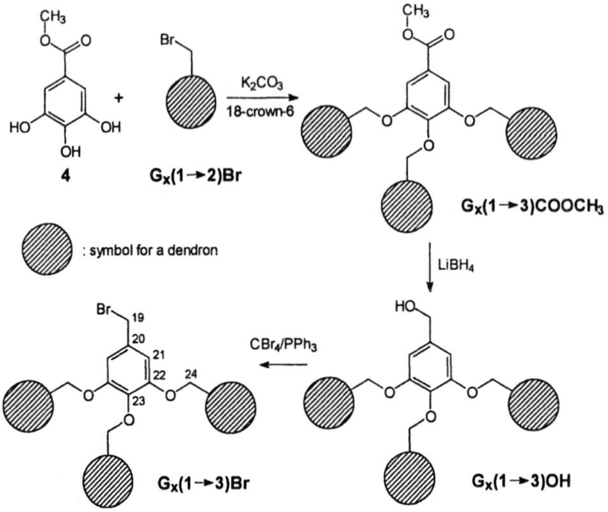

Scheme 2: Synthesis of $G_x(1\to3)$ dendons; the numbering given for $G_1(1\to3)Br$ is used for NMR signal assignment

While yields of 60% could be realized for $G_1(1\to3)Br$ when an improved workup procedure was used, the yields were significantly lower for the higher generations. We assume this is the result of excessive steric demand within the dendron molecules which affects its preparation – at least according to the procedures published in the literature.

Further proof of structure was carried out for all products using ^{1}H and ^{13}C NMR, 2D NMR experiments and mass spectrometry. As a representative example, Figure 3a shows the ^{1}H NMR spectrum of $G_1(1\to3)Br$ together with the signal assignment. As one can see, all observed resonances support to the expected constitution, and the lack of extraneous resonances proves the high purity of the material. Analogous results were obtained for most other dendrons which were used subsequently to derivatize the hydroxyphenyl-functionalized *o*-phenanthroline-based ligands.

Dendronized ligands

Using $G_x(1\to2)Br$ and $G_x(1\to3)Br$, the dendronized auxiliary ligands $5[G_x(1\to2/3)]$ – required to stabilize complex polymer **1** – and the dendronized ligand monomers **6[G$_x$(1→2/3)]** – required for the synthesis of coordination polymers **2** – were prepared. Auxiliary ligands **5** were dendronized according to Scheme 3:

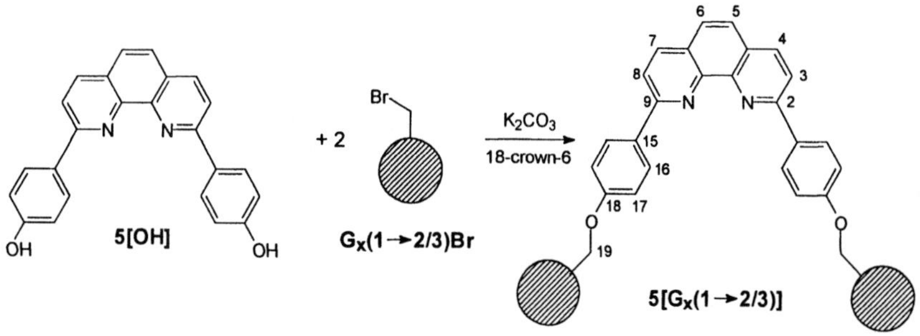

Scheme 3: Synthesis of dendronized o-phenanthrolines 5; the numbering given for 5[G$_x$(1→3)] is used for NMR signal assignment

Because of the low solubility of starting material **5[OH]**, the reactions were carried out in DMF. In the presence of potassium carbonate and 18-crown-6, the desired ligands were formed in yields of 30 – 60 % after careful purification: it was obvious that the rate of the conversions as well as the obtained yields were strongly dependent on the size of the dendron used. All products were readily soluble in a variety of organic solvents. As a representative example, Figure 3b shows the ^{1}H NMR of **5[G$_1$(1→3)]**. Analogously, we dendronized ligand mono-mer **6[OH]** (see Scheme 4 for the **G$_x$(1→3)** series).

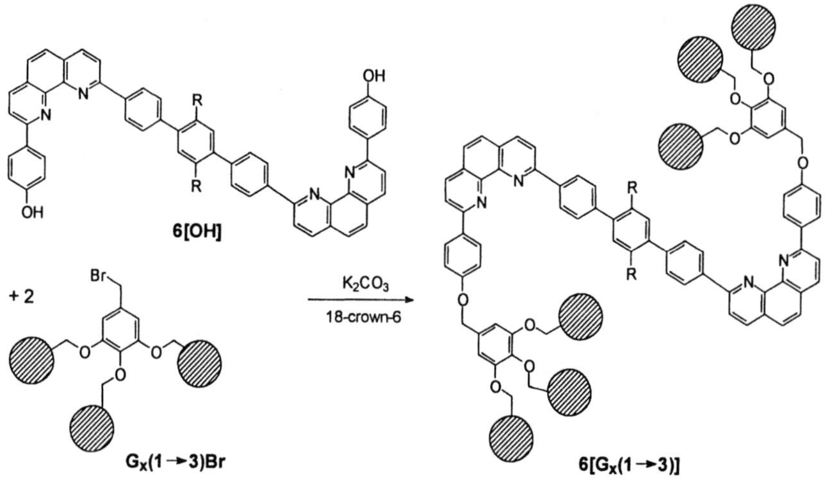

Scheme 4: Synthesis of G$_x$(1→3)-dendronized ligand monomers 6

The hydroxyl-functionalized ligand monomer **6[OH]** required was obtained by treating **6[OCH₃]** with pyridinium hydrochloride for 5 h at 230 °C. The pure, yellow material **6[OH]** was obtained in 95% yield. Subsequently, dendronization leading to **6[Gₓ(1→2/3)]** was possible in approx. 30% yield. Full characterization was done using NMR and MALDI MS techniques. As a representative example, Figure 6a shows the ¹H NMR spectrum of **6[G₁(1→3)]**.

Mononuclear model complexes

In order to analyze the complex formation behavior of the dendronized ligands, systematic series of mononuclear complexes **7** were prepared (Scheme 5). Most of the experiments were carried out in NMR tubes. The evolution of the color and the characteristic chemical shifts in the ¹H NMR spectra allowed estimating the rate and the completeness of the complex formation processes.

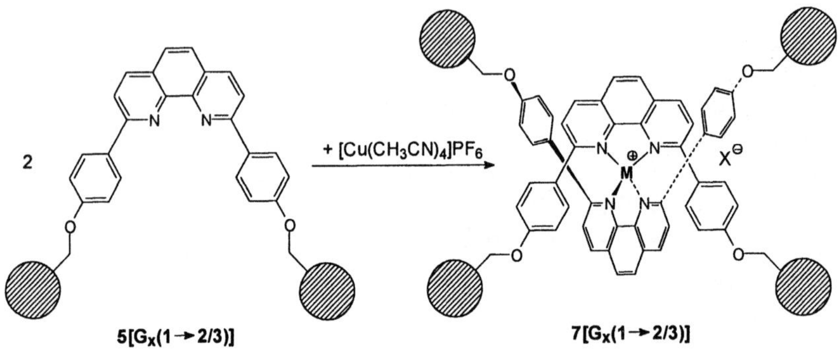

5[Gₓ(1→2/3)]   **7[Gₓ(1→2/3)]**

Scheme 5: Synthesis of dendronized mononuclear model complexes 7

As a representative example, Figure 3b shows the ¹H NMR spectrum of the free ligand **5[G₁(1→3)]**, and Figure 3c shows the corresponding spectrum of the resulting copper(I) complex **7[G₁(1→3)]**. Despite of the considerable steric demand of the dendrons, rapid complex formation was observed in all cases. Obviously, the dendrons are flexible enough to allow for complex formation.

Moreover, we show that ligand exchange is possible also in the dendronized complexes **7[G₁(1→2/3)]**. These studies were carried out by adding a second chelating ligand such as **5[OCH₃]** to solutions of the complexes **7** (Scheme 6) followed by monitoring the changes in the NMR spectra of the mixtures (Figure 4).

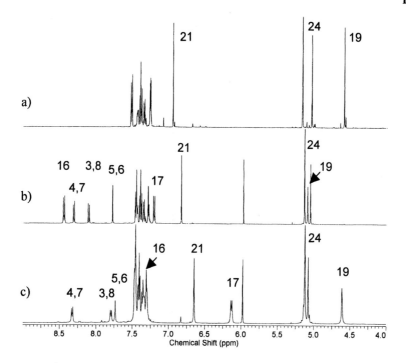

Figure 3: ^{1}H NMR spectra of (a) $G_1(1 \rightarrow 3)Br$ (acetone-d_6), (b) $5[G_1(1 \rightarrow 3)]$ and (c) $7[G_1(1 \rightarrow 3)]$ (1,1,2,2-tetrachloroethane-d_2)

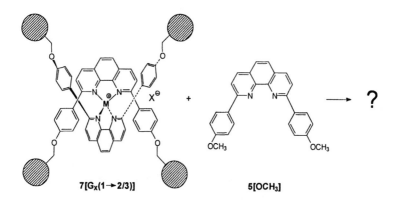

Scheme 6: Ligand exchange experiments on mononuclear complexes 7

The equilibria found in these ligand-exchange experiments allow the conclusion that there is a slight preference of the sterically less demanding ligands to be incorporated in the complexes.

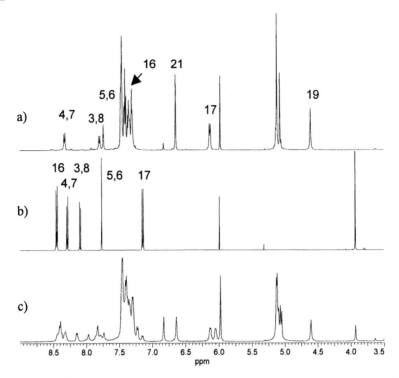

Figure 4: ^1H NMR spectra of (a) $7[G_1(1{\to}3)]$, of (b) free chelating ligand $5[OCH_3]$, and (c) of a 1:1 mixture thereof (tetrachloroethane-d_2)

Synthesis of complex polymers $1[G_x(1{\to}2/3)]$

Using the previously described copper(I) precursor polymer $1[L_2]$, the step-by-step introduction of the dendronized ligands $5[G_x(1{\to}2/3)]$ was studied (Scheme 7). $1[L_2]$ was obtained via dissolving the metal-free precursor polymer 1 in tetrachloroethane and addition of equimolar amounts of $[Cu(CH_3CN)_4]PF_6$. In the resulting precursor metallopolymer $1[L_2]$, two coordinating sites of each copper(I) are occupied by acetronitrile (L) as the ligands. Subsequently, the bidentate dendronized auxiliary ligands $5[G_x(1{\to}2/3)]$ were added. The progress of these conversions could be monitored using NMR. Figure 5 displays a representative spectrum of the metal-free precursor polymer 1, metallopolymer $1[L_2]$ and the dendronized complex polymer $1[G_1(1{\to}3)]$.

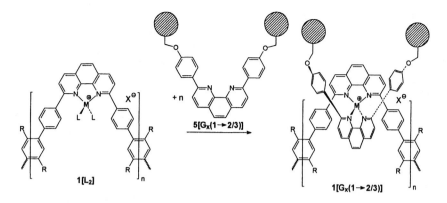

Scheme 7: Synthesis of complex polymers 1 bearing dendronized ligands

The observed chemical shifts, and here in particular the considerable shift of the α-CH$_2$ protons (*) of the *n*-hexyl side chains show clearly that all dendronized ligands are coordinated to the polymer.

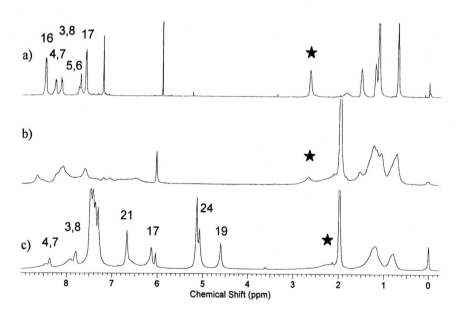

Figure 5: ^1H NMR spectra of (a) the metal-free precursor polymer 1, (b) 1[L$_2$], and (c) 1[G$_1$(1→3)] (1,1,2,2-tetrachloroethane-d$_2$)

Synthesis of coordination polymers 2[G$_x$(1→2/3)]

Finally, we broadened the concept of dendronized copper(I)-metallopolymers to well-defined coordination polymers **2**. For polymer synthesis, we used ligand monomers **6[G$_x$(1→2/3)]**. Again, the experiments were carried out in NMR tubes by stepwise addition of the metal monomer to the solution of the ligand monomer.

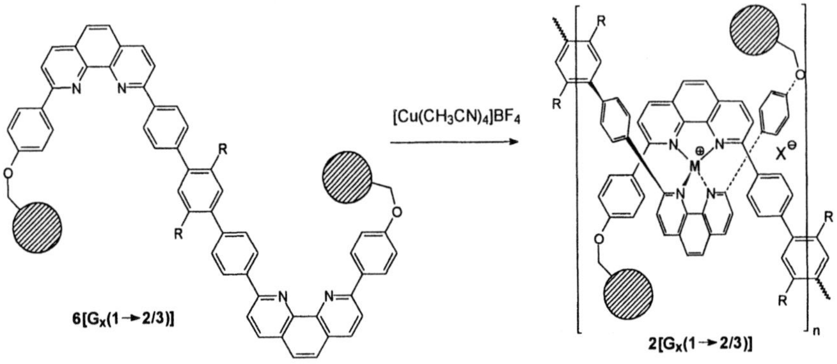

6[G$_x$(1→2/3)]

[Cu(CH$_3$CN)$_4$]BF$_4$

2[G$_x$(1→2/3)]

Scheme 8: Synthesis of copper(I) coordination polymers 2[G$_x$(1→2/3)]

After each addition of [Cu(CH$_3$CN)$_4$]PF$_6$, NMR spectra were recorded until precise 1:1 equivalence of ligand- and metal monomer was achieved – manifested by the disappearance of all endgroup absorptions. As an example, Figure 6 displays the ^{1}H NMR spectra of (*a*) ligand monomer **6[G$_1$(1→3)]**, (*b*) an oligomeric complex **2[G$_1$(1→3)]**, and (*c*) a high-moleclar-weight polymer **2[G$_1$(1→3)]**. The latter NMR spectrum – where no endgroup resonances can be detected – together with the increased solution viscosity support the formation of a very high-molecular-weight coordination polymer **2**. This polymer is formed despite of the sterically very demanding dendritic substituents. On the other hand, the dendrons gave the polymers an improved solubility in conventional organic solvents. This will allow extensive analysis of the chain molecules in solution using various methods in the future – despite their polyelectrolyte character.

Conclusions

We showed that dendrons attached to phenanthroline-based chelating ligands are well-appropriate to increase the solubility of kinetically labile copper(I) coordination polyelectrolytes in apolar, non-coordinating organic solvents.

Moreover, there is no evidence that the complex formation processes are affected by the sterically demanding dendrons. Therefore, in future, a more detailed analysis of the chain conformation and the solution properties of these metallopolymers will be possible.

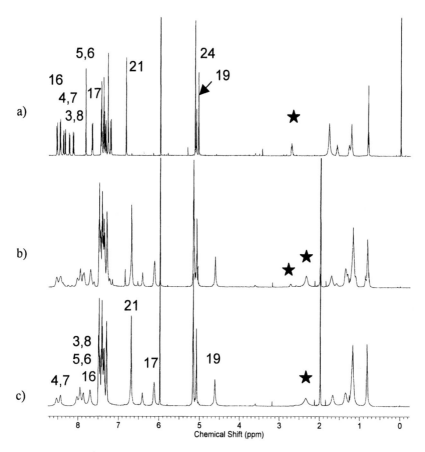

Figure 6: ¹H NMR spectra (a) of ligand monomer 6[G₁(1→3)], (b) of an oligomeric- and (c) of a high-molecular-weight coordination polymer 2[G₁(1→3)] (1,1,2,2-tetrachloroethane-d₂)

Acknowledgments

We thank the German Science Foundation (Deutsche Forschungsgemeinschaft, DFG) for financial support of this work.

References

1   Sauvage, J.-P.; Dietrich-Buchecker, C. (Eds.), *Catenanes, Rotaxanes and Knots*, Wiley VCH, Weinheim, **1999**.

2   Lehn, J.-M., *Supramolecular Chemistry – Concepts and Perspectives*, Wiley VCH, Weinheim, **1995**

3   Wöhrle, D.; Pomogaile, A. D., *Metal Complexes and Metals in Macromolecules*, Wiley VCH, Weinheim, **2003**.

4   Eisenbach, C. D.; Schubert, U. S., *Macromolecules* **1993**, *26*, 7372.

5   Eisenbach, C. D.; Göldel, A.; Terskan-Reinold, M.; Schubert, U. S., *Colloid Polym.Sci.* **1998**, *276*, 780.

6   Velten, U.; Rehahn, M., *J. Chem. Soc., Chem. Commun.* **1996**, 2639.

7   Velten, U.; Lahn, B.; Rehahn, M., *Macromol. Chem. Phys.* **1997**, *198*, 2789.

8   Velten, U.; Rehahn, M., *Macromol. Chem. Phys.* **1998**, *199*, 127.

9   Lahn, B.; Rehahn, M., *Macromol. Symp.* **2001**, *163*, 157.

10   Lahn, B.; Rehahn, M., *e-Polymers* **2002**, *no. 001*.

11   Kubasch, J.; Rehahn, M., *Polym. Preprints* **2004**, *45(1)*, 480.

12   [0]Albrecht-Gary, A.-M.; Saab, Z.; Dietrich-Buchecker, C. O.; Sauvage, J.-P., *J. Am. Chem. Soc.* **1985**, *107*, 3205.

13   Dietrich-Buchecker, C. O.; Sauvage, J.-P., *Tetrahedron Lett.* **1983**, *24*, 5091.

14   Hawker, C. J.; Fréchet, J. M. J., *J. Am. Chem. Soc.* **112**, 7638 (1990).

15   Nystrom, R. F.; Chaikin, S. W.; Brown, W. G., *J. Am. Chem. Soc.* **1949**, 3245.

16   Brown, H. C.; Narasimhan, S.; Choi, Y. M., *J. Org.Chem.* **1982**, *47*, 4702.

Chapter 2

A Comparative Study of Polymer Composition, Molecular Weight, and Counterion Effects on the Chelation of Bipyridine Macroligands to Iron(II)

Robert M. Johnson[1,2], Anne Pfister[1], and Cassandra L. Fraser[1,*]

[1]Department of Chemistry, University of Virginia,
Charlottesville, VA 22904
[2]Current address: Lubrizol Corporation, 29400 Lakeland Boulevard,
Wickliffe, OH 44092

Labile iron(II) tris(bipyridine)-centered star polymers are responsive materials. Bpy-centered macroligands based on polystyrene, poly(methyl methacrylate), poly(ε-caprolactone), poly(DL-lactic acid)) of different molecular weights were combined with FeX_2 salts ($X^- = I^-$, Br^-, $CClO_4^-$, BF_4^-) in 3:1 CH_2Cl_2:CH_3OH. Iron tris(bpy) centered star formation is facile for macroligands <15 kDa; however, both the rate and extent of chelation are polymer dependent and decrease with increasing molecular weight. Counterion ion (i.e. X^-) effects are minimal.

17

Introduction

Iron(II) complexes of α,α'-diimines have been studied extensively for the past 125 years[1] because of their optical properties,[2] electroactivity,[2,3] and in some cases, antibacterial activities.[3,4] Iron(II) tris(bipyridine) is an intensely red complex, which accounts for its use as a reagent for colorimetric iron determinations.[5,6] Iron(II) undergoes a transition from high to low spin upon chelation of a third bipyridine (bpy) ligand and the tris complex is significantly more stable than the mono or bis complexes.[7] Iron tris(bpy) complexes are labile; the ligands may dissociate with decreasing pH[8-10] or exchange with more nucleophilic ligands (CN⁻, OH⁻).[11,12]

The properties associated with [Fe(bpy)₃]²⁺ complexes make them interesting to incorporate into polymers. For example, in [Fe(bpy)₃]²⁺-centered star polymers, the inner sphere effects of iron tris(bpy) are combined with systematic tuning of the polymeric outer sphere. Previous work in our group has focused on iron tris(bipyridine)-centered polyoxazolines,[13-15] polylactides),[16] poly(ε-caprolactones),[16] polyacrylates,[17] and polystyrene,[17] as well as block copolymer combinations of these materials.[16-19] Related polymeric iron complexes have also been described by others.[20] The lability of the metal-centered polymer may potentially be exploited for controlled drug release by incorporating a hydrophobic guest molecule (i.e. a drug) in iron tris(bipyridine) centered star block copolymers (e.g. [Fe{bpy(PLA-PEG)₂}₃]²⁺ based on poly(lactide), PLA, and poly(ethylene glycol), PEG). Upon exposure to certain chemical stimuli (e.g. acid, base, oxidants), one or more ligands may dissociate, altering rates of drug release.[21] Moreover, iron tris(bipyridine) thin films could be useful in sensor technologies. For example, iron tris(bipyridine)-centered polymers are thermochromic, with their red-violet color fading when heated under nitrogen, but returning upon cooling.[14]

Since iron bipyridine complexes are not stable to many polymerization conditions, these labile polymeric complexes are usually accessed via the convergent chelation method which involves reaction of a macroligand with a metal salt (e.g. Equation 1). An exception to this generalization is iron tris(bipyridine)-centered polyoxazoline, which is prepared by metalloinitiation.

$$ (1) $$

These polyoxazoline complexes can be demetallated (for e.g. with OH⁻), and reconstituted by the addition of fresh iron(II) salt. Extinction coefficients for $[Fe(bpyPEOX_2)_3]^{2+}$ (PEOX = poly(2-ethyl-2-oxazoline)) generated by metalloinitiation were ~9500 $M^{-1}cm^{-1}$, while those reformed via chelation were ~8200 $M^{-1}cm^{-1}$, providing a comparison between the two synthetic strategies.[14] Similarly, biodegradable polyesters exhibited a range of chelation efficiencies, defined as $\varepsilon_{PMC}/\varepsilon_{bpy}$, which depended upon macroligand molecular weight and structure. For example, chelation of low molecular weight bpy-centered poly(ε-caprolactone) (bpyPCL$_2$) and poly(DL-lactide) (bpyPLA$_2$) to iron(II) resulted in molar absorptivities comparable to non-polymeric $[Fe(bpy)_3]^{2+}$. However, as the molecular weights were increased, the chelation efficiency decreased more markedly for bpyPCL$_2$ than for bpyPLA$_2$.[16] Additionally, it was noted that for certain polymer/metal salt/solvent combinations, namely bpyPLA$_2$, $Fe(NH_4)_2(SO_4)_2$ and CH_2Cl_2/MeOH, unexpectedly low yields of the corresponding polymeric iron tris(bpy) product were formed as determined by UV/vis spectrophotometric analysis.

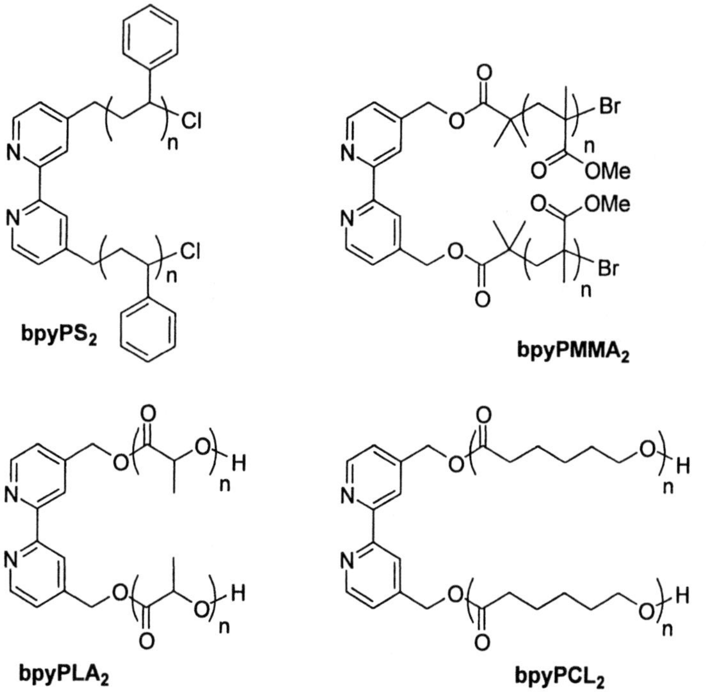

Figure 1. Bipyridine macroligands based on polystyrene (PS), poly(methyl methacrylate) (PMMA), polylactide (PLA), and poly(ε-caprolactone) (PCL).

These observations raised a number of questions which the present study was designed to address. Namely, how sensitive are polymeric Fe tris(bpy) complex extinction coefficients, ε, to different kinds of macromolecular outer spheres? Does a drop in calculated ε reflect polymer compositional effects on chromophore optical properties, or incomplete product formation and a shift in equilibria toward reactants, possibly mono or bis bpy intermediates in a given solvent system? How do the rate and extent of iron tris(bpy) complex formation (ascertained by absorbance at λ_{max}(MLCT) ~530 nm) vary with increasing molecular weight for different polymers? Additionally, are counterion effects significant and to what extent are they polymer dependent for a given solvent system? Spectrophotometric titrations and kinetics experiments were performed in a comparative study of polystyrene (PS), poly(methyl methacrylate) (PMMA), poly(lactic acid) (PLA) and poly(ε-caprolactone) (PCL) macroligands (Figure 1) of different molecular weights. Moreover, the influence of metal salt counterions on polymeric iron(II) tris(bpy) formation was also explored. These investigations enhance understanding of the ways in which coordination chemistry with macroligands is alike and different from analogous ligand systems lacking polymeric outer spheres.

Experimental

Materials. Bpy-centered macroligands were synthesized by previously described procedures.[16,17,22] Iron salts were purchased from Aldrich, stored in a N_2 dry box, and used without further purification. **CAUTION: Perchlorates are potentially explosive!** Dichloromethane used in titrations was purified via solvent columns.[23] All other reagents were used as received from Fisher Scientific.

Methods. UV-vis spectra were obtained on a Hewlett-Packard 8452A diode array spectrophotometer. Molecular weights were determined by gel permeation chromatography (GPC) ($CHCl_3$, 25 °C, 1.0 mL/min) using multiangle laser light scattering (MALLS) ($\lambda = 633$ nm, 25 °C), refractive index ($\lambda = 633$ nm, 40 °C) and diode-array UV-vis detection. Polymer Labs 5μ mixed C columns along with Wyatt Technology Corporation (Optilab DSP Interferometric Refractometer, Dawn DSP Laser Photometer) and Hewlett-Packard instrumentation (Series 1100 HPLC) and software (ASTRA) were used in GPC analysis.

Spectrophotometric Method.[24] Representative procedure for [Fe^{2+}] determination. A 3.4 mM stock solution of iron(II) tetrafluoroborate in CH_3OH was prepared in a 25 mL volumetric flask. The iron concentration was determined experimentally by titrating with a bipyridine stock solution (24.9 mg,

0.159 mmol) in 75% CH_2Cl_2/25% CH_3OH. To a 1-cm path-length cuvette was added 0.150 mL Fe^{2+} stock solution, 0.475 mL CH_3OH, and 1.875 mL CH_2Cl_2, resulting in approximately a 0.1 mM Fe^{2+}, 75% CH_2Cl_2/25% CH_3OH v/v solution. The iron solution was spectrophotometrically titrated with the bipyridine solution via a microliter syringe in 10 μL aliquots. Absorbances were corrected for dilution and plotted versus equivalents added.

Representative Procedure for Spectrophotometric Determination of $[Fe\{bpy(polymer)_2\}_3]^{2+}$ Formation. Fe^{2+} stock solution (0.150 mL), CH_3OH (0.475 mL), and CH_2Cl_2 (1.875 mL), was added to a 1-cm path-length cuvette resulting in a 0.1 mM Fe^{2+}, 3:1 CH_2Cl_2/CH_3OH v/v solution. The iron solution was spectrophotometrically titrated with the bipyridine-centered macroligand solution via microliter syringe in 50 μL aliquots. Absorbances were corrected for dilution and plotted versus equivalents added.

Representative Procedure for $[Fe(bpyX_2)_3]^{2+}$ Synthesis. One equivalent of titrated iron stock solution was added to a 5 mL round bottom flask (under nitrogen), then three equivalents of bpy-centered macroligand dissolved in dichloromethane was added, resulting in a 3:1 CH_2Cl_2/CH_3OH solution (~ 5 mL). The reaction were stirred and monitored by UV-vis until the absorbance reached a maximum value.

Representative Procedure for Chelation Experiments with Heating. Using a spectrophotometrically determined iron stock solution, one equivalent of iron was combined with three equivalents of macroligand in a 3:1 CH_2Cl_2/CH_3OH mixture under nitrogen in a tared round bottom flask. The reaction mixture was refluxed for 8-12 h, concentrated, dried in vacuo ~12 h and the mass was determined. The solid residue was then redissolved in the 3:1 CH_2Cl_2/CH_3OH solvent mixture and the absorbance at $\lambda_{max} \approx 530$ nm measured after equilibrium was reached.

Results & Discussion

Generating iron-centered star polymers by a chelation method requires careful balancing of conditions. Those favorable for the preparation of iron tris(bipyridine) are not suitable for chelating non-polar polymeric bpy ligands to iron(II). For example, iron salts require polar media such as water or methanol, while many polymeric macroligands require less polar solvents such as dichloromethane or tetrahydrofuran. Moreover, the resultant metal complex may undergo ligand dissociation in various solvents due to acidity ($CHCl_3$) or the presence of peroxides (in ethers such as THF).[16] A mixed solvent system consisting of 3:1 CH_2Cl_2/CH_3OH was selected for its compatibility with the library of macroligands prepared in our laboratory, while not adversely affecting the formation of the metal complexes, and their stabilities once formed.

To measure spectrophotometrically the number of equivalents of macroligand necessary to titrate the iron in solution as tris complex, the free iron concentration in a salt solution was first determined by titration with known ligands 2,2′-dipyridyl, and 4,4′-dimethyl 2,2′-bipyridine. A stock solution of $Fe(BF_4)_2 \cdot 6H_2O$ or another salt was prepared in methanol. The ligand titrant was prepared in a 3:1 CH_2Cl_2/CH_3OH solution in order to retain the CH_2Cl_2/CH_3OH solution composition as more ligand was added. A fraction of the Fe(II) stock solution was added to a cuvette, and dichloromethane and methanol were added to result in a 3:1 solvent system. The ligand solution was added to the iron solution and the mixture was stirred after each aliquot until the absorbance stabilized (i.e. typically within 10 min, although sometimes longer for high molecular weights). Absorbances were corrected for dilution, plotted versus equivalents added, and the end point was determined graphically as the intercept of the two linear segments of the titration curve (e.g. as shown for $bpyPS_2$ in Figure 3). The end point was then used to determine the concentration of Fe(II) in solution. This procedure was particularly important for hygroscopic Fe salts.

Effect of Polymer Composition on Iron(II) Tris(bipyridine) Synthesis

Once the concentration of Fe^{2+} in FeX_2 salt solutions was determined by titration, the effect of polymer composition on tris(bpy) complex formation was explored. A small bathochromic shift in the λ_{max} of the metal-to-ligand charge-transfer (MLCT) was observed for all polymeric metal complexes compared to the unsubstituted tris(bpy) complex: $[Fe(bpy)_3]^{2+}$: 522 nm; $[Fe\{bpy(polymer)_2\}_3]^{2+}$: ~530-533 nm. Spectral data for a $bpyPCL_2$ macroligand and the corresponding iron(II) complex are shown in Figure 2. Extinction coefficients for complexes prepared with different polymers (< ~15 kDa) varied little from each other for a given counterion, and were comparable to data for the corresponding iron(II) tris(bpy) complex (Table 1). This suggests that for low molecular weight materials, the MLCT transitions are not very sensitive to the polymeric outer spheres, and extinction coefficients can be useful for estimating and comparing molecular weight, counter ion, and other effects.

Spectrophotometric titrations were performed with $Fe(BF_4)_2 \cdot 6H_2O$ and other FeX_2 salts, with various low molecular weight bipyridine macroligands (Figure 1) in a manner analogous to the bipyridine calibrants (Figure 3). For the majority of these lower molecular weight polymers and metal salts evaluated, the number of equivalents of macroligand required to reach the end point of a spectrophotometric titration is slightly above three, indicative of the quality of the polymeric ligands synthesized (i.e. efficient polymerization from initiator sites on the bpy reagent in the macroligand preparation) (Table 2).

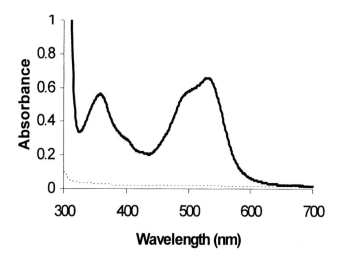

Figure 2. UV-vis spectra of [Fe(bpyPCL$_2$)$_3$](BF$_4$)$_2$ (—) (M$_n$star = 13.4 kDa, ε = 9583 M^{-1} cm^{-1}) and bpyPCL$_2$ (—) (M$_n$ = 4.4 kDa).

Table 1. Extinction Coefficients for Low Molecular Weight Polymeric Metal Complexes Determined from the End Points of Spectrophotometric Titrations in 3:1 CH$_2$Cl$_2$/CH$_3$OH.

Macroligand	M_n[a]	Extinction Coefficients[b]			
		BF$_4^-$	ClO$_4^-$	I$^-$	Br$^-$
bpyPS$_2$	8390	9616	10025	9369	9833
bpyPMMA$_2$	11810	9610	10470	9452	10470
bpyPLA$_2$	6630	9860	10449	9656	9708
bpyPCL$_2$	4380	9583	10139	9551	10178
bpy	156	9450	9750	9300	9650

[a] Molecular weight (Da) determined by GPC analysis (RI/MALLS detection) in CHCl$_3$.
[b] ε (M^{-1}cm^{-1}) measured at [Fe(bpy)$_3$]$^{2+}$ MLCT λ_{max} ~ 530-533 nm.

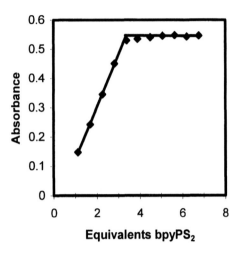

Figure 3. Spectrophotometric titration curve for bpyPS$_2$ (M$_n$ = 18.2 kDa) with Fe(BF$_4$)$_2$ in CH$_2$Cl$_2$/CH$_3$OH (3:1) to form [Fe(bpyPS$_2$)$_3$](BF$_4$)$_2$. Equivalents required to reach the end point of the titration = 3.3.

Table 2. Equivalents of Macroligand Required to Reach the Spectrophotometric Titration End Points for Various Iron Salts, FeX$_2$.

Macroligand	M_n*	Equivalents of Macroligand			
		BF$_4^-$	ClO$_4^-$	I$^-$	Br$^-$
bpyPS$_2$	7620	3.02	3.02	3.04	3.06
bpyPMMA$_2$	11810	3.03	3.02	3.03	3.05
bpyPLA$_2$	6630	3.03	3.05	3.04	3.08
bpyPCL$_2$	4380	3.02	3.07	3.09	3.03

*Molecular weight determined by GPC in CHCl$_3$ using MALLS/RI detection.

Molecular Weight Effects on Rates and Extent of Tris Complex Formation

With low molecular weight macroligands, polymeric metal complexes form readily in approximately five minutes, presumably because of the accessibility of the donor group and ease with which these systems can adopt the optimal bpy conformation for metal binding.[16] However, kinetics experiments show that reactions take longer for high molecular weight macroligands and sometimes do not go to completion, as indicated by calculated extinction coefficients that are depressed relative to Fe tris(bpy) and analogous lower molecular weight Fe tris(bpy)-centered star polymers (Table 3). To ensure that reactions were complete and thermodynamic not kinetic products were obtained, some reactions were heated at 50 °C for ~16 hours, however this still did not lead to samples with the anticipated extinction coefficients (i.e. ~9500 $M^{-1}cm^{-1}$).

Chelation efficiency, $\varepsilon_{PMC}/\varepsilon_{bpy}$, provides a relative measure of the extent to which a polymeric tris bpy complex forms, allowing for ready comparison between different polymers and reaction conditions.[25] For example, an Fe star prepared from bpyPS$_2$ (M_n = 8.39 kDa; MW star (calcd) = 25.4 kDa) reached a maximum absorbance after five minutes, with ε = 9616 $M^{-1}cm^{-1}$; chelation efficiency = 1.02). However, for a larger macroligand (M_n = 50.8 kDa; MW star (calcd) = 153 kDa), the reaction time to reach maximum absorbance increased to 130 minutes and the calculated extinction coefficient dropped significantly (ε = 7180 $M^{-1}cm^{-1}$; chelation efficiency = 0.76).

End points of spectrophotometric titrations performed with Fe(BF$_4$)$_2$·6H$_2$O and high molecular weight macroligands are often greater than 3:1, suggesting that the equilibrium may not be shifted entirely toward product formation. For example, titration with bpyPS$_2$ (M_n = 18,220) required 3.3 equivalents of ligand to reach the end point (Figure 3). Likewise, the end point for a bpyPCL$_2$ sample (M_n = 36,510) was reached at 3.5 equivalents of macroligand per Fe^{2+}. This contrasts with lower molecular weight polymers requiring ~3.0 equivalents (Table 2).

The extent to which rates increase and tris(bpy) formation decreases with increasing molecular weight is polymer dependent. For example, while chelation of bpyPCL$_2$ (53 kDa) resulted in an extinction coefficient of 4572 $M^{-1}cm^{-1}$, bpyPS$_2$ (50.8 kDa) correlated with ε = 7180 $M^{-1}cm^{-1}$, implying chelation efficiencies of 0.48 and 0.76 respectively (Table 3). This can be a result of a number of factors. One possibility is that bpy content varies with molecular weight for different polymerization mechanisms; higher molecular weights may contain less bpy per polymer, requiring more macroligand to reach the titration end point. Alternatively, if the error in molecular weight measurements varies

Table 3. Summary of UV-vis Data of Iron(II) Tris(bipyridine)-Centered Star Polymers

[Fe(bpyX$_2$)$_3$](BF$_4$)$_2$	M_n star (calcd) (kDa)[a]	ε_{PMC} (M^{-1} cm^{-1})[b]	Time (min)[c]	Chelation Efficiency[d]
[Fe(bpyPS$_2$)$_3$]$^{2+}$	25.4	9616	5	1.02
	82.0	9206	70	0.97
	153.0	7180	130	0.76
[Fe(bpyPMMA$_2$)$_3$]$^{2+}$	35.7	9610	5	1.02
	176.6	7450	60	0.79
[Fe(bpyPLA$_2$)$_3$]$^{2+}$	20.1	9860	5	1.04
	117.3	9145	60	0.97
	196.5	8129	105	0.86
[Fe(bpyPCL$_2$)$_3$]$^{2+}$	13.4	9583	5	1.01
	29.9	9067	10	0.96
	109.8	5845	55	0.62
	160.0	4572	90	0.48

[a] Calculated molecular weight of the polymeric metal complex with two BF$_4^-$ counterions. [b] Extinction coefficient calculated at maximum absorbance. [c] Time required to reach maximum absorbance.

with molecular weight, this too, may account for differences in titration data. Furthermore, differential polymer conformation and solvation could be playing a role. While tris complex formation may be favored thermodynamically from the standpoint of the metal, polymer molecular weight effects can prevent complete tris product formation. For example, inaccessibility of the bpy donor group embedded in a polymer matrix, steric bulk around the metal, or hindered rotation about the bpy C-C bond can slow or perhaps even prevent a third polymer chain from binding. Nonetheless, for the majority of macroligands explored to date, chelation efficiencies are >75%, even at higher molecular weights (Table 3).

Counterion Effects on Extinction Coefficients

Originally it was observed that polymeric metal complexes prepared in a CH$_2$Cl$_2$/CH$_3$OH using ferrous ammonium sulfate exhibited decreased molar absorptivities compared to those prepared from this salt followed by metathesis with hexafluorophosphate. Calculated extinction coefficients were on the order of 50-80% lower when no NH_4PF_6 was added in the preparation,[16] indicative of a sharp decrease in tris complex formation. Initially it was hypothesized that this stemmed from poor solvation of the hard SO_4^{2-} counterions in the non-polar polymeric outer sphere relative to PF_6^-. This prompted us to study the effects of different counterions on chelation reactions for a series of polymers. Spectrophotometric titrations were carried out on a variety of iron(II) salts. Because the iron (II) halide salts are readily oxidized, all reactions were performed under nitrogen. Ferrous chloride was also tested, but exhibited poor solubility in the reaction solvents. The end points of the titrations were reached after ~3.0 equivalents of macroligand were added; no significant variation was observed for different counterions (Table 2). These results suggest that incomplete tris complex formation for $Fe(NH_4)_2(SO_4)_2$ may result from some other factor, perhaps low solubility of the salt in methanol. By way of comparison, the salt is completely insoluble in ethanol and apparently is not solubilized by complexation to bpy donor sites.[26] Though coordination can sometimes drag poorly soluble metal salts into solution, here the amount of polymeric Fe tris(bpy) product remains low, even for longer reaction times.

The extinction coefficients of the polymeric metal complexes were calculated for different counterions (Table 1). For a given salt, the calculated extinction coefficients for polymeric complexes were generally slightly higher than that of $[Fe(bpy)_3]^{2+}$ prepared under the same conditions. Some differences were also observed for the same polymer with different metal salts. For example, metal complexes prepared from iron(II) perchlorate typically yielded slightly higher extinction coefficients than for other iron salts. Differences in extinction coefficients due to counterion effects have been noted previously, and were said to reflect the ability of anions to be accommodated in "pockets" between the ligands. Smaller anions are better able to situate in these pockets, however strongly hydrated ions such as Cl^- are unable to approach as nearly, thus decreasing their ability to engage in electrostatic attractions.[27,28] Some counterions, particularly I^-, may gain increased stability because of stronger interactions with the diimine ligands.[29] The extent to which such an argument applies to systems with polymeric outer spheres of different compositions and steric bulk in solution is not known.

Conclusions

The preparation of iron-centered star polymers via chelation methods is influenced by the chemical structure of the polymer as well as its size. Lower molecular weight polymers result in faster reaction rates and higher chelation efficiencies, while competing factors in higher molecular weight macroligands can prevent complete tris(bpy) formation. While the generation of the tris complex may impart stability to the metal core, for large macroligands of greater steric bulk, bpy sites may be less accessible and hindered from adopting the energetically less favorable cis conformation that is required for metal bonding. Although the solvent system was kept constant for this study, solvent effects are also expected to vary for different polymers. With the exception of ferrous ammonium sulfate, the choice of iron(II) salt had little impact on product formation.

Acknowledgments

We thank the National Science Foundation and the donors of The Petroleum Research Fund administered by the American Chemical Society for support of this work. We are also grateful to Prof. Frederick S. Richardson for helpful discussions.

References

1   Blau, F. *Monatsh.* **1888**, *19*, 647.
2   Bernhard, S.; Goldsmith, J. I.; Takada, K.; Abruna, H. D. *Inorg. Chem.* **2003**, *42*, 4389.
3   Brandt, W. W.; Dwyer, F. P.; Gyarfas, E. D. *Chem. Rev.* **1954**, *54*, 959.
4   Dwyer, F. P.; Gyarfas, E. C.; Koch, J.; Rogers, W. P. *Nature* **1952**, *170*, 190.
5   Woods, J. T.; Mellon, M. G. *Ind. Eng. Chem. Anal. Ed.* **1941**, *13*, 551.
6   Cagle, F. W.; Smith, G. F. *Anal. Chem.* **1947**, *19*, 384.
7   Stepwise formation constants for iron mono, bis and tris bpy formation in aqueous solution (25.0 °C) are: $K_1 = 1.6 \times 10^4$, $K_2 = 5.0 \times 10^3$, $K_3 = 3.5 \times 10^9$. Irving, C. H.; Mellor, D. H. *J. Chem. Soc.* **1962**, 5222.
8   Seiden, L.; Basolo, F.; Neumann, H. M. *J. Am. Chem. Soc.* **1959**, *81*, 3809.
9   Raman, S. *J. Inorg. Nucl. Chem.* **1976**, *38*, 781.
10 Krumholz, P. *J. Phys. Chem.* **1956**, *60*, 87.
11 Schilt, A. A. *J. Am. Chem. Soc.* **1960**, *82*, 3000.
12 Nord, G.; Pizzinno, T. *J. Chem. Soc., Chem. Commun.* **1970**, 1633.
13 McAlvin, J. E.; Fraser, C. L. *Macromolecules* **1999**, *32*, 1341.
14 McAlvin, J. E.; Scott, S. B.; Fraser. C. L. *Macromolecules* **2000**, *33*, 6953.
15 Park, C.; McAlvin, J. E.; Fraser, C. L.; Thomas, E. L. *Chem. Mater.* **2002**, *14*, 1225.

16 Corbin, P. S.; Webb, M. P.; McAlvin, J. E.; Fraser, C. L. *Biomacromolecules* **2001**, *2*, 223.

17 Fraser, C. L.; Smith, A. P. *J. Polym. Sci, Part A: Polym. Chem.* **2000**, *38*, 4704.

18 Johnson, R. M.; Fraser, C. L. *Macromolecules* **2004**, *37*, 2718.

19 Smith, A. P. Fraser, C. L. *Macromolecules* **2002**, *35*, 594.

20 For e.g., see the following and references therein: (a) Schubert, U. S.; Eschbaumer, C. *Angew. Chem. Int. Ed.* **2002**, *41*, 2892. (b) Andres, P. R.; Schubert, U. S. *Synthesis* **2004**, *8*, 1229. (c) Hofmeier, H.; Schubert, U. S. *Chem. Soc. Rev.* **2004**, *33*, 373. (d) El-ghayoury, A.; Hofmeier, H.; de Ruiter, B.; Schubert, U. S. *Macromolecules* **2003**, *36*, 3955. (e) Chujo, Y.; Sada, K.; Saegusa, T. *Macromolecules* **1993**, *26*, 6315.

21 For an interesting example of stimuli responsive metallopolymers, see: Beck, J. B.; Rowan, S. J. *J. Am. Chem. Soc.* **2003**, *125*, 13922.

22 Wu, X.; Fraser, C. L. *Macromolecules* **2000**, *33*, 4053.

23 Pangborn, A. B.; Giardello, M. A.; Grubbs, R. H.; Rosen, R. K.; Timmers, F. J. *Organometallics* **1996**, *15*, 1518.

24 For e.g., see: Goddu, R. F.; Hume, D. N. *Anal. Chem.* **1954**, *26*, 1740.

25 Chelation efficiency is a rough estimate of extent of reaction. This parameter is most useful if ε values for polymeric and the non-polymeric bpy complexes are comparable, as in the case of low molecular weight macroligands (e.g. Table 1). Bpy values for a given counterion are typically slightly lower than those for polymeric complexes, so chelation efficiency slightly overestimates the extent of reaction. Ways in which ε values are affected by high molecular weight materials is not known.

26 Macintyre, J. E. Ed.; *Dictionary of Inorganic Compounds*; Chapman & Hall: New York, 1992; Vol 3.

27 Johannson, L. *Chemica Scripta* **1976**, *9*, 30.

28 Johannson, L. *Chemica Scripta* **1976**, *10*, 72.

29 Johansson, L. *Chemica Scripta* **1976**, *10*, 149.

Chapter 3

Aqueous Metallosupramolecular Micelles with Spherical or Cylindrical Morphology

Jean-François Gohy[1], Bas G. G. Lohmeijer[2], Alexander Alexeev[2], Xiao-Song Wang[3], Ian Manners[3,4*], Mitchell A. Winnik[3,*], and Ulrich S. Schubert[2,*]

[1]CMAT Unit and CERMIN, Université Catholique de Louvain, Place L. Pasteur 1, B–1348 Louvain-la-Neuve, Belgium
[2]Laboratory of Macromolecular Chemistry and Nanoscience, Center for Nanomaterials, Eindhoven University of Technology and Dutch Polymer Institute, P.O. Box 513, 5600 MB Eindhoven, The Netherlands
[3]Department of Chemistry, University of Toronto, 80 St. George Street, Toronto, Ontario M5S 3H6, Canada
[4]Current address: The School of Chemistry, University of Bristol, Bristol BS8 1TS, United Kingdom

Supramolecular AB diblock copolymers have been prepared by the sequential self-assembly of terpyridine end-functionalized polymer blocks using Ru(III)/Ru(II) chemistry. Since this synthetic strategy has been used to attach hydrophobic polystyrene, PS, or poly(ferrocenylsilane), PFS, to a hydrophilic poly(ethylene oxide) block, amphiphilic metallo-supramolecular diblock copolymers have been obtained. These compounds form micelles in water, that were characterized by a combination of dynamic and static light scattering, transmission electron and atomic force microscopies. These techniques revealed the formation of spherical micelles for the PS-based systems while rod-like micelles are observed for the PFS-containing copolymer. This difference is attributed to the propensity of PFS to crystallize in the micellar core.

Background

Block copolymer micelles have been the topic of intense research during the last 30 years due to their potential application in various fields, including nanotechnology, biomedical applications, emulsion stabilization, viscosity regulation, catalyst support, surface modification, etc.[1] Aqueous block copolymer micelles are resulting from the dissolution of amphiphilic block copolymers in water and consist of a core formed by the hydrophobic polymer blocks surrounded by a corona containing the water-soluble blocks. Water-soluble blocks are typically either polyelectrolytes or non-ionic blocks such as poly(ethylene oxide), PEO.

The most commonly observed morphology for block copolymer micelles is the spherical one. This is resulting from the fact that the investigated amphiphilic block copolymers are generally containing a major hydrophilic block to be directly soluble in water. However, Eisenberg has shown that a variety of morphologies including rods, tubules, vesicles and many other intricate morphologies could be obtained from "crew-cut" micelles, formed by highly asymmetric block copolymers containing a large hydrophobic block.[2] Since these block copolymers could not be directly solubilized in water, the temporary use of an organic non-selective co-solvent was required. Aqueous micelles with rod-like or vesicular morphologies were also obtained by direct dissolution of amphiphilic block copolymers in water, as illustrated by the work of Antonietti et al.,[3] Möller et al.,[4] Meier et al.[5] and Bates and coworkers.[6,7]

One unifying rule accounting for block copolymer micelle morphology has been proposed by Discher and Eisenberg.[8] This rule should be considered for coil-coil block copolymer readily soluble in water and is expressed as a function of the mass fraction of the hydrophilic block, f, to total mass of the copolymer: spherical micelles are observed for $f > 45\%$; rod-like micelles are observed for $f < 45\%$; vesicles are observed for $f \sim 35\%$; inverted microstructures such as large compounds micelles are observed for $f < 25\%$. Sensitivity of these rules to the chemical composition and to the MW of the copolymer chains have however not yet been fully probed.

Another possibility to trigger the formation of non-spherical micelles is to use rod-coil copolymers. The introduction of a stiff segment results in an increase of the Flory-Huggins parameters in comparison with coil-coil copolymers.[9] As a result, non-spherical morphologies can be observed at higher f for rod-coil copolymer micelles compared to coil-coil ones. Block copolymer micelles in which the core-forming polymer blocks are able to crystallize are relatively similar to rod-coil copolymers and non-spherical micellar morphologies could also be obtained from such copolymers even if f is high.[10]

Amphiphilic block copolymers generally consist of a hydrophobic block, A, covalently linked to a hydrophilic, B, one. We have recently introduced a new class of amphiphilic block copolymers, in which the junction between the two blocks is a *bis*-2,2':6',2''-terpyridine-ruthenium(II) metal-ligand complex. These copolymers have been called metallo-supramolecular amphiphilic block copolymers. Terpyridine is used as the ligand in our approach. Briefly, a sequential two-step self-assembly process has been utilized to prepare these copolymers.[11] Terpyridine-ended polymer blocks, represented as A-[and B-[have been prepared, as described in detail elsewhere.[12] Ru(III)Cl$_3$ can only bind to one terpyridine ligand and has been reacted with e.g. the A-[block, resulting in a A-[RuCl$_3$ *mono*-complex. In a second step, A-[RuCl$_3$ has mixed with the other terpyridine-functionalized block B-[, and Ru(III) has been reduced to Ru(II). Under these conditions, the Ru(II) ions can coordinate two terpyridine ligands, and a A-[Ru(II)]-B *bis*-complex is exclusively formed. Compared to other transition metals, *bis*-2,2':6',2''-terpyridine-Ru(II) complexes are extremely stable. For example, they are not affected by pH changes from 0 to 14 (in water) and they do not open with time (on a several years timescale).[13] These A-[Ru]-B complexes are thus ideal candidates to be used instead of classical A-*b*-B covalent block copolymers for the formation of block copolymer micelles.

Objective

The objectives of this study were to obtain aqueous block copolymer micelles with either spherical or rod-like morphology from amphiphilic metallo-supramolecular copolymers. In order to reach this goal, we have synthesized two copolymers, both of them containing the same major PEO$_{70}$ block (the number in subscript being the number average degree of polymerization, the polydispersity index, PDI, of this block is 1.07) and either an amorphous polystyrene, PS$_{20}$ (PDI = 1.1), or a crystallizable poly(ferrocenylsilane), PFS$_{12}$ (PDI = 1.03), hydrophobic block. Formation of spherical micelles is expected from the PS$_{20}$-[Ru]-PEO$_{70}$ copolymers while rod-like micelles are anticipated from the PFS$_{12}$-[Ru]-PEO$_{70}$. Dynamic light scattering (DLS), transmission electron microscopy (TEM) and atomic force microscopy (AFM) have been used to characterize the resulting aqueous micelles.

Experimental

The synthetis and complete characterization of the PS_{20}-[Ru]-PEO_{70} and PFS_{12}-[Ru]-PEO_{70} amphiphilic copolymers have been previously reported.[12,14]

The micellar solutions were prepared by dissolving 50 mg of PS_{20}-[Ru]-PEO_{70} or PFS_{12}-[Ru]-PEO_{70} in N,N-dimethylformamide, DMF (1 mL). Distilled water was dropwise added with a calibrated pipette (50 µL increments). The scattered light intensity was measured after each addition of water and plotted versus the added water volume. The critical water concentration (cwc) was accordingly determined, following the procedure previously described by Eisenberg et al.[15] The dropwise addition was continued until 1 mL of water were added. The micellar solution was then placed in a dialysis bag (Spectra-Por membrane with a cut-off limit of 6000-8000 daltons). DMF was then eliminated by dialysis against regularly replaced distilled water. The concentration of the micelles in pure water was adjusted to 1 g/L.

Dynamic light scattering (DLS) measurements were performed on a Brookhaven Instruments Corp. BI-200 apparatus equipped with a BI-2030 digital correlator with 136 channels and a Spectra Physics He-Ne laser with a wavelength λ of 633 nm. A refractive index matching bath of filtered decalin surrounded the scattering cell, and the temperature was controlled at 25 °C. Toluene was used as a reference to determine the Rayleigh ratio. Data were analyzed by the method of the cumulants, as described elsewhere.[16] The Z-average diffusion coefficient over the whole set of aggregates was calculated from the first cumulant and the PDI of the aggregates was estimated from the Γ_2/Γ_1^2 ratio, where Γ_i is the ith cumulant. The diffusion coefficient extrapolated to zero concentration was related to the hydrodynamic diameter (D_h) by the Stokes-Einstein equation. The DLS data were also analyzed by the CONTIN routine, a method which is based on a constrained inverse Laplace transformation of the data and which gives access to a size distribution histogram for the aggregates.

Transmission electron microscopy (TEM) was carried out with a JEOL 2000 FX working at 200 kV and equipped with a CCD camera. The samples were prepared by drop-casting of a diluted (0.01 g/L) micellar solution on a Formvar-coated copper grid. No contrasting agent was used for the observation of the micelles. The Kontron KS 100 software was used to collect and analyze the TEM images.

Atomic force microscopy (AFM) measurements were performed with a Digital Instruments Nanoscope IIIa Multimode operated in air with the tapping mode. A diluted micellar solution (0.01 g/L) has been casted on a silicon wafer for the measurements.

Results & Discussion

Two amphiphilic metallo-supramolecular copolymers have been successfully synthesized, purified and fully characterized, as described in ref. 12 for the PS_{20}-[Ru]-PEO_{70} copolymer and in ref. 14 for the PFS_{12}-[Ru]-PEO_{70} sample. It was indeed demonstrated that the final samples were not contaminated by uncomplexed PS_{20}-[, PFS_{12}-[or PEO_{70}-[polymer blocks, that no *mono*-complexes PS_{20}-[Ru(III)], PFS_{12}-[Ru(III)] or PEO_{70}-[Ru(III)] were present, and that no *homo*-dimers PS_{20}-[Ru]-PS_{20}, PFS_{12}-[Ru]-PFS12 or PEO_{70}-[Ru]-PEO_{70} were present in the final samples. The chemical structures of these two copolymers are depicted in Figure 1.

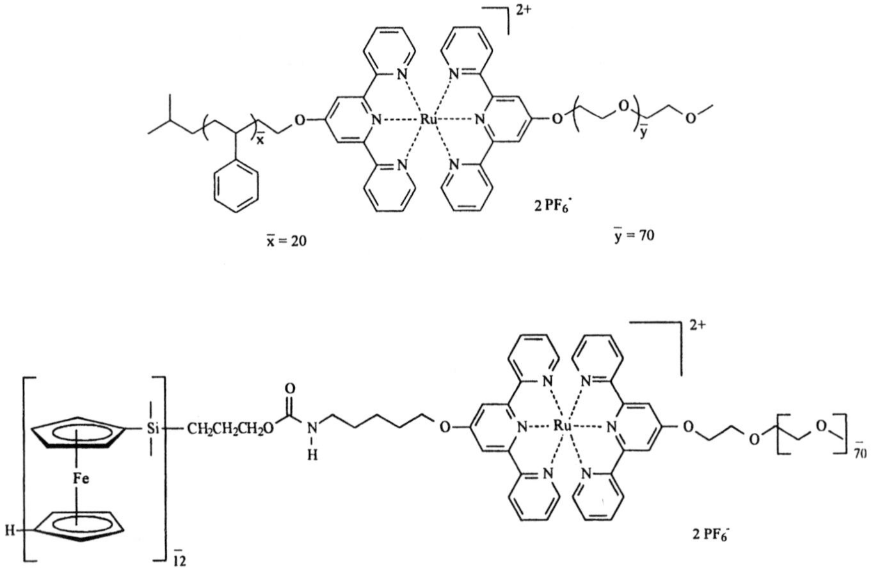

Figure 1. Chemical structure of the PS_{20}-[Ru]-PEO_{70} (top) and PFS_{12}-[Ru]-PEO_{70} (bottom) amphiphilic metallo-supramolecular copolymers.

Micelle Formation

Since the two investigated samples were poorly soluble in water at room temperature, the preparation method previously introduced by Eisenberg et al. was used (see Experimental).[15] Water was added dropwise to solutions of the bulk samples in DMF and the scattered intensity was measured after each water addition (see Figure 2). The cwc was accordingly determined and expressed as the percentage of water that has to be reached to observe micellization in the water/DMF mixture. The cwc was found to be 21 % for the PS_{20}-[Ru]-PEO_{70} sample and 35 % for the PFS_{12}-[Ru]-PEO_{70} one (for solutions obtained by dissolution of 50 mg of bulk sample in 1 mL DMF). The slightly lower cwc observed for the PS_{20}-[Ru]-PEO_{70} sample can be attributed to the longer hydrophobic block for this sample. After DMF elimination by dialysis, the two aqueous micellar solutions were measured by DLS. The results are shown in Table 1 and are in agreement with the formation of colloidal particles rather than block copolymer micelles. In order to obtain more information, the experimental intensity auto-correlation function was analyzed with the CONTIN routine.

The CONTIN distribution of D_h revealed the presence of two populations for the PS_{20}-[Ru]-PEO_{70} sample, in agreement with the large PDI.[17] The population with the smaller size (D_h~65 nm) could not be attributed to individual micelles since the expected D_h for micelles formed by such low molecular weight amphiphilic block copolymers should be in the 15-20 nm range.[1]

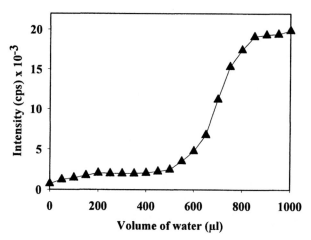

Figure 2. Determination of the critical water concentration (cwc) for PFS_{12}-[Ru]-PEO_{70} aqueous micelles. The scattered light intensity (in counts per second, cps) is plotted as a function of the added volume of water. The initial solution contains 50 mg of PFS_{12}-[Ru]-PEO_{70} dissolved in 1 mL of DMF.

The population with the larger size ($D_h \sim 200$ nm) is thought to result from the clustering of a large number of micelles. In order to gain more insight about the exact structure of these different populations, complementary experiments have been carried out. In this respect, cryo-TEM measurements were performed (see ref. 18 for details) in order to minimize perturbation of the micellar solution. Cryo-TEM pictures systematically showed individual micelles in which the PS core is imaged as a darker sphere with a diameter of 10 nm, and small aggregates containing a limited number of closely packed micelles and having a characteristic size in the 40-70 nm range.[18] Finally, analytical ultra-centrifugation studies revealed the presence of three aggregated species in the PS_{20}-[Ru]-PEO_{70} sample, that could respectively correspond to micelles, small clusters and large aggregates.[19] The mean aggregation number for PS_{20}-[Ru]-PEO_{70} individual micelles has been estimated by analytical ultracentrifugation and found to be 85.[19]

Combination of the results obtained from DLS, TEM, cryo-TEM and analytical ultracentrifugation allows to draw the picture depicted in Figure 3. Spherical micelles are formed in water by the PS_{20}-[Ru]-PEO_{70} sample. The diameter of the PS core is ~10 nm, and the D_h is in the 20-30 nm range. These micelles however aggregate into polydisperse small clusters, whose average size is around 60-70 nm. Finally, a very small number of large clusters is also observed. The resolution of the DLS Contin analysis is not sufficient enough to visualize these three species. Indeed, the individual micelles and the small aggregates are seen as a single broad population.

Table 1. DLS data obtained on the two investigated copolymers

Sample	Mean D_h (nm) obtained from the first cumulant of a cumulants analysis. PDI in parenthesis	Mean D_h's (nm) of the different populations obtained after CONTIN analysis
PS_{20}-[Ru]-PEO_{70}	150 (0.4)	65 and 202
PFS_{12}-[Ru]-PEO_{70}	89 (0.21)	95

In contrast to PS_{20}-[Ru]-PEO_{70} micelles, only one distribution is observed in the case of the PFS_{12}-[Ru]-PEO_{70} sample. Moreover, the PDI is much lower for the PFS-containing sample. The value of the D_h measured for the PFS_{12}-[Ru]-PEO_{70} sample is in agreement with the one previously obtained for cylindrical micelles from PFS-based copolymers in organic solvents.[20] The observation of a single population in the CONTIN histogram can be explained on the basis of a high flexibility of the cylindrical micelles in solution. The mean aggregation number for PFS_{12}-[Ru]-PEO_{70} micelles has been measured by static light scattering and found to be 2500.[20]

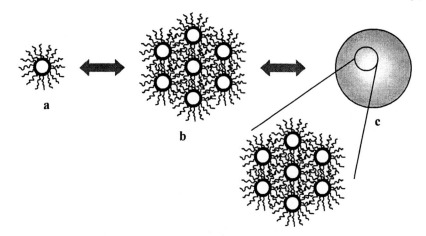

Figure 3. Scheme for PS$_{20}$-[Ru]-PEO$_{70}$ metallo-supramolecular micelles. The black ring around the PS core represents the bis-2,2':6',2''-terpyridine-Ru(II) complexes, located at the interface between the PS core (white sphere) and the PEO chains in black (a). These micelles aggregate into small (b) and large clusters (c, scale is reduced for clarity).

Micellar morphology

TEM and AFM investigations have been conducted on the micelles formed by the PS$_{20}$-[Ru]-PEO$_{70}$ and PFS$_{12}$-[Ru]-PEO$_{70}$ copolymers.

The formation of spherical micelles has been confirmed for the PS$_{20}$-[Ru]-PEO$_{70}$ sample as shown in Figure 4. Since, the electronic contrast is thought to originate from the ruthenium ions at the PS/PEO interface, it can be assumed that PS core are essentially visualized in the TEM images. The PS$_{20}$-[Ru]-PEO$_{70}$ have a tendency to stick together leading to grape structure. The aggregation of the micelles can however be overestimated while compared to the real situation in solution. Actually, the aggregation phenomenon shown in Figure 4 could be an artifact due to the drying process used in TEM specimen preparation. The formation of spherical micelles has been confirmed by AFM. The micelles now appear to be quite polydisperse in size. The biggest spheres could correspond to aggregates of a limited number of micelles but their internal structures could not be visualized by AFM.

The TEM and AFM pictures of the PFS$_{12}$-[Ru]-PEO$_{70}$ micelles are shown in Figure 5. No contrasting agent has been added and the contrast is thought to originate from both the ruthenium ions and the iron atoms of the PFS blocks. Rod-like micelles are clearly observed for this sample, whose diameter is constant (~6 nm) and whose length is highly variable.

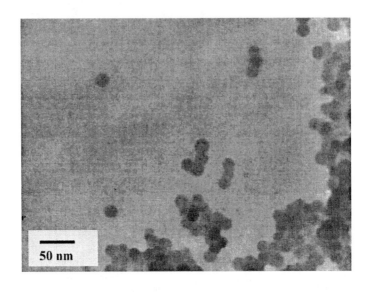

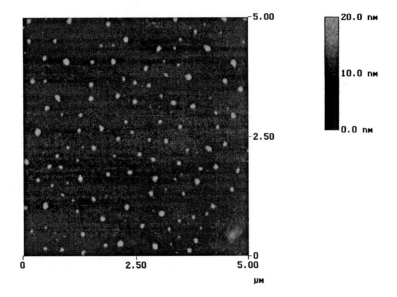

Figure 4. TEM (top) and height contrast AFM (bottom) pictures for PEO$_{20}$-[Ru]-PEO$_{70}$ micelles cast on on a formvar-coated copper grid (TEM) or on a silicon wafer (AFM).

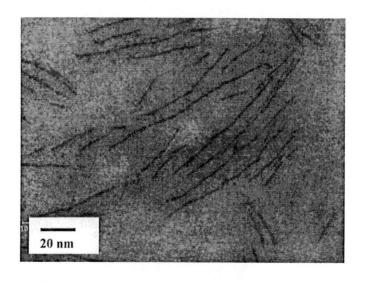

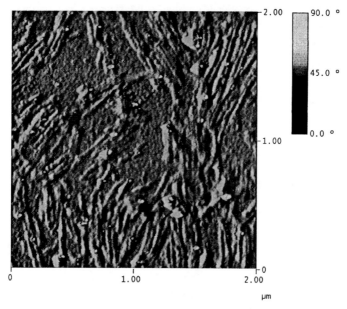

Figure 5. TEM (top) and phase contrast AFM (bottom) pictures for PFS$_{12}$-[Ru]-PEO$_{70}$ micelles cast on a formvar-coated copper grid (TEM) or on a silicon wafer (AFM).

These rod-like micelles have a tendency to lie parallel to each other. This kind of nematic order is thought to be an artifarct resulting from micelle preparation.

Spherical micelles are formed by the PS_{20}-[Ru]-PEO_{70} copolymer. Such a morphology is expected to be observed for micelles from a coil-coil amphiphilic diblock copolymer with $f \sim 60\%$.[8] For the PFS_{12}-[Ru]-PEO_{70} sample, f is ca. 50% and spherical micelles are expected. However PFS is a core-forming block that is characterized by a strong propensity to crystallize. This results in the formation of cylindrical micelles, as previously demonstrated for PFS-containing block copolymer in alkane solvents.[20] The presence of a crystallized PFS core has been previously ascertained in PFS_{12}-[Ru]-PEO_{70} micelles by differential scanning calorimetry experiments,[14] and is thought to explain why rod-like micelles are formed by this copolymer in water. However, it should be noted that large spherical micelles were prepared by a comparable PFS_{12}-b-PEO_{70} amphiphilic copolymer, but with a covalent bond at the block junction.[21] Nevertheless, the D_h measured by DLS (~160 nm) and the micellar diameter measured by TEM (~50 nm) for these spherical micelles are too large to be in agreement with equilibrium spherical dense micelles. It can be tentatively concluded that the *bis*-terpyridine ruthenium(II) complex have a beneficial effect, although unclear, on the crystallization of the PFS blocks.

Effect of added salt

Addition of salt has a deep effect on the micellar characteristic features. Whenever KCl has been added to a PS_{20}-[Ru]-PEO_{70} aqueous micellar solution and the final salt concentration has been adjusted to 1 mol/L, a sharp decrease in the characteristic size associated to the first population of the CONTIN histogram has been noted Actually, the mean D_h of the small-size population shifted from 65 to 43 nm. The reverse situation was observed for the large-size population. The effects can be rationalized on the basis of two different phenomenon. On the one hand, the cations of the added salt are known to interact with PEO.[22] As a result the PEO corona are carrying charged species and the local osmotic pressure is increasing. The clusters of micelles have thus a tendency to disintegrate in order to decrease the local osmotic pressure and the electrostatic repulsion forces. This could explain why the mean D_h of the first population decreases upon addition of salt. On the other hand, addition of salt results in the desolvation of the PEO chains, leading to the so-called "salting-out" effect. Therefore, the PEO coronal chains have a tendency to collapse which decreases the colloidal stability of the micelles. This effect is in line with the increase in size noted for the large aggregates.

The same experiment was conducted on the PFS_{12}-[Ru]-PEO_{70} aqueous micelles. When KCl concentration was adjusted to 1 mol/L, the mean D_h was found to be 86 nm. In this case, the slight decrease in micellar size can be attributed to the salting-out effect.

Upon addition of higher amounts of salt, flocculation is observed for the two investigated micellar solutions as a result of the reduced steric stabilization of the poorly solvated PEO chains.

Conclusions

Aqueous micelles have been prepared from metallo-supramolecular amphiphilic copolymers containing a PEO hydrophilic block and either a PS or a PFS hydrophobic block. Spherical micelles have been obtained from the PS-containing copolymer while flexible rod-like micelles have been observed for the PFS-based one. These micelles have been characterized by DLS, TEM and AFM. Clustering has been evidenced for the PS_{20}-[Ru]-PEO_{70} micelles, which is sensitive to the ionic strength of the aqueous medium. The rod-like micelles formed by the PFS_{12}-[Ru]-PEO_{70} copolymer are characterized by a constant diameter but are polydisperse in length. The formation of rod-like micelles is thought to be related to the crystallization of the PFS blocks in the micellar core.

Acknowledgments

JFG is grateful to the Communauté française de Belgique for the Action de Recherches Concertées NANOMOL 03/08-300. BGGL, AA and USS thank ir. B.F.M. de Waal (SyMOChem) for a generous gift of di-*tert*butyltricarbonate, Dipl.-Chem. M.A.R. Meier for MALDI-TOF MS measurements and the Dutch Council for Scientific Research (NWO), the Dutch Polymer Institute (DPI) and the Fonds der Chemischen Industrie for financial support. I.M. and M.A.W. are grateful to Materials and Manufacturing Ontario for an EMK Grant. I.M. thanks the Canadian Government for a Canada Research Chair.

References

1 Riess, G., *Prog. Polym. Sci.*, **2003**, *28*, 1107.
2 Zhang, L.; Eisenberg, A., *Science* **1995**; *268*, 1728.
3 Antonietti, M.; Heinz, S.; Schmidt, M.; Rosenauer, C., *Macromolecules* **1994**; *27*, 3276.
4 Spatz, J. P.; Mössmer, S.; Möller, M., *Angew. Chem. Int. Ed.* **1996**; *35*, 1510.
5 Grumelard, J.; Taubert, A.; Meier, W. *Chem Commun* **2004**, 1462.
6 Won,Y. Y.; Davis, H. T. ; Bates, F. S. *Science* **1999**, *283*, 960.
7 Discher, B. M. ; Won, Y. Y. ; Ege, D. S. ; Lee, J. C. M. ; Bates, F. S.; Discher, D. E.; Hammer, D. A. *Science* **1999**, *284*, 1143.
8 Discher, D. E.; Eisenberg, A. *Science* **2002**, *297*, 967.
9 Klok, H. A.; Lecommandoux, S. *Adv. Mater.* **2001**, *13*,1217.
10 Massey, J. A.; Power, K. N.; Manners, I.; Winnik, M. A. *J. Am. Chem. Soc.* **1998**, *120*, 9533.
11 Schubert, U. S.; Eschbaumer, C. *Angew. Chem. Int. Ed.* **2002**, *41*, 2892.
12 Lohmeijer, B. G. G.; Schubert, U. S. *Angew. Chem. Int. Ed.* **2002**, *41*, 3825.
13 Gohy, J.-F.; Lohmeijer, B. G. G.; Varshney, S. K.; Schubert, U. S. *Macromolecules* **2002**, *35*, 7427
14 Gohy, J.-F.; Lohmeijer, B. G. G.; Alexeev, A.; Wang, X.-S.; Manners, I. W.; Winnik, M. A.; Schubert, U. S. *Chem. Eur. J.* **2004**, *10*, 4315.
15 Zhang, L.; Shen, H.; Eisenberg, A. *Macromolecules* **1997**, *30*, 1001.
16 Berne, B. J.; Pecora, R. J. *Dynamic Light Scattering*; John Wiley and Sons: Toronto, **1976**.
17 Gohy, J.-F.; Lohmeijer, B. G. G.; Schubert, U. S. *Macromolecules* **2002**, *35*, 4560.
18 Regev, O.; Gohy, J.-F.; Lohmeijer, B. G. G.; Varshney, S. K.; Hubert, D. H.W.; Schubert, U. S. *Colloid Polym. Sci.* **2004**, *282*, 407.
19 Vogel, V.; Gohy, J.-F.; Lohmeijer, B. G. G.; van den Broek, J. A.; Haase, W.; Schubert, U. S.; Schubert D. *J. Polym. Sci., Part A*, **2003**, *41*, 3159.
20 Massey, J. A.; Power, K. N.; Manners, I.; Winnik, M. A. *J. Am. Chem. Soc.* **1998**, *120*, 9533.
21 Resendes, R.; Massey, J.; Dorn, H.; Winnik, M. A.; Manners, I. *Macromolecules* **2000**, *33*, 8.
22 Kjellander, R.; Florin, E. *J. Chem. Soc., Faraday Trans.* **1981**, *177*, 2053.

Chapter 4

Thermodynamics of 2,2′:6′,2″-Terpyridine–Metal Ion Complexation

Rainer Dobrawa[1], Pablo Ballester[2], Chantu R. Saha-Möller[1], and Frank Würthner[1,*]

[1]Institut für Organische Chemie, Universität Würzburg,
Am Hubland, D–97070 Würzburg, Germany
[2]Institució Catalana de Recerca i Estudis Avançats (ICREA)
and Institut Català d'Investigació Química (ICIQ), Avgda.
Països Catalans, s/n, 43007 Tarragona, Spain
*Corresponding author: wuerthner@chemie.uni-wuerzburg.de

The complexation of 2,2':6',2"-terpyridine (tpy), which is an important building block in metallosupramolecular chemistry, has been studied by titration experiments with the perchlorate hexahydrate salts of Fe^{2+}, Co^{2+}, Ni^{2+}, Cu^{2+} and Zn^{2+} in acetonitrile solution applying UV-vis and NMR spectroscopy and isothermal titration calorimetry (ITC). UV-vis titrations showed characteristic spectral changes upon addition of metal ions to tpy solutions indicating the formation of the $M(tpy)_2^{2+}$ complexes at a metal/ligand ratio of 1:2. ^1H NMR experiments revealed the exact amounts of 1:1 and 1:2 species during the titrations based on characteristic changes of the chemical shifts of tpy protons. ITC experiments were performed in both titration directions and gave ΔH^0 values between -14.5 and -22.2 kcal/mol for the complexation of the metal ions to tpy ligand together with estimates for the lower limit of the binding constants, which are in the range of 10^8 M^{-1} and higher.

43

Background

With the evolution of supramolecular chemistry[1] the use of metal complexes became popular for the construction of noncovalently bound architectures.[2,3] In particular the 2,2':6,'2"-terpyridine (tpy) ligand became an often used receptor unit for metal-ligand mediated self-assembly due to its easy availability and its predictable complexation behavior. Terpyridine derivatives with residues in the 4'-position are of special importance as they allow the supramolecular arrangement of two units in an exactly collinear fashion.

In the last decade an increasing number of supramolecular structures based on the tpy ligand have been published combined with an increase in complexity and functionality of the systems. A number of groups are working on the construction of coordination polymers based on the tpy ligand,[4] a field which was pioneered by Constable and coworkers.[5] Further classes of supramolecular systems constructed by tpy-metal complexes are dendrimers,[6] grid-like structures[7] and helicates.[8] Due to their luminescence properties, especially the Ru^{2+} and Os^{2+} complexes of terpyridine have been widely used to construct functional supramolecular systems, for example molecular rods with tpy complexes as energy donor and/or acceptor units. Extensive research has been done in this field by the groups of Sauvage, Balzani, and Ziessel.[9] Various ruthenium-terpyridine compounds have been studied for application as photoactive dyes in dye sensitized solar cells.[10] Furthermore, terpyridine is used as a receptor unit in fluorescence sensors for the recognition of metal ions[11,12] and amino acids.[13,14] Recently even more elaborate supramolecular systems, so-called molecular muscles, based on catenanes and rotaxanes have been introduced.[15]

Objective

In the context of our own research interests on metal ion-directed formation of multichromophoric architectures[16] and materials[17] we became recently interested in this highly useful terpyridine receptor unit for the preparation of metallosuramolecular polymers.[18] In this paper we like to report on our studies on the characterization of tpy complexes of some frequently used first row transition metal ions in a nonaqeous solvent, especially with regard to its application under proper thermodynamic control. A series of titration experiments has been conducted applying UV-vis and NMR spectroscopy and isothermal titration calorimetry (ITC). The results are reported in the following sections.

Experimental

General. Solvents were purified and dried according to standard procedures.[19] Spectroscopy grade solvents were used for UV-vis spectroscopy. All
compounds are commercially available and were used without further purification. Metal(II) perchlorate hexahydrate salts were supplied by Sigma-Aldrich,
Steinheim, Germany, except for iron(II) perchlorate hexahydrate (Alfa Aesar,
Karlsruhe, Germany). ^{1}H NMR spectra were recorded on a Bruker AMX400
spectrometer (400 MHz) in d_3-acetonitrile and chemical shifts δ (ppm) are calibrated against the residual CH_3CN signal ($\delta = 2.96$ ppm).

Isothermal Titration Calorimetry (ITC). Experiments were performed
using a MicroCal isothermal titration calorimeter system and analyzed using the
software (Origin 7) supplied with the instrument. In this technique the parameter
of interest for comparison with the experiment is the change in heat content
$\Delta Q(i)$ from the completion i-1 injection $Q(i$-1) to the completion of the i
injection $Q(i)$. The expression for the heat release $\Delta Q(i)$, from the ith injection is:

$$\Delta Q(i) = Q(i) + \frac{dV_i}{V_o}\left[\frac{Q(i) + Q(i-1)}{2}\right] - Q(i-1)$$

V_o = volume of the active cell; dV_i = volume of the ith injection.

The process of fitting experimental data relies on the selection of an
appropriate binding model, in the present case a model for two sites, and
involves 1) initial guesses of Ks (stability constants K_{11} and K_{11-21}) and ΔHs
(interaction enthalpies ΔH_{11} and ΔH_{11-21}); 2) calculation of $\Delta Q(i)$ for each
injection. This implies solving numerically the equation (done in Origin) to
obtain the values for the fractional saturation of each binding assuming the
parameters values of Ks and ΔHs assigned in step 1; 3) comparison of these
values with the measured heat for the corresponding experimental injection; 4)
improvement in the initial values of Ks and ΔHs by standard Marquardt methods;
and 5) iteration of the above procedure until no further significant improvement
in fit occurs with continued iteration.

Titrations were performed in both directions either by addition of metal salt
solution (5 mM) into terpyridine solution (0.5 mM) or by addition of terpyridine
solution (10 mM) into metal salt solution (0.5 mM). Solutions were prepared in
HPLC grade acetonitrile. Measurements were conducted at 25.3–25.5 °C, an
aliquot of 4 μL was applied per addition using a 250 μL syringe, injection
duration was 5 s, the interval between two subsequent injections was 100 s. All
calculated ΔH values are corrected against the heat of dilution and the
concentration changes during the titration. The obtained isotherms are
rectangular in shape with the height corresponding exactly to ΔH and with a
sharp drop occurring precisely at the stoichiometric equivalence point. The

rectangular shape is due to the fact that the titration was carried out under tight binding conditions. This shape of the curves is adequate for the determination of the enthalpy of binding and the stoichiometry of the complex. On the other hand, an accurate determination of the stability constants by non-linear least square fitting of calorimetric titration data requires the use of isotherm curves obtained under intermediate binding strength conditions (see page 37 Microcal User's Manual MAU130030 Rev. A 10/29/03). For this reason the reported stability constants are lower limit best estimates.

Spectroscopic titrations. ^{1}H NMR and UV-vis titrations were performed by adding solutions of the metal perchlorate hexahydrate (guest) to a solution of the tpy (host) in either a 5 mm NMR tube or a 1 cm path cuvette using microliter syringes. In both types of titration experiments the tpy was present in the guest solution at the same concentration as that in the NMR tube or cuvette to avoid dilution effects.

^{1}H NMR Titration Experiments with tpy. Aliquots of a zinc perchlorate hexahydrate solution (62 mM in d_3-acetonitrile) were added to a tpy solution (31 mM in d_3-acetonitrile) and ^{1}H NMR spectra were recorded after each addition. The quantities of the different species were calculated from the integrated areas of NMR signals.

UV-vis Titration Experiments with tpy and Metal Perchlorates. Titrations were performed at 25.0 °C by addition of aliquots of the respective metal perchlorate hexahydrate solution (0.5 mM in acetonitrile) to the host solution (0.05 mM in acetonitrile). UV-vis spectra were recorded after each addition.

Results & Discussion

Ligand Conformation and Complex Types

The solid state conformation of the uncomplexed and unprotonated form of tpy is *trans/trans* with respect to the N-C-C-N bonds, which is explained by the steric repulsion of the *ortho*-hydrogen atoms (3/3' and 5'/3") in combination with an electrostatic repulsion of the nitrogen lone pairs. The conformation of tpy in solution follows the same trend but is dependent on the hydrogen-bonding properties of the solvent. In acidic media, stabilization of the *cis/cis* conformation can be achieved by protonation of one pyridine unit, which leads to a favorable hydrogen-bonding between the proton and the lone pairs of the remaining free nitrogen atoms.

Two predominant types of tpy complexes can be found, depending on the metal ion and on the stoichiometry applied in the crystallization process: a 2:1 complex showing a slightly distorted octahedral coordination with the two tpy units arranged perpendicular to each other, and a 1:1 complex with the metal ion in a trigonal bipyramidal or a square-based pyramidal five-coordinate environment.[20] Figure 1 shows both the 1:1 and 2:1 complexes.

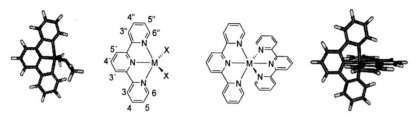

Figure 1. Structures and molecular models[a] of 1:1 (left) and 2:1 (right) tpy complexes. The 1:1 complex is modeled with Zn^{2+} and two acetonitrile molecules, X represents any counterion or coordinating solvent molecule. Depending on the metal ion also hexacoordinated species (distorted octahedral environment) have to be considered for the 1:1 complex.

UV-vis Titration Studies

UV-vis titrations have been performed to study the complexation of terpyridine with the metal ions of the first transition metal row from iron(II) to zinc(II). All metals ions were applied as the metal perchlorate hexahydrate salts, since they offer some advantages: the salts are all commercially available and have already been used in the construction of supramolecular systems. Due to their hydrated form they are not hygroscopic, which facilitates the precise preparation of stock solutions, and all salts are readily soluble in acetonitrile, as is the ligand itself. Acetonitrile was chosen since it acts as a competitive ligand for the complexation, which should lead to a decreased binding constant and additionally facilitate ligand exchange. Furthermore, it is not protic, thus eliminating any effects due to protonation or hydrogen bonding to the pyridine units.

The UV-vis spectrum of terpyridine in acetonitrile shows a broad, unstructured band at 235 nm and a second broad band at 280 nm with two shoulders at 300 nm and 310 nm (Figure 2, solid line). Both bands are attributed to π-π* transitions and the broadness of the bands and the absence of vibronic finestructure indicates the population of various conformations.[21] Apart from some special features, which will be discussed for each metal ion in the following paragraph, all metal ions used in this series cause similar changes in the tpy spectrum which are closely related to those caused by protonation of tpy.[21,22] Figure 2 shows the UV-vis spectrum of tpy in acetonitrile solution before (solid line) and after (dashed line) addition of trifluoroacetic acid. Both absorption bands are sharpened considerably pointing at a more coplanar chromophore. In addition, a new

---

[a] Fujitsu Quantum CAChe 5.0 (AM1)

band is coming up between 300-350 nm. The latter process is characteristic for the conformational change in the tpy from the *trans/trans* conformation in solution to the perfectly planarized *cis/cis* conformation in the complex. In the protonated species, the *cis/cis* conformation is stabilized by hydrogen bonds from the lateral pyridine units to the proton.

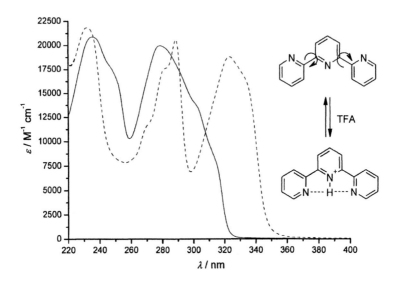

Figure 2. Spectra of tpy in acetonitrile (0.5 mM, solid line) and same solution after protonation with one equivalent of trifluoroacetic acid (dashed line).

Figure 3 shows the UV-vis titrations of terpyridine with the perchlorate hexahydrate salts of iron(II), cobalt(II), nickel(II) and zinc(II). The characteristic feature of the Fe(tpy)$_2$$^{2+}$ complex (Figure 3A) is the quite intense ($\varepsilon = 12000$ M^{-1} cm^{-1}) metal-ligand charge transfer (MLCT) band with a maximum at 550 nm, which is responsible for its deep violet color. The Co(tpy)$_2$$^{2+}$ complex (Figure 3B) also shows absorbance in the visible region but only with a small extinction coefficient of approx. 1400 M^{-1}cm^{-1}. A very weak absorption band can also be observed for Ni(tpy)$_2$$^{2+}$ (Figure 3C) at approx. 550 nm and 725 nm ($\varepsilon \approx 400$ M^{-1}cm^{-1}). The respective zinc(II) complex with the closed shell (d^{10}, Figure 3D) shows no absorption at all at wavelengths > 350 nm, confirming the absence of charge transfer transitions arising from metal-ligand coordination. The titration curves, which were evaluated for characteristic bands, show that all spectral changes occur at metal/ligand ratios below 0.5 indicating the formation of the M(tpy)$_2$$^{2+}$ complex. The straight slope and its abrupt saturation at a ratio of 0.5 indicate a high binding constant which cannot be determined from this set of data. All titrations show very clear isosbestic points suggesting that only two species, namely the uncomplexed tpy and the metal-complexed tpy, are present during the titration process. $M(tpy)^{2+}$ and $M(tpy)_2^{2+}$ cannot be distinguished, but based on the saturation at a ratio of 0.5 it can be concluded that all added metal ions are directly converted to $M(tpy)_2^{2+}$ up to a ration of 0.5.

The absence of spectral change at higher metal/ligand ratios does not necessarily mean that there is no further change in the system. As it has been mentioned, the most significant change in the UV-vis absorption spectrum of terpyridine is due to the *cis*/*trans* conformations in the tpy and concomitant formation of a planar chromophore, which is identical in the 1:1 and the 2:1 complex. Therefore, UV-vis spectroscopy is not suited to examine the reversibility of tpy complexes and to reveal the preference for the system at ratios > 0.5.

The titration of terpyridine with copper(II) is the only example in this series where a significant spectral change can be observed also in the range of metal/ligand ratio between 0.5 and 1. Figure 4 (left) shows the titration up to a molar ratio of 0.5 with the same features as mentioned previously for Fe^{2+}, Co^{2+}, Ni^{2+} and Zn^{2+}. In contrast, a sharp change is observed when the ratio of 0.5 is exceeded, characterized by the increasing intensity of three sharp bands between 260 nm and 280 nm and the shift of the long-wavelength band by ca. 10 nm combined with a splitting into two separate maxima. This second process terminates immediately when 1:1 stoichiometry is reached indicating the formation of a defined 1:1 complex. The fact that the spectrum of the $Cu(tpy)_2^{2+}$ and the $Cu(tpy)^{2+}$ species are not identical suggests two different terpyridine coordination modes in the two species. This hypothesis is supported by the results of ITC titrations and a model will be discussed in the respective paragraph.

^{1}H NMR Titration Experiments with Iron(II) and Zinc(II) Salts

To assess the reversibility of terpyridine-metal complex formation, ^{1}H NMR titrations of terpyridine have been performed with iron(II) and zinc(II) perchlorate. These two metals afford diamagnetic complexes and therefore allow the straightforward measurement of ^{1}H NMR spectra. We have excluded cobalt(II), nickel(II) and copper(II) complexes because of their paramagnetic high-spin nature, although it has been recently reported that cobalt(II) complexes can be studied by NMR under particular conditions.[23] ^{1}H NMR proved to be an ideal method to confirm ligand exchange (reversibility) in the Zn–tpy complex system as all species (tpy, $Zn(tpy)^{2+}$, $Zn(tpy)_2^{2+}$) present during a titration can be identified by their characteristic changes of the chemical shifts. This is possible because ligand exchange processes are slow on the NMR timescale but fast on the laboratory timescale. Figure 5 (left) shows the NMR spectra at different metal/ligand ratios, whereas Figure 5 (right) depicts a titration curve presenting the relative amount of the three species against the Zn^{2+}/tpy ratio. The signals of the protons H6/H6" (*) and H3'/H5' (#) are highlighted, since they

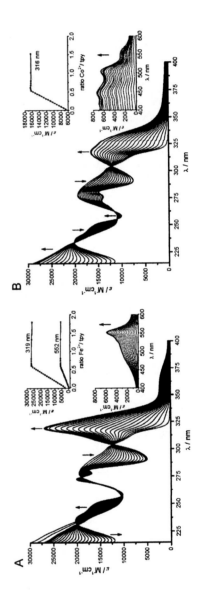

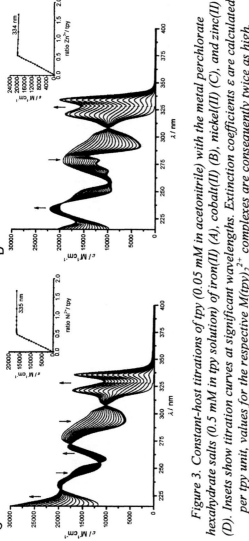

Figure 3. Constant-host titrations of tpy (0.05 mM in acetonitrile) with the metal perchlorate hexahydrate salts (0.5 mM in tpy solution) of iron(II) (A), cobalt(II) (B), nickel(II) (C), and zinc(II) (D). Insets show titration curves at significant wavelengths. Extinction coefficients ε are calculated per tpy unit, values for the respective M(tpy)$_2^{2+}$ complexes are consequently twice as high.

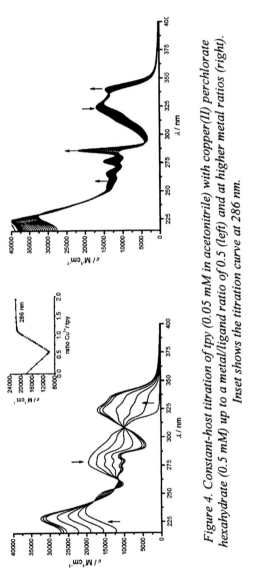

Figure 4. Constant-host titration of tpy (0.05 mM in acetonitrile) with copper(II) perchlorate hexahydrate (0.5 mM) up to a metal/ligand ratio of 0.5 (left) and at higher metal ratios (right). Inset shows the titration curve at 286 nm.

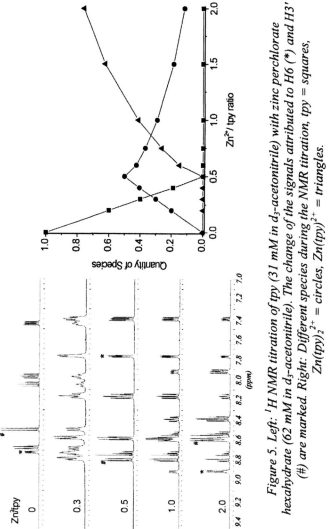

Figure 5. Left: ^1H *NMR titration of tpy (31 mM in d_3-acetonitrile) with zinc perchlorate hexahydrate (62 mM in d_3-acetonitrile). The change of the signals attributed to H6 (*) and H3' (#) are marked. Right: Different species during the NMR titration, tpy = squares, $Zn(tpy)_2^{2+}$ = circles, $Zn(tpy)^{2+}$ = triangles.*

are most sensitive to the coordination mode of the tpy complex. The protons H6 and H6", which are positioned next to the nitrogen atoms of the lateral pyridine rings, are of particular importance, since their chemical shift exhibit a drastic change on transformation of the free ligand to 2:1 and subsequently to the 1:1 complex species. In the 2:1 complex, H6/H6" are positioned directly above the central pyridine unit of the second tpy and experience shielding by the aromatic ring current. Therefore the signal is shifted to high field by approximately 1 ppm. Consequently, the absence of a second tpy unit in the 1:1 complex leads to a downfield shift. A significant change is also observed for the signal of H3', which is shifted downfield upon complexation. This signal shift is even more characteristic in 4'-substituted terpyridine derivatives where it appears as a sharp singlet whose change can easily be monitored.[18a]

The spectra and the titration curve in Figure 5 show that at a Zn^{2+}:tpy ratio >0.5 an equilibrium between 2:1 and 1:1 complexes is present. At a metal/ligand ratio of 2, still 25% of the tpy units are bound in a 2:1 complex. This equilibrium is dependent on the nature of the counterion and the solvent. For an identical titration applying zinc trifluoromethane sulfonate (triflate, OTf), the percentage of 2:1 complex present at a molar ratio of 2 was only 10% compared to 25% in the case of the perchlorate complex. Che and coworkers[24] could not observe reversibility for zinc terpyridine complexes when applying zinc acetate and zinc chloride in related ^{1}H NMR titrations in d_6-dmso, but observed reversible complexation when zinc nitrate hexahydrate was applied. In conclusion, the complexation behavior of tpy with Zn^{2+} is strongly dependent on the interplay between the metal-tpy interaction, the coordination ability of the counterion and the nature of the solvent, which can act as a competitive ligand or, if alcohols are applied, can interact with the counterions and tpy ligands as hydrogen bond donors.

Isothermal Titration Calorimetry (ITC)

To further investigate the binding energy for tpy-metal bonding, ITC titrations[25,26,27] were performed with the same series of metal ions that have been already characterized by UV-vis spectroscopy. The interesting advantage of ITC method is that the titration direction is not restricted to the addition of metal ions to a tpy solution. The latter is the case for ^{1}H NMR and UV-vis spectroscopy, as the metal ions do not show NMR signals or UV-vis absorption, which could be monitored during the titration. In contrast, ITC is only sensitive to the heat of reaction, independent on the nature of the compound present in the cell and added in aliquots through the syringe. Again, all experiments were performed with the perchlorate hexahydrate salts in acetonitrile solution to ensure comparability.

ITC results for the zinc/tpy system in acetonitrile are presented in and Figure 6. The addition of tpy to a zinc perchlorate solution (Figure 6 A) yields a constant reaction heat of -14.5 kcal/mol up to a molar ratio of 2. This stoichiometry is anticipated for the formation of a 2:1 complex. The fact that no difference in the heat flow can be detected for the whole titration allows two interpretations: either only 2:1 complexes are formed from the beginning and no 1:1 species are present in considerable amounts in the course of both experiments, or the enthalpy detected for the complexation is identical, independent whether tpy forms a complex with a free Zn^{2+} or alternatively with an already formed 1:1 complex. ^{1}H NMR titrations indicate an equilibrium between 2:1 and 1:1 species. If this information is considered, ΔH^0 of -14.5 kcal/mol represents the enthalpy of the first as well as the second process.

Figure 6 B shows the inverted situation, with Zn^{2+} being added to a tpy solution. As expected based on our UV-vis studies, the endpoint of the titration is reached at a molar ratio of Zn^{2+}/tpy = 0.5. In this case, the resulting ΔH^0 gives the sum of the values for the first and the second binding event. A ΔH^0 of -28.4 kcal/mol corresponds to two tpy-Zn^{2+} coordination events of -14.2 kcal/mol in good accordance with the former results. As the identical titration was also done by ^{1}H NMR, it is known that after the ratio of 0.5 is exceeded, the $Zn(tpy)_2^{2+}$ complex is opened to form the 1:1 species. ITC results indicate that this process is not connected with a significant calorimetric signal. Therefore, the only process yielding a detectable enthalpy change is the terpyridine complexation, independent whether this process takes place with a free Zn^{2+} ion or a monocomplexed $Zn(tpy)^{2+}$ species. From these results the absence of a special thermodynamic stabilization of the octahedral $Zn(tpy)_2^{2+}$ complex can be concluded.

The titrations with iron(II), cobalt(II) and nickel(II) perchlorates gave similar curves which only differ in the absolute values for ΔH^0. In all titrations, the first and the second binding event cannot be distinguished. As the concentration of the solutions was optimized for an acceptable signal/noise ratio and therefore had to be kept in the order of 0.5 mM, it was not possible to determine the exact binding constants for the systems under investigation. Comparison with the UV-vis titrations (Figure 3), which have been performed with tenfold dilute solutions, show that also in this concentration range the binding strength is too high to determine K precisely. Curve-fitting tests with different binding constants could be applied to estimate a lower limit for the binding constant, which is in the order of $>10^8$ M^{-1}. Table 1 and Table 2 summarize the resulting enthalpies and estimated binding constants for all metal ions studied.

ITC study with copper perchlorate. The UV-vis titration study for copper perchlorate in acetonitrile (Figure 4) revealed that the copper-tpy bond exhibits pronounced reversibility and that the UV-vis spectra of the 1:1 and the 2:1 species are different. ITC studies have been used in order to obtain a deeper insight into the Cu-tpy binding process. In analogy to the previously discussed ITC stud-

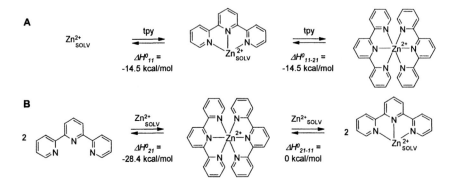

Scheme 1. Thermodynamic scheme for the titration of zinc perchlorate with tpy in acetonitrile and the corresponding ΔH^0 values determined by ITC for addition of tpy to the Zn^{2+} solution (A) and vice versa (B).

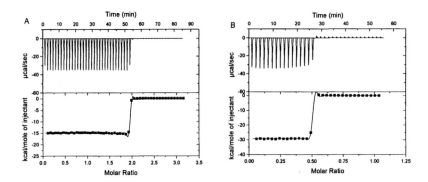

Figure 6. (A) ITC titration curves for the addition of tpy (10 mM) to a Zn^{2+} solution (0.5 mM). (B) Reversed titration of Zn^{2+} (5 mM) to a tpy solution (0.5 mM).

Table 1. Summary of the ITC results for the titration of tpy to the metal perchlorate hexahydrate solutions in acetonitrile.[a]

	$M^{2+} + tpy \rightleftarrows M(tpy)^{2+}$			$M(tpy)^{2+} + tpy \rightleftarrows M(tpy)_2^{2+}$		
	$\Delta H^0{}_{11}$ [kcal/mol] [kJ/mol]		$K_{11}{}^{(b)}$ [M^{-1}]	$\Delta H^0{}_{11\text{-}21}$ [kcal/mol] [kJ/mol]		$K_{11\text{-}21}{}^{(b)}$ [M^{-1}]
Fe^{2+}	-19.1	-79.9	$>10^8$	-19.1	-79.9	$>10^8$
Co^{2+}	-14.7	-61.5	$>10^8$	-14.7	-61.5	$>10^8$
Ni^{2+}	-16.0	-66.9	$>10^8$	-16.0	-66.9	$>10^8$
Cu^{2+}	-22.2	-92.2	$>10^8$	**-13.0**	**-54.4**	$\approx 10^6$
Zn^{2+}	-14.5	-60.7	$>10^8$	-14.5	-60.7	$>10^8$

[a] Errors for ΔH^0 are ±0.5 kcal/mol. Indices of ΔH^0 values refer to schemes 1 and 2. [b] Estimated values. Note that some data presented in our earlier publication (ref. 18c, table 1) are corrected here.

Table 2. Summary of the ITC results for the titration of metal perchlorate hexahydrates to a tpy solution in acetonitrile.[a]

	$2\,tpy + M^{2+} \rightleftarrows M(tpy)_2^{2+}$			$M(tpy)_2^{2+} + M^{2+} \rightleftarrows 2\,M(tpy)^{2+}$		
	$\Delta H^0{}_{21}$ [kcal/ mol] [kJ/mol]		$K_{21}{}^{(b)}$ [M^{-2}]	$\Delta H^0{}_{21\text{-}11}$ [kcal/ mol] [kJ/mol]		$K_{21\text{-}11}$
Fe^{2+}	-38.4	-160.7	$>10^{16}$	± 0	± 0	$4^{(c)}$
Co^{2+}	-28.2	-118.0	$>10^{16}$	± 0	± 0	$4^{(c)}$
Ni^{2+}	-31.2	-130.5	$>10^{16}$	± 0	± 0	$4^{(c)}$
Cu^{2+}	-33.8	-141.4	$>10^{16}$	**-8.7**	**-36.4**	$100^{(d)}$
Zn^{2+}	-28.4	-118.8	$>10^{16}$	± 0	± 0	$4^{(c)}$

[a] Errors for ΔH^0 are ±0.5 kcal/mol. Indices of ΔH^0 values refer to schemes 1 and 2. [b] Estimated values. [c] Calculated assuming cooperativity factor equals to 1. [d] Calculated as the ratio $K_{11}/K_{11\text{-}21}$.

ies, Figure 7 (left) shows the result of the titration of terpyridine to a copper(II) solution. In this curve, two different binding events can be distinguished, one with saturation occurring at a molar ratio of 1:1, the second at a molar ratio of 1:2 with a decreased ΔH^0. Since the stoichiometries fit to the binding of the first and the second tpy unit, the released enthalpy for the binding of the second tpy is smaller.

A complementary result is obtained for the reverse titration, the addition of tpy to a copper(II) solution. For all other metal ions, only a residual baseline signal was detected after exceeding the ratio of 0.5 in this type of titration, meaning that in the case of a reversible complexation there is no enthalpy change when the 2:1 species is changed into the 1:1 complex. For copper, exactly the opposite behavior is observed: when the 2:1 complex, which is present in solution at the molar ratio of 0.5, is opened to form two 1:1 species, this process is strongly exothermic, indicating that a thermodynamically more favorable species is formed.

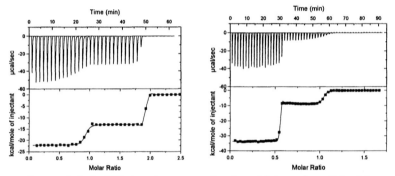

Figure 7. Left: ITC titration curves for the addition of tpy (10 mM)
to a Cu^{2+} solution (0.5 mM). Right: Reverse titration of Cu^{2+} (5 mM)
into a tpy solution (0.5 mM).

The ΔH^0 values measured for the cupper(II)-tpy binding processes by ITC titration indicate that in 1:1 and 2:1 complexes the tpy ligands are differently bound which is in corroboration with the results of UV-vis titration studies presented previously. For the coordination of the first tpy ligand with Cu^{2+} to form $Cu(tpy)^{2+}$ species, a ΔH^0 value of -22.2 kcal/mol is measured (corresponds to -7.4 kcal/mol for each pyridine unit of the tridentate tpy), but only -13.4 kcal/mol are obtained for the coordination of the second tpy ligand leading to the $Cu(tpy)_2^{2+}$ complex (Scheme 2A). In other words, the enthalpy gain by bonding of three pyridine units of the second tpy corresponds to that for two pyridines of the first tpy. This thermodynamic outcome can be rationalized in terms of different binding of the two tpy ligands in $Cu(tpy)_2^{2+}$ due to the well-known Jahn-Teller distortion in octahedral copper(II)-tpy complexes.[28] The crystallographic data reported for a $Cu(tpy)_2^{2+}$ complex revealed a strong distortion in CuN_6 octahedron with two C-N bond lengths of ca. 2.3 Å and other four C-N bond distances of ca. 2.0 Å.[28] Accordingly, in our 2:1 complex (Scheme 2A) two pyridine units are more weakly bound to Cu^{2+} (represented by dotted bonds in structure) compared to the other four pyridines and, therefore, the former bonds contribute less to the enthalpy gain. This rational is also supported by the fact that the reverse titration by addition of Cu^{2+} ions to the $Cu(tpy)_2^{2+}$ system (Scheme 2B) provides a considerable enthalpy gain of $\Delta H^0 = -8.7$ kcal/mol.

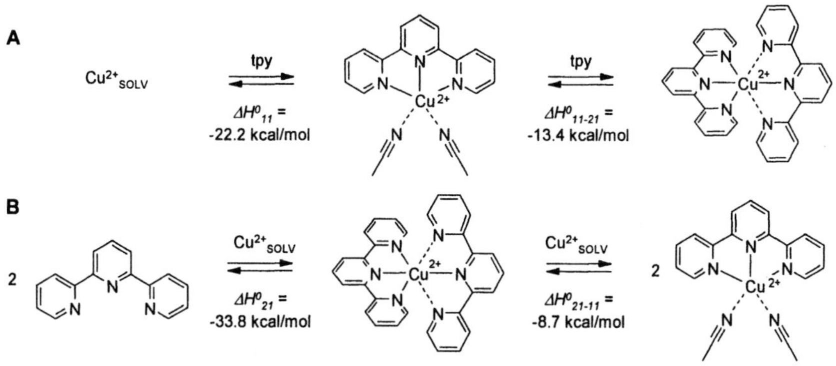

Scheme 2. Proposed mechanism for the titration of copper perchlorate with tpy in acetonitrile and the corresponding ΔH values determined by ITC for addition of tpy to the Cu^{2+} solution (A) and vice versa (B). The values suggest that the second tpy ligand in the 2:1 complex is bound less tightly.

All results of the ITC titrations are summarized in Tables 1 and 2 for the two titration modes respectively. The attachment of the first tpy unit to the copper(II) ion gives the largest binding enthalpy with a ΔH^0 of -22.2 kcal/mol, followed by iron(II) with -19.1 kcal/mol. The value for nickel(II) is situated in between with -16.0 kcal/mol, whereas the measured enthalpies of cobalt(II) and zinc(II) are the lowest with >-15 kcal/mol. With the exception of copper(II), all values determined by the addition of metal to the tpy solution, meaning an immediate formation of the 2:1 complex, are approximately twice the value for the sequential measurement by titration of tpy to the metal ion solution. The assessment of the reported complexation enthalpies by values obtained by other groups[29] is difficult, since those experiments have been performed in aqueous solution and with various counterions, so that no direct comparison is possible. In addition, it should be mentioned that our measurements in acetonitrile did not take into account the influences of ionic strength.

Conclusion

In conclusion, the complexation of terpyridine with a series of transition metal ions has been studied in acetonitrile as a nonaqueous, aprotic solvent by UV-vis and ^{1}H NMR spectroscopy and isothermal titration calorimetry (ITC). A collection of UV-vis spectra is available for all investigated metal complexes. The characteristic changes in the ^{1}H NMR spectrum have been discussed for the Zn^{2+}-terpyridine system showing reversible ligand exchange. ITC studies provide binding enthalpies which could be determined for both titration directions, i.e. also the addition of tpy to the metal ion solution, which is not possible in UV-vis and NMR studies. The knowledge of the basic binding parameters, i.e. the binding constant, enthalpy and reversibility is essential for the rational construction of more elaborate supramolecular architectures. Zn^{2+} and Cu^{2+} complexes seem to be most rewarding candidates for this research.[30]

Acknowledgments

The authors acknowledge financial support by the *Deutsche Forschungsgemeinschaft* (DFG, grant project WU 317/3-1) and the *Deutscher Akademischer Austausch Dienst* (DAAD, Acciones Integradas Hispano-Alemanas).

References

1   (a) Lehn, J.-M. Supramolecular Chemistry: Concepts and Perspectives; VCH: Weinheim, **1995**; (b) Steed, J. W.; Atwood, J. L. Supramolecular Chemistry: A Concise Introduction; John Wiley & Sons, Ltd.: Chichester, **2000**.

2   Swiegers, G. F.; Malefetse, T. J. *Chem. Rev.* **2000**, *100*, 3483-3537.

3   Würthner, F.; You, C.-C.; Saha-Möller, C. R. *Chem. Soc. Rev.* **2004**, *33*, 133-146.

4   (a) Schütte, M.; Kurth, D. G.; Linford, M. R.; Cölfen, H.; Möhwald, H. *Angew. Chem. Int. Ed.* **1998**, *37*, 2891-2893; (b) Kelch, S.; Rehahn, M. *Macromolecules* **1999**, *32*, 5818-5828; (c) Schubert, U. S.; Eschbaumer, S. *Angew. Chem. Int. Ed.* **2002**, *41*, 2892-2926; (d) Lohmeijer, B. G. G.; Schubert, U. S. *Macromol. Chem. Phys.* **2003**, *204*, 1072-1078; (e) Andres, P. R.; Schubert, U. S. *Adv. Mater.* **2004**, *16*, 1043-1068.

5   Constable, E. C.; Cargill Thompson, A. M. W. J. *Chem. Soc., Dalton Trans.* **1992**, 3467-3475.

6   Newkome, G. R.; He, E.; Moorefield, C. N. *Chem. Rev.* **1999**, *99*, 1689-1746.

7   Ziener, U.; Lehn, J.-M.; Mourran, A.; Möller, M. *Chem. Eur. J.* **2002**, *8*, 951-957.

8   Piguet, C.; Bernardinelli, G.; Hopfgartner, G. *Chem. Rev.* **1997**, *97*, 2005-2062.

9   Reviews: (a) Sauvage, J.-P.; Collin, J.-P.; Chambron, J.-C.; Guillerez, S.; Coudret, C.; Balzani, V.; Barigelletti, F.; De Cola, L.; Flamigni, L. *Chem. Rev.* **1994**, *94*, 993-1019; (b) Ziessel, R.; Hissler, M.; El-Ghayoury, A.; Harriman, A. *Coord. Chem. Rev.* **1998**, *178-180*, 1251-1298; (c) Baranoff, E.; Collin, J.-P.; Flamigni, L.; Sauvage, J.-P. *Chem. Soc. Rev.* **2004**, *33*, 147-155.

10 Zakeeruddin, S. M.; Nazeeruddin, M. K.; Pechy, P.; Rotzinger, F. P.; Humphry-Baker, R.; Kalyanasundaram, K.; Graetzel, M.; Shklover, V.; Haibach, T. *Inorg. Chem.* **1997**, *36*, 5937-5946.

11 Goze, C.; Ulrich, G.; Charbonniere, L.; Cesario, M.; Prange, T.; Ziessel, R. *Chem. Eur. J.* **2003**, *9*, 3748-3755.

12 Barigelletti, F.; Flamigni, L.; Calogero, G.; Hammarstrom, L.; Sauvage, J.-P.; Collin, J.-P. *Chem. Commun.* **1998**, 2333-2334.

13 Aiet-Haddou, H.; Wiskur, S. L.; Lynch, V. M.; Anslyn, E. V. *J. Am. Chem. Soc.* **2001**, *123*, 11296-11297.

14 Wong, W.-L; Huang, K.-H.; Teng, P.-F.; Lee, D.-S.; Kwong. H.-L. *Chem. Commun.* **2004**, 384-385.

15 Jimenez-Molero, M. C.; Dietrich-Buchecker, C.; Sauvage, J.-P. *Chem. Commun.* **2003**, 1613-1616.

16 (a) Würthner, F.; Sautter, A.; Schmid, D.; Weber, P. J. A. *Chem. Eur. J.* **2001**, *7*, 894-902; (b) Sautter, A.; Schmid, D.; Jung, G.; Würthner, F. *J. Am. Chem. Soc.* **2001**, *123*, 5424-5430; (c) Würthner, F.; Sautter, A. *Org. Biomol. Chem.* **2003**, *1*, 240-243 (d) You, C.-C.; Würthner, F. *J. Am. Chem. Soc.* **2003**, *125*, 9716-9725.

17 (a) Würthner, F.; Thalacker, C.; Sautter, A. *Adv. Mater.* **1999**, *11*, 754-758; (b) Würthner, F.; Thalacker, C.; Diele, S.; Tschierske, C. *Chem. Eur. J.* **2001**, *7*, 2245-2253; (c) Würthner, F.; Chen, Z.; Hoeben, F. J. M.; Osswald, P.; You, C.-C.; Jonkheijm, P.; van Herrikhuyzen, J.; Schenning, A. P. H. J.; van der Schoot, P. P. A. M.; Meijer, E. W.; Beckers, E. H. A.; Meskers, S. C. J.; Janssen, R. A. J. *J. Am. Chem. Soc.* **2004**, *126*, 10611-10618.

18 (a) Dobrawa, R.; Würthner, F. *Chem. Commun.* **2002**, 1878-1879; (b) Dobrawa, R.; Kurth, D. G.; Würthner, F. *Polym. Preprints* **2004**, *45*, 378-379; (c) Dobrawa, R.; Lysetska, M.; Ballester, P.; Grüne, M.; Würthner, F. *Macromolcules* **2005**, *38*, 1315-1325; (d) Dobrawa, R.; Würthner, F. *J. Polym. Sci. Part A: Polym. Chem.*, in press.

19 Perrin, D. D.; Armarego, W. L. F. Purification of Laboratory Chemicals, 2nd ed., Pergamon Press: Oxford, **1980**.

20   (a) Corbridge, D. E. C.; Cox, E. G. *J. Chem. Soc.* **1956**, 594-603; (b) Harris, C. M.; Lockyer, T. N.; Stephenson, N. C. *Aust. J. Chem.* **1966**, *19*, 1741-43; (c) Constable, E. C.; Phillips, D.; Raithby, P. R. *Inorg. Chem. Commun.* **2002**, *5*, 519-521.

21   Nakamoto, K. *J. Phys. Chem.* **1960**, *64*, 1420-1425.

22   Offenhartz, P. O'D.; George, P.; Haight, G. P. Jr. *J. Phys. Chem.* **1963**, *67*, 116-118.

23   (a) Rao, J. M.; Macero, D. J.; Hughes, M. C. *Inorg. Chim. Acta* **1980**, *41*, 221-226; (b) Waldmann, O.; Hassmann, J.; Müller, P.; Volkmer, D.; Schubert, U. S.; Lehn, J. M. *Phys. Rev. B.* **1998**, *58*, 3277-3285; (c) Chow, H. S.; Constable, E. C.; Housecroft, C. E.; Kulicke, K. J.; Tao, Y. *Dalton Trans.* **2005**, 236-237.

24   Yu, S.-C.; Kwok, C.-C.; Chan, W. K.; Che, C.-M. *Adv. Mater.* **2003**, *15*, 1643-1647.

25   Freire, E.; Mayorga, O. L.; Straume, M. *Anal. Chem.* **1990**, *62*, 950A-959A.

26   Blandamer, M. J.; Cullins, P. M.; Engberts, J. B. F. N. *J. Chem. Soc., Faraday Trans.* **1998**, *94*, 2261-2267.

27   Jelesarov, I.; Bosshard, H. R. *J. Mol. Recognit.* **1999**, *12*, 3-18.

28   Allmann, R.; Henke, W.; Reinen, D. *Inorg. Chem.* **1978**, *17*, 378-382.

29   (a) Holyer, R. H.; Hubbard, C. D.; Kettle, S. F. A.; Wilkins, R. G. *Inorg. Chem.* **1966**, *5*, 622-62; (b) Kim, K.Y.; Nancollas, G. H. *J. Phys. Chem.* **1977**, *81*, 948-952; (c) Calì, R.; Rizzarelli, E.; Sammartano, S.; Siracusa, G. *Transition Met. Chem.* **1979**, *4*, 328-323; (d) Bullock, J. I.; Simpson, P. W. G. *J. Chem. Soc., Faraday Trans.* **1981**, *77*, 1991-1997.

30   (a) Chichak, K. S.; Cantrill, S. J.; Pease, A. R.; Chiu, S.-H.; Cave, G. W. V.; Atwood, J. L.; Stoddart, J. F. *Science* **2004**, *304*, 1308-1312; (b) Poleschak, I.; Kern, J.-M.; Sauvage, J.-P. *Chem. Commun.* **2004**, 474-476.

Chapter 5

Grid Forming Metal Coordinating Macroligands: Synthesis and Complexation

Richard Hoogenboom[1], Jurriaan Huskens[2], and Ulrich S. Schubert[1,*]

[1]Laboratory of Macromolecular Chemistry and Nanoscience, Eindhoven University of Technology and the Dutch Polymer Institute (DPI), P.O. Box 513, 5600 MB Eindhoven, The Netherlands
[2]Laboratory of Supramolecular Chemistry and Technology, MESA[+] Institute for Nanotechnology, University of Twente, P.O. Box 217, 7500 AE, Enschede, The Netherlands
*Corresponding author: email: u.s.schubert@tue.nl; Internet: www.schubert-group.com

The synthesis of poly(L-lactide) bispyridylpyridazine macroligands from a hydroxyl functionalized ligand by controlled ring-opening polymerization is described. The resulting macroligands were characterized by ^{1}H-NMR spectroscopy, GPC and MALDI-TOF MS. The metal coordination of those macroligands with copper(I) resulting in polymeric [2×2] metal grids was investigated by UV-vis titration studies. In addition, the complexation behavior obtained from these experiments was modeled in order to explore the complexation parameters and the corresponding mechanism.

Background

In recent years, the incorporation of supramolecular moieties into well-defined polymer structures was established as attractive method for the construction of novel materials that combine the mechanical properties of the polymers and the physical and reversible properties of the supramolecular moieties.[1,2] For directed self-assembly, the most promising types of supramolecular interactions are hydrogen bonding and metal–ligand interactions because of their high directionality. In this contribution, metal-coordinating units are utilized because their self-assembly can be easily tuned from very labile to inert by varying the metal ion. In addition, the complexation can be triggered by addition of metal ions and the formed complexes can be addressed and manipulated by changes in pH, electrochemistry, temperature or concentration. Well-defined polymers bearing a metal coordinating end-group can be prepared by post-functionalization of polymers[3,4,5] or by the utilization of functional inititiators and/or terminating agents for living/controlled polymerization techniques.[6-10] We have chosen for the functional initiator approach since it provides easier isolation of the macroligands: After precipitation each polymer chain has a ligand attached, whereas post-functionalization often requires difficult chromatographic purification steps since the functionalized and unfunctionalized polymers have very similar properties. Up to now, this functional initiator approach has been mainly applied to incorporate bipyridine[6-8,10,11] and terpyridine[6,9,12] ligands as polymer end-groups. Those macroligands assemble upon the addition of metal ions with two or three ligands, respectively. However, in literature many ligands have been described that form larger grid-like metal complexes with various metal ions.[13] Figure 1 depicts some examples of ligands that form [1×1], [2×2] or [3×3] metal complexes with copper(I) or silver(I) ions.

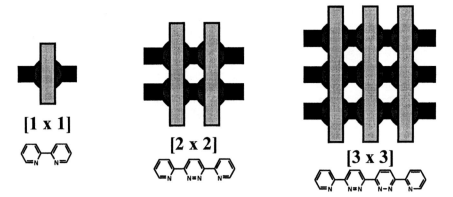

Figure 1. Examples of metal coordinating ligands that self-assemble into grid-like architectures with copper(I) and (sliver(I) ions.

Objective

The objective of this study was to extend the supramolecular initiator approach from bipyridine and terpyridine initiators to grid-forming initiators to construct larger polymeric architectures (schematically depicted in Figure 2). For this purpose, the synthetically easy accessible 3,6-*bis*(2-pyridyl)-pyridazine ligand was chosen, which forms [2×2] metal grids with copper(I)[14] and silver(I)[15] ions. In this contribution, we describe the synthesis and characterization of poly(*L*-lactide) macroligands from a hydroxyl functionalized 3,6-*bis*(2-pyridyl)-pyridazine.[16,17] In addition, extensive UV-vis titration studies and subsequent modeling of the complexation behavior is discussed for macroligands with different molecular weigths.

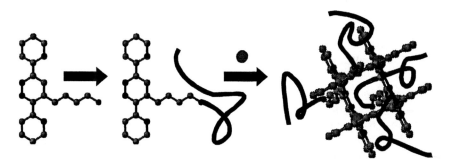

Figure 2. The construction of polymeric [2×2] metal grids: Synthesis of bis-pyridylpyridazine macroligands followed by self-assembly with copper(I) ions.

Results & Discussion

Synthesis of the poly(*L*-lactide) macroligands

The synthesis of *bis*pyridylpyridazine macroligands via the functional initiator approach requires a functional handle on the ligand. Therefore, a hydroxyl functionalized ligand was synthesized by an inverse-electron demand Diels-Alder reaction between 3,6-*bis*(2-pyridyl)-tetrazine and 5-hexyne-1-ol, which resulted in the formation of 3,6-*bis*-(2-pyridyl)-4-hydroxybutylpyridazine **1** after the elimination of a nitrogen molecule. A similar reaction between 3,6-*bis*(2-pyridyl)-tetrazine and ethyne was first reported by Butte and Case.[18]

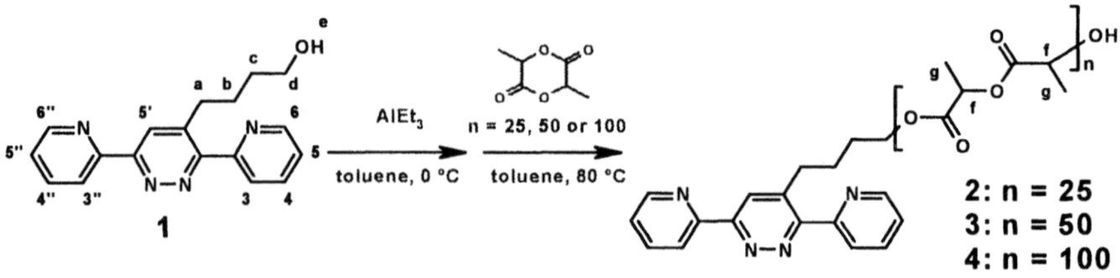

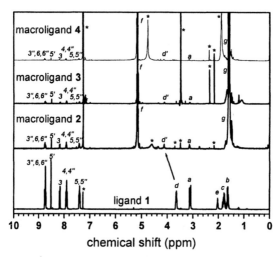

Scheme 1. Synthesis of the poly(L-lactide) macroligands 2-4 from 3,6-bis(2-pyridyl)-4 hydroxybutylpyridazine 1.

In the next step towards the construction of polymeric [2×2] metal grids, an aluminum alkoxide was generated *in situ* from the hydroxyl functionalized 3,6-*bis*(2-pyridyl)-pyridazine by reacting it with triethyl aluminum. This aluminum alkoxide was utilized as initiator for the controlled ring-opening polymerization of *L*-lactide resulting in macroligands **2-4** as depicted in Scheme 1. The utilization of aluminum alkoxides for the controlled ring-opening polymerization of *L*-lactides was first described by Kricheldorf and coworkers.[19] After purification of the polymer by precipitation in methanol, the resulting poly(*L*-lactide)s were characterized with ¹H-NMR spectroscopy, gel permeation chromatography (GPC) and MALDI-TOF-MS. The ¹H-NMR spectra clearly shows the corresponding proton resonances for both the ligand and the polymer backbone (f and g at 5.2 ppm and 1.6 ppm) as depicted in Figure 3. Moreover, the ligand resonances decreased with increasing polymer length as expected.

Figure 3. ¹H-NMR spectra obtained for 3,6-bis-(2-pyridyl)-4-hydroxybutylpyridazine 1 and the macroligands 2-4 (in CDCl₃).

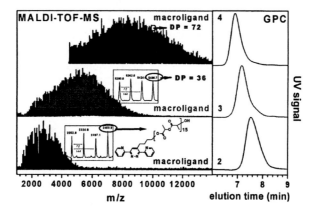

Figure 4. MALDI-TOF-MS spectra with end-group analysis (left) and GPC signals obtained by UV-detector for the macroligands 2-4 (right).

GPC analysis also clearly demonstrated the incorporation of the ligand into the polymer chains since the polymer signal was detected with both refractive index and ultraviolet detectors (Figure 4 right). The UV-detector was set to 254 nm, where the polymer does not absorb. Moreover, end-group analysis from the MALDI-TOF-MS spectra corresponded exactly to a certain amount of monomer units and the ligand as depicted in Figure 4 left. The number average molecular weight (M_n) for the macroligands was calculated from ^{1}H-NMR spectroscopy (from integrating the ligand and polymer signals), GPC (against polystyrene standards) and from MALDI-TOF-MS. The resulting M_n's are plotted against the initial monomer to initiator ([M]/[I]) ratios in Figure 5.

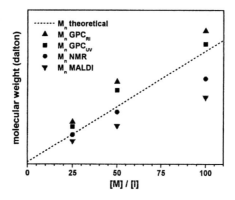

Figure 5. Dependence of molecular weight on the initial monomer to initiator ratio.

The molecular weights correspond well to the theoretical molecular weights, whereby only the molecular weight of the largest macromonomer is a bit off due to insufficient reaction time. Moreover, the $M_{n,GPC}$ is too high due to its polystyrene calibration and the $M_{n,MALDI}$ is too low due to the higher ionization probability of the lower molecular weight poly(L-lactide) chains. MALDI-TOF-MS and GPC characterization of the macroligands 2-4 revealed narrow molecular weight distributions with polydispersity indices equal or lower than 1.20. The good correspondence between the theoretical and obtained molecular weights and these low PDI values clearly demonstrated that the polymers were synthesized in a controlled manner.

Complexation studies of the poly(L-lactide) macroligands

The complexation of the macroligands 2-4 was investigated by UV-vis titration studies. To a solution of the macroligand in diochloromethane (DCM), a *tetrakis*acetonitrile copper(I) hexafluorophosphate solution in DCM was added step-wise. After each titration step a UV-vis spectrum was recorded resulting in the titration graphs as depicted in Figure 6 left for the three macroligands 2-4. The possibility of performing those UV-titration studies in DCM solution already demonstrated that the properties of both the ligand and the polymers are combined in the macroligands: Adding the same copper(I) salt solution to a solution of the unreacted 3,6-*bis*-(2-pyridyl)-4-hydroxybutyl-pyridazine 1 in DCM instantaneously led to precipitation of the self-assembled copper(I) metal grids. Upon addition of copper(I)ions to the macroligands, a metal to ligand charge transfer (MLCT) band appears with λ_{max} at 467 nm that during the course of the titration shifts toward 437 nm (see insets of the titration graph Figure 6 left). The basis of this shift could not be unambiguously proven. However, it most likely results from a transition from complexes with two ligands and one copper(I) ion into complete polymeric [2×2] grid-like complexes. Remarkably, this shift in λ_{max} already occurs with the first titration step (~ 0.1 equivalents of copper(I) ions) for the smallest macroligand 2, whereas λ_{max} only shifts with the addition of ~ 0.5 equivalents of copper(I) ions for the other macroligands 3 and 4. To further elucidate the ongoing complexation processes, the increase in absorption at 437, 467 and 567 nm was plotted against the added equivalents of copper(I) ions (Figure 6 right). The difference in complexation behavior of the smaller macroligand 2 compared to the other macroligands 3 and 4 is also clearly noticeable from these plots.

To gain further insight into the complexation mechanism, the titration data were modeled (Figure 6 right, dotted lines). To simplify the modeling of this complexation that consist of various equilibria (over ten different processes), it

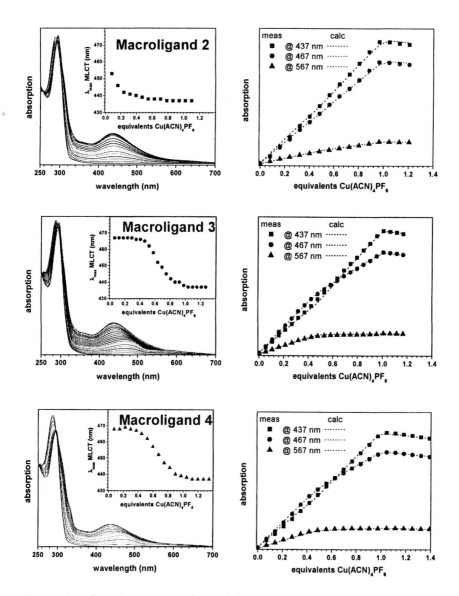

Figure 6. Left: UV-vis spectra obtained during the titration of Cu(1)(ACN)₄PF₆
*to a solution of the macroligands **2** (top), **3** (middle) and **4** (bottom); the insets*
show the observed shifts in MLCT from 467 nm to 437 nm; Right: Observed
(symbols) and modeled (dotted lines) change in absorption at 437, 467 and
*567 nm for the macroligands **2** (top), **3** (middle) and **4** (bottom).*

was assumed that the grid-formation only consists of two discernible complexation steps (L = ligand):

Step 1: $$2 \cdot L + Cu(I) \xrightarrow{k_1} L_2 Cu(I)$$

Step 2: $$2 \cdot L_2 Cu(I) + 2 \cdot Cu(I) \xrightarrow{k_2} L_4 Cu(I)_4$$

The experimental data obtained for macroligands **3** and **4** can be nicely fitted with only these two equilibria. Unfortunately, no reliable values for the complexation constants could be extracted from these fits: The two sharp transitions in the data at 0.5 and 1.0 equivalents of copper(I) ions render this impossible in this concentration regime. However, the positions of the inflection points confirm the overall stoichiometries of the complexes. Moreover, the experimental data obtained during the titration of macroligand **2** could be fitted with the two-step process as well. However, also modeling utilizing only a one-step complexation event to the copper(I) grids resulted in an adequate fit to the data. This is likely to reflect an increased stability of the grid relative to the ML_2 complex for this particular ligand, which is attributed to less steric hindrance in this case compared to **3** and **4** leading to efficient cooperativity and thus favoring the direct formation of the grid instead of the intermediate ML_2 complex. In conclusion, no quantitative complexation constants could be determined from the modeling but the stoichiometries are supported by the inflection points. Moreover, the possibility of fitting the titration data of macroligand **2** with a one-step process, which is not possible for macroligands **3** and **4**, suggests that the formation of complete copper(I) grids is stronger for this smaller macroligand **2**.

Conclusions

Poly(*L*-lactide) macroligands containing a telechelic 3,6-*bis*(2-pyridyl)-pyridazine moiety were successfully synthesized by introducing a hydroxyl group to the ligand and subsequent polymerization utilizing an aluminum alkoxide initiated ring-opening polymerization. UV-vis complexation studies demonstrated different complexation behavior for the smallest macroligand **2** compared to macroligands **3** and **4**. The experimental data for the complexation process were successfully modeled assuming only two important complexation steps. However, this simplification of the complexation mechanism prohibited the determination of complexation constants.

Acknowledgments

Richard Hoogenboom and Ulrich S. Schubert thank the *Dutch Scientific Organization* (NWO), the *Dutch Polymer Institute* (DPI) and the *Fonds der*

Chemischen Industrie for financial support. Jurriaan Huskens thanks the *Council for Chemical Sciences of the Dutch Organization for Scientific Research* (NWO-CW) for the Vidi Vernieuwingsimpuls grant (number 700.52.423).

References

1 Brunsveld, L.; Folmer, B. J. B.; Meijer, E. W.; Sijbesma, R. P. *Chem. Rev.* **2001**, *101*, 4071-4097.

2 Schubert, U. S.; Eschbaumer, C. *Angew. Chem.* **2002**, *114*, 3016-3050; *Angew. Chem. Int. Ed.* **2002**, *41*, 2892-2962.

3 Schubert, U. S.; Eschbaumer, C. *Macromol. Symp.* **2001**, *163*, 177-187.

4 Lohmeijer, B. G. G.; Schubert, U. S. *Angew. Chem.* **2002**, *114*, 3980-3984; *Angew. Chem. Int. Ed.* **2002**, 41, 3825-3829.

5 Hoogenboom, R.; Andres, P. R.; Kickelbick, G.; Schubert, U. S. *SynLett* **2004**, *10*, 1779-1783.

6 Schubert, U. S.; Heller, M. *Chem. Eur. J.* **2001**, *7*, 5252-5259.

7 Schubert, U. S.; Hochwimmer, G. *Polym. Preprints* **1999**, *40*, 340-341.

8 Wu, W.; Collins, J. E.; McAlvin, J. E.; Cutts, R. W.; Fraser, C. L. *Macromolecules* **2001**, *34*, 2812-2821.

9 Heller, M.; Schubert, U. S. *Macromol. Symp.* **2002**, *177*, 87-96.

10 Corbin, P. S.; Webb, M.P.; McAlvin, J. E.; Fraser, C. L. *Biomacromolecules* **2001**, *2*, 223-232.

11 Marin, V.; Holder, E.; Hoogenboom, R.; Schubert, U. S. *J. Polym. Sci.: Part A: Polym. Chem.* **2004**, *42*, 4153-4160.

12 Lohmeijer, B. G. G.; Schubert, U. S. *J. Polym. Sci.: Part A: Polym. Chem.* **2004**, *42*, 4016-4027.

13 For a review on grid-like metal complexes, see: Ruben, M.; Rojo, J.; Romero-Salguero, F. J.; Uppadine, L. H.; Lehn, J.-M. *Angew. Chem.* **2004**, *116*, 3728-3747; *Angew. Chem. Int. Ed.* **2004**, *43*, 3644-3662.

14 Youinou, M.-T.; Rahmouni, N.; Fischer, J.; Osborn, J. A. *Angew. Chem.* **1992**, *104*, 771-773; *Angew. Chem. Int. Ed.* **1992**, *31*, 775-778.

15 Baxter, P. N. W.; Lehn, J.-M.; Kneisel, B. O.; Fenske, D. *Angew. Chem.* **1997**, *109*, 2067-2070; *Angew. Chem. Int. Ed.* **1997**, *36*, 1978-1981.

16 For the synthesis of functionalized 3,6-*bis*(2-pyridyl)-pyridazines, see: Hoogenboom, R.; Kickelbick, G..; Schubert, U. S. *Eur. J. Org. Chem.* **2003**, 4887-4896.

17 For the synthesis of the poly(*L*-lactide) macroligands, see: Hoogenboom, R.; Wouters, D.; Schubert, U. S. *Macromolecules* **2003**, *36*, 4743-4749.

18 Butte, W. A.; Case, F. H. J. Org. Chem. **1961**, *26*, 4690-4692.

19 Kricheldorf, H. R.; Mang, T.; Jonté, J. M. *Macromolecules,* **1984**, *17*, 2173-2181.

Chapter 6

Light-Active Metal-Based Molecular-Scale Wires

Anthony Harriman[1,*], Abderahim Khatyr[2], Sarah A. Rostron[1], and Raymond Ziessel[2]

[1]Molecular Photonics Laboratory, School of Natural Sciences (Chemistry), Bedson Building, University of Newcastle, Newcastle upon Tyne NE1 7RU, United Kingdom
[2]Laboratoire de Chimie Moléculaire, Ecole Européenne de Chimie, Polymères et Matériaux, Université Louis Pasteur, 25 rue Becquerel, F–67087, Strasbourg Cedex 02, France

Several disparate systems have been examined as putative molecular-scale wires able to conduct charge over relatively long distances under illumination. The key feature of such systems concerns the level of electronic coupling between the terminals, which are themselves formed from photoactive ruthenium(II) or osmium(II) poly(pyridine) complexes. Photophysical properties have been recorded in fluid solution and, where appropriate, the mechanism for long-range electron exchange has been clarified. In order to ascertain the importance of excited states localized on the organic framework, the photophysical properties of the free ligands have also been determined. The ligands bind zinc(II) cations in solution and thereby affect the absorption and emission spectra. This allows determination of the binding constants.

Background

The future development of effective systems for use as molecular photonic devices is contingent on the availability of molecular materials able to conduct charge over reasonable distances and along preferred pathways.[1] Ideally, the mechanism for charge migration will involve through-bond electron transfer (or exchange) and will facilitate efficient transport, storage and retrieval under illumination with visible light.[2] It will be necessary to transfer such information over distances in excess of 100 Å with high efficiency and excellent directionality. In recent years, it has become clear that there are two major mechanisms that favor long-range information transfer at the molecular level; namely, electron tunneling[3] and charge hopping.[4] The difference between these two extreme mechanisms is illustrated in Figure 1. Thus, electron tunneling occurs via superexchange interactions and makes use of virtual orbitals on the connecting bridge. On the other hand, charge hopping occurs over a series of short-range steps and involves states on the bridge as real intermediates. The design of the molecular system, especially the energetics of the bridge, depends on the choice of mechanism. Although early indications suggested that charge hopping provides the more likely route to success there are strong suggestions[5] that electron tunneling might prove to be more effective in the long run.

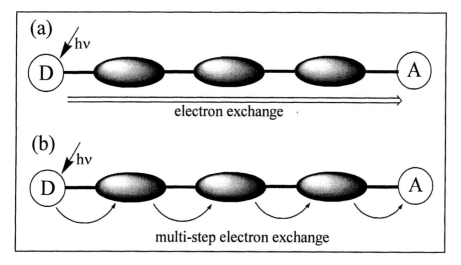

Figure 1. Pictorial representation of long-range electron exchange occurring via (a) electron tunneling and (b) electron hopping processes.

Understanding and optimizing such systems requires detailed knowledge of the precise role of the bridge. This is particularly important with light-activated devices and it is essential that we obtain a proper understanding of the excited-state behavior of all components in the system. Since there exists the possibility for π,π^*, intramolecular charge-transfer, metal-to-ligand and ligand-to-metal charge transfer excited states,[6] of both singlet and triplet parentage, it is clear that the excited state manifold will be heavily congested. Add to this the realization that the energetics of π,π^* and charge-transfer states are likely dependent on the degree of conjugation[7] and it becomes obvious that resolving the various excited-state identities will not be straightforward. In fact, there are several examples where the various excited states reside in thermal equilibrium at ambient temperature, thereby further complicating the situation.[6]

As the bridge becomes longer there is a greater propensity for its involvement as a real intermediate in the overall chemistry. It is possible that the mechanism will evolve with increasing length of the bridge – moving from electron tunneling to charge hopping as the bridge becomes more extended.[8] It is also important to consider the flexibility of the bridge under operating conditions since few molecular scaffolds are really rigid in solution at room temperature. Segments of the bridge might act together rather than blend into continuous states. This is a rich and varied field that provides opportunities for the design of unusual materials.

Objective

The aim of this study is to explore the excited state manifold for a series of oligomeric metal complexes of increasing nuclearity. These materials are designed to operate as prototypic molecular-scale wires that conduct electronic charge. The system comprises a series of 2,2'-bipyridine ligands separated by diethynylene-phenoxyl groups. The key issue is to resolve excited states localized on the metal complex and on the organic framework. It is recognized that the energy of the latter excited states will likely change on attachment of the metal cation. The emergence of next-generation photonic devices depends upon our ability to resolve these issues.

Experimental

The various ligands were prepared and purified as before.[9] All metal complexes were prepared by heating gently the free ligands with the required amount of [Ru(bipy)$_2$Cl$_2$] in ethanol. The course of reaction was followed by thin-layer chromatography and the target complexes were isolated as the hexafluorophosphate salts by column chromatography using alumina as support and a mixture of dichloromethane and methanol as the mobile phase. All complexes were unambigously characterized by ^{1}H-NMR, FT-IR, FAB-MS or ES-MS, cyclic voltammetry and elemental analysis. Solvents were spectroscopic grade and distilled from suitable drying agents immediately before use. All photophysical measurements were made with dilute solutions after purging with nitrogen. Absorption spectra were recorded with a Hitachi U3310 spectrophotometer whilst emission studies were made with either a Yvon-Jobin Fluorolog tau-3 spectrophotometer. Low temperature studies were made with an Oxford Instruments Optistat DN cryostat. Luminescence lifetimes were measured with the Fluorolog spectrophotometer.

Deconvolution of the emission spectra into the minimum number of Gaussian components was made with PEAKFIT. The spectra were reconstituted using purpose-written software. Quantum mechanical calculations were made using GAUSSIAN-03 or TITAN, running on a fast PC. Fluorescence quantum yields were measured relative to 9,10-dimethylanthracene in cyclohexane[10] and corrections were made for changes in refractive index of the solvent.

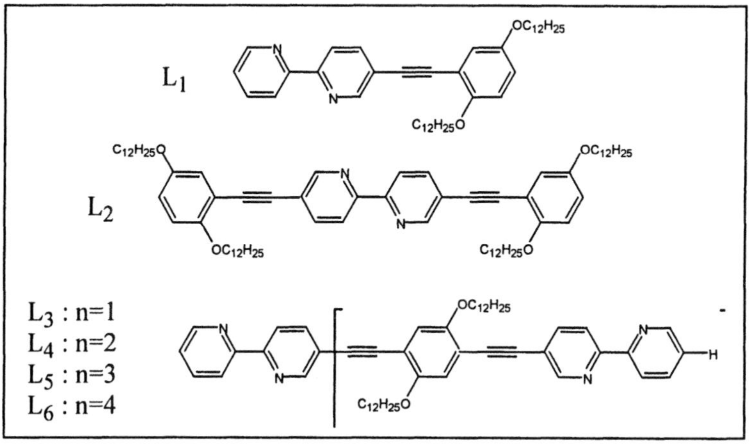

Figure 2. Structural formulae for the free ligands used in this work.

Results & Discussion

The photophysical properties of the various systems were measured, usually as a function of temperature, in an effort to identify the nature of the emitting species. A particular concern with this work is to determine the triplet energy of the briding unit with respect to that of the corresponding metal complex. It has to be considered that triplet states localized on the bridge might be π,π^* or intramolecular charge transfer in origin.[6] These different configurations can be resolved by virtue of their response to solvents of different polarity and to changes in temperature.

Photophysical properties of the free ligands L_1-L_6

A set of polytopic ligands containing 2,2'-bipyridine units in the backbone was synthesized and characterized.[9] The structures are shown in Figure 2. In order to provide for adequate solubility in organic solvents, the polytopic ligands were built around 1,4-dialkoxybenzene groups. These latter groups were intended to increase the solubility but also to isolate the individual bipyridine units, at least in an electronic sense. One way to assess the success of this strategy is to measure the photophysical properties of the ligands prior to attaching the metallo-fragments. Thus, each of the free ligands displays pronounced absorption in the near-UV region that can be assigned to a combination of π,π^* transitions.[9] The wavelength corresponding to the lowest-energy transition (λ_{MAX}) shows a modest dependence on the degree of oligomerization (Table 1). Strong fluorescence is observed for each compound but the fluorescence maximum (λ_{FLU}) shows a pronounced Stokes shift (Table 1) in N,N-dimethylformamide solution. The fluorescence quantum yields (Φ_F) are very high, approaching unity in some cases, but the emission lifetimes (τ_F) are quite short (Table 1). The radiative rate constants depend only on the average frequency of the fluorescence spectrum and are otherwise independent of the nature of the compound. This latter finding holds true in a range of solvents of different polarity. Despite the high fluorescence yields, weak phosphorescence could be detected in an ether:pentane:alcohol 5:5:2 glass at 77 K. The phosphorescence maxima (λ_{PHO}) show a slight dependence on the molecular length (Table 1).

These results imply that the degree of electronic communication along the molecular axis is modest. At first sight, this seems surprising but it was found that the lowest-energy excited singlet state possesses considerable charge-transfer character. This effect was evidenced by the strong dependence of the Stokes shift on the polarity of the surrounding solvent. For L_3, this dependence corresponds to a change in dipole moment upon excitation of 22.2 D. In contrast, the excited triplet state appears to retain similar polarity to that of the ground state. This means that the singlet-triplet energy gap depends markedly on solvent polarity. At room temperature, nonradiative deactivation of the excited singlet state is primarily via intersystem crossing to the triplet manifold. By correlating the derived rate constants with the measured energy gap, in accordance with classical Marcus-Jortner theory,[11] the reorganization energy associated with intersystem crossing was calculated to be ca. 1,500 cm^{-1}.

Table 1. Photophysical properties recorded for the free ligands L_1-L_6 in N,N-dimethylformamide solution at room temperature. NB The phosphorescence spectra were recorded in an EPA glass at 77 K.

Cpmd	λ_{MAX} / nm	λ_{FLU} / nm	Φ_F	τ_F / ns	λ_{PHO} / nm
L_1	347	477	0.64	4.1	545
L_2	355	504	0.19	2.0	540
L_3	388	448	0.88	1.5	547
L_4	399	478	0.44	1.6	545
L_5	397	478	0.37	1.6	545
L_6	401	480	0.35	1.7	545

On the basis of quantum chemical calculations, it appears that the electron donor in these ligands is the dialkoxybenzene unit while the electron acceptor is the bipyridine group. The molecular structure favors short-range interactions and it is this realization that inhibits the creation of a giant dipole along the molecular axis. Since increasing the degree of oligomerization merely adds incremental numbers of charge-transfer units to the chain, there is only a weak correlation between excited state energy and the molecular length. The energy of the triplet state, in particular, is relatively insensitive to increasing degree of oligomerization. This is good news for the future development of molecular-scale photoactive wires.

Attachment of zinc(II) cations

Each of the ligands L_1-L_6 was titrated with a stock solution of $Zn(ClO_4)_2$ in chloroform solution and the course of reaction was followed by UV-visible absorption and fluorescence spectroscopies. The presence of the cation caused significant changes in the UV-visible absorption spectrum (Figure 3) and led to a serious extinction of fluorescence from the ligand (Figure 4). It is clear that the cation coordinates to the vacant bipyridine sites available on the ligand. For both L_1 and L_2, coordination of zinc(II) could be analysed satisfactorily in terms of the stepwise formation of 1:2 and 1:3 (metal:ligand) complexes. The derived stability constants are collected in Table 2. More complicated binding behavior is evident, however, for the higher-order ligands. Indeed, molecular mechanics studies indicated that these polytopic ligands facilitate formation of metallo-helicates, despite the bulky dialkoxybenzene units.

The effect of added zinc(II) cations on the absorption spectrum of the free ligand is to induce formation of a new absorption band at longer wavelength. This band is characteristic of the metal complex and is most likely an intramolecular charge-transfer transition. The fluorescence spectra show progressive loss of emission characteristic of the free ligand and the concomitant

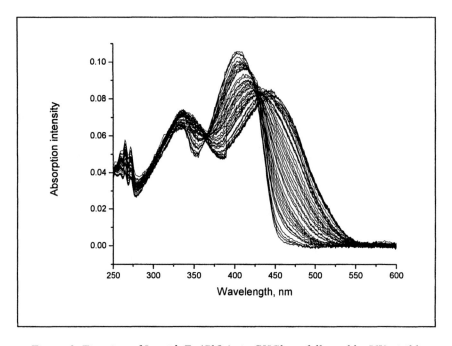

Figure 3. Titration of L$_5$ with Zn(ClO$_4$)$_2$ in CHCl$_3$ as followed by UV-visible absorption spectroscoy; the molar ratio of cation:L$_5$ increases progressively from 0 to 70.

growth of a new fluorescence profile at long wavelength. Again, this new band is attributed to the metal complex and possesses the usual features of an intramolecular charge-transfer transition. Clearly, isosbestic points are not preserved throughout the titrations; indicating the involvement of several species. Stability constants (β) were determined using SPECFIT to analyse the data collected over many different concentrations of ligand and cation. Comparable values were found from fitting the absorption and fluorescence data.

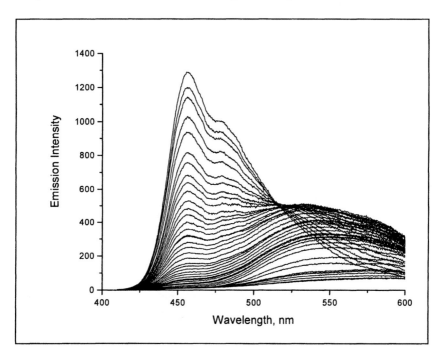

Figure 4. Spectrofluorescence titration of L_5 with $Zn(ClO_4)_2$ in $CHCl_3$, carried out over the same cation:L_5 ratio as described for Figure 3.

In fitting the titration data we have assumed that the zinc(II) cation tries to maximize the coordination geometry to 6. We did not see clear evidence for formation of the 1:1 complex, except with L_2 and a large excess of cation. For ligands of intermediate length, namely L_3 and L_4, there is a propensity to wrap three ligands around two or three cations, each being coordinated to a bipyridine site. These are stable complexes. With L_5, both 3:3 and 4:3 complexes are formed, together with the 2:3 complex. However, the longest ligand, L_6, does not form the expected 5:3 (presumed!) helicate-type complex and prefers to assemble complexes with lower stoichiometry. This might be because of electrostatic repulsion or steric crowding. [NB The stoichiometry, as written here, refers to the number of metal cations to the total number of polytopic ligands wrapped around those cations.]

Table 2. Overall stability constants derived from UV-visible absorption spectroscopy for attachment of zinc(II) cations to the various ligands in chloroform solution at room temperature. The stoichiometries refer to metal:ligand ratios.

log β	1:2	1:3	2:3	2:2	3:3	4:3	5:3
L_1	10.9	15.3	-	-	-	-	-
L_2	9.8	12.1	-	-	-	-	-
L_3	-	11.8	32.3	23.8	-	-	-
L_4	-	11.9	16.6	-	21.3	-	-
L_5	-	-	25.7	-	31.7	37.8	-
L_6	-	13.9	19.0	-	33.0	7.1	-

The presence of the zinc(II) cation lowers the energy of the first-excited singlet state, which is of intramolecular charge-transfer character, by raising the reduction potential of the 2,2'-bipyridine ligand to a less negative value. There is no obvious dependence on chain length and each metal complex appears to act independently. Phosphorescence could not be detected at 77 K but this does not mean that the corresponding triplet excited states are not formed.

Attachment of ruthenium(II) bis(2,2'-bipyridine) fragments

Each of the vacant 2,2'-bipyridine groups was coordinated with a single Ru(bipy)$_2$ metallo-fragment to give the corresponding tris complex (RL where L refers to a particular ligand and the nuclearity of the resulting complex depends on the number of bipyridine subunits present on the bridging ligand). Apart from the dimer, the polynuclear complexes studied here contain two different types of substituted bipyridine ligand. The terminal unit has only a single ethynylene attachment but the central or connecting units are each substituted with two ethynylene groups (a typical example is given in Figure 5). It is likely that the number of such groups will affect the photophysical properties of the metal complex due to a modification of the triplet energy or by a perturbation of the extent of electron delocalisation at the triplet level. In order to assess the importance of this structural change, the photophysical properties of the relevant metal complexes were evaluated in deoxygenated acetonitrile solution at room temperature. Additional studies were made at lower temperature and in solvents in different polarity. In all cases, the effect of molecular oxygen on the emission properties were examined.

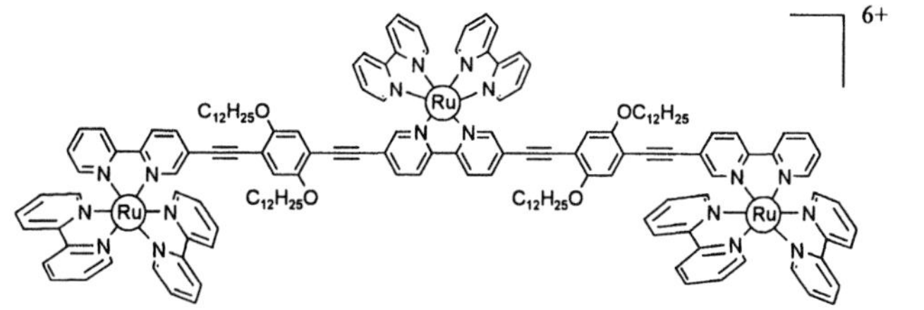

Figure 5. Idealized representation of the trinuclear complex RL₄

For RL_1, the absorption spectrum shows the expected metal-to-ligand, charge-transfer (MLCT) transitions centered around 460 nm. There is a weak tail stretching beyond 600 nm that can be attributed to the spin-forbidden MLCT transitions. The parent 2,2'-bipyridine ligands show prominent absorption centered around 290 nm. There are two obvious absorption bands associated with the ethynylene-substituted ligand. One band appears around 320 nm whilst the second band is apparent between 350 and 420 nm. This latter band overlaps to some degree with the MLCT transition.

The absorption spectrum recorded for RL_2 is qualitatively similar to that described above and, in particular, π,π^* transitions associated with the parent ligands are seen clearly at 290 nm. The disubstituted 2,2'-bipyridine ligand displays a relatively intense absorption band at 320 nm and a broad, structureless transition centered around 400 nm. The MLCT transition appears as a weak shoulder at 460 nm on the low-energy side of this latter band, although the spin-forbidden MLCT transition is well resolved. The ligand-centered (LC) absorption transition seen around 400 nm is similar to that found for the free ligand. It is possible that RL_1 and RL_2 also display an intramolecular charge-transfer transition where the dialkoxybenzene unit acts as donor and the acceptor is the 2,2'bipyridine group bearing the ethynylene substituent.

Both complexes exhibit luminescence in deoxygenated acetonitrile solution at ambient temperature. The spectral profiles are broad and centered in the far-red region. The emission maximum (λ_{LUM}) occurs at 628 nm for RL_1 but is red-shifted to 653 nm for RL_2. In both cases, the corrected excitation spectrum matched with the absorption spectrum across the visible and near-UV regions. The emission intensity was extensively quenched by molecular oxygen, indicating that luminescence arises from a triplet excited state. Quantum yields (Φ_{LUM}), measured in deoxygenated acetonitrile at 20 °C, indicate that the emission is relatively weak (Table 3). The measured Φ_{LUM} for RL_1 is essentially twice that found for RL_2. Under these conditions, the luminescence signal decays via first-order kinetics for both compounds. The derived lifetimes (τ_{LUM}) are collected in Table 3 and were found to be independent of laser intensity and monitoring wavelength. Again, it is notable that the triplet lifetime found for RL_1 exceeds that recorded for RL_2 under identical conditions.

The binuclear complex RL_3 is built around the unit characterized for RL_1. The absorption spectrum recorded in acetonitrile solution shows the expected transitions associated with the parent ligand at 290 nm (Figure 5). The polytopic ligand exhibits absorption peaks at 320 and 350 nm that appear to be similar to the higher-energy transition noted for the mononuclear complexes. The lower-energy transition appears as a broad, structureless band centered at 430 nm. This latter band is somewhat red shifted with respect to the mononuclear complexes and serves to obscure the MLCT transitions associated with the metal complexes. In fact, the spin-allowed MLCT transition can be recognised as a slight shoulder around 480 nm. The spin-forbidden MLCT transition is still apparent in the spectrum as a tail stretching beyond 600 nm. The absorption bands attributed to the polytopic ligand agree well with those characterized earlier for the free ligand,[9] except that the presence of the metal centers induces a modest red shift. These absorption bands are probably due to π,π^* transitions.

Table 3. Photophysical properties recorded for the various ruthenium(II) complexes in deoxygenated acetonitrile at ambient temperature.

Cmpd	λ_{LUM} / nm	Φ_{LUM}	τ_{LUM} / ns	k_{RAD} / 10^4 s^{-1}
RL_1	628	0.055	850	6.5
RL_2	653	0.026	370	7.0
RL_3	634	0.033	620	5.3
RL_4	653	0.022	400	5.5
RL_5	660	0.018	370	4.9
RL_6	665	0.014	360	3.9

Weak emission is observed with RL_3 in deoxygenated acetonitrile solution at room temperature (Figure 6). The luminescence profile is broad and centered around 634 nm, which is similar to that found for RL_1 under these conditions. Good agreement was noted between the corrected excitation spectrum and the absorption spectrum over the range 300 to 600 nm. It is interesting to note that whereas the emission lifetime found for RL_3 remains similar to that found for RL_1 there is a more substantial decrease in the quantum yield (Table 3). This suggests that the radiative rate constant is lower for the dimer than for the corresponding monomer but it should be noted that recent work has reported triplet quantum yields less than unity for closely-coupled binuclear complexes.[12] Comparing the various parameters with those recorded for the mononuclear complexes (Table 3) shows that the binuclear complex exhibits very similar behavior to that displayed by RL_1. Apart from the slightly lower Φ_{LUM}, there are no obvious effects due to the increase in nuclearity and it appears that the central dialkoxybenzene unit does not promote increased conjugation along the molecular axis in this case.

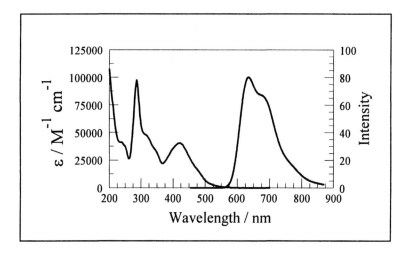

Figure 6. Absorption and emission spectra recorded for RL$_3$ in acetonitrile solution at room temperature.

The trinuclear complex RL$_4$ comprises a central unit comparable to RL$_2$ and terminal complexes reminescent of RL$_1$. The absorption spectrum recorded in acetonitrile is dominated by π,π^* transitions associated with the polytopic ligand. Absorption due to the parent ligand is seen clearly at 290 nm but the polytopic ligand absorbs over the range 300 to 500 nm. Some fine structure is apparent in the near-UV and the absorption maximum occurs at 430 nm. However, the spin-allowed MLCT transition is not obvious in the spectrum. As above, the spin-forbidden MLCT band can be seen as a tail on the low-energy side of the π,π^* transitions. Relative to the free ligand, the lowest-energy π,π^* transitions noted for RL$_4$ are red shifted by about 20 nm but the spectral profile remains the same. The origin of this red shift can be traced to the increased planarity of the ligand upon coordination and to the reduction of the LUMO energy that accompanies formation of the metal complex.

Weak luminescence occurs for RL$_4$ in deoxygenated acetonitrile at room temperature. The emission maximum is centered at 653 nm, as is that found for RL$_2$ under identical conditions. The emission quantum yield was relatively low but the lifetime remained similar to those found for the mononuclear complexes (Table 3). The decay kinetics remained strictly mono-exponential at all monitoring wavelengths.

The experimental data suggest that the three metal complexes on RL$_4$ act independently. Since the two terminal complexes, these being akin to RL$_1$, possess somewhat higher triplet energies it is reasonable to suppose that they will transfer energy to the central unit. This latter unit emits with the characteristic features of RL_2. The absence of any apparent temperature effect on the triplet lifetime rules out the possibility of LC triplet states of comparable energy. Thus, as with RL_3, it appears that the trinuclear complex is segmented into individual complexes that do not communicate in an electronic fashion.

The absorption spectrum recorded for the tetranuclear complex RL_5 shows the main features already recognised for RL_4. The parent 2,2'-bipyridine units absorb strongly at 290 nm and two distinct absorption bands can be seen for the polytopic ligand. The lower energy LC absorption band is centered at 438 nm and completely obscures the MLCT transition. It is difficult to resolve the spin-forbidden MLCT band from the overlapping π,π^* transitions. There is a small red shift relative to RL_4 whilst the absorption peak is red shifted by about 30 nm with respect to that of the free ligand. The absorption spectrum recorded for the pentanuclear complex RL_6 is very similar to that described for the tetranuclear complex except that the absorption peak is shifted to 445 nm. Again, the anticipated MLCT transitions cannot be resolved in the spectrum.

Luminescence could be detected from the tetranuclear complex in deoxygenated acetonitrile at room temperature. The emission maximum is red shifted by about 7 nm compared to that observed for the trinuclear complex (Table 3). Both the emission quantum yield and the lifetime are reduced somewhat with respect to RL_4, probably because of the lower triplet energy. Very similar behavior is found for the pentanuclear complex RL_6. In both cases, the emission is quenched by molecular oxygen and is hardly affected by modest changes in temperature. These latter studies provide no indication for the presence of a relatively low energy LC triplet state that could be populated by thermal equilibration from the lowest-energy triplet. Overall, the emission spectral properties remain similar to those recorded for the trinuclear complex. Again, the appearance is that the central dialkoxybenzene subunit acts to isolate a relatively small luminescent center. If so, we suppose that energy transfer occurs from the terminals to the central units.

Conclusions

The most interesting feature to emerge from this study is that the lowest-energy triplet is associated with the MLCT state in all the RuL-type complexes examined here. There is no obvious indication for a π,π^* triplet state localized on the polytopic ligand, even for the pentanuclear species. Such behavior might arise from the central dialkoxybenzene unit functioning as an insulator or because the pendant metal complexes interupt the electronic conjugation of the polytopic ligand. It is likely that intramolecular triplet energy transfer occurs from the terminal metal complexes to those complexes bearing two ethynylene substituents, because of the lower triplet energy. The segmented nature of these systems might be a useful facet for constructing artificial light-harvesting arrays.

Acknowledgment

This research was supported by the Engineering and Physical Sciences Research Council (GR/S00088/01), the CNRS, the University of Newcastle and ULP. We thank Johnson Matthey for the loan of precious metal salts.

References

1   Khondaker, S. I.; Yao, Z.; Cheng, L.; Henderson, J. C.; Yao, Y. X.; Tour, J. M. *Appl. Phys. Lett.* **2004**, *85*, 645-647.
2   Weiss, E. A.; Ahrens, M. J.; Sinks, L. E.; Gusev, A. V.; Ratner, M. A.; Wasielewski, M. R. *J. Am. Chem. Soc.* **2004**, *126*, 5577-5584.
3   Lin, J. P.; Beratan, D. N. *J. Phys. Chem. A* **2004**, *108*, 5655-5661.
4   Berlin, Y. A.; Hutchinson, G. R.; Rempala, P.; Ratner, M. A.; Michl, J. *J. Phys. Chem. A* **2003**, *107*, 3970-3980.
5   Harriman, A.; Khatyr, A.; Ziessel, R.; Benniston, A. C. *Angew. Chem. Int. Ed.* **2000**, *39*, 4287-4292.
6   Harriman, A.; Khatyr, A.; Ziessel, R. *Dalton Trans.* **2003**, 2061-2068.
7   Tan, C. Y.; Pinto, M. R.; Schanze, K. S. *Chem. Commun.* **2002**, 446-447.
8   Davis, W. B.; Ratner, M. A.; Wasielewski, M. R. *J. Am. Chem. Soc.* **2001**, *123*, 7877-7886.
9   Khatyr, A.; Ziessel, R. *J. Org. Chem.* **2000**, *65*, 7814-7824.
10 Murov, S. L.; Carmichael, I.; Hug, G. L. *Handbook of Photochemistry* Marcel Dekker, New York, 1993.
11 Van Hal, P. A.; Meskers, S. C. J.; Janssen, R. A. J. *Appl. Phys. A* **2004**, *79*, 41-46.
12 Benniston, A. C.; Grosshenny, V.; Harriman, A.; Ziessel, R. *Dalton Trans.* **2004**, 1227-1232.

Chapter 7

Triads Containing Terpyridine–Ruthenium(II) Complexes and the Perylene Fluorescent Dye

Harald Hofmeier[1,2], Philip R. Andres[1,2], and Ulrich S. Schubert[1,2,*]

[1]Laboratory of Macromolecular Chemistry and Nanoscience, Eindhoven University of Technology and the Dutch Polymer Institute (DPI), P.O. Box 513, 5600 MB Eindhoven, The Netherlands
[2]Center for Nanoscience, Photonics and Optoelectronics, Universität München, Amalienstrasse 54, 80799 München, Germany
*Corresponding author: email: u.s.schubert@tue.nl; Internet: www.schubert-group.com

A perylene unit was modified with two terpyridine moieties. Each terpyridine chelating moiety was subsequently complexed with ruthenium(III) trichloride to yield a dinuclear *bis*(monoterpyridine-ruthenium(III) chloride)perylene derivative. Alternatively, it is possible to obtain the corresponding perylene derivative bearing two *bis*-terpyridine-ruthenium(II) complexes via conversion of the of the terpyridine-substituted perylene with [Ru(tpy)Cl₃]. The compounds were characterized in detail by MALDI-TOF mass spectrometry, NMR, UV-vis as well as fluorescence spectroscopy.

© 2006 American Chemical Society

Introduction

N-Heterocyclic ligands have been used extensively during the last decades for the formation of transition metal complexes. The terpyridine ligand represents one of the most prominent examples, due to its outstanding complexation abilities with a wide range of metal ions and the specific electrochemical and photochemical properties.[1] In recent time this moiety has also become a key building block in polymer science[2] and supramolecular chemistry,[3] after important developments had been made concerning the accessibility of functionalized terpyridine building blocks.[4,5] In particular the ruthenium chemistry nowadays allows a directed access of unsymmetrical metal complexes via the terpyridine-Ru(III)/Ru(II) route.[6] We have just recently reported about the formation of a new class of nanomaterials utilizing such terpyridine-ruthenium chemistry.[7] In addition, the combination of energy/electron deficient moieties with energy/electron rich moieties on a molecular level has gained much interest over the last decade, especially in search of good charge separation for new organic solar cells. Also for these types of compounds, terpyridine-ruthenium complexes have been extensively used.[8]

Perylene units were already combined with terpyridine moieties and supramolecular polymers were reported after addition of zinc(II) ions.[9] In this contribution, we sought to synthesize a defined molecular triad containing a room temperature fluorescent perylene moiety and two *bis*-terpyridine-Ru(II) complexes. Such systems could be studied in future regarding electron and energy transfer processes and may find potential applications e.g. in organic solar cells.

Results & Discussion

Synthesis and characterization of the perylene-terpyridine system

Perylenetetracarboxilic dianhydride (PTCDA) **1** was reacted with the easily accessible amino-functionalized 2,2':6',2"-terpyridine **2** in quinoline yielding the *bis*-terpyridine-terminated perylene dye **3** in 87% yield (Scheme 3.5). Compound **3** could be characterized in detail using NMR, elemental analysis and MALDI-TOF mass spectrometry. The reaction with ruthenium(III) chloride in DMF led to a precipitation of the corresponding *bis*-ruthenium(III) complex **4** in 62% yield. This compound represents a key building block for the construction of complex supramolecular architectures. The addition of the preformed terpyridine-Ru(III) complex **5** to ligand **3** gave the ruthenium(II)-*bis*-terpyridine complex **6** in 44% yield (after anion exchange with hexafluorophosphate counter-ions) and precipitation into diethyl ether.

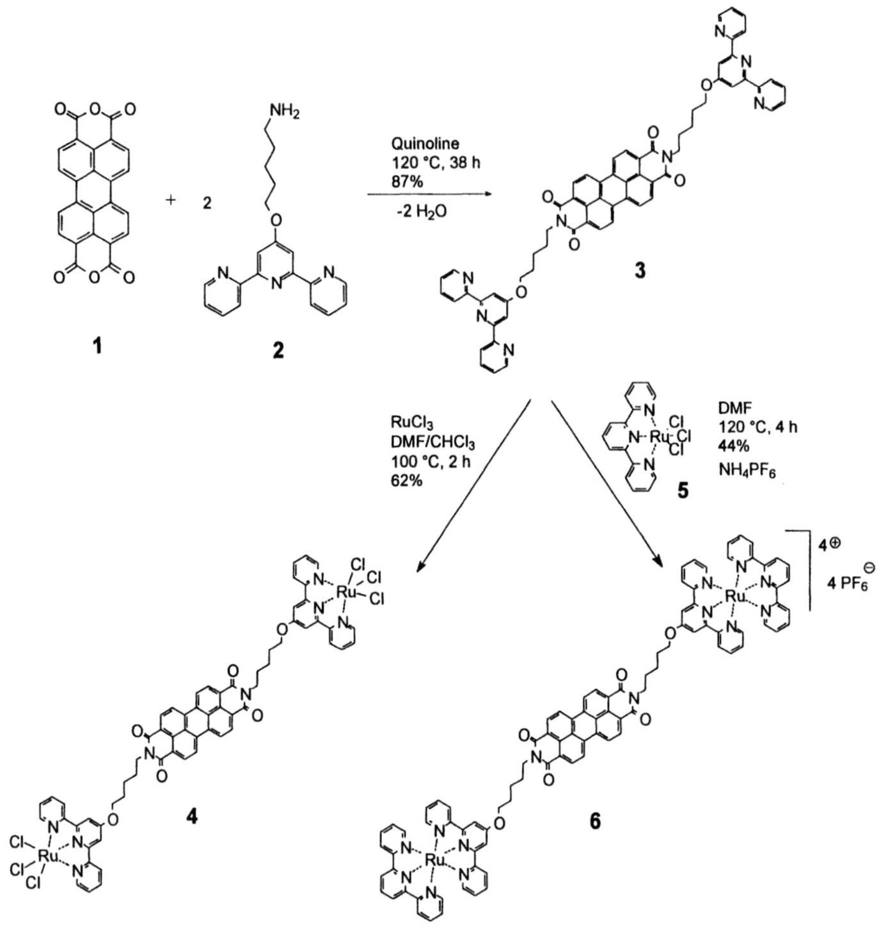

Scheme 1. Schematic representation of the synthesis of the bis-terpyridinyl perylene system and the subsequent complexation with RuCl₃ and terpyridine-Ru-fragments.

MALDI-TOF mass spectrometry has proven to be a mild and efficient method for the characterization of high-molecular weight molecules and metal complexes that cannot be detected by conventional MS-methods.[10] The mass spectrum of ligand 3 is shown in Figure 1 (top). The isotopic pattern fits to the simulated isotopic distribution. Also the terpyridine-ruthenium(III) mono-complex can be easily detected by this method (Figure 1, bottom). In this case, several distributions could be observed, corresponding to subsequent losses of chloride ions. Finally, complex 6 (M^+ without the four hexafluorophosphate counterions) with a molar mass of 2273 g/mol could be detected by MALDI-TOF mass spectrometry (Figure 2). In all cases, singly-charged species are observed, because only these ions survive long enough to be detected (ions carrying a higher charge are neutralized by the electrons that are also emitted in the desorption process).[11]

Due to the low solubility of **3** in chloroform (~ 2 × 10^{-5} mol/L), NMR spectroscopy was carried out in d-trifluoroacetic acid (d-TFA). This resulted in a protonation of the terpyridine rings. All expected signals with the corresponding integral values could be observed. The NMR spectrum of compound **3** is shown

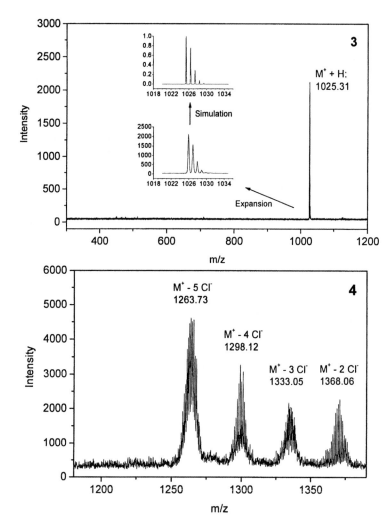

Figure 1. MALDI-TOF mass spectra of the perylene-terpyridine ligand 3 and of the corresponding ruthenium(III)-complex 4 (matrix: dithranol).

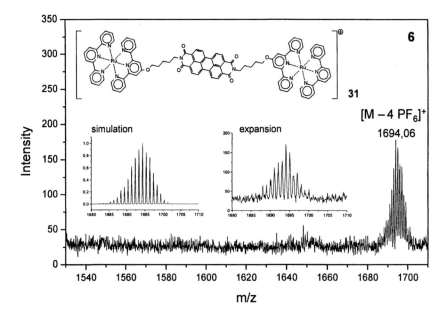

*Figure 2. MALDI-TOF mass spectrum of the perylene terpyridine ruthenium(II) complex **6** (matrix: dithranol).*

in Figure 3. The signals of the methylene groups of the spacer could be found at around 2 ppm and the CH_2-signals next to the oxygen of the imide group were detected at around 4.6 ppm. In the range between 8 and 9.4 ppm the aromatic signals were found. The resonances of the perylene protons were detected at 8.93 ppm.

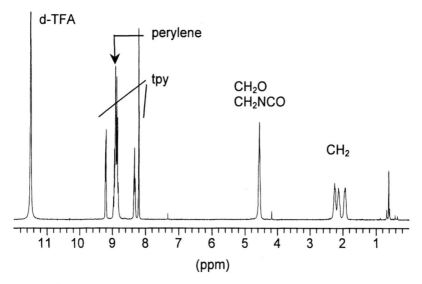

Figure 3. ^{1}H-NMR spectrum of the free tpy$_2$– perylene ligand 3 (impurities from the solvent d-TFA at: 0.6-0.9 ppm and at 4.1 ppm).

The UV-vis spectrum clearly revealed the characteristic absorption bands of perylene between 400 and 550 nm for all investigated compounds (Figure 4). In the UV spectral region **3** represents the well-known ligand-centered (LC) absorption band ($\pi^* \leftarrow \pi$ transition) of the 2,2':6',2"-terpyridine with a maximum around 280 nm. For the ruthenium(II)-complex **6**, a bathochromic shift of these absorption bands to 309 nm can be observed, which serves as an indication of complex formation. It should be mentioned that the spin-allowed metal-to-ligand charge-transfer transition (MLCT) of ruthenium terpyridine complexes at about 480 nm lies beneath the perylene absorptions. Moreover, NMR proved the existence of the *bis*-complex **6** (for a different approach towards such perylene-terpyridine compounds see ref.[9]).

Perylene is known to be fluorescent at room temperature. The emission spectrum of compound **6** (Figure 5) shows an emission maximum at 545 nm.

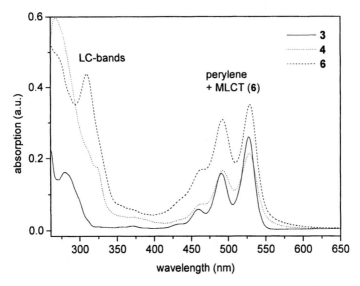

*Figure 4. UV-vis absorption spectra of the compounds **3** (in chloroform), **4** and **6** (in DMSO).*

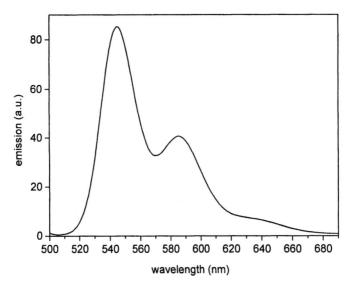

*Figure 5. Fluorescence emission spectrum of the perylene complex **6** (in DMSO).*

Conclusions and outlook

A *bis*(terpyridine-ruthenium(II))-perylene triad was prepared by complexing a *bis*-terpyridine functionalized perylene system with terpyridine-ruthenium(III) trichloride and characterized by NMR, UV-vis, MALDI-TOF MS and fluorescence spectroscopy. This system represents a model compound regarding functionalized systems that could be used e.g. for optical sensors. Further studies will include the preparation of similar systems with improved solubility characteristics as well as detailed fluorescence lifetime studies (including possible energy or electron transfer processes). Moreover, perylene will be introduced into polymeric systems.

Experimental

Basic chemicals were obtained from Sigma-Aldrich. The amino-functionalized terpyridine **2** was prepared as described in the literature[5,12] and the perylene-anhydride **1** was received from the BASF-AG (Ludwigshafen, Germany). MALDI-TOF mass spectra were measured with a Bruker Reflex II and an Applied Biosystems Voyager 6020 using dithranol as matrix. NMR-Spectra were measured with a Bruker ARX 300 (^{1}H: 300 MHz, ^{13}C: 75 MHz). UV-vis spectra were recorded on a Varian Cary 50 and on a Perkin Elmer Lamda-45 (1 cm cuvettes), fluorescence measurements were performed on a Perkin Elmer LS 50B luminescence spectrometer (emission wavelength: 486 nm). The samples were measured by using a 1-cm quartz cuvette at room temperature. All liquid samples had concentrations of around 5×10^{-6} mol/L.

Perylene ligand 3

To a mixture of 3,4:9,10-perylenetetracarboxilic dianhydride **1** (0.050 g, 1.127 mmol) and 5-(2,2':6',2''-terpyridin-4'-yloxy) pentylamine **2** (0.170 g, 0.508 mmol), quinoline (15 mL) was added. The mixture was stirred for 38 h at 120 °C and after cooling to room temperature the reaction mixture was suspended in ethanol and the dark green precipitate was filtered. After washing with ethanol (3 × 50 mL) and chloroform (3 × 50 mL) the product was dried *in vacuo*, yielding **3** (0.114 g, 87%) as a dark green solid: M.p. > 300 °C; ^{1}H NMR (d-TFA, 300 MHz, 298 K): δ = 1.95 (m, 4H, H$_\gamma$), 2.13 (m, 4H, H$_\beta$), 2.26 (m, 4H, H$_\delta$), 4.58 (m, 8H, H$_{\alpha,\varepsilon}$), 8.21, 8.34, 8.93, 9.21 (m, 20H, H$_{3,3''}$, H$_{3',5'}$, H$_{4,4''}$, H$_{5,5''}$, H$_{6,6''}$), 8.93 (m, 8H, H$_{IV,V}$); MS (MALDI-TOF, matrix: dithranol): *m/z* = 1025.31 (calcd for C$_{64}$H$_{48}$N$_8$O$_6$$^{+\cdot}$ 1025.37); UV-Vis (CH$_3$CN): λ_{max} (ε [Lmol^{-1} cm^{-1}]) 526.0 (3 730),

94

493.0 (4 450), 416.5 (7 880), 237.5 (8 650). $C_{64}H_{48}N_8O_6$ + 1 H_2O = 1043.39 g/mol) found (calc.): C 73.99 (73.69); H 4.58 (4.83); N 11.01 (10.74).

Ruthenium(III) monocomplex 4

To a hot solution of **3** (30 mg, 0.029 mmol) in 30 mL DMF $RuCl_3$ (24 mg, 0.12 mmol) was added. After heating for 4 hours to 100 °C 20 mL of ethanol were added, the mixture was cooled down and the greenish-brown precipitate was filtered and washed with water, ethanol and diethyl ether yielding **4** (26 mg, 62%): MS (MALDI-TOF, matrix: dithranol): m/z = 1368.06 (calcd for $C_{64}H_{48}N_8O_6Ru_2Cl_4^+$ 1368.05), 1333.05 (calcd for $C_{64}H_{48}N_8O_6Ru_2Cl_3^+$ 1333.08); UV-Vis (CH_3CN): λ_{max} (ε [$Lmol^{-1}cm^{-1}$]) 528.5 (17 680), 492.0 (12 990), 461.0 (6 120), 374.5 (2 950), 325.5 (8 050), 312.0 (7 190), 292.9 (9 180).

Ruthenium(II) complex 6

To a hot solution of terpyridine-$RuCl_3$ **5**[6] (26 mg, 0.06 mmol) in 30 mL DMF $AgBF_4$ (35 mg, 0.18 mmol) was added. The formed AgCl was removed by filtration. Compound **3** (30 mg, 0.029 mmol) was added and the mixture was heated to 120 °C for 4 hours. The crude product was precipitated by addition of an aqueous solution of NH_4PH_6 (500 mg). After filtration the product was recrystallized from acetonitrile/diethyl ether yielding **6** (28 mg, 44%): ^1H NMR (CD_3CN): δ = 1.82 (m, 4H, CH_2), 2.10 (m, 4H, CH_2), 4.20 (m, 2H, $C\underline{H}_2N$), 4.59 (t, 2H, J = 6.49 Hz, $C\underline{H}_2O$), 7.13 (m, 4H, $H_{5,5"}$), 7.29 (d, J = 4.96 Hz, 2H, $H_{6,6"}$), 7.43 (d, J = 5.34 Hz, 2H, $H_{6,6"}$), 7.9 (m, 6H, $H_{4,4"}$ + perylene), 8.30 (m, 2H, $H_{perylene}$), 8.32 (s, 2H, $H_{3',5'}$), 8.36 (t, J = 8.01 Hz, 1H, $H_{4'}$), 8.46 (d, J = 8.01 Hz, 4H, $H_{3,3"}$), 8.72 (d, J = 8.39 Hz, 2H, $H_{3',5'}$); MS (MALDI-TOF, matrix: dithranol): m/z = 1694.06 (calcd for $C_{94}H_{70}N_{14}O_6Ru_2^+$ 1694.37); UV-Vis (CH_3CN): λ_{max} (ε [$Lmol^{-1}cm^{-1}$]) 217 (196 500); 258 (110 600); 303 (89 330); 486 (62 250); 521 (66 540); elemental analysis ($C_{94}H_{70}N_{14}O_6Ru_2P_4F_{24}$ + 2 diethyl ether = 4971.96 g/mol) found (calc.): C 50.41 (50.80); H 4.01 (3.89); N 7.63 (7.97).

Acknowledgments

The authors wish to thank the *Dutch Scientific Organization* (NWO) and the *Fonds der Chemischen Industrie* for financial support.

References

(1) McWhinnie, W. R.; Miller, J. D. *Adv. Inorg. Chem. Radiochem.* **1969**, *12*, 135-215; Constable, E. C. *Adv. Inorg. Chem. Radiochem.* **1986**, *30*, 69-121.

(2) Schubert, U. S.; Eschbaumer, C. *Angew. Chem. Int. Ed.* **2002**, *41*, 2892-2926; Hanabusa, K.; Nakano, K.; Koyama, T.; Shirai, H.; Hojo, N.; Kurose, A. *Makromol. Chem.* **1990**, *191*, 391-396; Potts, K. T.; Usifer, D. A. *Macromolecules* **1988**, *21*, 1985-1991; Kelch, S.; Rehahn, M. *Macromolecules* **1999**, *32*, 5818-5828; Ng, W. Y.; Gong, X.; Chan, W. K. *Chem. Mater.* **1999**, *11*, 1165-1170; Schubert, U. S.; Hien, O.; Eschbaumer, C. *Macromol. Rapid Commun.* **2000**, *21*, 1156-1161; Heller, M.; Schubert, U. S. *Macromol. Rapid Commun.* **2001**, *22*, 1358-1363; Schubert, U. S.; Eschbaumer, C. *Macromol. Symp.* **2001**, *163*, 177-187.

(3) Constable, E. C. *Tetrahedron* **1992**, 10013-10059; Constable, E. C. In *Progress in Inorganic Chemistry*; Karlin, K. D., Ed.; John Wiley & Sons: New York, 1994; Vol. 42, pp 67-138; Potts, K. T.; Keshavarz-K, M.; Tham, F. S.; Abruña, H. D.; Arana, C. R. *Inorg. Chem.* **1993**, *32*, 4422-4435; Hasenknopf, B.; Lehn, J.-M. *Helv. Chim. Acta* **1996**, *79*, 1643-1650; Newkome, G. R.; Cardullo, F.; Constable, E. C.; Moorefield, C. N.; Thompson, A. M. W. C. *Chem. Commun.* **1993**, 925-927.

(4) Schubert, U. S.; Eschbaumer, C.; Hochwimmer, G. *Synthesis* **1999**, 779-782; Heller, M.; Schubert, U. S. *Eur. J. Org. Chem.* **2002**, *67*, 8269-8272; Heller, M.; Schubert, U. S. *Synlett* **2002**, 751-754.

(5) Schubert, U. S.; Eschbaumer, C.; Hien, O.; Andres, P. *Tetrahedron Lett.* **2001**, *42*, 4705-4707; Andres, P. R.; Lunkwitz, R.; Pabst, G. R.; Böhn, K.; Wouters, D.; Schmatloch, S.; Schubert, U. S. *Eur. J. Org. Chem.* **2003**, 3769-3776.

(6) Sullivan, B. P.; Calvert, J. M.; Meyer, T. J. *Inorg. Chem.* **1980**, *19*, 1404-1407; Constable, E. C.; Thompson, A. M. W. C.; Tocher, D. A.; Daniels, M. A. M. *New J. Chem.* **1992**, *16*, 855-867; Togano, T.; Nagao, N.; Tsuchida, M.; Kumakura, H.; Hisamatsu, K.; Howell, F. S.; Mukaida, M. *Inorg. Chim. Acta* **1992**, *195*, 221-225; Pickardt, J.; Staub, B.; Schäfer, K. O. *Z. Anorg. Allg. Chem.* **1999**, *625*, 1217-1224.

(7) Schubert, U. S.; Hofmeier, H. *Macromol. Rapid Commun.* **2002**, *23*, 561-566; Gohy, J.-F.; Lohmeijer, B. G. G.; Schubert, U. S. *Macromol. Rapid Commun.* **2002**, *23*, 555-560; Gohy, J.-F.; Lohmeijer, B. G. G.; Schubert, U. S. *Macromolecules* **2002**, *35*, 4560-4563; Schubert, U. S.; Schmatloch, S.; Precup, A. A. *Design. Monom. Polym.* **2002**, *5*, 211-221; Gohy, J.-F.; Lohmeijer, B. G. G.; Varshney, S. K.; Schubert, U. S. *Macromolecules* **2002**, *35*, 7427-7435; Lohmeijer, B. G. G.; Schubert, U. S. *Angew. Chem. Int. Ed.* **2002**, *41*, 3825-3829.

(8) Hofmeier, H.; Schubert, U. S. *Chem. Soc. Rev.* **2004**, *33*, 373-399.

(9) Dobrawa, R.; Würthner, F. *Chem. Commun.* **2002**, 1878-1879.

(10) Meier, M. A. R.; Lohmeijer, B. G. G.; Schubert, U. S. *J. Mass Spectrom.* **2003**, *38*, 510-516.

(11) Schubert, U. S.; Eschbaumer, C. *Polym. Prepr.* **2000**, *41*, 676-677; Karas, M.; Glückmann, M.; Schäfer, J. *J. Mass Spectrom.* **2000**, *35*, 1-12.

(12) Newkome, G. R.; He, E. *J. Mater. Chem.* **1997**, *7*, 1237-1244.

Chapter 8

The Preparation of Metallosupramolecular Polymers and Gels by Utilizing 2,6-*bis*-(1'-Methyl-benzimidazolyl)Pyridine–Metal Ion Interactions

J. Benjamin Beck and Stuart J. Rowan*

Department of Macromolecular Science and Engineering, Case Western Reserve University, 2100 Adelbert Road, Cleveland, OH 44106

Metallo-supramolecular polymers and gels have been prepared from ditopic monomer units, which consist of a 4-hydroxy-2,6-*bis*-(1'methyl-benzimidazolyl)pyridine unit attached to both ends of a polyether chain, mixed with either a transition metal ion (e.g. Co(II) or Zn(II)) or a combination of transition and lanthanoid metal (e.g. La(III), Eu(III)) ions. Such materials show dramatic reversible responses to a variety of stimuli, including thermal, mechanical, chemical and light.

Background

Stimuli-responsive polymers (SRPs) exhibit an ideally dramatic change in properties upon application of an external stimulus, such as a change in temperature, ionic strength, pH, electric or magnetic fields or by chemical or biological analytes. Examples include liquid crystal polymers,[1] polymer solutions and gels,[2] which can undergo a change in phase morphology,[3] electro- and magnetorheological fluids[4] and electro-active polymers (EAPs).[5] SRPs could play a role in a wide range of potential applications, including smart films, sensors, actuators, electro-optic devices, etc. Supramolecular chemistry,[6] with its use of weak reversible noncovalent interactions, has the potential to be a powerful tool in building SRPs which respond to environmental conditions that affect the degree of the molecular interactions. In order to maximize the effect that the environmental stress will have on the material's properties, systems which utilize noncovalent interactions in the construction of polymeric aggregates can be designed.[7,8,9] One way to access such a supramolecular polymer is the attachment of binding motifs onto the two ends of a core unit. These monomer units will then aggregate together, self-assembling into polymeric architectures which have a mixture of covalent bonds and noncovalent interactions along the polymeric backbone. For example, Figure 1 shows schematic representations of two homoditopic monomer units which have complementary binding sites that drive the self-assembly of an A-A/B-B type supramolecular polymer (Figure 1a) and a metallosupramolecular polymer (Figure 1b). The use of such a supramolecular polymerization process,[10-14] means that any change in the strength of the intermolecular noncovalent interactions can and will result in a dramatic modification of the supramolecular structure (e.g. degree of polymerization, DP) and thus effect significant changes in the properties of the material.

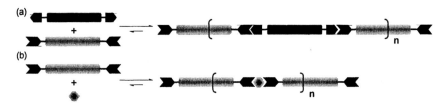

Figure 1. Schematic representations of two supramolecular polymerization processes; formation of (a) an A-A/B-B polymer and (b) a metallo-supramolecular polymer.

Supramolecular polymers have a number of interesting properties which help to make them attractive from the point of view of developing SRPs: 1) they form spontaneously, without the need for an initiation process (or catalyst); 2) they are 'dynamic' *i.e.,* are formed under reversible conditions; and 3) termination processes during the self-assembly (polymerization) process are limited. As a result, the degree of polymerization depends, to a large extent, on the strength of the supramolecular interaction between the monomers (K_a) and the total monomer (repeat unit) concentration [M]. If growth of the supramolecular polymer operates through a *Multi-Stage Open Association mechanism*, which is, simply put, a standard reversible step-growth process where the binding constant is independent of the molecular weight, then the size of the aggregate (DP) will approximately equal $2(K_a[M])^{1/2}$. This suggests that in order to obtain significant molecular weights of such self-assembled, reversible polymers then supramolecular motifs with large binding constants and/or large total monomer concentrations need to be employed. Most metallo-supramolecular polymers are comprised of a two-component A-A/B-B system (Figure 1) and as such the above prediction for DP would only hold up if the molar ratio of the components were exactly 1:1. Of course, as with all step-growth polymerizations, deviations from a 1:1 ratio of the complementary monomer units will have a significant negative effect on the overall DP.[15] Another major assumption of this model is that no cyclization occurs, therefore the DP is calculated for the formation of linear aggregates only. Thus ditopic monomer units which possess identical binding motifs can exhibit very different degrees of polymerization depending on the predisposition of the core to form thermodynamically stable macrocycles.

There are a wide range of noncovalent interactions than can be utilized to build supramolecular polymers, from hydrophobic interactions to hydrogen bonding and metal-ligand interactions. We have focused our attention in this area on two classes of noncovalent binding motifs, namely nucleobase interactions[16] and metal-ligand interactions. Metal-ligand interactions are of particular interest for a number of reasons. There are a myriad of different types of metal-ligand interactions, which offer a broad range in both their thermodynamic and kinetic stabilities, that can be employed in the construction of metallo-supramolecular polymers (Figure 1b). This diversity of possible metal/ligand combinations allows the researcher to tune the properties of these self-assembled materials by simply varying the metal ion and/or ligand. The thermodynamic stability of the metal/ligand interaction will influence the size of the aggregate (or degree of polymerization), while its kinetic stability should influence how responsive the material is to environmental factors. For example, kinetically labile metal-ligand systems should produce 'dynamic' polymers which are under continuous equilibrium (c.f. supramolecular polymers), while kinetically 'inert' metal-ligand combinations will produce polymers similar in nature to standard covalent systems. Furthermore, the incorporation of metal ions into polymeric systems opens up the possibility of imparting the functional properties of the metal ion, e.g. catalysis, light-emitting, conducting, gas binding etc., into the polymer. In recent years a number of groups have started to investigate the potential that this type of noncovalent bond has in the formation of such organic/inorganic polymer hybrids.[17]

Objective

The objectives of this study were to prepare ditopic monomer units which consist of polyether cores with the tridentate 2,6-*bis*-(1'-methyl-benzimidazolyl)-4-oxypyridine (**O-Mebip**) ligand attached to either end and examine the effect on the properties of these materials upon self-assembly with both transition metal ions and lanthanoid metal ions.

Experimental

NMR spectra were recorded on a Varian Gemini-300 or 600 MHz NMR spectrometer. Mass spectrometry was carried out on a Bruker BIFLEX III MALDI time-of-flight mass spectrometer using 2-(hydroxyphenylazo)-benzoic acid (HABA) as the matrix. UV-visible spectra were obtained by a Perkin Elmer Lambda 800 UV-VIS spectrometer. Fluorescence spectra were obtained with a SPEX Fluorolog 3 (Model FL3-12); corrections for the spectral dispersion of the Xe-lamp, the instrument throughput, and the detector response were applied. Variable temperature experiments were performed using a fiber optic detector placed in line with solution containing sealed vials atop a hot stage. The entire hot stage was enclosed so as to prevent ambient light from interfering. Temperatures reported are estimates based on the setting of the hot stage. VT fluorescence spectra described in this publication are the average of five scans using 1.000 second integration time with an interval of 1 nm. All emission spectra were excited at 340 nm.

Synthesis of 1. A mixture of **4** (0.990 g, 2.39 mmol) in dry DMF (10 mL) was stirred for 30 minutes until fully dissolved. Trimethylacetyl chloride (1.5 mL, 12.19 mmol) was added and the reaction mixture stirred for 10 minutes. *N*-methyl morpholine (3.00 mL, 27.26 mmol) and a solution of *bis*-(3-aminopropyl) terminated polytetrahydrofuran (**5**) in DMF (0.45 g, 0.32 mmol, in 5 mL) were then added and the reaction was allowed to stir for 48 hrs at room temperature. The DMF was removed under vacuum and the solids stirred in chloroform, filtered, and collected. The solution containing the organic fraction was evaporated to dryness and the resulting solids were purified by column chromatography (silica gel; CH$_2$Cl$_2$/MeOH 100:0, 98:2,…, 90:10) to yield 210 mg of **1** (35%). ^{1}H NMR (CDCl$_3$) δ 7.92 (4H, s), 7.85 (4H, d, J = 6.7 Hz), 7.38 (12H, m), 4.74 (4H, s), 4.23 (12H, s), 3.39 (68H, m), 1.61 (64H, m). ^{13}C NMR (CDCl$_3$) δ 166.7, 164.9, 151.9, 150.0, 142.7, 137.5, 124.0, 123.2, 120.5, 111.9, 110.2, 71.3, 70.9, 70.7, 69.6, 67.5, 38.0, 32.8, 29.5, 26.8, 26.7. MALDI-MS (matrix: HABA): M_n = 1725 m/z, M_w = 1884 m/z, PDI 1.09. FT-IR (cm^{-1}) 1674, 1544, 1595, 1571, 1446, 1477, 1369. UV-Vis: λ_{max} = 314 nm. PL: ($\lambda_{excitation}$ = 320 nm) $\lambda_{emission}$ = 365 nm.

Typical Sample Preparation of Metallo-Supramolecular Polymers. A solution containing 20.1 mg (0.01 mmoles) of **1** in 200 μL of chloroform was mixed with a stoichiometric amount of 3.72 mg (0.01 mmoles) of zinc perchlorate hexahydrate in 135 μL acetonitrile. This mixed solvent solution was then cast onto a glass slide to make a film. The complex was allowed to air dry and then was vacuum dried in an oven for several hours.

Results & Discussion

Monomer Synthesis

Two different monomers, which have 2,6-*bis*-(1'-methyl-benzimidazolyl)pyridine units attached to either end, have been prepared and investigated. The first is a polydispersed ditopic monomer based on a polytetrahydrofuran core (**1**) and the second a monodispersed ditopic monomer based on a pentaethylene glycol core (**2**). The synthesis of the ligand 4-hydroxy-2,6-*bis*-(1'-methyl-benzimidazolyl)pyridine (**3**) was achieved in one step via a Philips condensation of chelidamic acid and *N*-methyl-1,2-phenylenediamine heated in phosphoric acid (85%) at 220 °C for 10 hours.[18] The synthesis of **1** followed the synthetic methodology which we have previously employed[16a] for the attachment of supramolecular motifs to a polytetrahydrofuran chain. Thus we functionalized the hydroxyl group of **3** with benzylbromoacetate under basic conditions in ethanol to yield directly the acetic acid derivative **4** (Scheme 1). The synthesis of the supramolecular telechelic macromonomer **1** was then achieved by reacting **4** with the commercially available *bis*-(3-aminopropyl) terminated polytetrahydrofuran (**5**, M_n = 1,400 g mol^{-1}, PDI = 1.10) using mixed anhydride peptide coupling. The mono-dispersed ditopic monomer **2** was prepared in one step via the reaction of **3** and *bis*-iodopenta(ethylene glycol) under basic conditions (Scheme 1).[19]

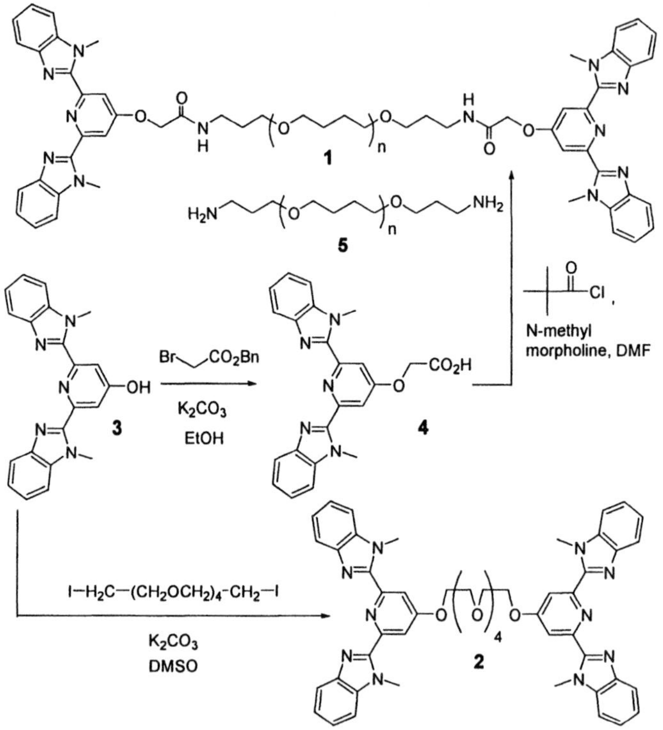

Scheme 1. The synthesis of the 2,6-bis-(1'-methylbenzimidazolyl)pyridine-terminated monomers.

The structure of **1** was confirmed by NMR and MALDI-TOF MS. End-group analysis of the ^{1}H NMR spectrum estimates the molecular weight (M_n) of **1** to be about 2,000 g·mol^{-1}. This value is in good agreement with the MALDI-TOF MS spectrum (Figure 2) which indicates that the molecular weight (M_n) of **1** is about 1,800 g·mol^{-1} with a PDI of 1.09. The major peaks correspond to [M+Na]$^+$ with the minor peaks corresponding to either [M+H]$^+$ or [M+K]$^+$. It is also important to note that the MALDI-TOF MS shows no evidence of any mono-substituted material or unreacted polymer **5**.

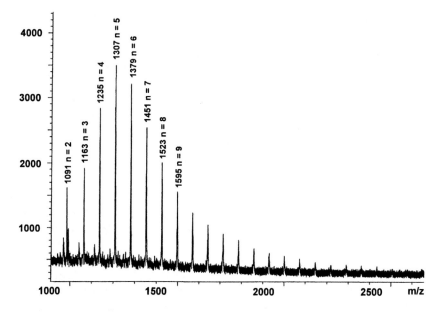

*Figure 2. MALDI-MS of **4** (NaCl + HABA matrix). All labeled peaks represent [M + Na] with varying number of -(C₄H₈O)- repeat units.*

Self-assembly of Metallo-Supramolecular Polymers

The formation of the metallo-supramolecular materials, $[1 \cdot MX_2]_n$ and $[2 \cdot MX_2]_n$, can be achieved by simple addition of one equivalent of the appropriate metal ion salt to a solution of the ditopic monomer, 1 or 2. We have found that a variety of ions (e.g. Zn^{2+}, Co^{2+}, Fe^{2+}) can be utilized to interact with the tridentate 2,6-*bis*-(1'-methyl-benzimidazolyl)pyridine (BIP) ligand. We have carried out some initial studies on the binding of the 2,6-*bis*-(1'-methylbenzimidazolyl)-4-oxypyridine (O-Mebip) ligands. Titration studies of 2 with Zn^{2+} ions revealed a cooperative effect in the formation of the 2:1 O-Mebip/metal complex as well as a strong overall binding constant (ca. 10^6 M^{-1}) in acetonitrile.[19] This is comparable to the binding of Zn^{2+} ions to other terdentate ligands such as terpy.[20]

To study the effect that the metal ions have on the properties of the ditopic monomers viscosity studies were carried out using a Cannon-Ubbelohde micro dilution viscometer. The intrinsic viscosity, $[\eta]$, of a sample is related to the molecular weight M of the polymer through the Mark-Houwink-Sakurada equation: $[\eta] = KM^a$, where K and a are experimentally determined polymer and environmentally specific constants. The K and a values for these materials are not known, however, we can draw some basic conclusions based on the comparison of the intrinsic viscosity values. The relative viscosities of both the starting ditopic ligands (1 or 2) and the metallo-supramolecular polymers $([1 \cdot Zn(ClO_4)_2]_n$ or $[2 \cdot Zn(ClO_4)_2]_n)$ were measured at a variety of different concentrations. Initial studies on the metallo-supramolecular polymers in organic solvents showed the presence of the polyelectrolyte effect, namely an increase in reduced viscosity at high dilution. Therefore, all subsequent viscosity studies were carried out in organic solvents which contain 0.1 M solution of tetrabutylammonium hexafluorophosphate to screen this effect. Most of the viscosity studies were carried out in a combination of 1:1 chloroform/acetonitrile as this solvent was found to dissolve all the materials of interest. The intrinsic viscosity, $[\eta]$, can be estimated by extrapolation of the reduced viscosity data (Figure 3) to where polymer concentration is zero. Intrinsic viscosity values of 2.5 mL/g are obtained for both 2 and $[2 \cdot Zn(ClO_4)_2]_n$ while values of 6.0 mL/g and 10.3 mL/g are obtained for 1 and $[1 \cdot Zn(ClO_4)_2]_n$, respectively. Given that optical spectroscopy indicates that most if not all of the ligands are complexed at these concentrations and the fact that little or no viscosity difference is observed between 2 and $[2 \cdot Zn(ClO_4)_2]_n$, then this suggests the formation of a significant amount of macrocycles in $[2 \cdot Zn(ClO_4)_2]_n$ at the concentrations studied. Studies could not be carried out at higher concentrations on account of the limited solubility of $[2 \cdot Zn(ClO_4)_2]_n$. A viscosity difference is observed between 1 and $[1 \cdot Zn(ClO_4)_2]_n$, at similar concentrations to those studies for $[2 \cdot Zn(ClO_4)_2]_n$, consistent with the formation of higher molecular weight aggregates upon metal ion complexation with this larger ditopic monomer. It should be stated that there

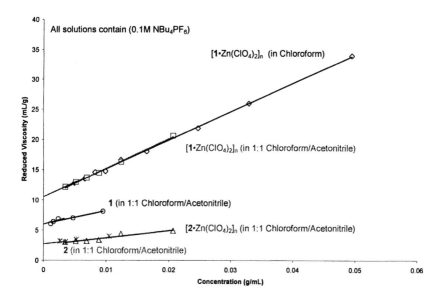

Figure 3. Reduced viscosities of 1, [1·Zn(ClO₄)₂]ₙ, 2 and [2·Zn(ClO₄)₂]ₙ at different concentrations (Cannon-Ubbelohde: chloroform/acetonitrile (1:1) + 0.1 M NBu₄PF₆ or chloroform + 0.1 M NBu₄PF₆).

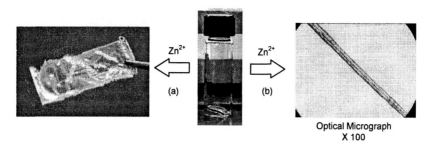

Figure 4. Optical Micrographs of 1, (a) a film of 1·Zn(ClO₄)₂ solution cast from dichloromethane and (b) a fiber of 1·Zn(ClO₄)₂ (x 100) which was melt processed.

is presumably the formation of some rings in this material, however the amount of macrocycles present would be expected to be less as a consequence of entropic effects reducing the likelihood of the chain ends in the macromonomer coming together.

As well as forming polymers in solution Figure 4 shows that the addition of metal ion can also enhance the mechanical properties of the material in the solid state. While **2** is an oil at room temperature addition of $Zn(ClO_4)_2$ to **2** does indeed result in a material that can be solution processed into films (Figure 4a) and melt processed into fibers (Figure 4b). It should be noted that no mechanically stable films of the 1:1 complex of **1** with $Zn(ClO_4)_2$ could be obtained from solution. However, brittle fibers could be obtained from the melt of this material.

Self-assembly of Metallo-Supramolecular Gels

It is known that 2,6-*bis*-(1'-methylbenzimidazolyl)pyridine ligands not only bind transition metal ions in a ratio of 2:1 but can also bind the larger lanthanoid ions in a 3:1 ratio.[21] Therefore, we reasoned that it should be possible to prepare metallo-supramolecular gels by simply mixing transition metal ions (> 95%) and lanthanoid metal ions (< 5%) with an appropriate *bis*-**Mebip** functionalized monomer (Scheme 2).

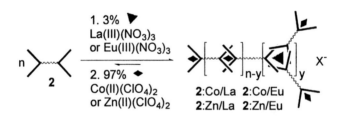

Scheme 2. Schematic representation of the formation of a metallo-supramolecular gel using a combination of lanthanoid and transition metal ions mixed with a ditopic monomer.

We have previously shown[19,22] that the use of the mono-dispersed monomer **2** in conjunction with either Co(II) or Zn(II) and Ln(III) or Eu(III) ions can result in the formation of gel-like materials. The self-assembly of the gel-like materials occurred spontaneously upon addition of the lanthanoid (III) nitrate (3 mol% based on the total number of **O-Mebip** ligands) followed by the transition metal ion perchlorate (97 mol% based on the total number of **O-Mebip** ligands) to a solution of **2** in $CHCl_3/CH_3CN$. The four possible metallo-supramolecular gel-like materials **2**:Co/La, **2**:Zn/La, **2**:Co/Eu, and **2**:Zn/Eu, were prepared in this way and upon removal of the solvent all four could be reswollen with pure CH$_3$CN (800% by wt.) by heating to the sol state and allowing to cool to room temperature. At these concentrations and using similar preperative procedures the 1:1 mixture of [**2**·Zn(ClO$_4$)$_2$]$_n$ forms precipitates from a CH$_3$CN solution but does not form a macroscopic gel. However, gels can be obtained from a 1:1 mixture of [**2**·Zn(ClO$_4$)$_2$]$_n$, which contains no lanthanoid ions, if after heating to the sol state the system was allowed to cool slowly by placing it in a bath at 30°C. This is interesting as it suggests that the mechanism for gel formation in these systems maybe something other than simple formation of a supramolecular crosslinked species. It is possible that theses materials self-assembly into a higher ordered structure that results in the formation of the gel. We are currently investigating the morphology of these systems in more detail to further elucidate the mechanism of gelation and gain a better understanding of the role of the lanthanoid ion in the changing the properties of the gel. Interestingly, addition of >10% Ln(III) does results in an intractable solid which is consistent with the formation of a highly crosslinked material. Both Co materials are orange in color, indicative of the binding of the ligand to the Co(II), while the Zn(II) systems are slightly off-white in color. All four gels show thermoresponsive behavior. For example, heating **2**:Zn/Eu to ca. 100 °C results in a reversible gel-sol transition (Figure 5). At these higher temperatures the orange color of the Co-materials persist in solution suggesting that it is the La/ligand interactions which are being thermally broken. Furthermore, these materials are also mechano-responsive, exhibiting a thixotropic (shear-thinning) behavior. Figure 5 shows how the shaking of **2**:Zn/Eu can result in the formation of a free-flowing liquid, which upon standing for ca. 20 seconds results in the reformation of the gel-like material. Initial rheological studies[23] on the **2**:Zn/La system demonstrate that this material has a yield point of approximately 155 Pa. The recovery time is very dependent on the amount of acetonitrile present in the system (material with less solvent recovers more quickly).

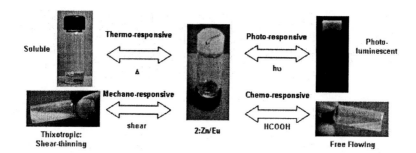

Figure 5. Mutli-stimuli responsive metallo-supramolecular gels of 2:Zn/Eu swollen in acetonitrile (800% by wt.).

The Eu (III) containing materials offer the possibility of utilizing the interesting spectroscopic properties of this lanthanide ion.[24] For example, Eu(III) ions can show an intense metal-centered luminescence in the presence of an appropriate UV absorbing ligand via the so-called "antenna effect".[25] This is in effect a light conversion process which occurs by absorption of the light by the ligand, followed by a ligand-to-metal energy transfer process finally resulting in the metal ion-based emission. **Mebip** ligands have been shown[26] to act as "antenna" for Eu(III) ions and as such this opens up the possibility of these material to be utilized as photo- or electroluminescent materials.[27] **2**:Zn/Eu was indeed photoluminescent showing the emission bands indicative of the lanthanide metal centered emission (581, 594, 616, 652 nm) as well as a ligand centered emission at 397 nm. Note the emission of the unbound ligand is at 365 nm, indicating that the ligand emission is sensitive to metal binding. While, not surprisingly, **2**:Zn/Ln displayed only the metal-bound ligand based emission (397 nm), **2**:Co/Eu did not show any fluorescent behavior, probably on account of the presence of low energy metal centered levels which facilitate radiationless decay processes.[28] As such the photo-responsive nature of these materials can be controlled by the nature of transition metal ion as well as lanthanide ion. The luminescence of **2**:Zn/Eu can also be used as a tool to examine the nature of metal binding in this system. For example, heating **2**:Zn/Eu shows a substantial reduction in the lanthanide-based emission (Figure 6) but no significant shift of the ligand emission, further supporting the fact that the Ln(III)/ligand bonds are being thermally broken.

Lanthanides are known to bind to carboxylic acids.[29] Therefore, we reasoned that these lanthanide-containing systems should also be chemo-responsive to such molecules, i.e. breakdown of the gel-like material should occur upon addition of a small amount of formic acid. Addition of 0.85% by weight of formic acid to a **2**:Zn/Eu gel (ca. 12% solid in acetonitrile) results not only in the loss in the mechanical stability of the material but also quenching of the Eu(III) emission (Figure 6). This is consistent with the formate anion displacing the **Mebip** ligand on the Eu(III) cation, resulting in a "switching off" of the aromatic ligands antenna effect. This process can be reversed by drying out the material in a vacuum oven for 8 hours at 40 °C. Upon reswelling the material with acetonitrile an increased Eu(III) emission is observed. To further demonstrate that it is the competitive binding of the carboxylate ion that results in the breakdown of the gel we added sodium acetate (ca. 0.12% by weight) to a preformed **2**:Zn/La gel (ca. 12% solid in acetonitrile). This results in a free flowing suspension, after mixing, which does not recover its gel-like state.

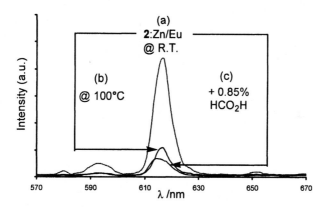

Figure 6. A section of the photoluminescent spectra of the acetonitrile swollen 2:Zn/Eu materials after being excited at 340 nm. (a) As prepared, (b) at 100 °C and (c) after addition of 0.85% HCO₂H.

110

Conclusions

In summary, we have used a combination of metal ions in conjunction with the *bis*-ligand monomers to produce supramolecular polyelectrolyte materials. We have been able to form stand alone films and fibers using a ditopic macromonomer 1 in conjunction with metal ions such as Zn(II). Furthermore, we have prepared self-assembled polyeletrolyte gels from the ditopic monomer 2, which exhibit, thermo-, chemo-, mechano-responses as well as light-emitting properties. The nature of the response exhibited by these systems depends upon the metal ion and the amount of swelling solvent. Given the wide variety of metal ions (with different metal/ligand kinetic stabilities and functional properties), counter ions and possible cores of the ligand-terminated monomer a wide variety of environmentally responsive metallo-supramolecular materials can be envisaged.

Acknowledgments

This material is based upon work supported by the National Science Foundation under Grant No. CAREER: CHE-0133164 and by the Case School of Engineering. The authors would like to thank Prof. A. M. Jamieson, Dr Y. Zhao and W. Weng for help with the rheological experiments and Jennifer Ineman for initial studies on polymer 1.

References

1. Picken, S. J. *Macromol. Symp.* **2000**, *154*, 95-104.
2. For some recent examples see (a) Nowak, A. P.; Breedveld, V.; Pakstis, L.; Ozbas, B.; Pine, D. J.; Pochan, D. J.; Deming, T. J. *Nature* **2002**, *417*, 424-428. (b) Elliott P. T.; Xing, L. L.; Wetzel W. H.; Glass, J. E. *Macromolecules* **2003**, *36*, 8449-8460. (c) Eastoe, J.; Sánchez-Dominguez, M.; Wyatt, P.; Heenan, R.K. *Chem. Commun.* **2004**, 2608-2609.
3. (a) Siegel, R. A.; Firestone, B. A. *Macromolecules* **1988**, *21*, 3254-3259. (b) Kwon, I. C.; Bae, Y. H.; Kim, S. W. *Nature* **1991**, *354*, 291-293. (c) Holtz, J. H.; Asher, S. A. *Nature* **1997**, *389*, 829-832. (e) Miyata, Y.; Asami, N.; Uragami, T. *Nature* **1999**, *399*, 766-769.
4. (a) Lengalova, A.; Pavlinek, V.; Saha, P.; Quadrat, O.; Stejskal J. *Coll. Surf. A* **2003**, *227*, 1-8. (b) Park, J. H.; Chin, B. D.; Park, O. O. *J. Coll. Interf. Sci.* **2001**, *240*, 349-354.
5. For example, see: *Electroactive Polymers [EAP] Actuators as Artificial Muscles: Reality, Potential, and Challenges* ed. Y. Bar-Cohen; SPIE Press, Bellingham, 2001.

6. For example, see: *Comprehensive Supramolecular Chemistry* (Eds. J. L. Atwood, J. E. D. Davies, D. D. MacNicol, F. Vögtle), Pergamon, Oxford, 1996, 11 vols.

7. Brunsveld, L.; Folmer, B.J.B.; Meijer, E.W.; Sijbesma, R.P. *Chem. Rev.* **2001**, *101*, 4071-4097.

8. Ciferri, A. *Macromol. Rapid Commun.* **2002**, *23*, 511-529.

9. *Supramolecular Polymers, Second Edition* Cifferi, A., Ed.; CRC Press, Taylor and Francis: Boca Raton, **2005**.

10. Kotera, M.; Lehn, J.-M.; Vigneron, J.-P.; *J. Chem. Soc. Chem. Commun.* **1994**, *2*, 197-199.

11. St. Pourcain C.; Griffin, A.C. *Macromolecules* 1995, **28**, 4116-4121.

12. (a) Shimizu, T.; Iwaura, R.; Masuda M.; Hanada T. Yase, K. *J. Am. Chem. Soc.* **2001**, *123*, 5947-5955. (b) Iwaura, R.; Yoshida, K.; Masuda, M.; Yase K.; Shimizu, T. *Chem. Mater.* **2002**, *14*, 3047-3053.

13. Yamauchi, K.; Lizotte, J.R.; Long, T.E. *Macromolecules* **2002**, *35*, 8745-8750.

14. Xu, J.; Fogleman, E.A.; Craig, S.L. *Macromolecules* **2004**, *37*, 1863-1870.

15. Vermonden, T.; van der Gucht, J.; de Waard, P.; Marcelis, A. T. M.; Besseling, N. A. M.; Sudhölter E. J. R.; Fleer, G. J.; Cohen Stuart, M. A. *Macromolecules* **2003**, *36*, 7035-7044.

16. (a) Rowan, S. J.; Suwanmala, P.; Sivakova, S. *J. Polym. Sci. A: Polym. Chem.* **2003**, *41*, 3589-3596. (b) Sivakova, S.; Rowan, S. J. *Chem. Commun.* **2003**, *19*, 2428-2429.

17. For recent examples of linear metallo-supramolecular polymers, see: (a) Schmatloch, S.; van den Berg, A. M. J.; Alexeev, A. S. Hofmeier, H.; Schubert, U. S. *Macromolecules* **2003**, *36*, 9943-9949. (b) Hinderberger, D.; Schmelz, O.; Rehahn, M.; Jeschke, G. *Angew. Chem. Int. Ed.* **2004**, *43*, 4616-4621. (c) Kurth, D. G.; Meister, A.; Thuenemann, A.F.; Foerster, G. *Langmuir* **2003**, *19*, 4055-4057. (d) Yount, W. C.; Juwarker, H.; Craig, S. L. *J. Am. Chem. Soc.* **2003**, *125*, 15302-15303. (e) Andres, P. R.; Schubert, U. S. *Adv. Mater.* **2004**, *16*, 1043-1068.

18. Froidevaux, P.; Harrowfield, J.M.; Sobolev, A.N. *Inorg. Chem.* **2000**, *39*, 4678-4687.

19. Rowan, S.J.; Beck, J. B. *Faraday Discuss.* **2005**, *128*, 43-53.

20. For example see: Holyer, R. H.; Hubbard, C. D.; Kettle, S. F. A.; Wilkins, R. G. *Inorg Chem*, **1966**, *5*, 622-625

21. Piguet, C.; Bünzli, J.-C. G.; Bernardinelli, G.; Hopfgartner, G.; Williams, A.F. *J. Alloys Comp.* **1995**, *225*, 324-330.

22. Beck, J.B.; Rowan, S.J. *J. Am. Chem. Soc.* **2003**, *125*, 13922-13923.

23. Zhao, Y.; Beck, J. B.; Rowan, S.J.; Jamieson, A.M. *Macromolecules* **2004**, *37*, 3529-3531.

24. Bender, J.L.; Corbin, P.S.; Fraser, C.L.; Metcalf, D.H.; Richardson, F.S.; Thomas, E.L.; Urbas, A.M. *J. Am. Chem. Soc.* **2002**, *124*, 8526-8527.
25. (a) Lis, S. *J. Alloys Comp.* **2002**, *341*, 45-50. (b) Sabbatini, N.; Guardigli, Lehn, J.-M. *Coord. Chem. Rev.* **1993**, *123*, 201-228.
26. Petoud, S.; Bünzli, J.-C. G.;Glanzman, T.; Piguet, C. Xiang, Q.; Thummel, P. *J. Lumin.* **1999**, *82*, 69-79.
27. Kido, J.; Okamoto, Y. *Chem. Rev.* **2002**, *102*, 2357-2368.
28. Armaroli, N.; De Cola, L.; Balzani, V.; Sauvage, J.-P.; Dietrich-Buchecker, C.O.; Kern, J.-M. Bailal, A. *J. Chem. Soc. Dalton Trans.* **1993**, 3241-3247.
29. For example, see: (a) Ma, L.; Evans, O.R.; Foxman, B.M.; Lin, W. *Inorg. Chem.* **1999**, *38*, 5837-5840. (b) Kawa, M.; Frechet, J.M.J. *Chem. Mater.* **1998**, *10*, 286-296.

Chapter 9

Thermal Stability, Rheology, and Morphology of Metallosupramolecular Polymers Based on *bis*-Terpyridine–Ruthenium(II) Complexes

Harald Hofmeier[1,2], Marielle Wouters[3], Daan Wouters[1,2], and Ulrich S. Schubert[1,2,*]

[1]Laboratory of Macromolecular Chemistry and Nanoscience, Eindhoven University of Technology, P.O. Box 513, 5600 MB Eindhoven, The Netherlands
[2]Center for Nanoscience, Photonics and Optoelectronics, Universität München, Amalienstrasse 54, 80799 München, Germany
[3]TNO Industries, 5600 MB Eindhoven, The Netherlands
*Corresponding author: email: u.s.schubert@tue.nl;
Internet: www.schubert-group.com

Supramolecular polymers, based on poly(ethylene glycol) and terpyridine-ruthenium(II) complexes, were investigated regarding their thermal stability by TGA. In addition, the temperature-dependent melt viscosity was studied using a rheometer and the results were compared to the corresponding classical poly(ethylene glycol). Finally, the morphology of an annealed film was investigated by AFM.

Background

Supramolecular polymers represent a new class of macromolecules that can show different properties than classical covalent polymers due to the reversibility of the non-covalent bond. Such polymers could find future applications in "smart materials" that might be "tuned" by addressing the supramolecular entities. The introduction of supramolecular moieties into polymers by end-group modification of oligomers and low-molecular-weight-polymers, leading to telechelics, has already been demonstrated for hydrogen bonding[1] and metal coordinating systems.[2] An important role for the construction of metal-coordination systems is played by 2,2':6',2"-terpyridine,[3] which acts as a chelating ligand for a variety of transition metal ions [e.g. Fe(II), Zn(II), Ru(II)]. A wide range of different architectures, such as AB and ABC block copolymers,[4] linear coordination polymers,[5-7] cross-linked systems,[8] and even star-shaped[9] and grafted structures[10] could be constructed. Linear terpyridine coordination polymers have so far been obtained by the use of metal ions like Fe(II), Zn(II) or Co(II),[5,11,12] some of them with chiral precursors leading to chiral polymers. Also rigid linkers were employed for the preparation of metallopolymers with Ru(II)[7,13] and Fe(II) ions,[14] resulting in rigid-rod-like structures.

Polymers, based on *bis*-terpyridine metal complexes, can also be reversibly decomplexed and recomplexed by external stimuli such as temperature, pH or electrochemistry ("switchable polymers").[15,16] An example is the thermal reversibility of a poly(ester) containing iron(II) *bis*-terpyridine complexes.[17] Furthermore, strong competing ligands are able to open terpyridine-metal complexes.[5,15] Ruthenium(II) is one of the most favorable metal centers in the engineering of supramolecular polymers since it forms very stable metal-coordination systems and allows both the directed synthesis of asymmetric as well as symmetric complexes. In addition, ruthenium complexes possess interesting optical and photophysical properties.[3,18]

The synthesis of linear metallopolymers built from terpyridine-functionalized small organic (diethylene glycol) as well as polymeric [poly(ethylene glycol)] telechelics with ruthenium(II) ions has already been described in the literature (Figure 1).[19-21] The solution viscosity was studied in detail including the polyelectrolyte effect (increase of the reduced viscosity of charged polymers by dilution), suggesting the formation of high-molecular-weight species.[7,22] Furthermore, the morphology of these materials was studied by AFM.

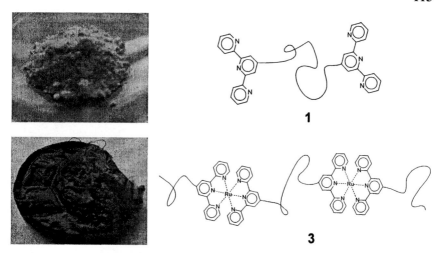

Figure 1. Photographic picture and schematic representation of the telechelic 1 (top right) and the ruthenium metallopolymer 3 (bottom right).

Objective

Metallo-supramolecular polymers containing terpyridine complexes have rarely been investigated regarding their thermal stability and their rheological properties. To the best of our knowledge, no examples of such studies of this type of polymer has appeared. In this contribution, the stability of such metallopolymers (Scheme 1) regarding thermal degradation under inert as well as oxidative atmosphere is investigated and the results compared to the corresponding uncomplexed telechelics as well as the analogous covalent polymer. Furthermore, the melt viscosity of the bulk material over a wide temperature range is described. AFM studies were already performed on drop-casted films, showing a lamellar morphology. Since a double lamella structure was found on analogous iron(II)-metallopolymers[12] after annealing of a film, the morphology of the ruthenium(II) system was investigated.

Results & Discussion

Thermal stability of the metallopolymer

To investigate the thermal as well as oxidative stability of **1**, **2** and **3**, thermal gravimetric analysis (TGA) was performed under an inert gas (nitrogen)

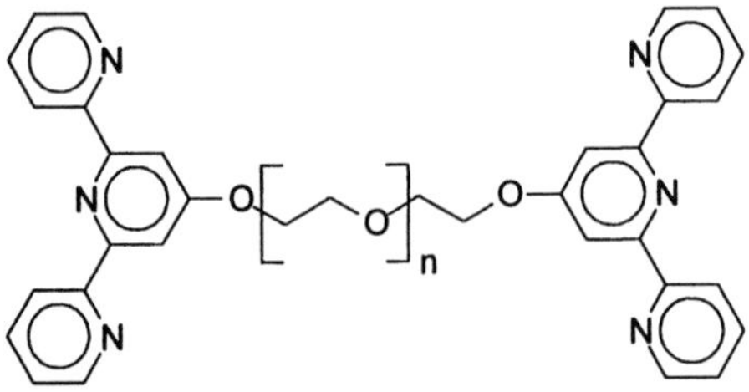

1: $\overline{n}$ = 179

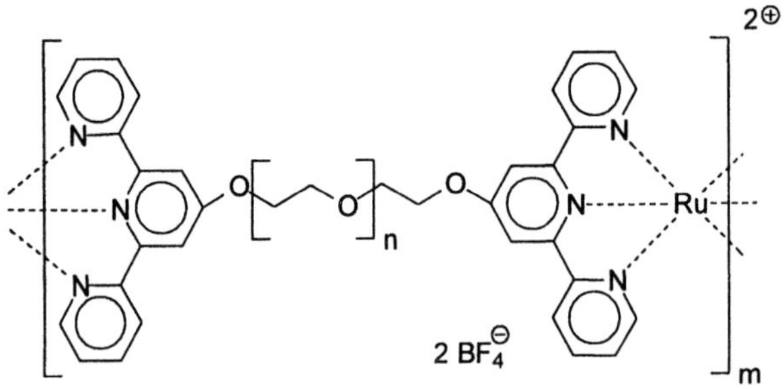

2: n = 1
3: $\overline{n}$ = 179

Scheme 1. Schematic representation of the investigated polymers 1-3.

and air atmosphere, respectively (Figure 2, for preliminary results see also ref.[21]). For comparison, also α,ω-*bis*hydroxy-poly(ethylene glycol)$_{180}$ (the precursor of **1**) and a high molecular weight poly(ethylene glycol) (200 000 daltons; abbreviated as PEG 200K) was investigated. In nitrogen atmosphere, the metal-free polymers revealed the fastest decompositions with a 5% weight loss between 228 and 250 °C (Table 1), while the coordination polymer **3** showed an increased stability (5% onset at 321 °C). Polymer **2** revealed an even higher decomposition temperature with a 5%-onset at 331 °C due to the large percentage of complex component within the polymer. In addition, a flatter slope was found. As a result of the higher metal content in this polymer, a large quantity of residue (20%) remained. The weight-percentage of ruthenium in the initial polymer **3** was 12%; the residue could contain compounds possessing boron and phosphorus (from the counterions) besides ruthenium oxides.

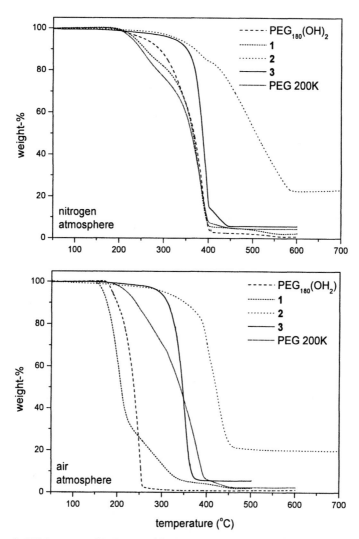

Figure 2. TGA traces of 1, 2, 3 and hydroxy-terminated PEG₁₈₀ as well as high molecular weight PEG (200 000 daltons) in nitrogen (top) and air (bottom) atmospheres.

The mechanism of degradation of poly(ethylene glycol)s was found to be random scission of the polymer chains involving free radical species.[23] Therefore, the stabilization of the coordination polymer could be ascribed to trapment of the radicals by the metal complex moieties. Repetition of the TGA measurements in air showed that compound **1** is oxidized more easily with a 5%-onset of 168 °C (67 K less than the nitrogen-measurement) than coordination polymer **2** (295 °C, 26 K less than in N_2). The mechanism probably involves the addition of oxygen to form peroxy groups, which subsequently form radicals. Enhanced radical formation would logically result in a faster decomposition. Another effect that may play a role for the decomposition profile could be the different volatilities of the formed fragments: Fragments bearing a metal complex moiety would evaporate more difficult. This has to be considered for polymer **2** (with short linkers), in the case of **3**, the length of the PEG chains should allow the formation of easily evaporable fragments.

The curve for **1** revealed a flattening slope between 230 and 320 °C, which was not found for the terpyridine-free analogue. An explanation could be an influence of the terpyridine moieties on the oxidative decomposition (terpyridines with short PEG-chains are formed, which have a decreased probability of chain scission). Finally, poly(ethylene glycol) of high molecular weight (200 000 daltons) was investigated to study the influence of the molecular weight on the decomposition behavior. Whereas thermal degradation of this polymer is similar to the low molecular weight compound **1**, the oxidative decomposition is slower than for the low-molecular weight compounds, but it is still faster than for coordination polymer **3**. In the case of additional oxidative decomposition, the molecular weight shows a significant influence on the decomposition, because more scission steps are necessary to fully decompose the polymer. This effect should then be also present for thermal decomposition, however, since the process is much slower, it may play a minor role. Additionally, the rate of heating can play a role. Finally, the metal-rich coordination polymer **2** revealed a decreased onset (306 °C) of decomposition, indicating a faster decomposition in an oxidative atmosphere.

Table 1. 5% onsets (°C), as found by TGA for compounds 1, 2, 3, PEG$_{180}$(OH)$_2$ and PEG (200 000 daltons) in nitrogen and air atmospheres.

	nitrogen	air
PEG$_{180}$(OH)$_2$	255	188
1	235	168
3	321	295
PEG 200K	228	222
2	337	307

The conclusion, drawn from the TGA results, is that the metallopolymers possess an increased stability towards thermal and oxidative decomposition due to the stable *bis*-terpyridine-ruthenium moieties. Moreover, oxidative processes have a less drastic influence on the coordination polymer compared to the metal-free polymers potentially due to radical trapping. However, this is just a hypothesis that has to be proven.

Rheological properties

The melt viscosities [complex viscosity $\eta^*(\omega)$] of precursor **1** and coordination polymer **3** were studied using a rheometer (Figure 3). The viscosity of telechelic polymer **1** decreased by five orders of magnitude during the melting transition (from 1.5×10^6 to 20 Pa.s); whereas, the viscosity of coordination polymer **3** stayed at higher levels; 4 orders of magnitude higher than for its precursor. Only a decrease of less than two powers of ten was observed (from 7×10^7 to 10^5 Pa.s). A covalent poly(ethylene glycol) with an $\overline{M}_n$ of 200 000 daltons (see TGA investigations) was also investigated for comparison. Its complex viscosity has the same order of magnitude as the coordination polymer **3** before and after the transition. These findings provide further indication for the presence of high molecular weight polymer chains in **3**. The location of the melting transition is in accordance with the DSC results. To investigate the stability of the coordination polymer during the rheometry measurement (to exclude any chain rupture), the solution viscosity was measured before and after the rheometry measurement, revealing the same viscosities. Therefore, rupture of the polymer chains can be excluded, as expected for ruthenium(II) complexes.

Furthermore, the storage and loss moduli were calculated. The storage modulus G' is a measure of the elasticity and describes the solid behavior of the substance; whereas, the loss modulus G'' is a measure of the fluid behavior. They are in relation to the dynamic viscosity as follows, where ω denotes the angular frequency.

$$\eta^*\omega = G' + iG'' \qquad (1)$$

A comparison of the moduli shows that the storage modulus for **3** is always larger than the loss modulus; whereas for the covalent PEG, the lines (moduli in dependence of temperature) cross after melting, as expected for conventional linear polymers (Figure 4). This behavior of the coordination polymer suggests the presence of intermolecular interactions (ionic interactions of the charges), which could remain largely intact after the melting transition (in agreement to the polyelectrolyte behavior of **3**). This corresponds to the behavior of ionomers:[24] Before melting, the charges are clustered within the polymer matrix, acting as cross-linkers. During the melting transition, these clusters can "dissolve". However, the electrostatic interactions are still present but are acting in a dynamic fashion.

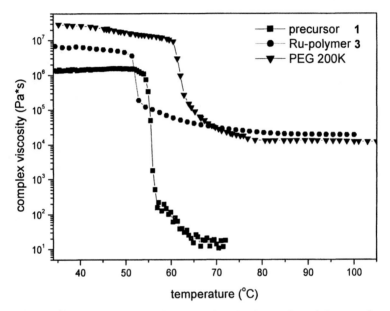

Figure 3. Temperature dependence of the absolute value of the complex viscosity of precursor **1**, coordination polymer **3**, and poly(ethylene glycol) ($\overline{M}_n$ = 200 000 daltons) at ω = 1 Hz.

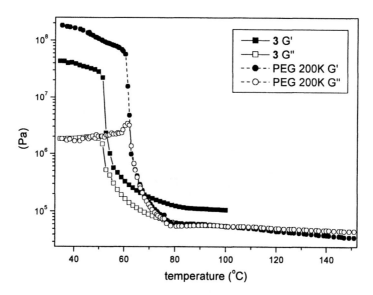

Figure 4. G' and G"-values of coordination polymer 3 and PEG
($\overline{M}_n$ = 200 000 daltons).

Morphology

It is known that poly(ethylene glycol) forms a lamellar morphology due to folding of the polymer chains.[25] A comparable structure of lamellae was also found by atomic force microscopy for the metallopolymer **3**.[19] Subsequently, a film was annealed at elevated temperature (on the heating stage of the AFM) and measured after cooling to ambient temperature. Now a double lamella structure was visible (Figure 5). The measurements could be repeated with a different AFM-tip to exclude a tip-convolution effect as a cause for the observed structures (however, it has to be mentioned that in some regions of the sample, these features were not found).

A potential explanation for this morphology could be the presence of once-folded poly(ethylene glycol) chains. For low molecular-weight PEGs, also double lamellae were found, which was explained by an arrangement of the chains by hydrogen-bonding of the terminal hydroxy groups.[26] In the present case, the complex units could be responsible for this arrangement. Scheme 2 displays the basic concept. The polymer in the neighborhood of the complexes is expected to be amorphous because the complex moieties are much bulkier than hydroxy endgroups. Smaller crystalline domains could be the result, which is in accordance to the DSC data.[19] The current findings for the ruthenium polymer are in agreement with the morphology found for an iron(II) coordination polymer of a similar constitution.[12]

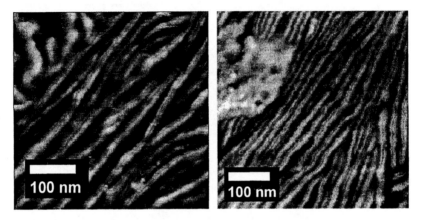

Figure 5. AFM phase images (two different experiments) of an annealed sample of 3, revealing a double lamella structure.

Scheme 2. Schematic representation of a potential poly(ethylene glycol) chain folding model, initiated by the terpyridine ruthenium(II) complex units.

Conclusions

It could be shown that the introduction of *bis*-terpyridine-ruthenium(II) complexes into a well-known polymer system can drastically change its properties. The metallopolymer showed an increased stability to thermal and oxidative degradation, originating from the complex units. In bulk, an increased melt viscosity was found due to strong intermolecular interactions of the charges. Finally, the complexes induce a folding pattern that is expressed in a double lamella structure.

Experimental details

The synthesis and characterization of the described compounds can be found in the literature.[19,20] The schematic structure of the polymers is shown in Scheme 1. TGA measurements have been performed on a Perkin Elmer Pyris 6 with a heating rate of 5 °C/min. Atomic force microscopy (AFM) has been performed in tapping mode on a DI Multimode with a Nanoscope IIIa controller (Digital Instruments, Santa Barbara, CA, USA). The used cantilever was a "golden" silicon cantilever type NSG11-A (NT-MDT). Samples were prepared from dropcasting solutions (10 mg/mL) of **3** (in chloroform) onto silicon wafers. Furthermore, a Solver47H (NT-MDT, Moscow), equipped with a heating stage, was used. Rheological experiments were conducted on a Paar Physica UDS 200 rheometer. A parallel plate with a diameter of 8 mm was used and the measurements were performed within the linear viscoelastic regime at an oscillating frequency of 1 Hz and a strain amplitude of 1%. Temperature scans were conducted from 30 °C to 72 °C (**2**) respective 100 °C (**3** and "PEG 200K") with a heating rate of 1 °C/min

Acknowledgments

Richard Hoogenboom, Daan Wouters and Ulrich S. Schubert thank the *Dutch Scientific Organization* (NWO), the *Dutch Polymer Institute* (DPI) and the *Fonds der Chemischen Industrie* for financial support.

References

(1) Brunsveld, L.; Folmer, B. J. B.; Meijer, E. W.; Sijbesma, R. P. *Chem. Rev.* **2001**, *101*, 4071-4097; Rieth, L. R.; Eaton, R. F.; Coates, G. W. *Angew. Chem. Int. Ed.* **2001**, *40*, 2153-2156; Yamauchi, K.; Lizotte, J. R.; Hercules, D. M.; Vergne, M. J.; Long, T. E. *J. Am. Chem. Soc.* **2002**, *124*, 8599-8604;

Ilhan, F.; Gray, M.; Rotello, V. M. *Macromolecules* **2001**, *34*, 2597-2601; Cooke, G.; Rotello, V. M. *Chem. Soc. Rev.* **2002**, *31*, 275-286; Yamauchi, K.; Lizotte, J. R.; Long, T. E. *Macromolecules* **2003**, *36*, 1083-1088.

(2) Schubert, U. S.; Eschbaumer, C. *Angew. Chem. Int. Ed.* **2002**, *41*, 2892-2926; Kelch, S.; Rehahn, M. *Macromolecules* **1998**, *31*, 4102-4106; McAlvin, J. E.; Scott, S. B.; Fraser, C. L. *Macromolecules* **2000**, *33*, 6953-6964; Fraser, C. L.; Smith, A. P. *J. Polym. Sci., Part A: Polym. Chem.* **2000**, *38*, 4704-4716; Lahn, B.; Rehahn, M. *e-Polymers* **2002**, *001*, 1-33.

(3) Schubert, U. S.; Eschbaumer, C.; Andres, P.; Hofmeier, H.; Weidl, C. H.; Herdtweck, E.; Dulkeith, E.; Morteani, A.; Hecker, N. E.; Feldmann, J. *Synth. Met.* **2001**, *121*, 1249-1252.

(4) Gohy, J.-F.; Lohmeijer, B. G. G.; Schubert, U. S. *Macromolecules* **2002**, *35*, 4560-4563; Gohy, J.-F.; Lohmeijer, B. G. G.; Varshney, S. K.; Decamps, B.; Leroy, E.; Boileau, S.; Schubert, U. S. *Macromolecules* **2002**, *35*, 9748-9755; Gohy, J.-F.; Lohmeijer, B. G. G.; Varshney, S. K.; Schubert, U. S. *Macromolecules* **2002**, *35*, 7427-7435; Lohmeijer, B. G. G.; Schubert, U. S. *Angew. Chem. Int. Ed.* **2002**, *41*, 3825-3829.

(5) Schmatloch, S.; González, M. F.; Schubert, U. S. *Macromol. Rapid Commun.* **2002**, *23*, 957-961.

(6) Kelch, S.; Rehahn, M. *Macromolecules* **1997**, *30*, 6185-6193.

(7) Kelch, S.; Rehahn, M. *Macromolecules* **1999**, *32*, 5818-5828.

(8) El-Ghayoury, A.; Hofmeier, H.; de Ruiter, B.; Schubert, U. S. *Macromolecules* **2003**, *36*, 3955-3959; Hofmeier, H.; Schubert, U. S. *Macromol. Chem. Phys.* **2003**, *204*, 1391-1397; Calzia, K. J.; Tew, G. N. *Macromolecules* **2002**, *35*, 6090-6093.

(9) Wu, X.; Collins, J. E.; McAlvin, J. E.; Cutts, R. W.; Fraser, C. L. *Macromolecules* **2001**, *34*, 2812-2821; Fraser, C. L.; Smith, A. P.; Wu, X. *J. Am. Chem. Soc.* **2000**, *122*, 9026-9027; Corbin, P. S.; Webb, M. P.; McAlvin, J. E.; Fraser, C. L. *Biomacromolecules* **2001**, *2*, 223-232; Smith, A. P.; Fraser, C. L. *Macromolecules* **2002**, *35*, 594-596; Schubert, U. S.; Heller, M. *Chem. Eur. J.* **2001**, *7*, 5252-5259.

(10) Schubert, U. S.; Hofmeier, H. *Macromol. Rapid Commun.* **2002**, *23*, 561-566.

(11) Bernhard, S.; Takada, K.; Diaz, D. J.; Abruña, H. D.; Mürner, H. *J. Am. Chem. Soc.* **2001**, *123*, 10265-10271; Bernhard, S.; Goldsmith, J. I.; Takada, K.; Abruña, H. D. *Inorg. Chem.* **2003**, *24*, 4389-4393; Schubert, U. S.; Hien, O.; Eschbaumer, C. *Macromol. Rapid Commun.* **2000**, *21*, 1156-1161.

(12) Schmatloch, S.; van den Berg, A. M. J.; Alexeev, A. S.; Hofmeier, H.; Schubert, U. S. *Macromolecules* **2003**, *36*, 9943-9949.

(13) Hjelm, J.; Constable, E. C.; Figgemeier, E.; A. Hagfeld, R. H.; Housecroft, C. E.; Mukhtar, E.; Schofield, E. *Chem. Commun.* **2002**, 284-285.

(14) Schütte, M.; Kurth, D. G.; Linford, M. R.; Cölfen, H.; Möhwald, H. *Angew. Chem. Int. Ed.* **1998**, *37*, 2891-2893.

(15) Gohy, J.-F.; Lohmeijer, B. G. G.; Schubert, U. S. *Macromol. Rapid Commun.* **2002**, *23*, 555-560.

(16) Lohmeijer, B. G. G.; Schubert, U. S. *Macromol. Chem. Phys.* **2003**, *204*, 1072-1078.

(17) Heller, M.; Schubert, U. S. *Macromol. Rapid Commun.* **2001**, *22*, 1358-1363.

(18) Hofmeier, H.; Schubert, U. S. *Chem. Soc. Rev.* **2004**, *33*, 373-399; Sauvage, J. P.; Collin, J. P.; Chambron, J. C.; Guillerez, S.; Coudret, C.; Balzani, V.; Barigelletti, F.; De Cola, L.; Flamigni, L. *Chemical Reviews (Washington, DC, United States)* **1994**, *94*, 993-1019.

(19) Hofmeier, H.; Schmatloch, S.; Wouters, D.; Schubert, U. S. *Macromol. Chem. Phys.* **2003**, *204*, 2197-2203.

(20) Schmatloch, S.; van den Berg, A. M. J.; Hofmeier, H.; Schubert, U. S. *Design. Monom. Polym.* **2004**, *7*, 191-201.

(21) Hofmeier, H.; Schmatloch, S.; Wouters, D.; Schubert, U. S. *Trans. Mater. Res. Soc. Jpn.* **2004**, *29*, 203-206.

(22) Pals, D. T. F.; Hermans, J. J. *Rec. Trav. Chim.* **1952**, *71*, 433-457; Dautzenberg, H.; Jaeger, W.; Kötz, J.; Philipp, B.; Seidel, C.; Stscherbina, D. *Polyelectrolytes*; Carl Hanser Verlag: Munich, 1994.

(23) Madorsky, S. L.; Straus, S. *J. Polym. Sci., Part A: Polym. Chem.* **1959**, *36*, 183-194; Cameron, C. G.; Ingram, M. D.; Qureshi, M. Y.; Gearing, H. M.; L Costa, G. C. *Eur. Polym. J.* **1989**, *25*, 779-784.

(24) Han, S.-I.; Im, S. S.; Kim, D. K. *Polymer* **2003**, *44*, 7165-7173; Kang, H.; Lin, Q.; Armentrout, R. S.; Long, T. E. *Macromolecules* **2002**, *35*, 8738-8744; Kim, J.-S.; Hong, M.-C.; Nah, Y. H. *Macromolecules* **2002**, *35*, 155-160.

(25) Beekmans, L. G. M.; Meer, D. W. v. d.; Vansco, G. J. *Polymer* **2002**, *43*, 1887-1895; Snetivy, D.; Vancso, G. J. *Polymer* **1992**, *33*, 432-434.

(26) Barnes, W. J. F.; Price, P. *Polymer* **1964**, *5*, 283-292; Cheng, S. Z. D.; Bu, H. S.; Wunderlich, B. *Polymer* **1988**, *29*, 579-583.

Chapter 10

Novel Block Copolymers with Terpyridine Pendant Groups

Gregory N. Tew[*], Khaled A. Aamer, and Raja Shunmugam

Department of Polymer Science and Engineering,
University of Massachusetts at Amherst, 120 Governors Drive,
Amherst, MA 01003
[*]Corresponding author: tew@mail.pse.umass.edu

Polymer architectures containing metal-ligands in their side chain represent a diverse and highly functional approach to hybrid organic-inorganic materials. The ability to synthesize block copolymers has recently been demonstrated and is facilitated by advances in controlled radical polymerization techniques. Using a combination of the three most common methods, a variety of block copolymers have been prepared. Both a direct and indirect approach to ligand incorporation was demonstrated. Subsequent functionalization with metal ions leads to a cornucopia of properties illustrated here by metal induced gelation and solvochromic sensors.

Background

Integration of organic and inorganic components into the same material can lead to a wide variety of advanced materials with unique properties.[1] In some cases, an increase in structural complexity gives rise to new properties, which cannot be foreseen on the basis of the single constituting moieties.[2] The assembling of disparate components, or molecular fragments, may give rise to new materials that exhibit useful physical and chemical properties in the condensed phase. One such approach is the creation of hybrid organic macromolecules and inorganic metal-ligand complexes. Metal-ligands impart many properties including luminescence, electro- and photo-chemistry, catalysis, charge, magnetism, and thermochromism.[3] The synthesis of metal-ligand polymers has typically focused on incorporating metal-ligands at the chain terminus.[4-9] A significant amount of work has focused on the use of metal complexation to drive polymerization and build dendrimers.[10,11] A similar approach used reversible metal ligand interactions to build metallo-supramolecular gels.[8,12] Alternatively, using macroligands at the core of radical initiators produced novel star polymers with heteroarms as well as diblock and triblock copolymers from a set of unsymmetrical, difunctional bipyridine reagents.[13,14]

Much less work has focused on the incorporation of metal ligands into the polymer side chain.[15-21] Polyoxazolines were prepared containing bipy that gelled in the presence of transition metal ions and were thermally reversibility.[15-17] These workers assumed network formation based on the observation of a solid after metal addition. Upon further dilution in water, the solid swells and then dissolves. It was suggested that the solid dissolved because intermolecular bonds were replaced with intramolecular ones. Potts and Usifer first reported polymers with terpyridine units in the side chain.[18,19] They showed that 4'-vinyl-2,2':6',2"-terpyridinyl readily formed homopolymers as well as styrene copolymers using AIBN initiation. Similar polymers and their transition metal complexes were investigated a few years later.[20,21]

The preparation of polymeric systems containing terpyridine (terpy) in the side chain has been limited due, in general, to the lack of emission properties from transition metal complexes.[22,23] In addition, there has been a limited number of commercially available functionalized terpy. Despite the lack of emission from transition metal complexes, other useful properties including electrochemical, photochemical, magnetism, and thermochromism result from the complexes. Interestingly, terpy binds to a range of lanthanide ions resulting in excellent luminescence, which we have studied.[24] As a result, a little more than two years ago, our lab and Schubert's reported methylmethacrylate (MMA) polymers containing terpy in the side chain.[22,23] We showed the solution viscosity of these polymers increased upon addition of copper (II) ions.[23] These reports explored only random copolymers until our recent work on block-random copolymers[25] which localized the metal ligand to one segment of a block copolymer for the first time. The ability to generate block copolymer architectures with metal-ligands confined to one block will have important applications in the field of supramolecular polymer science.

Access to block copolymers required the application of controlled or living polymerization techniques which also lead to very well controlled polymerizations in terms of molecular weight (MW), polydispersity (PDI), architecture, and monomer composition. Our approach has focused on living controlled radical polymerization (CRP) methods including atom transfer radical polymerization (ATRP), nitroxide mediated radical polymerization (NMP), and reversible addition-fragmentation chain transfer polymerization (RAFT). In addition to these synthetic methods, a direct approach that involves the polymerization of vinyl functionalized terpy was developed as well as an indirect approach which focuses on the polymerization of active ester monomers and subsequent conversion to incorporate terpy after polymerization. For this indirect method, the N-methacryloxysuccinimide (OSu) ester was selected since these esters are more hydrolytically stable than other commonly used active esters.[26] In addition, to the best of our knowledge, the controlled polymerization of OSu monomer has only been reported twice for the copolymerization of acrylate based polymers.[27,28] Our work extends the use of this monomer to very well defined block copolymers with monomodal size exclusion chromatography (SEC) curves and narrow molecular weigth distribution (MWD).

Objective

This chapter describes our efforts to create block copolymers using NMP, RAFT, and ATRP. Following the synthetic efforts, the properties of various polymer-metal ion complexes are described including reversible viscosity changes, gel formation, and solvochromic sensors. These properties are generated by the addition of transition metal ions like Cu (II), Zn (II), and Co(II).

Experimental

Materials

Methyl methacrylate (MMA), n-butyl methymethacrylate (nBMA) and Styrene (S) were vacuum-distilled and stored in an air free flask in the freezer. Azobis(isobutyronitrile) (AIBN) was recrystallized from methanol and stored in the freezer. Poly(ethylene glycol) methyl ether methacrylate macromonomer (PEGMA); M_n = 480) was obtained from Aldrich and purified by passing through a neutral alumina. 4'-Chloro-2, 2': 6', 2''-terpyridine was purchased from Lancaster and all other chemicals were used as received from Aldrich. Reagent grade DMF or freshly distilled THF was used for GPC. All other solvents were used as received. CuBr (98%) was obtained from Fischer Scientifics, pentamethyl diethylenetriamine (PMDETA) (99%), ethyl 2-bromoisobutyrate (2-EiBBr, 98%), and anisole (99.7%) were obtained from Aldrich and used without further purification. PMMA macroinitiator, terpyridine amine, and other polymers were synthesized following the procedure as reported earlier.[23-25]

Results & Discussion

Direct Method

Following successful synthesis of random copolymers[23], we turned our attention to block architectures. Using the direct method, we determined ATRP would be incompatible with the terpy unit and so we explored NMP and RAFT methods.[25,29] NMP successfully synthesized poly(S-b-(S-ran-S_{terpy}) copolymers using a polystyrene macroinitiator with M_n of 44 kDa. The final block copolymer contained 10 mol % terpy, had a M_n value of 67 kDa against poly(S) standards, and MWD of 1.4. The presence of the terpy causes broadening of the peak, which is consistent with previous reports on pyridine containing polymers. The high mol % incorporation of terpy obtained at 56 % conversion is consistent with observations that the terpy functional monomer appears to add to the growing chain in preference to styrene. These results show NMP allows block architectures, based on styrene, to be prepared in which the terpy unit is confined to one segment. The application of NMP is limited to styrenic and acrylate monomers and so we explored the synthesis of block-random MMA structures by RAFT. Shown in Scheme 1 is the successful preparation of a block-random copolymer using the poly(MMA) macro-chain transfer agent in the presence of

AIBN (0.4:1 AIBN:RAFT agent), MMA (90 mol %), and the functionalized terpy monomer (10 mol%) in benzene at 60 °C as shown in Scheme 1. Figure 1 shows the overlaid GPC traces for the poly(MMA) macro-chain transfer agent and block copolymer. The total mol % incorporation of terpy is 2.5 mol % based on elemental analysis of the nitrogen content and ^{1}H NMR at 95% conversion. This indicates Sty$_{Terpy}$ has a lower tendency to incorporate than MMA in the copolymer under these RAFT conditions. The GPC trace shows little broadening in the peak, most likely due to the decreased mol % incorporation of terpy compared to the previous block copolymers. Figure 2 shows the aromatic region of the ^{1}H-NMR spectrum for copolymer confirming the presence of terpy in the backbone.

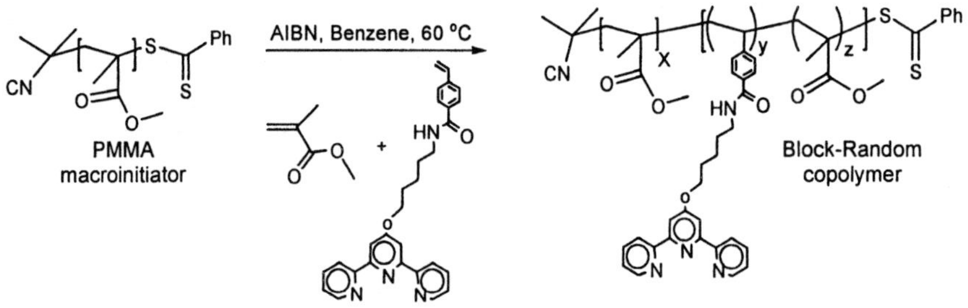

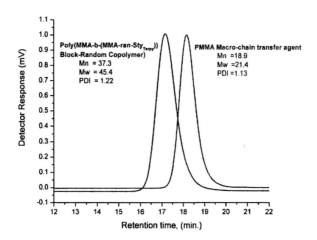

Scheme 1. Synthesis of poly(MMA-b-(MMA-ran-Sty$_{Terpy}$) synthesized via RAFT.

Figure 1. Overlaid GPC trace of α, ω-(thiobenzoyl thio) polymethylmethacrylate macro-chain transfer agent, and poly(MMA-b-(MMA-ran-Sty$_{Terpy}$)), synthesized via RAFT in THF as mobile phase.

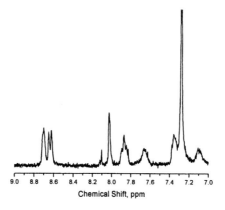

Figure 2. ^{1}HNMR of the aromatic region of poly(MMA-b-(MMA-ran-Sty$_{Terpy}$) in CDCl$_3$

Block-Block

Diblock copolymers are very interesting macromolecules because of their remarkable microphase separation properties and hierarchical ordering of the two dissimilar blocks into different morphologies. The incorporation of terpy units in one of the blocks is very tempting to synthesize since the terpy block can be functionalized with various transition metal ion to provide additional chemical functions or by lanthanides to generate highly luminescent polymeric system confined within the morphology.

As shown in Scheme 2, the alkoxyamine end functionalized polystyrene macroinitiator was used to grow a second block of Sty$_{Terpy}$ in diglyme at 125 °C in the presence of catalytic acetic anhydride. The polymer was isolated by partitioning between cyclohexane/methanol. ^{1}H-NMR, shown in Figure 3, confirmed the presense of polymeric terpy units and elemental analysis of the nitrogen content, gave 27 mol% Sty$_{Terpy}$ which provides some evidence that a second block of Sty$_{Terpy}$ was generated. However, a polymer with such high terpy percentage could not characterized by any GPC methods available to us. Several attempts to remove the terpy units from the polymer backbone were not successful and included treating the polymer in refluxing methanol/NaOH as well as transamidation of the amine functionalized terpy with benzylamine in the presence of Sc(OTf)$_3$ and toluene at 90 °C.

The diblock copolymer was finally characterized by optical methods in which a thin film of the diblock was spun on a silicon wafer, annealed for 3 days at 170 °C and the reflection optical micrograph was recorded. Figure 4 shows the optical micrograph of the diblock at the film edge where sharp color changes are observed. These color changes, along with no observable macrophase separation, strongly support microphase separation into layers that would most likely originate from a diblock copolymer. It is the presence of this microphase separation that generates the color changes as a result of refractive index difference leading to light refraction.

Indirect Method

The direct approach has some advantages but we have observed multiple limitations especially in the size exclusion chromatography (SEC) characterization as mentioned above.[30] Therefore, we considered alternative routes to these macromolecules including a post-polymerization process that would allow easy characterization of the pre-polymer and convinent montoring of the reaction to incorporate terpy. At the same time, this approach, which is illustrated here with amine functionalized terpy, could be easily extended to incorporate many different chemical functions including bioactive moieties like proteins. This approach also overcomes batch to batch differences that occur in polymerization reactions including monomer sequence heterogeneity and tacticity. Further, one batch of active ester polymer would yield a family of terpy containing molecules in which only the concentration of terpy varies. Generating reactive polymers for subsequent modification has been studied extensively[31] but the use of activated ester monomers has gained favor recently due to their chemical versatility.[27,28] We have focused on N-methacryloxysuccinimide (OSu) since these esters are more hydrolytically stable than other commonly used active esters and the conversion can easily be followed by IR. We developed optimized ATRP conditions for the homopolymerization of OSu in nonpolar solvent[24] and below we describe our efforts toward block copolymer formation.

Using macroinitiators of MMA or S, block copolymers of OSu were prepared by optimized ATRP conditions (Scheme 3). Formation of the diblock copolymers was confirmed by NMR, IR and overlaying SEC traces of the macroinitiator and resulting copolymer. A typical example is shown in Figure 5 for a PMMA macroinitiator and P(MMA-b-OSu) block copolymer. All block copolymers showed monomodal Gaussian-shaped SEC peaks suggesting very good chain extension from the macroinitiator. The process appears to work well regardless of whether the macroinitiator belongs to the same monomer class

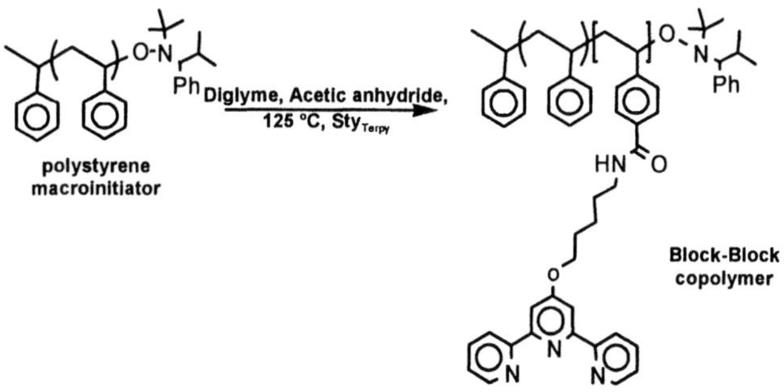

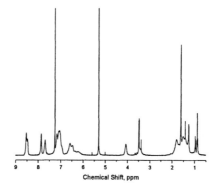

Scheme 2. Synthesis of Poly(sty-b-Sty$_{Terpy}$) via NMP.

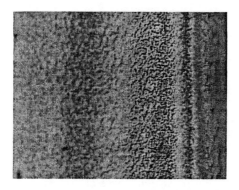

Figure 3. ^{1}HNMR spectrum of Poly(Sty-b-Sty$_{Terpy}$) in CDCl$_3$.

Figure 4. Optical micrograph of annealed diblock film on oxidized surface of silicon wafer.

(methacrylate) or not. Here, a p(MMA) macroinitiator was used to polymerize OSu leading to very good block copolymer formation which is an improvement over an earlier report in which a macroinitiator based on OSu was used to generate poly(OSu-b-MMA) copolymers.[27] These workers reported less efficient initiation based on a substantial macroinitiator peak present after copolymer formation. This is an interesting comparison and further work is necessary to understand if the monomer order or differences in chemistry control good initiation.

After block copolymer formation, the active ester functions were converted to terpy. The conversion was easily accomplished by reacting an amine functionalized terpy with the polymer in anhydrous DMSO and triethylamine at 60 °C for 2 h as shown in Scheme 3. By ^{1}H NMR and IR spectroscopy, the reaction proceeds to greater than 95 % conversion. Covalent attachment of terpy to the polymer backbone is supported by the shift in the methylene protons adjacent to the amine. In the terpy molecule, these protons come at 2.6 ppm but shift to 3.3 ppm when bound to the polymer backbone while the methylene protons adjacent to oxygen have the same integration as those at 3.3 ppm indicating that all signals from the terpy molecule are associated with backbone attachment. In agreement with these observations, the complete disappearance of the signal at 2.8 ppm, corresponding to the methylene protons of the succinamide ring, was observed. The active ester has characteristic IR stretches at 1808, 1781, and 1672 cm^{-1} which disappear completely upon reaction, suggesting high conversion of the active esters.

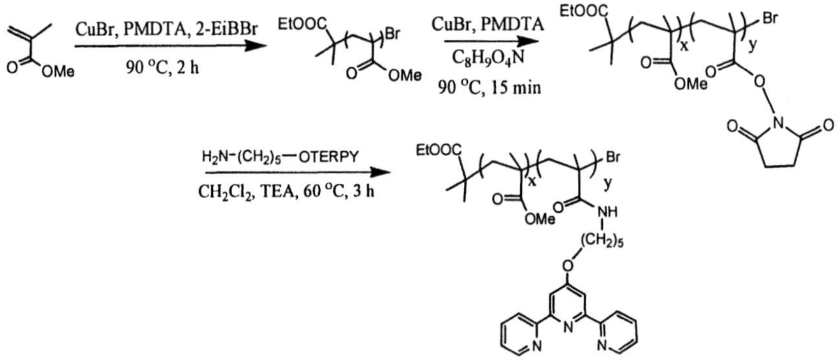

Scheme 3. ATRP synthesis of p(MMA-b-OSu) and post-polymerization functionalization with terpyridine.

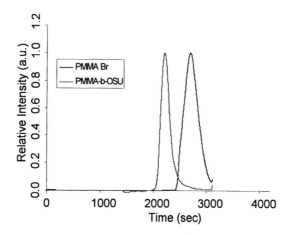

Figure 5. SEC chromatograms of a p(MMA) macroinitiator, centered at 2750 sec, and p(MMA-b-OSu) block copolymer, centered at 2200 sec.

Intramolecular Cross-linking

Addition of Cu(II) ions to a MMA random copolymer containing terpy units results in the expected UV-Vis spectroscopy changes; however, the observation of metal-ligand bonds in the polymer does not indicate if they are inter- or intramolecular bonds. Therefore, solution viscosity measurements were taken to determine if interchain cross-links were formed. A solution of polymer was monitored at a single concentration as Cu(II) ions were added. For the experiments represented in Figure 6, a polymer solution of 4 mg/mL (0.35 wt %) in 1:1 CHCl$_3$:MeOH was used for both polymers, terpy copolymer and p(MMA) homopolymer. Polymer solutions were stirred for 30 min after each addition of metal ions. As seen in this figure, there is a clear difference in the response of the terpy copolymer and homopolymer to metal ion addition. As more metal is added, the viscosity of the copolymer increases rapidly while the homopolymer remains flat. Such an increase in viscosity is consistent with increased polymer molecular weight, suggesting intermolecular cross-links are formed. When the final solution was allowed to stand for an extended period (>2 days), the viscosity remained constant (near 1.5) suggesting the intermolecular interactions in this systems are stable. Stable cross-links were not reported with bipy functionalized polyoxazolines and may be related to the different ligands used. In comparison to hydrogen bonding assemblies, no difference in relative viscosity was observed at concentrations of 4 mg/mL.[32] Finally, the addition of PMDETA to the viscous solution resulted in the immediate decrease in viscosity as shown by the larger red triangle.

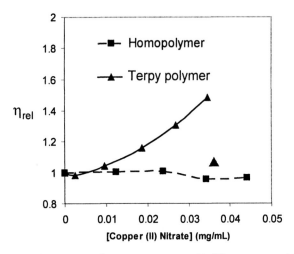

Figure 6. The change in relative viscosity as Cu(II) ions are added to the solution. Only the terpyridine containing polymer shows an increase in viscosity, while the PMMA homopolymer remains essentially unchanged. At the end point, a better ligand, PMDETA, is added and the viscosity decreases as shown with the larger red triangle.

A direct comparison of the mol % terpy and concentration in the backbone was conducted and the experiments are shown in Figure 7. The square and diamond data points were collected from solutions containing 5 mg/mL and 4 mg/mL of polymer, respectively. Although the overall concentration of polymer is almost identical, the mol % of terpy in the polymer is higher for the diamonds, at 12.8 mol % compared to just 4.5 mol % for the squares. Thus an increased change in viscosity is observed for the higher mol % terpy sample. This result is consistent with our earlier observations that interchain crosslinking is likely to be responsible for the increase in viscosity because the presence of more terpy units would improve the likelihood of cross-links. Similarly, a more concentrated sample should provide a larger change in relative viscosity. When solutions containing polymer with 4.5 mol % terpy were studied at 5 mg/mL and 20 mg/ml, it was observed that the more concentrated solution provided a large overall change in viscosity as shown by the square and triangle data points in Figure 7.

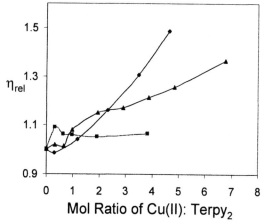

Mol Ratio of Cu(II): Terpy$_2$

Figure 7. Comparison of mol % terpyridine and polymer concentration. The square points were obtained from a 4.5 mol % terpy polymer while the diamond data points were collected from a 12.8 mol % polymer. The overall polymer concentrations were 5 mg/mL and 4 mg/mL, respectively. At the same time, an increase in polymer concentration leads to a larger change in relative viscosity as by the square and triangle data points. These data were collected from a polymer with 4.5 mol % terpy at 5 mg/mL and 20 mg/ml, respectively.

A single, low concentration of polymer was used in these experiments to determine the response to metal ions. In fact, the initial concentration of polymer in the solution does not detectably change the viscosity of the solvent. In contrast, if the experiment is repeated with a significantly higher concentration of terpy functionlized polymer then a solid precipitates upon metal ion addition as shown in Figure 8. This solid is insoluble in any solvent, including DMF which usually dissolves both the apo-polymer (no metal ion) and metal complexed polymer. This insolubility suggests the precipitate is highly cross-linked by the metal-ligand complexes.

Figure 8. Picture of a highly cross-linked gel produced by adding Cu(II) ions to a more concentrated polymer solution. The addition of DMF, a good solvent for the apo- and metal-containing polymer, swells but does not dissolve the sample.

Polymer Based Sensors

The incorporation of metal-complexes into macromolecular architectures produces multi-functional materials. Beyond the use of these interactions to drive polymerizations and reversible cross-links, applications in sensors should be enabled. Toward this end, we are exploring these systems for sensing of other metal ions, solvent, temperature, and chemical agents. We have already reported thermochromic materials based on lanthanides and recently discovered Hg(II) selective sensors. Here, we report on Co(II) films that serve as solvent sensors based on observable solvochroism from green to brown. Figure 9 shows pieces of a polymer film produced from a Co(II) complexed macromolecule based on PMMA. When these films are prepared at room temperature, they are green in color. However, upon exposure to methanol or water they change to brown within 100 secs. The color returns to green upon removal of the agent and the process is completely repeatable. The overall time scale to sense and return to the initial green color is temperature depended, but approximately 2 mins. These results have not been optimized and thinner films would be expected to generate much faster time scales for recovery. In addition, the use of these polymeric materials to develop methanol or water sensors may be limited by other technology; however, this simple experiment represents a critical proof of principle.

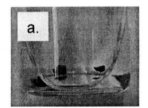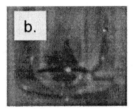

Figure 9. (a) Green film before and (b) brown film after exposure to methanol.

Conclusions

Creating hybrid materials by integrating metal ligand complexes into polymer architectures, including block copolymers, is a versatile approach to novel supramolecular materials. Extensive work at this early stage needs to focus on synthetic methods that provide access to a wide variety of polymer chemistries and architectures. This report describes significant advances toward block copolymers using CRP methods combined with a direct or indirect approach to ligand incorporation. Each approach, direct or indirect, has advantages; however, the indirect approach allows access to well characterized polymers with high terpy content. Until the problems of SEC chromatography are solved, this indirect approach remains the top choice in our laboratory. Preliminary results demonstrate the array of properties that will be realized from these unique macromolecules. Applications in areas such as sensors, 'smart,' and self-healing materials are expected.

Acknowledgments

We thank Kevin Calzia for his work early in this project. We thank the ARO Young Investigator and PECASE programs for generous support of this work. Initial support from the ARL Center of Excellence on Polymers (W911NF-04-1-0191) is greatly acknowledged. G.N.T thanks the ONR Young Investigator, NSF-CAREER, 3M Nontenured faculty grant, and Dupont Young Faculty Award programs as well as NIH for support.

References

(1) Stupp, S. I.; Braun, P. V., *Science* **1997**, *277*, 1242-1248.
(2) Lehn, J.-M. *Supramolecular chemistry-Concepts and prospectives; VCH: Weinheim*, 1995.
(3) Balzani, V.; Juris, A.; Venturi, M.; Campagna, S.; Serroni, S., *Chem. Rev.* **1996**, *96*, 759-833.
(4) Lamba, J. J. S.; Fraser, C. L., *J. Am. Chem. Soc.* **1997**, *119*, 1801-1802.
(5) Schubert, U. S.; Kersten, J. L.; Pemp, A. E.; Eisenbach, C. D.; Newkome, G. R., *Eur. J. Org. Chem.* **1998**, 2573-2581.
(6) Johnson, R. M.; Corbin, P. S.; Ng, C.; Fraser, C. L., *Macromolecules* **2000**, *33*, 7404-7412.
(7) Schubert, U. S.; Nuyken, O.; Hochwimmer, G., *J. Macromol. Sci.-Pure Appl. Chem.* **2000**, *37*, 645-658.
(8) Beck, J. B.; Rowan, S. J., *J. Am. Chem. Soc.* **2003**, *125*, 13922-13923.
(9) Schubert, U. S.; Hien, O.; Eschbaumer, C., *Macromol. Rap. Commun.* **2000**, *21*, 1156-1161.
(10) Newkome, G. R.; He, E., *J. Mater. Chem.* **1997**, *7*, 1237-1244.
(11) Schubert, U. S.; Eschbaumer, C., *Angew. Chem. Int. Ed.* **2002**, *41*, 2893-2926.
(12) Zhao, Y. Q.; Beck, J. B.; Rowan, S. J.; Jamieson, A. M., *Macromolecules* **2004**, *37*, 3529-3531.
(13) Johnson, R. M.; Fraser, C. L., *Macromolecules* **2004**, *37*, 2718-2727.
(14) Smith, A. P.; Fraser, C. L., *Macromolecules* **2003**, *36*, 2654-2660.

140

(15) Chujo, Y.; Sada, K.; Saegusa, T., *Macromolecules* **1993**, *26*, 6315-6319.
(16) Chujo, Y.; Sada, K.; Saegusa, T., *Macromolecules* **1993**, *26*, 6320-6323.
(17) Chujo, Y.; Sada, K.; Saegusa, T., *Polymer J.* **1993**, *25*, 599-608.
(18) Hanabusa, K.; Nakano, K.; Koyama, T.; Shirai, H.; Hojo, N.; Kurose, A., *Makromol. Chem.* **1990**, *191*, 391.
(19) Hanabusa, K.; Nakamura, A.; Koyama, T.; Shirai, H., *Makromol. Chem.* **1992**, *193*, 1309-1319.
(20) Potts, K. T.; Usifer, D. A.; Guadalupe, A.; Abruna, H. D., *J. Am. Chem. Soc.* **1987**, *109*, 3961-3967.
(21) Potts, K. T.; Usifer, D. A., *Macromolecules* **1988**, *21*, 1985-1991.
(22) Schubert, U. S.; Hofmeier, H., *Macromol. Rapid Comm.* **2002**, *23*, 561-566.
(23) Calzia, K. J.; Tew, G. N., *Macromolecules* **2002**, *35*, 6090-6093.
(24) Shunmugam, R.; Tew, G. N., **2005**, submitted.
(25) Aamer, K.; Tew, G. N., *Macromolecules* **2004**, *37*, 1990-1993.
(26) Hermanson, G.; *Bioconjugate techniques*, Ed.: New York, 1996, pp 139-140.
(27) Monge, S.; Haddleton, D. M., *Eur. Polym. J.* **2004**, *40*, 37.
(28) Godwin, A.; Hartenstein, M.; Muller, A. H. E.; Brocchini, S., *Angew. Chem. Int. Ed.* **2001**, *40,*, 595.
(29) Lohmeijer, B. G. G.; Schubert, U. S., *J Polym. Sci. Part A: Poly. Chem.* **2004**, *42*, 4016-4027.
(30) Meier, M. A. R.; Lohmeijer, B. G. G.; Schubert, U. S., *Macromolecular Rapid Communications* **2003**, *24*, 852-857.
(31) Heilmann, S. M.; Rasmussen, J. K.; Krepski, L. R., *J Polym. Sci. Part A: Poly. Chem.* **2001**, *39*, 3655.
(32) Rieth, L. R.; Eaton, R. F.; Coates, G. W., *Angew Chem Int Edit* **2001**, *40*, 2153-2156.

Chapter 11

Complexation Parameters of Terpyridine–Metal Complexes

Philip R. Andres[1], Harald Hofmeier[1], and Ulrich S. Schubert[1,2,*]

[1]Laboratory of Macromolecular Chemistry and Nanoscience,
Eindhoven University of Technology, P.O. Box 513,
5600 MB Eindhoven, The Netherlands
[2]Center for Nanoscience, Ludwig-Maximiliams-Universität München,
Geschwister-Scholl Platz 1, 80333 München, Germany
*Corresponding author: email: u.s.schubert@tue.nl;
Internet: www.schubert-group.com

Terpyridine ligands were used for the construction of model complexes, which were characterized by NMR, UV-vis and MALDI-TOF mass spectrometry. Subsequently, the kinetics of the complex formation was investigated by isothermal titration microcalorimetry (ITC) to obtain the complexation constants. In addition, the complexation reaction was studied by UV-vis titration experiments.

Introduction

Metal complexes of 2,2':6',2"-terpyridines have been known since the isolation of 2,2':6',2"-terpyridine by Morgan and Burstall[1] and have been studied to a wide extent regarding synthesis and characterization.[2] More recently such complexes have been applied in fields such as metallo-supramolecular chemistry, where the metal centers are used to control self organization processes[3] or used for photophysical purposes, from which energy or electron transfer processes to or from the photoactive complex leads to molecular light to energy conversion systems.[4] Metal-ions suitable for the build-up of such metallo-assemblies are transition metal ions such as iron(II), ruthenium(II) or nickel(II), which lead to pseudo-octahedral *bis*-terpyridine complexes with high stability constants.[5,6]

Concerning the reversibility of such metallo-supramolecular complex structures, only little is known about the exact thermodynamic and kinetic processes up to date. Complexes with stronger binding metals such as iron(II)-*bis*-terpyridine complexes start to exchange at higher temperatures of over 100 °C;[7] whereas the metal-ions, which have significantly lower association constants with terpyridines, are zinc(II), cadmium(II) or manganese(II).[5,6] Therefore these metal-ions should prove to be interesting for metallo-supramolecular structures, which can be reversed by applying less severe conditions, such as high temperatures as mentioned above for the iron(II) case.

In this contribution the preparation and characterization of such complexes are presented, followed by a study of the kinetics and thermodynamics of the complex formation using UV-vis spectroscopy and ITC microcalorimetry.

Results and discussion

Synthesis and characterization of 4'-functionalized *bis*-terpyridine metal complexes

Two complexes of a heptyl-functionalized terpyridine were prepared, as models. Both complexes presented here could be characterized by means of NMR, UV-vis spectroscopy and MALDI-TOF mass spectrometry. First the zinc(II) *bis*-terpyridine complex **2** was synthesized by reacting the terpyridine ligand **1** and zinc(II) acetate in a 2:1 ratio (Scheme 1).

The reaction conditions were chosen according to standard literature procedures involving this type of *bis*-terpyridine complexation with transition metal ions.[2,8] First, complexation takes place in methanol solution at 70 °C using the soluble acetate salt of zinc(II). Addition of a large excess of NH_4PF_6 caused the *bis*-complex to precipitate. The pure product was then obtained by crystallization of the crude filtrate by diffusion of diethyl ether into acetonitrile.

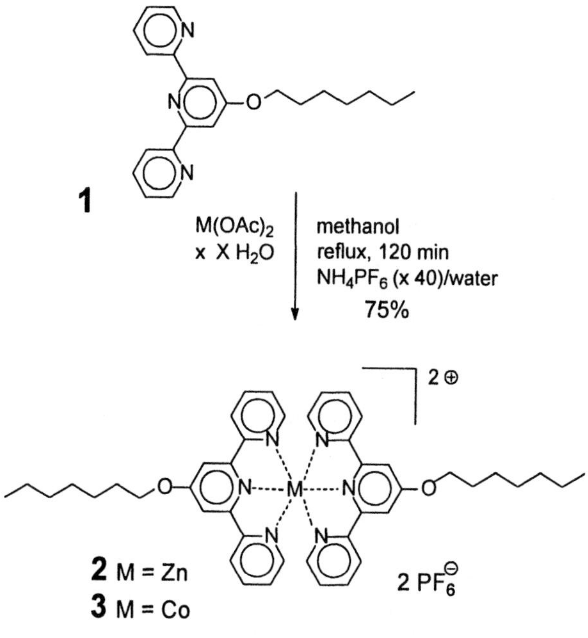

1

M(OAc)$_2$
x X H$_2$O

methanol
reflux, 120 min
NH$_4$PF$_6$ (x 40)/water

75%

2 ⊕

2 M = Zn
3 M = Co

2 PF$_6^{\ominus}$

Scheme 1. Synthesis of the metal complexes 2 and 3.

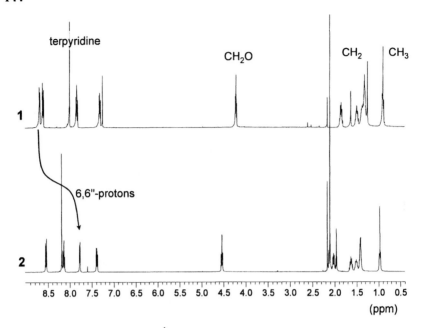

Figure 1. Comparison of the ^{1}H-NMR spectra of the ligand 1 (top spectrum, in chloroform) with its zinc(II) complex 2 (bottom spectrum, in acetonitrile).

The diamagnetic nature of zinc(II) allows for characterization of the complex with NMR spectroscopy. The ^{1}H-NMR spectrum shows significant differences when compared with the free ligand (Figure 1).

The shifts of the signals are mainly due to the influence of the metal center on the terpyridine ligands and to the fact that the terpyridines are locked in a pseudo-octahedral position in which some of the ligand protons "feel" the magnetic fields created by the opposite lying terpyridine. The protons in the 6,6"-positions especially show a significant shift: because of their position directly above the aromatic ring of the other ligand, the ring current causes their upfield shift. The UV absorption will be discussed later.

Apart from *bis*-complexes, mono-complexes are also readily formed by applying one equivalent of the metal-salt, which was shown for cadmium (Cd(tpy)Cl$_2$) by Pickardt *et al.*[9] This fact should also reflect in the binding constants, which will be discussed later herein.

The cobalt(II) complex **3** was formed in a similar fashion by the reaction of two equivalents 4'-(heptyloxy)-terpyridine with one equivalent of Co(II)acetate. The addition of excess NH$_4$PF$_6$ then immediately produced quantitative precipitation of the crude product. After filtration and washing with MeOH as well as water the pure complex was obtained after recrystallization from acetonitrile/diethyl ether to give brown crystals. The shifts in the ^{1}H-NMR are influenced by the paramagnetism of Co(II) (Figure 2). Due to the hyperfine interaction of the unpaired Co(II) electrons with the ligand protons a shift to low field is observed with the signals still being sharp enough for integration.[10] Through comparison with similar complexes reported by Constable *et al.* the spin-state of the Co(II) could be assigned as low-spin.[11]

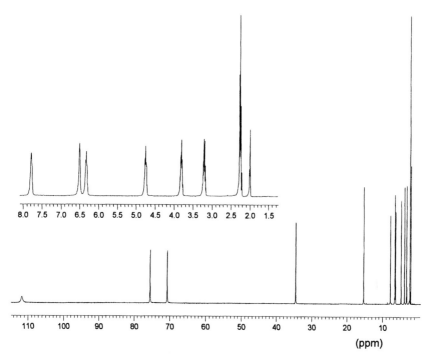

Figure 2. Knight shift [1]*H-NMR spectrum of cobalt complex 3 (in acetonitrile).*

The existence of the complexes **2** and **3** was proven by MALDI-TOF mass spectrometry. The mass spectra revealed several isotope distributions, which could be assigned. All fragments refer to singly positive charged compounds.[12] An ion pair of the complex cation with one PF_6 counterion (only for **3**) as well as the cation without counter-ions were detected. These fragment types are often observed for *bis*-terpyridyl complexes.[13] In addition, matrix adducts (loss of one terpyridine ligand and binding of a dithranol molecule) are formed during the MALDI process.

The shift of the ligand centered (LC) absorption band in the UV-vis spectrum to lower energy is typical for terpyridine complexation and can also be seen for the above discussed zinc(II) and cobalt (II) complex (for the spectra see the UV-titration parts discussed later). In the case of the cobalt(II) complex a weak metal-to-ligand charge transfer (MLCT) absorption becomes visible, when compared to the spectrum of the free ligand. Especially regarding future applications including Co(II)-terpyridine complexation the appearance of such a MLCT band could be of importance for the detection of successful complexation reactions.

Investigation of the complexation papameters with ITC and UV-vis titrations

In the following section, a more detailed insight regarding the thermodynamics and kinetics of the complexation of 4'-chloroterpyridine with cobalt(II), zinc(II), and manganese(II) will be presented.

Thermodynamic and kinetic data of the complexation of terpyridine-ligands with transition metal ions have been determined mainly by stopped flow measurements.[5,6,14,15] Isothermal titration microcalorimetry (ITC) has been proven to be a useful method for the study of complexation parameters of other metal-to-ligand complexation systems, such as: human serum albumin with nickel(II)[16] or a lignin derivative with Cu, Pb, and Mg.[16,17] Here, it was attempted to determine stability constants for the commonly known terpyridine manganese(II), zinc(II), and cobalt(II) complexes by ITC.

Briefly, the method works as follows. Aliquots of one binding partner (here the metal salt) are titrated with another binding partner (here the 4'-chloro-terpyridine). The heat evolution upon each titration in comparison to a reference cell containing only the solvent is measured at a given temperature (26 °C). The titration syringe acts at the same time as a stirrer in order to minimize diffusion effects in the cell. Most known attempts in literature, which concern thermodynamics and kinetics of terpyridine-metal complexation, focus on one reaction step at a time, using e.g. a large excess of the metal-ion, with the focus on mono-complexation (pseudo-first order condition). With ITC it should be possible, especially for the metal-ions with rather low K's as the ones mentioned above, to determine multiple binding constants in one experiment. Concerning the statistics of complexation, the following assumptions were made. Based on what is known from literature, the investigated metal-ions are believed to coordinate in a hexacoordinate fashion with the terpyridine ligand. Thus, there is the possibility of coordinating one and/or two terpyridines per metal-ion. It should be mentioned, that lower coordination modes could also be possible, which would reflect the dependency of the binding-sites towards each other. Since this question cannot be answered with complete certainty, the question of the sites being dependant or independent will not be discussed here. For using the sequential binding sites model, as described above, it has to be assumed that both complexation sites in the apo-metal ions are identical. The following equations are of importance for fitting the experimental data according to a sequential binding sites model, which, in this case, takes into account two sequential binding sites on the apo-metal ion. In equation 1, F_n is the fraction of total apo-metal having n bound ligands. F_n contains the fitting parameters K_n and the free concentration of ligand (for details see manual of MicroCal ITC).[18] The heat content after the ith injection is then determined from equation 1.

Equation 1: $Q = M_t V_0 (F_1 \Delta H_1 + F_2[\Delta H_1 + \Delta H_2])$

Equation 2: $\Delta Q(i) = Q(i) + \dfrac{dV_i}{V_0}\left[\dfrac{Q(i) + Q(i-1)}{2}\right] - Q(i-1)$

with: Q = total heat content of the solution contained in V_0;
      M_t = bulk concentration of metal-salt in V_0;
      V_0 = active cell volume;
      ΔH_n = free enthalpy for reaction step n;
      F_n = fraction of metal-salt having n bound ligands;
      Q_i = total heat released after the ith injection;
      ΔQ_i = heat released from completion of the $(i-1)^{th}$ to the ith injection.

This heat content is then used for the expression of the heat released ΔQ_i, from the ith injection (Equation 2), which then leads to a Marquardt minimization routine.

In parallel, UV-vis titrations were performed for these systems. The fitting of the data in this case was attempted using the standard mass balance equations and the Lambert-Beer equation. Additionally, the different epsilons for different complex-species in one system (e.g. mono-complex or *bis*-complex) were also allowed to vary as fitting parameters. For more detailed information, the reader is referred to the literature.[19] The choice of the ligand used was 4'-chloro-terpyridine, for which no thermodynamic and kinetic investigations have been performed yet, and which represents an important building block in metallo-supramolecular science. Cobalt(II) acetate-tetrahydrate, zinc(II) chloride, and manganese(II) chloride were the metal-salts used for the investigations. The choice of metal-salts had to do mainly with purity, availability, and past experiences. All experiments were performed in methanol.

Concerning the use of ITC for terpyridine to metal interactions, it has to be taken into account that this method is limited especially in terms of determining formation constants in the range of $K = 10^4 - 10^8$ L/mol ("K-window"). This is mainly due to the limited minimal amount of titrant, which can be used and which then leads to too few data points to obtain a good fit for the experimental data.

The first system discussed here is cobalt(II) acetate-tetrahydrate / 4'-chloro-terpyridine. As known from literature the complexation constants for the reaction of unfunctionalized terpyridine with excess cobalt(II) acetate (pseudo-first order) in water gave a complexation constant of 2.5×10^8 L/mol, measured by stopped flow. This already indicates that cobalt-terpyridine complexation may lie just out of "K-window" for the ITC method. Nevertheless, a measurement was attempted (Figure 3).

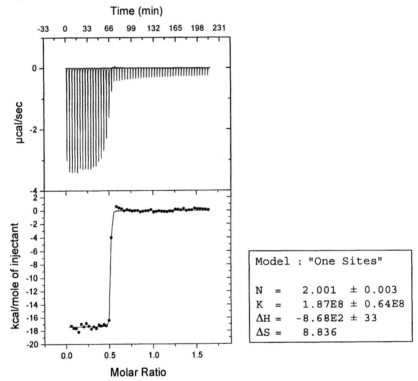

Figure 3. ITC-titration of cobalt(II) acetate to 4'-chloro-terpyridine (cell conc.: 0.09 mM, in MeOH).

On the upper left of Figure 3 the actual titration can be seen as recorded. The graph below shows the integral heat per injection plotted against the molar ratio. The raw data have been corrected for heat of dilution, by subtracting a heat-dilution measurement, where the metal-salt solution is titrated to the pure solvent methanol (titration of pure solvent to the ligand solution does not have to be taken into account, because there is no heat effect). As expected, obtaining a good fit was not possible. Nevertheless, when using a simpler fitting model, which only considers one type of identical sites (in principal the same as the above described site model, without the condition of sequential order), a suitable fit can be obtained. This result gives, of course, only one $K = K_1 = K_2$, but nonetheless a rough estimate of the range of the binding constants can be obtained. This result also matches the above-mentioned findings in literature. On the other hand, this fit clearly shows a sharp transition at a molar ratio of 0.5 cobalt(II) ions to chloroterpyridine ligands, which clearly proves the formation of a *bis*-complex. This sharp transition also indicates "cooperative binding" where K_2 should prove to be K_1, meaning that in equilibrium for all ratios almost no mono-complex, rather only *bis*-complex, is stable. However, this latter indication cannot be proven beyond doubt, since the sequential order binding model does not give results in this case. This also means that the values obtained for the enthalpy, which is as expected positive and the entropy, which is also clearly positive (4 hydrated waters are freed), are only average estimates. These findings can now be compared with data obtained from a UV-vis titration. Generally, it has to be noted that the UV-vis titrations require much smaller concentrations (between 1 and 2 orders of magnitude). This has to be carefully taken into account when comparing with the results obtained by ITC. Figure 4 shows the UV-vis titration as well as the rise in absorption at 329 nm upon addition of cobalt(II) acetate-tetrahydrate including the fitting curve.

A good fit could be achieved using the standard mass balance equations for a 2:1 ligand to metal binding model, however, the K-values obtained by this method cannot be trusted because of too few data points around the metal to ligand ratio of 0.5. The most similar compound to the one investigated here, which is known from literature, especially regarding the acetate counter-ions, is a *bis*-4'-*para*-tolyl-terpyridine cobalt(II) di-acetate complex.[20]

The next system investigated was zinc(II) chloride / 4'-chloro-terpyridine. Already from qualitatively looking at the ITC measurement, one can observe a transition, which fits the measurement requirements (Figure 5). Again, the data points have been corrected for dilution, as described above.

The experimental data can be fitted very well with the above described sequential binding sites model. The value of 4.6×10^6 L/mol for K_1 is in good agreement with the value of 1.0×10^6 obtained for unfunctionalized terpyridine with excess zinc(II) by stopped flow measurements (pseudo-first order) in water.[6] Furthermore, the ITC measurement provides a value for K_2, which lies significantly below the value for K_1. This negative cooperativity would suggest that the mono-complex is the favored product upon reacting metal and ligand in a 1:1 ratio as opposed to the above discussed cobalt case. The enthalpies are, as to be expected, negative for both reaction steps. The entropy for the first step is slightly negative, which seems reasonable considering the formation of one product (the mono-complex) out of two educts (chloroterpyridine and zinc(II) chloride) and indicates that the chlorides are tightly bound to the zinc(II). On the

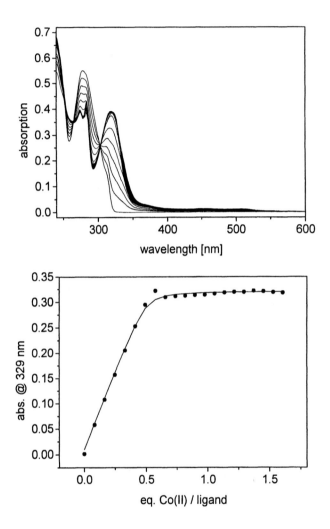

Figure 4. Top: UV-vis-titration of cobalt(II) acetate to 4'-chloro-terpyridine (conc.: 2.5 × 10^{-5} mM, in MeOH); Bottom: rise in absorption at 329 nm as a function of the molar ratio.

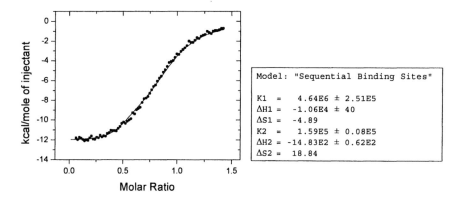

Figure 5. ITC-titration of zinc(II) chloride to 4'-chloro-terpyridine (cell conc.: 0.10 mM, in MeOH).

other hand, the entropy for the second step is highly positive, which can be illustrated by the displacement of the two chloride anions by the second terpyridine.

The UV-vis spectrum for the zinc(II) system indicates a more complicated behavior than for the case of cobalt(II). Up to the equivalence of zinc(II) to chloro-terpyridine of 1 to 2, a linear increase in the arising ligand centered (LC) band is observed. However, after further titration, the maximum of the LC band starts to shift significantly and remains unchanged after a 1:1 ratio is reached (Figure 6).

Qualitatively, this observation is in good accordance with the finding of negative cooperativity from the ITC experiment. Upon further titration after a 1:2 zinc to ligand ratio is reached, the formation of mono-complexes immediately starts taking place. Hence, it can be concluded, that the final maximum of the LC band belongs to the mono-complex, in contrast to the maximum observed up to a titration of a 1:2 (zinc to ligand) ratio, which belongs to the *bis*-complex. Because of this shift, no fit of the rise to absorbance could be performed. A crystal structure for the 1:1 complex with unsubstituted terpyridine as the chloride salt has been reported.[21] Also a *bis*-complex of a 5,5"-methyl substituted terpyridine with PF_6^- - counter-ions was reported by our group.[8]

Lastly, manganese(II) chloride to chloro-terpyridine was investigated. The fit of the experimental data obtained by ITC results in a value of 4.95×10^4 L/mol for K_1 (Figure 7), which can be compared to the value 1.0×10^5 L/mol found for terpyridine with excess manganese(II) perchlorate (pseudo-first order) measured by the stopped flow method in anhydrous methanol.[15]

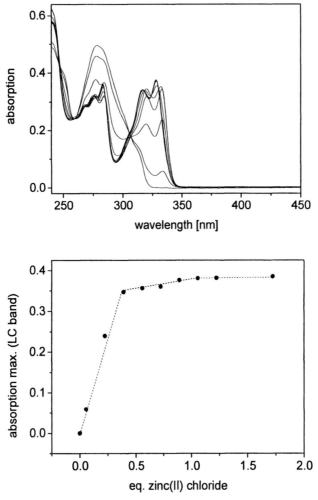

Figure 6. Top: UV-vis-titration of zinc(II) chloride to 4'-chloro-terpyridine (conc.: 10^{-5} M, in MeOH). Bottom: rise in maximum absorption of the LC band between 325 nm and 335 nm as a function of the molar ratio.

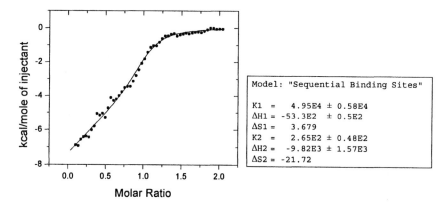

Figure 7. ITC-titration of manganese(II) chloride to 4'-chloro-terpyridine (cell conc.: 1.0 mM, in MeOH).

Using the same methods Holyer et al.[6] found K_1 to be 2.5×10^4 L/mol for terpyridine and manganese(II) measured by stopped flow in water; however, a recent study using the ITC method and 4'-hydroxy-terpyridine / manganese(II) chloride in water solution gave a much higher value for K_1 of 2.0×10^7 L/mol.[22] Except for the last mentioned system, where the hydroxy-group in the 4'-position shows a considerable influence on complexation process, all other known values from literature for similar systems lie in the same order as the value measured in this work. The value obtained for K_2 lies over two orders of magnitude lower at 2.6×10^2 L/mol. Again, this means negative cooperativity and this does seem reasonable when taking into account that mono-complexes with manganese(II) chloride have been prepared and their X-ray crystal structures have been determined.[23].

Conclusions

Through addition of transition metal salts to terpyridine ligands, *bis*-terpyridine metal complexes were prepared and characterized. The complexes could be characterized by ^{1}H-NMR, UV-vis spectroscopy as well as MALDI-TOF mass spectrometry. Detailed studies of the complexation parameters (kinetics and thermodynamics) of the weaker binding metal-ions cobalt(II), zinc(II) and manganese(II) were performed using ITC and UV-vis spectrophotometric titrations for the ligand 4'-chloro-terpyridine. The results obtained are comparable to what is known from literature for unfunctionalized terpyridine. Furthermore, there is strong evidence from ITC that zinc(II) chloride complexes 4'-chloro-terpyridine with negative cooperativity, meaning that K_1 for the mono-complexation is larger than K_2 for the *bis*-complexation.

Experimental part

General remarks and instruments used

Basic chemicals were obtained from Sigma-Aldrich. 4'-Heptyloxy-terpyridine **1** was synthesized as described in ref.[24]. NMR spectra were measured on a Varian Mercury 400 spectrometer. The chemical shifts were calibrated to the residual solvent peaks or TMS. UV-vis spectra were recorded on a Perkin Elmer Lamda-45 (1 cm cuvettes, acetonitrile). MALDI-TOF mass spectra were measured on a BioSystems Perseptive Voyager 2000 instrument with dithranol as matrix. ITC experiments were carried out in methanol on a VP-ITC isothermal titration calorimeter form MicroCal™.

Bis(4'-heptyloxy-2,2':6',2''-terpyridine) zinc(II) hexafluorophosphate 2

To a stirred suspension of **1** (122 mg, 0.37 mmol) in MeOH (10 mL), Zn(OAc)$_2$ × 2 H$_2$O (40.4 mg, 0.18 mmol) was added and then the mixture was heated for 60 min to 70 °C, followed by addition of NH$_4$PF$_6$ (150 mg, 0.92 mmol) in 10 mL water. The resulting white precipitate was filtered and washed with MeOH (3 × 30 mL). Solvent residues were evaporated under reduced pressure and the crude product was recrystallized by diffusion of diethyl ether into an acetonitrile solution and then dried under high vacuum yielding 136 mg (71%) of a white solid.

UV-vis (CH$_3$CN): λ_{max} (ε) = 322 (27400), 309 (25400), 274 (44200), 244 nm (56200) nm (L mol^{-1}cm^{-1}).

^{1}H-NMR (400 MHz, CH$_3$CN): δ = 0.95 (3 H, t, J = 7.32 Hz, CH$_3$), 1.39 (4 H, m, CH$_2$), 1.49 (2 H, m, CH$_2$), 1.61 (2 H, m, CH$_2$), 2.00 (2 H, m, CH$_2$), 4.52 (2 H, t, J = 6.59 Hz, CH$_2$O), 7.37 (2 H, dd, J = 7.82, 4.89 Hz, H$_{5,5''}$), 7.76 (2 H, d, J = 4.89 Hz, H$_{6,6''}$), 8.12 (2 H, dd, J = 7.82, 7.82 Hz, H$_{4,4''}$), 8.17 (2 H, s, H$_{3',5'}$), 8.53 (2 H, d, J = 7.82 Hz, H$_{3,3''}$).

MS (MALDI-TOF, matrix: dithranol): m/z = 636.3 [M − ligand − 2 PF$_6$ + dithranol − H]$^+$, 759.4 [M − 2 PF$_6$]$^+$.

Bis(4'-heptyloxy-2,2':6',2''-terpyridine) cobalt(II) hexafluorophosphate 3

To a stirred suspension of **1** (100 mg, 0.29 mmol) in MeOH (10 mL), Co(OAc)$_2$ × 4 H$_2$O (35.8 mg, 0.15 mmol) was added at room temperature. After 30 min NH$_4$PF$_6$ (95 mg, 0.58 mmol) in 10 mL MeOH was added upon which an orange-brown precipitate was formed. The mixture was stirred for another 10 min and the precipitate was filtered and washed with MeOH (3 × 30 mL) and CHCl$_3$ (3 ×

30 mL). Residual solvent was evaporated *in vacuo*. The crude product was then recrystallized by diffusion of diethyl ether into an acetonitrile solution and then dried under high vacuum yielding 113 mg (72%) of an orange solid.

UV-vis (CH_3CN): λ_{max} (ε) = 306 (25800), 274 (38800), 244 nm (49400) nm (L $mol^{-1}cm^{-1}$).

IR (ATR): $1/\lambda$ (cm^{-1}): 1615, 1603, 1571, 1558 (C=C, C=N, terpyridine); 827 (P-F, PF_6).

^{1}H-NMR (300 MHz, CH_3CN): δ = 2.20 (3 H, t, J = 7.31 Hz, CH_3), 3.15 (2 H, m, CH_3), 3.74 (2 H, m, CH_3), 4.69 (2 H, m, CH_3), 6.26 (2 H, m, CH_3), 6.45 (2 H, m, CH_3), 7.73 (2 H, m, CH_3), 15.20, 34.20, 70.65, 75.42, 111.59 (10 H, H-terpyridine).

MS (MALDI-TOF, matrix: dithranol): m/z = 631.3 [M − ligand − 2 PF_6 + dithranol − H]$^+$, 753.4 [M − 2 PF_6]$^+$, 898.2 [M − PF_6]$^+$.

Anal. Calcd for $C_{44}H_{50}N_6O_2P_2F_{12}Co$ (1043.78): C, 50.63; H, 4.83; N, 8.05. Found: C, 50.45; H, 4.71; N, 7.97.

Acknowledgments

The authors thank the *Dutch Scientific Organization* (NWO) and the *Fonds der Chemischen Industrie* for financial support.

References

(1)   Morgan, S. G.; Burstall, F. H. *J. Chem. Soc.* **1931**, 20-30.

(2)   Constable, E. C. *Adv. Inorg. Chem. Radiochem.* **1986**, *30*, 69-121; McWhinnie, W. R.; Miller, J. D. *Adv. Inorg. Chem. Radiochem.* **1969**, *12*, 135-215.

(3)   Lehn, J.-M. *Supramolecular Chemistry, Concepts and Perspectives*; VCH: Weinheim, 1995; Hofmeier, H.; Schubert, U. S. *Chem. Soc. Rev.* **2004**, *33*, 373-399.

(4)   Nazeeruddin, M. K.; Pechy, P.; Renouard, T.; Zakeeruddin, S. M.; Humphry-Baker, R.; Comte, P.; Liska, P.; Cevey, L.; Costa, E.; Shklover, V.; Spiccia, L.; Deacon, G. B.; Bignozzi, C. A.; Grätzel, M. *J. Am. Chem. Soc.* **2001**, *123*, 1613-1624; Yu, S.-C.; Kwok, C.-C.; Chan, W.-K.; Che, C.-M. *Adv. Mater.* **2003**, *15*, 1643-1647.

(5)   Hogg, R.; Wilkins, R. G. *J. Chem. Soc.* **1962**, 341-350.

(6)   Holyer, R. H.; Hubbard, C. D.; Kettle, S. F. A.; Wilkins, R. G. *Inorg. Chem.* **1966**, *5*, 622-625.

(7)   Heller, M.; Schubert, U. S. *Macromol. Rapid Commun.* **2001**, *22*, 1358-1363.

(8)   Schubert, U. S.; Eschbaumer, C.; Andres, P.; Hofmeier, H.; Weidl, C. H.; Herdtweck, E.; Dulkeith, E.; Morteani, A.; Hecker, N. E.; Feldmann, J. *Synth. Met.* **2001**, *121*, 1249-1252.

(9)   Pickardt, J.; Staub, B.; Schafer, K. O. *Z. Anorg. Allg. Chem.* **1999**, *625*, 1217-1224.

(10) Bertini, I.; Luchinat C. *Coord. Chem. Rev.* **1996**; *150*.

(11) Constable, E. C.; Housecroft, C. E.; Kulke, T.; Lazzarini, C.; Schofield, E. R.; Zimmermann, Y. *J. Chem. Soc., Dalton Trans.* **2001**, 2864-2871; Constable, E. C.; Hart, C. P.; Housecroft, C. E. *Appl. Organomet. Chem.* **2003**, *17*, 383-387.

(12) Karas, M.; Glückmann, M.; Schäfer, J. *J. Mass Spectrom.* **2000**, *35*, 1-12.

(13) Schubert, U. S.; Eschbaumer, C. *J. Incl. Phenom. Macrocycl. Chem.* **1999**, *35*, 101-109; Meier, M. A. R.; Lohmeijer, B. G. G.; Schubert, U. S. *J. Mass Spectrom.* **2003**, *38*, 510-516.

(14) Buck, D. M. W.; Moore, P. *J. Chem. Soc., Dalton Trans.* **1976**, 638-642.

(15) Benton, D. J.; Moore, P. *J. Chem. Soc., Dalton Trans.* **1973**, 399-404.

(16) Saboury, A. A. *J. Chem. Thermodyn.* **2003**, *35*, 1975-1981.

(17) Garcia-Valls, R.; Hatton, T. A. *Chem. Eng. J.* **2003**, *94*, 99-105.

(18) http://www.microcalorimetry.com.

(19) Huskens, J.; van Bekken, H.; Peters, J. A. *Comput. Chem.* **1995**, *19*, 409-416.

(20) Kurth, D. G.; Schütte, M.; Wen, J. *Colloids Surf., A* **2002**, *198-200*, 633-643.

(21) Vlasse, M.; Rojo, T.; Beltran-Porter, D. *Acta Crystallogr., Sect. C: Cryst. Struct. Commun.* **1983**, *C39*, 560-563.

(22) Wieprecht, T.; Xia, J.; Heinz, U.; Dannacher, J.; Schlingloff, G. *J. Mol. Catal. A: Chem.* **2003**, *203*, 113-128.

(23) Harris, C. M.; Lockyer, T. N.; Stephenson, N. C. *Aust. J. Chem.* **1966**, *19*, 1741-1743.

(24) Andres, P. R.; Hofmeier, H.; Lohmeijer, B. G. G.; Schubert, U. S. *Synthesis* **2003**, 2865-2871.

Chapter 12

Poly[2-(Acetoacetoxy)Ethyl Methacrylate]-Based Hybrid Micelles

Theodora Krasia and Helmut Schlaad[*]

Colloid Department, Max Planck Institute of Colloids and Interfaces,
Am Mühlenberg 1, 14476 Potsdam-Golm, Germany

Block copolymers based on n-butyl methacrylate (BMA) and 2-(acetoacetoxy)ethyl methacrylate (AEMA), a commercially available monomer bearing a β-dicarbonyl ligand, are used for the solubilization of metal ions (Fe^{3+}, Co^{2+}, Pd^{2+}) in organic media. Stable colloidal dispersions of inverse hybrid micelles can be obtained in cyclohexane. Even though not soluble in the continuous phase, the metal ions are evenly distributed among aggregates. Due to their narrow size distribution, the hybrid micelles tend to form ordered domains when being deposited on a solid substrate.

Background

Coordination compounds play an extraordinary role in diverse areas like supramolecular chemistry,[1] chemistry of living matter and biominerals, catalysis, hydrometallurgy, and waste-water treatment.[2] Over the past decades, polymer science has strongly been entering the field of inorganic-organic hybrid materials, aiming to combine potential applications of metal compounds with the special properties of block copolymers, namely formation of nanometer-scale structured materials, electrosteric stabilization of colloids, and good mechanical performance.[3,4]

Most studies in the field of metal-coordinating polymers and functional hybrid materials have been done with poly(acrylic acid)- and polyvinylpyridine-based block copolymers, see for example the work of Antonietti et al.,[5-14] Möller et al.,[15-23] and Eisenberg et al.[24-27] Missing so far in the list of available block copolymers are those carrying β-dicarbonyl moieties.[2] β-Dicarbonyls are known to act as strong bi-dentate ligands, capable of coordinating a wide range of metal ions with different geometries and oxidation states. Whereas keto tautomers are rather "soft" ligands, only replacing neutral molecules in metal ion salts like water or ethers, enolates can exchange acetate or halide anions.[28]

The empty spot in the list of copolymers was filled in 2001, when Schlaad et al.[29] described the synthesis of block copolymers based on 2-(acetoacetoxy)ethyl methacrylate (AEMA, see structure in Figure 1). Acetoacetoxy moieties of PAEMA are preferentially in the keto tautomeric form (~92%), but can be converted into the enolate upon addition of a base. First results[29] showed that PAEMA-based block copolymers promote a solubilization of ferric salts in very hydrophobic organic media and produce stable colloidal dispersions.

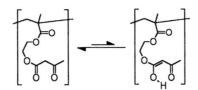

Figure 1. Chemical structure of PAEMA.

At least two major advantages are apparent when using PAEMA instead of polyvinylpyridines. First, it might be used in applications where amine-free ligands are preferable (e.g., drug delivery) and second, its glass transition temperature is well below room temperature (+3 °C;[30] compare: polyvinylpyridine, > +100 °C), thus promoting the formation of monodispersed equilibrium structures rather than "frozen" ones.

Objective

The objective of this work was to screen the potential of PBMA-PAEMA block copolymers (PBMA = poly(n-butyl methacrylate)) in the production of colloidal hybrid materials. Colloids containing different metal ions (Fe^{3+}, Co^{2+}, and Pd^{2+}) have been prepared in cyclohexane solution and characterized by means of spectroscopy (^{1}H NMR and UV/visible), static and dynamic light scattering (SLS and DLS), analytical ultracentrifugation (AUC), and scanning force microscopy (SFM).

Experimental

Chemicals. Chemicals and solvents were purchased from Aldrich with the highest purity grade available and were used as received.

Block copolymer synthesis. PBMA$_{342}$-PAEMA$_{39}$ (the subscripts denote the number of repeating units) was synthesized according to a procedure described elsewhere.[29] In a first step, a PBMA$_{342}$-b-PHEMA$_{39}$ (HEMA = 2-hydroxyethyl methacrylate) (polydispersity index, PDI = 1.05) was prepared by sequential group transfer polymerization (GTP) of BMA and TMSHEMA at r.t. in tetrahydrofuran (THF) using 1-methoxy-1-trimethylsiloxy-2-methyl-prop-1-ene as the initiator and tetrabutylammonium bibenzoate as the catalyst; TMS protecting groups were removed by HCl-catalyzed hydrolysis. This material was then converted into PBMA$_{342}$-PAEMA$_{39}$ by exhaustive azeotropic acetoacetylation with *tert*-butyl acetoacetate in a mixture of benzene and water.

Analytical instrumentation and methods. (i) ^{1}H NMR spectra were recorded at 25 °C on a Bruker DPX-400 spectrometer operating at 400.1 MHz. Signals were referenced to that of traces of non-deuterated solvent arising at δ = 7.24 (chloroform) or 1.44 ppm (cyclohexane). (ii) UV/visible spectra were recorded at r.t. using a UVIKON 940/941 dual-beam grating spectrophotometer (Kontron Instruments) equipped with a 1-cm quartz cell. (iii) AUC experiments were performed on an Optima XL-1 ultracentrifuge (Beckman-Coulter, Palo Alto, CA) with Rayleigh interference and UV/visible absorption optics. Sedimentation velocity runs were done with ~0.4 wt % polymer solutions at 25 °C and 60000 rpm. Time-dependent concentration profiles were evaluated using the SEDFIT 5 software (P. Schuck; http://www.analyticalultracentrifugation. com/). (iv) SLS experiments were carried out at 20 °C with a frequency-doubled Neodym-YAG laser light source (Coherent DPSS532, intensity 300 mW, λ = 532 nm), an ALV goniometer, and an ALV-5000 multiple-tau digital correlator (ALV GmbH, Langen, Germany). Measurements were performed on 0.1-0.4 wt % polymer solutions at scattering angles from (15°) 30°-150° at 3° intervals. Data were evaluated by a standard Zimm analysis to provide the molecular weight (M_w) of aggregates, radius of gyration (R_g), and second virial coefficient (A_2). Refractive index increments dn/dc were measured using an NFT-Scanref differential refractometer operating at λ = 633 nm. **(v)** DLS experiments were carried out on a spectrometer consisting of an argon ion laser (λ = 633 nm, 30-600 mW; Coherent Innova 300), a self-constructed goniometer, a single photon detector (ALV SO-SIPD), and a multiple-tau digital correlator (ALV 5000/FAST). DLS autocorrelation functions were measured at different polymer concentrations (0.1-0.4 wt %) and scattering angles (30°, 50°, 70°, and 90°) and were evaluated with the program FASTORT.EXE.[31] From the obtained diffusion coefficients, hydrodynamic radii (R_h) were calculated via the Stokes-Einstein equation. **(vi)** SFM experiments were performed with a Nanoscope Multimode IIIa (Digital instruments, Santa Barbara, CA) employing silicon cantilevers (k = 42 N·m^{-1}; Olympus Optical Co. Ltd., Japan). Specimens were prepared by spin-coating or slowly drying 0.15 wt % polymer solutions on mica or graphite and were scanned in the tapping mode at a resonance frequency of 300 khz.

Results & Discussion

First experiments with DLS showed that PBMA$_{342}$-PAEMA$_{39}$ dissolves in chloroform without forming aggregates. Accordingly, the ^{1}H NMR signals of both block segments are observed in CDCl$_3$. Upon the addition of FeCl$_3$·6H$_2$O (f_{Fe} = [Fe]/[AEMA] = 0.33), the characteristic signals of PBMA become considerably broader and those of PAEMA vanish completely (see Figure 2). Obviously, the coordination of the metal ion salt reduces solubility of the PAEMA segment in chloroform, thus leading to the formation of aggregates with a PAEMA-metal core and a PBMA solvating corona. Formation of the hybrid aggregates is accompanied by a color change from colorless to purple. Due to the fact that the metal ion salt itself is soluble in chloroform, the metal ion salt is released from the PAEMA core when the polymer concentration is lower than ~2 wt %. The color of the solution then is yellow-orange instead of purple.

For a better localization of the salt, cyclohexane was chosen as the solvent. Cyclohexane is a non-solvent for FeCl$_3$·6H$_2$O as well as PAEMA and a good solvent for PBMA. As indicated by DLS, PBMA$_{342}$-PAEMA$_{39}$ forms spherical micelles in cyclohexane with an average hydrodynamic radius R_h = 32 nm. ^{1}H NMR confirmed that micelles have a PBMA solvating corona and a PAEMA core. Added FeCl$_3$·6H$_2$O was found to dissolve readily within a few minutes, and the color of the dispersion changed from colorless to purple. The existence of the PAEMA-metal complex was confirmed by UV/visible spectroscopy (see Figure 3). It is supposed that the octahedral structure of the ferric salt is preserved, the four water molecules adjacent to the iron atom being replaced by two keto tautomeric acetoacetoxy ligands. The two chloride substituents, on the other hand, should have not been touched by the acetoacetoxy groups. In fact, solubilization of anhydrous $FeCl_3$ failed when applying the same experimental protocol. Right after small amounts of a water/methanol mixture had been added to the dispersion, transforming the iron(III) chloride into a hydrate, a red-purple complex was instantly formed.

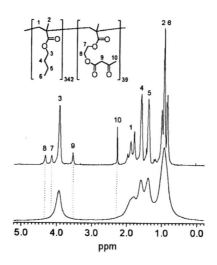

Figure 2. ^1H NMR spectra of $PBMA_{342}$-$PAEMA_{39}$ in the absence (top) and presence (bottom) of $FeCl_3 \cdot 6H_2O$ (f_{Fe} = 0.33) in $CDCl_3$.

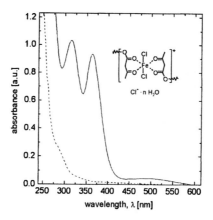

Figure 3. UV/visible spectra of $PBMA_{342}$-$PAEMA_{39}$ (dashed line) and $PBMA_{342}$-$PAEMA_{39}$ + $FeCl_3 \cdot 6H_2O$ (f_{Fe} = 0.33; solid line) in cyclohexane.

As can be seen from AUC sedimentation-velocity experiments (see Figure 4), the sedimentation-coefficient distribution, $g^*(s)$, of the $PBMA_{342}$-$PAEMA_{39}$

micelles shifts to higher s values after being loaded with $FeCl_3 \cdot 6H_2O$. This behavior is due to the different densities of the organic and inorganic components: $\rho = 1.1$ ($PBMA_{342}$-$PAEMA_{39}$) and 1.8 $g \cdot cm^{-3}$ ($FeCl_3 \cdot 6H_2O$). The original shape of the distribution is maintained, which indicates that the ferric salt is evenly distributed among the aggregates. Thus, since the complexation reaction was a heterogeneous process, there must have taken place a dynamic intermolecular exchange of salt molecules between the aggregates, supposedly via a fusion-fission mechanism.[32] The polymer-metal colloids might therefore be considered as equilibrium structures.

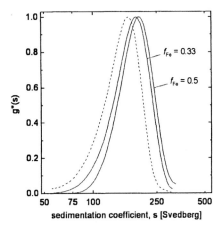

Figure 4. Sedimentation-coefficient distributions of the micellar solutions of PBMA$_{342}$-PAEMA$_{39}$ (dashed line) and PBMA$_{342}$-PAEMA$_{39}$ + FeCl$_3 \cdot 6H_2O$ (f$_{Fe}$ = 0.33 and 0.5, solid lines) in cyclohexane.

DLS and SLS indicate that neither the size nor the shape of micelles is affected by the addition of $FeCl_3 \cdot 6H_2O$ ($R_h \sim 30$ nm, $R_g/R_h \sim 0.85$; R_g: radius of gyration) (see Table 1). Micelles are spherical in shape, which is also seen in SFM (see Figure 5). The average number of polymer chains per micelle, $Z = M_w^{micelle}/M_w^{polymer}$, was found to decrease with increasing amount of ferric salt loaded into the PAEMA core. At $f_{Fe} = 0.85$, the aggregation number drops to about 60% of the initial Z value of $PBMA_{342}$-$PAEMA_{39}$ micelles ($Z \sim 342$). This aggregation behavior is rather unexpected, having in mind the work of Förster et al.[33] on the micellization of strongly segregated block copolymers. The addition of the salt should have increased the incompatibility of the two segments and thus led to a growing of the aggregates and eventually to a transition from spheres to cylinders.[27] Seemingly, the PAEMA core shrinks rather than swells upon the addition of $FeCl_3 \cdot 6H_2O$.

Colloidal dispersions are only stable if $f_{Fe} < 1$ (cf. values of A_2 of in Table 1), otherwise coagulation is observed. The steric stabilization of colloids by the PBMA layer is then not sufficient to shield the attractive forces between the PAEMA-metal centers and to avoid coalesence.[32]

Table 1. Results of SLS/DLS analysis of PBMA$_{342}$-PAEMA$_{39}$ micelles in the presence of different amounts of FeCl$_3$·6H$_2$O, solvent: cyclohexane.

Fe^{3+}, f_{Fe}	R_h [nm]	R_g [nm]	R_g/R_h	$A_2\ 10^9$ [mol·dm^3·g^{-2}]	Z
0	32	26	0.81	1.5	342
0.25	33	29	0.88	0.9	274
0.33	33	28	0.85	0.9	269
0.45	31	27	0.87	1.3	297
0.60	30	27	0.90	1.1	266
0.85	28	24	0.86	-0.6	207
1	---------------------------- coagulation ----------------------------				

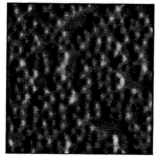

$1 \times 1\ \mu m^2$

Figure 5. SFM phase image of PBMA$_{342}$-PAEMA$_{39}$ micelles loaded with FeCl$_3$·6H$_2$O (f_{Fe} = 0.33), spin-coated from cyclohexane solution on mica.

As mentioned earlier, AEMA in the keto tautomeric form is only capable of replacing neutral ligands in metal ion salts. Exchange of acetate substituents requires deprotonation of the AEMA units with e.g. triethylamine. Assisted by the addition of the amine, the metal ions of cobalt(II) and palladium(II) acetate as well as FeCl$_3$·6H$_2$O could be transferred into the core of PBMA$_{342}$-PAEMA$_{39}$ micelles in cyclohexane. Complexation of the metal ions was accompanied by a precipitation of triethylammonium acetate (see Scheme 1) and a change of color of the solution from colorless to either pink (Co^{2+}), yellow (Pd^{2+}), or orange

(Fe^{3+}); UV-visible spectra are shown in Figure 6. Similar results for the corresponding metal acetoacetonates have been reported elsewhere.[28,34-36]

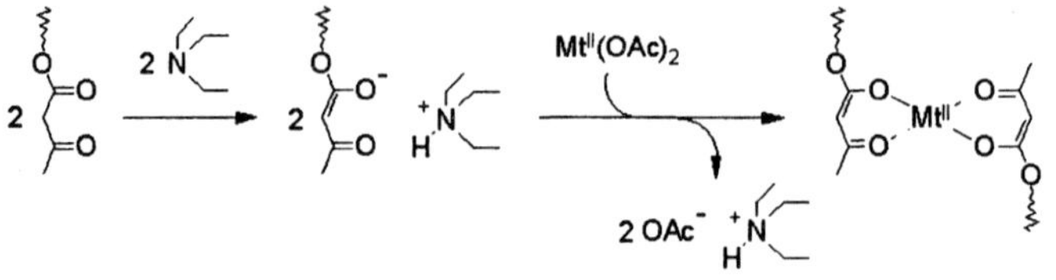

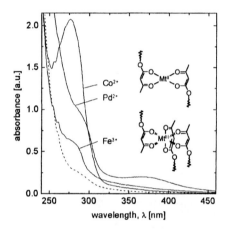

Scheme 1. *Formation of a β-ketoesterenolate by the deprotonation of AEMA acetoacetoxy units with triethylamine and complexation of metal(II) acetates.*

Figure 6. *UV/visible spectra of PBMA$_{342}$-PAEMA$_{39}$ (dashed line) and PBMA$_{342}$-PAEMA$_{39}$ + Co^{2+}/Pd^{2+}/Fe^{3+} (f$_{metal}$ ~ 0.5; solid lines) in cyclohexane.*

As indicated by DLS/SLS and SFM, aggregates maintained their spherical shape and size upon complexation with Co^{2+} and Pd^{2+} (R_h ~ 32 nm, cf. Table 2 and Figure 7). As a matter of the narrow particle size distribution, micelles tend to form ordered domains on a solid substrate upon evaporation of the solvent. It is noteworthy that the aggregation numbers Z slightly increase with increasing amount of Pd^{2+} added to the PAEMA micellar core (+ 20% at f_{Pd} ~ 0.5). This effect is much more pronounced when the metal ion is Fe^{3+} (+ 140% at f_{Fe} ~ 0.5, see Table 2). In addition, the value of R_g/R_h raises from 0.8 to 1.2, indicating that these micelles are not spherical but cylindrical in shape (cf. above). Hence, the aggregation behavior of PAEMA-based hybrid micelles is vastly affected by the tautomeric structure of the acetoacetoxy ligand as well as the nature of the metal ion.

Table 2. Results of SLS/DLS analysis of $PBMA_{342}$-$PAEMA_{39}$ (enolate) micelles loaded with different amounts of Pd^{2+}/Fe^{3+}, solvent: cyclohexane.

Pd^{2+}, f_{Pd}	R_h [nm]	R_g [nm]	R_g/R_h	$A_2\ 10^9$ [mol·dm³·g⁻²]	Z
0	35	28	0.80	1.5	323
0.12	31	29	0.94	0.7	361
0.23	32	28	0.88	0.8	398
0.40	32	29	0.90	0.6	367
0.55	33	30	0.91	0.7	393
Fe^{3+}, f_{Fe}					
0	35	28	0.80	1.5	323
0.35	45	50	1.11	2.2	412
0.52	58	69	1.19	0.3	784

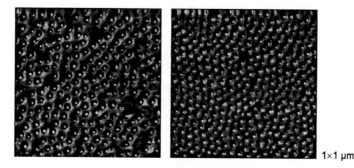

1×1 µm²

Figure 7. SFM phase images of $PBMA_{342}$-$PAEMA_{39}$ (enolate) micelles loaded with Co^{2+} ($f_{Co} \sim 0.5$, left) and Pd^{2+} ($f_{Pd} \sim 0.5$, right) on graphite.

In summary, it has been shown that PAEMA-based block copolymers can be used to produce well-defined colloidal hybrid materials. Dimension of the aggregates can be controlled by the tautomeric structure of AEMA ligands and the nature of the metal ion. However, more systematic studies are needed to get a comprehensive understanding of the system. It will also be interesting to see the performance of these materials, for example in the production of ordered hybrid films, in comparison to polyvinylpyridines.

166

Acknowledgments

Markus Antonietti, Reinhard Sigel, Birgit Schonert, Helmut Cölfen, Antje Völkel, Anne Heilig, Marlies Gräwert, Olaf Niemeyer, Ines Below, Costas S. Patrickios, and Erich C. are thanked for their contributions to this work. Financial support was given by the *Max Planck Society* and the *German Academic Exchange Service* (DAAD).

References

1   Lehn, J.-M. *Supramolecular Chemistry - Concepts and Perspectives*; VCH: Weinheim, 1995.

2   Kaliyappan, T.; Kannan, P. *Prog. Polym. Sci.* **2000**, *25*, 343-370.

3   Förster, S.; Antonietti, M. *Adv. Mater.* **1998**, *10*, 195-217.

4   Cölfen, H. *Macromol. Rapid. Commun.* **2001**, *22*, 219-252.

5   Antonietti, M.; Wenz, E.; Bronstein, L.; Seregina, M. *Adv. Mater.* **1995**, *7*, 1000-1005.

6   Antonietti, M.; Förster, S.; Hartmann, J.; Oestreich, S. *Macromolecules* **1996**, *29*, 3800-3806.

7   Antonietti, M.; Förster, S.; Hartmann, J.; Oestreich, S.; Wenz, E. *Nachr. Chem. Tech. Lab.* **1996**, *44*, 579-586.

8   Antonietti, M.; Thünemann, A.; Wenz, E. *Coll. Polym. Sci.* **1996**, *274*, 795-800.

9   Seregina, M. V.; Bronstein, L. M.; Platonova, O. A.; Chernyshov, D. M.; Valetsky, P. M.; Hartmann, J.; Wenz, E.; Antonietti, M. *Chem. Mater.* **1997**, *9*, 923-931.

10 Platonova, O. A.; Bronstein, L. M.; Solodovnikov, S. P.; Yanovskaya, I. M.; Obolonkova, E. S.; Valetsky, P. M.; Wenz, E.; Antonietti, M. *Coll. Polym. Sci.* **1997**, *275*, 426-431.

11 Antonietti, M.; Förster, S.; Oestreich, S. *Macromol. Symp.* **1997**, *121*, 75-88.

12 Klingelhöfer, S.; Heitz, W.; Greiner, A.; Oestreich, S.; Förster, S.; Antonietti, M. *J. Am. Chem. Soc.* **1997**, *119*, 10116-10120.

13 Breulmann, M.; Förster, S.; Antonietti, M. *Macromol. Chem. Phys.* **2000**, *201*, 204-211.

14 Steunou, N.; Förster, S.; Florian, P.; Sanchez, C.; Antonietti, M. *J. Mater. Chem.* **2002**, *12*, 3426-3430.

15 Spatz, J. P.; Sheiko, S.; Möller, M. *Adv. Mater.* **1996**, *8*, 513-517.

16 Spatz, J. P.; Mossmer, S.; Möller, M. *Chem. Eur. J.* **1996**, *2*, 1552-1555.

17 Selvan, S. T.; Spatz, J. P.; Klok, H. A.; Möller, M. *Adv. Mater.* **1998**, *10*, 132-134.

18 Spatz, J. P.; Herzog, T.; Mossmer, S.; Ziemann, P.; Möller, M. *Adv. Mater.* **1999**, *11*, 149-153.

19 Spatz, J. P.; Mossmer, S.; Hartmann, C.; Möller, M.; Herzog, T.; Krieger, M.; Boyen, H. G.; Ziemann, P.; Kabius, B. *Langmuir* **2000**, *16*, 407-415.

20 Mossmer, S.; Spatz, J. P.; Möller, M.; Aberle, T.; Schmidt, J.; Burchard, W. *Macromolecules* **2000**, *33*, 4791-4798.

21 Kastle, G.; Boyen, H. G.; Weigl, F.; Ziemann, P.; Riethmuller, S.; Hartmann, C. H.; Spatz, J. P.; Möller, M.; Garnier, M. G.; Oelhafen, P. *Phase Transitions* **2003**, *76*, 307-313.

22 Boyen, H. G.; Kastle, G.; Zurn, K.; Herzog, T.; Weigl, F.; Ziemann, P.; Mayer, O.; Jerome, C.; Möller, M.; Spatz, J. P.; Garnier, M. G.; Oelhafen, P. *Adv. Func. Mater.* **2003**, *13*, 359-364.

23 Haupt, M.; Miller, S.; Glass, R.; Arnold, M.; Sauer, R.; Thonke, K.; Möller, M.; Spatz, J. P. *Adv. Mater.* **2003**, *15*, 829-831.

24 Moffitt, M.; McMahon, L.; Pessel, V.; Eisenberg, A. *Chem. Mater.* **1995**, *7*, 1185-1192.

25 Moffitt, M.; Eisenberg, A. *Macromolecules* **1997**, *30*, 4363-4373.

26 Moffitt, M.; Vali, H.; Eisenberg, A. *Chem. Mater.* **1998**, *10*, 1021-1028.

27 Sidorov, S. N.; Bronstein, L. M.; Kabachii, Y. A.; Valetsky, P. M.; Lim Soo, P.; Maysinger, D.; Eisenberg, A. *Langmuir* **2004**, *20*, 3543-3550.

28 Mehrotra, R. C.; Bohra, B.; Gaur, D. P. *Metal β–Diketonates and Allied Derivatives*; Academic Press: New York, 1978.

29 Schlaad, H.; Krasia, T.; Patrickios, C. S. *Macromolecules* **2001**, *34*, 7585-7588.

30 Krasia, T.; Soula, R.; Börner, H. G.; Schlaad, H. *Chem. Commun.* **2003**, 538-539.

31 Schnablegger, H.; Glatter, O. *Appl. Opt.* **1991**, *30*, 4889-4896.

32 Evans, D. F.; Wennerström, H. *The Colloidal Domain - Where Physics, Chemistry, Biology, and Technology Meet*; VCH: New York, 1994.

33 Förster, S.; Zisenis, M.; Wenz, E.; Antonietti, M. *J. Chem. Phys.* **1996**, *104*, 9956-9970.

34 Teyssié, P.; Smets, G. *Makromol. Chem.* **1958**, *26*, 245-251.

35 De Wilde-Delvaux, M. C.; Teyssié, P. *Spectrochimica Acta* **1958**, *12*, 280-293.

36 Corain, B.; Zecca, M.; Mastrorilli, P.; Lora, S.; Palma, G. *Makromol. Chem., Rapid Commun.* **1993**, *14*, 799-805.

Chapter 13

Designing Supramolecular Porphyrin Arrays for Surface Assembly and Patterning of Optoelectronic Materials

Jayne C. Garno[1], Chang Xu[2], Griorgio Bazzan[2], James D. Batteas[3], and Charles Michael Drain[2,4,*]

[1]Department of Chemistry, Louisiana State University, Baton Rouge, LA 70803
[2]Department of Chemistry and Biochemistry, Hunter College and The Graduate Center, City University of New York, New York, NY 10021
[3]Chemical Science and Technology Laboratory, National Institute of Standards and Technology, 100 Bureau Drive, Gaithersburg, MD 20899
[4]The Rockefeller University, 1230 York Avenue, New York, NY 10021

Designing nanoscale devices requires a clear understanding of the hierarchical structural organization of the functional material – from the design of the molecule, to how the molecules self-assemble into supramolecular structures, to how these structures organize on various surfaces. Investigations comparing the structures in solution to the structures found on surfaces such as in thin films, nanoparticles, and nanoarrays, reveal the complex interplay between the energetics of self-organization and the energetics of interactions of supramolecular systems with surfaces. A perspective on the design of porphyrins that self-assemble into discrete supramolecular structures that subsequently self-organize into nanoparticles and films, and some of the factors dictating solution and surface morphology is presented.

Introduction

The advantages and disadvantages of self-assembled/organized, supramolecular systems[†] as components of materials and/or devices are well delineated.[1-12] The synthesis of the molecular components is oftentimes straightforward with good yields, and the formation of complex, multicomponent systems by self-assembly can also proceed in remarkably high yields. The disadvantages of self-assembled systems largely stem from the complex equilibria inherent to supramolecular entities that make both characterization and material/device stability keystone issues in real-world applications. However, the principles and strategies for the *de novo* design of multicomponent supramolecular systems that self-assemble into solid-state materials such as crystals, or self-organize into predictably sized aggregates such as nanoparticles, are far less understood. Furthermore, the complex interactions of these materials with support/substrate surfaces have more recently become a topic of interest.

For self-assembled materials in devices one may consider four levels of structural organization.[13] (1) The primary structure is that of the molecular components, for which we have exquisite control. (2) The secondary structure is that of the supramolecule, and while there remains much to be discovered, supramolecular synthesis is a mature field. (3) The tertiary structure pertains to how the supramolecular systems self-assemble or self-organize into solid-state materials, and predicting the tertiary structure of both molecular and supramolecular systems remains a major challenge, but some progress has been made in the last few years especially in terms of what is often called crystal engineering. (4) The quaternary structure describes how solid-state supramolecular systems self-assemble and/or self-organize onto surfaces as components of materials and devices – and may include the interconnections between the macroscopic and nanoscaled realms. This last level of structural order is important because the complex interactions between supporting substrate surfaces and the molecules, supramolecules, and supramolecular materials can significantly alter the structure, therefore the function. The nature of the electronic interactions between surfaces and molecules (adsorbed or covalently attached) is of major importance and the subject of much theoretical

---

[†] Though there is leeway, for the purposes of this perspective the following definitions are used. Nanoscaled is < 200 nm; self-assembly is the formation of a discrete supramolecular entity wherein there is no tolerance for error since a different supramolecular system results from errors; self-organization is the formation of a non-discrete supramolecular system, such as a multimer or a monolayer, wherein there is a threshold below which errors are non-critical.

and experimental study. From another standpoint, the surface can be considered an additional design parameter to exploit in the design, assembly/organization and function of these materials.[14-17]

The porphyrinoids [porphyrins (Por), phthalocyanines (Pc), porphyrazines (Pz), and corroles (Cor)] are exemplary molecules to construct supramolecular photonic materials because of their remarkable stability in view of their diverse photophysical and chemical properties.[14-20] The photonic properties such as: excited state lifetimes, redox potentials, catalytic activities, magnetism, optical cross sections, etc. can be systematically varied both by the metalation with nearly every metal in the periodic table, and by the substituents on the macrocycle. The structural or architectural organization of the chromophores in a material, as well as other environmental factors such as solvent or matrix, has a profound influence on the functionality of photonic materials. The importance of both the supramolecular structure and the matrix are notably typified by the photosynthetic reaction centers and antenna complexes, wherein the former serves as a conduit for electrons and the latter a conduit of energy. These are self-assembled and self-organized systems, respectively. Since there are no covalent bonds between chromophores, and there is evidence that some antenna complexes do not need a protein scaffold, both can have quantum yields near unity. It has been demonstrated that materials composed of porphyrinoids can be very robust to real-world conditions[3-5] such as elevated temperatures, and the presence of dioxygen and/or water, and that they are quite versatile tectons that afford rich variations in supramolecular topologies.

The number of reports on the self-organization and self-assembly of porphyrinoids has grown nearly exponentially[15] since the late 1980s when we reported the formation of a photogated transistor that functioned via an ion-chain composed of transiently formed porphyrin cations self-organized by electrostatic interactions with hydrophobic anions.[12] Numerous supramolecular systems[20-27] using electrostatics,[11,12] hydrogen bonding,[20] metal ion coordination,[20-26] and reversible covalent bonds such as disulfides have been reported and the subject has been well reviewed.[14,15,21] The formation of 3-dimensional crystal lattices of porphyrins by designed supramolecular chemistry has also been reviewed.[21] The reason for this wide interest in the supramolecular chemistry porphyrins and their analogues arise from the aforementioned photonic properties, the robustness of the chromophores, and the supramolecular design flexibility. Potential applications include catalysts, sensors, molecular electronics, molecular sieves, and solar energy conversion – all of which have been demonstrated to various degrees – that have superior properties or new functions compared to the individual, non-supramolecular components.[25,28-38]

Self-Assembled Porphyrin Arrays

The self-assembly of porphyrins using specific intermolecular interactions[2] has been reviewed[14,15,21] and salient features are outlined herein. Hydrogen-bonding provides a variety of recognition motifs and is reversible thus potentially affording good yields of supramolecular arrays, but simple recognition motifs with 3-4 H-bonds are unlikely to be suitable for deposition of discrete arrays on surfaces because the intermolecular interactions are too weak to maintain the nanoarchitecture as solvent evaporates. Coordination chemistry has the advantages of more robust but reversible bonds, and the metal ion linkers can dictate a variety of geometries and provide further functionality. Therefore, self-assembly via coordination chemistry is widely studied in terms of photonic materials. There are well over 30 possible different geometric topologies that metallopyridylporphyrins can afford,[15] and combined with the various topologies of the transition metals, the possible porphyrinic arrays are limited only by imagination. Secondly, hard-soft metal-ligand interactions can be exploited to accomplish multistep assembly of more complex supramolecular systems held together by different metal ions.[15] Note that in designing photonic materials, the heavy atom effect must be considered. Herein, we discuss the self-organization of simple trapezoidal tetrameric arrays of pyridylporphyrins with 180^0 topologies assembled with *cis*PtCl$_2$ at the corners, Figure 1. This results in a neutral, organic soluble array. The pyridyl-Pt(II) bond is more robust than other square planar metals such as Ni(II) and Pd(II),[14,20,21] and other 90^0 topologies on metal ions require pre-chelation of open coordination sites other than the two ligands to be exchanged for the pyridyl moieties.

Self –Organized Porphyrinic Materials

Secondary, non-specific intermolecular interactions[2] such as pi-stacking, van der Waals, and electrostatic interactions of all kinds can be used to self-organize the supramolecular porphyrin arrays in the solid state.[39-41] These secondary interactions tend to be non-specific or directional, and tend to result in self-organized aggregates such as porphyrin nanoparticles, columnar stacks, nanocrystals, and films. Therefore, in order to investigate the commingled roles of supramolecular structure, dynamics, and stability as well as the effects of peripheral substitution on the surface organization of supramolecular arrays, we have appended dodecyloxy groups onto the peripheral phenyl groups of the tetrameric array, Figure 1.

Organization of Self-Assembled Arrays on Surfaces

We use one recent example to illustrate the manifold factors that influence supramolecular materials morphology, and present three proof-of-concept examples on the quaternary organization of these materials on surfaces. The orientation of porphyrins on surfaces is determined by factors such as the nature of the peripheral substituents, R, and their position on the macrocycle. For example, small substituents on the 4-position of tetraaryl porphyrins favor π-stacking, whereas those on 2 or 3 positions inhibit significant π-stacking and generally weaken surface-porphyrin interactions.[17,26]

The role of peripheral groups in the self-organization of self-assembled multiporphyrinic arrays on surfaces was examined for Pt(II)-linked trapezoidal tetrameric porphyrin arrays with peripheral tert-butylphenyl or dodecyloxyphenyl functionalities.[16]

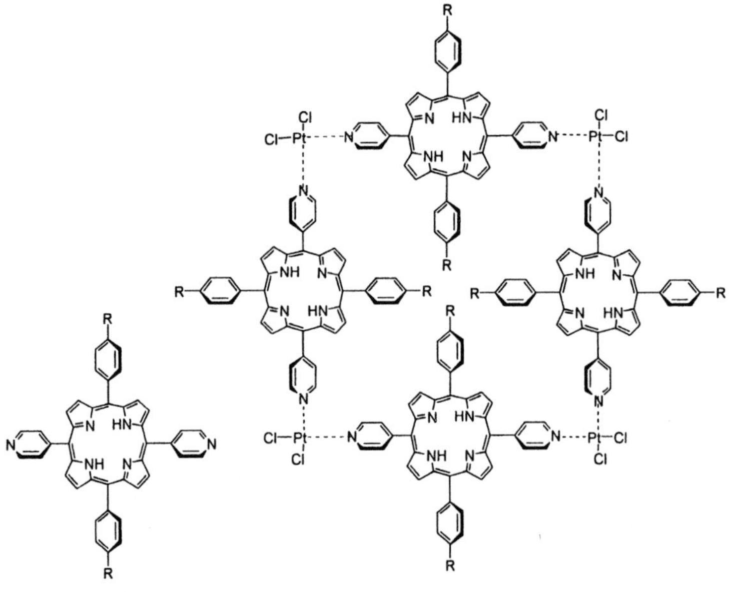

1 R = tert-butyl
2 R = dodecyloxy

3 R = tert-butyl
4 R = dodecyloxy

Figure 1. Porphyrin building blocks and self-assembled array 3, 4 (Adapted from Ref. 16).

AFM investigations reveal that the supramolecular architecture of the Pt(II) assembled trapezoids remain intact when cast on glass and that the orientation and length of peripheral alkyl substituents influence the resulting structures on surfaces. The *tert*-butylphenyl substituted porphyrin arrays form small aggregates which organize in a vertical direction via π-stacking interactions among the macrocycles. In contrast, tetrameric porphyrin arrays with dodecyloxyphenyl groups form a continuous film via van der Waals interactions between the peripheral hydrocarbon chains. By appending peripheral dodecyloxyphenyl groups to porphyrin building blocks, the strategies were both to minimize conformational flexibility compared to simple self-assembled dimers[20,22] and to increase inter-array interactions via the self-organizing properties of long chain hydrocarbons to yield organized films on surfaces.

^{1}H NMR data suggest that there is some flexibility in the supramolecular structure of the trapezoidal tetramers and steric energy minimization calculations (MM2) indicate arrangements schematically depicted in Figure 2.

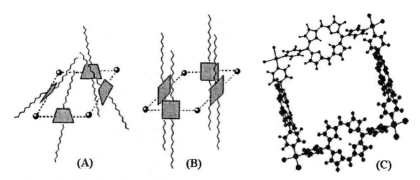

Figure 2. Models of possible arrangements of the porphyrins in tetramer 4 (Adapted from Ref. 16).

Using this model, the distance from side to side (porphyrin face to porphyrin face) of the tetramer is approximately 1.8 nm, and the diagonal distance between opposite platinum atoms is approximately 2.6 nm. A similar structure was proposed for tetramers of porphyrin with Re(II) corners.[24,25] The square pyramidal (*C4v*) arrangement shown in Figure 2A has four identical porphyrins with pyridyl, pyrrole and phenyl protons facing into or away from the center of the structure.

Surface Organization of Tetrameric Porphyrin Arrays. When samples were prepared by solvent evaporation via drop deposition, a variety of surface morphologies are observed that are largely a consequence of the concentration of the solution. Representative surface structures of **3** and **4** assayed by AFM are shown in Figures 3 and 4. Array **3** forms discrete nanoclusters or porphyrin stacks of various heights on glass surfaces. The aggregates are organized randomly across the surface, and it is clear that the porphyrin stacks do not merge and are separated by distances of at least 50 nm. These are observed over a supramolecular concentration range from 1 to ~50 μM. Cursor measurements indicate that the columnar stacks have variable heights, ranging from 1.5 to 18 nm. Thus, for the surface organization of the tert-butylphenyl tetramer **3**, the π-π stacking interactions in solution direct the assembly into columnar structures. With surface deposition, the porphyrin planes maximize the interactions with the substrate.

In contrast, the dodecyloxyphenyl functionalized porphyrin supermolecule **4** forms films of different morphologies on glass depending on the concentration of the deposition solution. At low <10 μM concentrations, this array forms small islands of variable horizontal dimensions (50 – 700 nm) and heights that range between 5 and 20 nm. At higher concentrations, ~100 μM, array **4** forms continuous 10 nm thick films (Figure 3), and at intermediate concentrations form films with large circular defects. At still greater concentrations, ~200 μM, nanocrystalline domains are observed.[16]

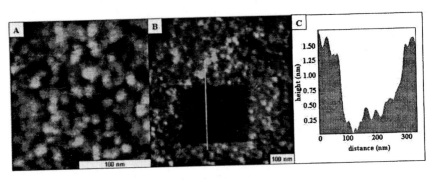

*Figure 3. AFM image of the continuous film formed from supramolecular porphyrin array **4** with dodecyloxyphenyl substituents on glass. (A) Close-up view in ethanol; (B) hole fabricated in the same porphyrin film; and (C) representative line profile for the nanopattern (Adapted from Ref. 16).*

The tetramers are packed closely in random arrangements, touching neighboring clusters to form a densely aggregated structure. Within the films, the individual supramolecular clusters are nearly spherical with a uniform geometry and size (Figure 3). Nanoshaving was used to measure the thickness of the film (Figure 3B and 3C).[27] These results indicate that the dispersion interactions among the peripheral long alkyl chains on tetramer **4** affect both the supramolecular structure and the self-organization of the supermolecule on surfaces. The dodecyloxy groups significantly enhance the horizontal intermolecular interactions during surface deposition, and columnar stacks are not observed in solution or on surfaces.

The proposed intermolecular interactions in which the sides of the porphyrin trapezoids assemble in a side-on arrangement on the surface are represented schematically in Figure 4. This ensures the maximal interaction between the hydrocarbon chains in and between arrays of **4**, minimizes interactions with the hydrophilic surface, and allows one porphyrin face and two $PtCl_2$ units to interact with the surface. We find that the organization of these arrays depends on the surface energetics, *vide infra*.

Quarternary Organization: Patterning Arrays on Surfaces

The ability to incorporate self-assembled materials into devices requires that the materials be organized on surfaces in predefined patterns. There are a variety of means to accomplish this task including both bottom-up and top-down strategies. Using widely available PDMS stamps is a way to pattern the arrays of **4** on surfaces rapidly and reproducibly. The dodecyloxy groups make array **4** highly soluble compared to most porphyrins and porphyrinic arrays, and this can be exploited as part of the quaternary organization of the materials (Figure 5). Though the films are not yet perfect, double deposition and/or annealing the system may provide the route to patterns of high quality thin films of these arrays.

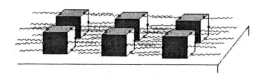

Figure 4. Surface organization of 4 due to horizontal interactions between supramolecular arrays (Adapted from Ref. 16).

The peripheral R groups dictate not only solubility of the multiporphyrin array but also play crucial role in the self-organization of the tetramers on surfaces and influence the surface binding energetics. Therefore, for surface patterning of porphyrin assemblies, we focused on the deposition and properties of tetramer **4**. To prepare patterned microstructures of the porphyrin arrays, a clean PDMS stamp was placed onto a gold substrate. Next, a drop (5 μL) of porphyrin solution (2.4×10^{-6} M of the Pt(II)-linked dodecyloxy array in toluene) was introduced at the end of the channels, and guided to selected areas of the substrate by capillary action, (Figure 5) as previously described for protein solutions using microfluidic networks.[42] After drying, the PDMS stamp was removed, and the samples were imaged in air (Figures 5B and 5C). The heights of the microstructures ranged from 40 to 160 nm along the lengths of the microstructures, depending on the shape of the channels and the uniformity of

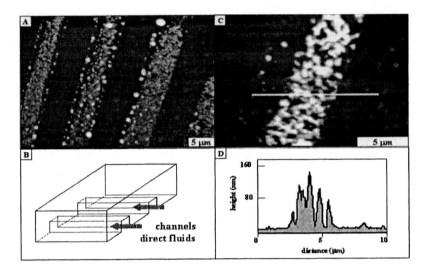

Figure 5. Microprinting of dodecyloxy array 4 on Au. (A) Channels of nanoparticles were prepared by placing a PDMS stamp against the surface. Solutions are pulled into the channels via capillary action. (B) AFM topograph of 5 μm lines of porphyrin spaced 5 μm apart. (C) Zoom-in view of a single channel; (D) cursor for line trace in B.

deposition (Figure 5D). This lithographic approach is a rapid and practical means to chemically pattern substrates for microscopy characterizations.

Organizing porphyrin materials on surfaces provides a means of generating potential device structures and for controlling the dimensions and density of materials by confinement in narrow (microns) channels of PDMS. In our initial work we have patterned Pt(II)-linked dodecyloxy material on Au and glass surfaces using microfluidic channels formed with polydimethyl siloxane (PDMS). Using this approach, patterned arrays can be formed over large areas which conform to the dimensions of the PDMS mold. Figure 5B demonstrates the patterning of the Pt(II)-linked dodecyloxy material over large areas of a Au surface. In this example, the microchannels were 5 μm wide. The flow of the solution through the channels is found to be semi-continuous with patterns extending beyond 50 microns. This suggests that at higher concentrations, the lateral interactions of the side chains as well as significant π stacking can promote organization into larger structures, potentially with liquid crystalline behavior.[18,19,22] Other stamping methods will likely yield similar results.[43]

Porphyrin Nanoparticles

Porphyrin nanoparticles are promising new materials for nanotechnology,[35,36] because of their rich photochemistry, stability, and catalytic activity.[28] Analogous to inorganic and other organic nanostructures,[21,37] it is expected that nanoparticles of porphyrins will have unique photonic properties in comparison to large-scale bulk materials. The increased surface area and aggregate structure of nanoparticles of catalytic porphyrins provide enhanced stability and catalytic rates.[28] Porphyrin nanoparticles were prepared by mixed solvent techniques using polyethyleneglycol (PEG) derivatives.[28] As previously reported with other nanoparticle preparations, the PEG stabilizer prevents agglomeration[38] and is a contributing factor in determining nanoparticle size.

	R	M	Average radius (nm)
1	(pyridyl-O-O- structure) Fe(III)		27 (±6)
2	4-carboxylphenyl	$2H^+$	46 (±40)
3	4-carboxylphenyl, methyl ester	$2H^+$	58 (±47)
4	Phenyl	$2H^+$	11 (±4)
5	4-methoxyphenyl	$2H^+$	34 (±30)
6	4-pyridyl	$2H^+$	70(±50)
7	5,15-(mesityl)-10,20-(4-bromo-phenyl)	$2H^+$	24(±6)

Table 1: DLS measurements for nanoparticles formed from meso-substituted tetraphenylporphyrins (Adapeted from Ref. 28).

Dynamic light scattering (DLS) was used to characterize the size of porphyrin nanoparticles in solution (Table 1). Since applications often require surface-bound nanoparticles, AFM was also used to characterize nanoparticles supported on glass substrates, in the absence of solvent. For example, in solution, the average diameter for Fe(III) nanoparticles, **1**, measured 54 nm while AFM height measurements of the Fe(III) nanoparticles **1** ranged from 30-65 nm. However, this was not necessarily a consistent trend for all of the structures in Table 1. These porphyrin nanoparticles are likely held together by both hydrophobic and π-stacking effects.[26,34,39-41] In polar solvents such as water, these intermolecular forces become stronger, and regardless of the substituted moiety, the porphyrin core remains hydrophobic. The size of the nanoparticle is somewhat controllable via the types and ratios of the host and guest solvents.

As might be predicted, the UV-Vis spectra of porphyrin nanoparticles are significantly different compared to spectra of the corresponding porphyrin solutions. For nanoparticles, the Soret (B) bands were significantly broadened and split. These spectra are consistent with other porphyrin nanoscaled aggregates and are well-understood to be indicative of electronic coupling of the chromophores.[17-20,23,34-36,39-41]

Preliminary results on the catalytic activity of Fe porphyrin nanoparticles, Fe(III) 1 reveal that they have a ~70-fold greater turnover number and a 10-fold

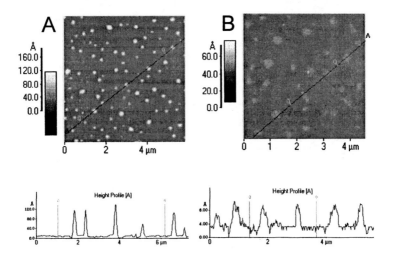

Figure 6. Structures of porphyrin nanoparticles from evaporation of a Fe(III) 1 nanoparticle solution on a hydrophilic oxidized silicon surface. (A) deposited nanoparticles with heights on the order of 5 – 10 nm; (B) disassembled particles on the surface only one molecular layer thick (~ 0.5 nm).

greater rate than the individual porphyrin in solution using a standard catalytic epoxidation of cyclohexene.[44] Most porphyrin nanoparticle solutions are exceptionally stable as judged by the unchanging optical spectrum and DLS after months of storage.

Organization of porphyrin nanoparticles on surfaces

Nanoparticles of Fe(III) **1** are stable and can be deposited on substrates wherein the surface morphology of the resultant material is further dictated not only by the nature of the surface, but also by the rate of solvent evaporation. When the Fe(III) **1** nanoparticle solution was deposited on UV-ozone cleaned silicon surface, which renders the native oxide hydrophilic, and the solvent was slowly evaporated, porphyrin aggregates with heights much less than the diameter of original nanoparticles in solution were observed. With these conditions, two types of surface morphologies were typically observed for porphyrin aggregates. One type of aggregate was particle-like, and the other exhibits a thin disk shape. The thinnest disk-shaped structure observed with AFM was about ~ 0.5 nm thick (Figure 6), which corresponds to only a single layer of porphyrin molecules laying flat on the surface. The disks spread across several hundred nanometers of the surface, which is larger than the average diameter of the original nanoparticles in solution. The presence of solvent is essential to assist the transformation of porphyrin nanoparticle on surfaces, as the structure of the final porphyrin aggregates are stable in air and remain the same over weeks. The small porphyrin aggregates originate form the breaking down of larger size nanoparticles in the original solution. The break down and collapse of Fe(III)TPEGPP nanoparticles on hydrophilic surface can be attributed to the hydrophilic-hydrophilic interaction between Fe(III) nanoparticles, 1, and the surface. Conversely, if the solvent evaporation rate is fast and the surface is hydrophobic, the average height of the final aggregate structure on the surface corrolates well with the size of the colloidal particles in solution.

Electrostatic Deposition

In addition to the drop casting of self-assembled porphyrin arrays such as the squares, and the self-organized porphyrinic nanoparticles, we have investigated the formation of thin films of porphyrinic materials using electrostatic forces. We have found that 20 ± 4 nm films can be formed by successively dipping a freshly cleaved mica surface into a 0.1 mM solution of a tetracationic porphyrin, followed by a 0.1 mM solution of a teraanionic europium polyoxometalate (POM), and repeating this for eight cycles (Figure 7). These electrostatically organized films are stable in air and to removal from the surface by solvents such as water and toluene. The same procedure on a mica substrate patterned with ~1 μm wide by ~150 nm tall polyethylene polymer lines yields ~60 nm porphyrin-POM films in the ~ 15 μm valleys between the polymer lines (Figure 7).

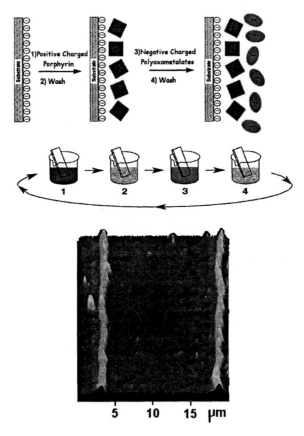

Figure 7. Eight cycles of dipping a mica substrate patterned with a polyethylene polymer (1 μm wide x 150 nm tall lines) in 0.1 mM solutions of tetracationic 5,10,15, 20-tetrakis(N-methyl-4-pyridinium)porphyrin and tetra-anionic polyoxometalate (POM) [EuPW$_{11}$O$_{39}$]$^{4-}$ results in ~ 60 nm ± 6 nm porphyrin-POM hybrid films on a substrate separated by a pattern of 150 nm tall polymer lines spaced 15 μm apart. This proof-of-principle experiment demonstrates that no specific surface chemistry or capillary stamp is required to form patterns of porphyrinic materials on surfaces. The square in the center is the result of a nanoshaving experiment to determine film thickness.

Summary

The three proof-of-principle methods to pattern arrays on surfaces illustrate a few of the general considerations for the quarternary organization of porphyrinic materials on surfaces. (1) Using a neutral, hydrophobic, self-assembled porphyrin array on a neutral surface allows the formation of patterns by capillary action under a PDMS stamp. (2) Self-organized nanoparticles of a pentacationic porphyrin remain in tact on hydrophobic surfaces, but fall apart on hydrophilic surfaces driven by surface-molecule interactions. (3) A polymer patterned onto an anionic mica substrate, allow the formation of patterns of thin films of tetracationic porphyrins by electrostatic interactions using a layer-by-layer method, and a tetra anionic polyoxometalate counter ion.

Future Directions

The hierarchical structural organization of porphyrins – from molecules to supramolecules to nanoscaled aggregates to patterns on surfaces – has been demonstrated. The physical chemical properties and the architecture of the supramolecular array are important design criteria for the deposition of these materials on surfaces; however, supramolecular dynamics and/or conformation flexibility are also factors that determine the final structure and morphology on surfaces. Just as importantly, the surface properties can be expoited as a means to control both tertiary and quarternary structure. The formation of hierarchical structures using both specific and nonspecific interactions may or may not be cooperative, but imparts a high degree of stability and control of size not found in most supramolecular systems.[26] The the ability to synthesize many porphyrins in good yields[45,46] and the variety of functions of porphyrinic systems on surfaces warrents further investigations into applications as nanoscaled photonic materials. Three proof-of-principle methodologies are presented that afford patterned arrays of porphyrinic materials on surfaces: transport via capillary action in microchannels under a stamp, reorganization of colloidal nanoparticles via surface energetics, and electrostatic absorption on mica surfaces with nanostructured polymer patterns.

Acknowledgement. The authors gratefully acknowledge funding from the National Science Foundation (CHE-0135509, IGERT-9972892), the National Institutes of Health (SCORE S06GM60654), and the Israel-United States Binational Science Foundation (1999082). Hunter College Chemistry infrastructure is supported by the National Institutes of Health (including the RCMI program G12-RR-03037), the National Science Foundation, and the City University of New York.

References

1. Ball, P. *Nature* **2001**, *409*, 413-416.
2. Alivisatos, A. P.; Barbara, P. F.; Castleman, A. W.; Chang, J.; Dixon, D. A.; Klein, M. L.; McLendon, G. L.; Miller, J. S.; Ratner, M. A.; Rossky, P. J.; Stupp, S. I.; Thompson, M. E. *Adv. Mater.* **1998**, *10*, 1297-1336.
3. Fox, M. A. *Acc. Chem. Res.* **1999**, *32*, 201-207.
4. Aviram, A.; Ratner, M. *Ann. N. Y. Acad. Sci.* **1998**, *852*, 1-21.
5. Reed, M. A. *MRS Bull.* **2001**, *26*, 113-120.
6. Lehn, J.-M. *Angew. Chem.* **1990**, *102*, 1347-1362.
7. Lehn, J.-M. *Angew. Chem., Int. Ed. Engl.* **1990**, *102*, 1304-1319.
8. Lindsey, J. S. *New. J. Chem.* **1991**, *15*, 153-180.
9. Stang, P. J.; Olenyuk, B. *Acc. Chem. Res.* **1997**, *30*, 507-518.
10. Fujita, M. *Structure and Bonding*, 2000; Vol. 96.
11. Drain, C. M.; Mauzerall, D. *Bioelectrochem. Bioenerg.* **1990**, *24*, 263-266.
12. Drain, C. M.; Christensen, B.; Mauzerall, D. *Proc. Natl. Acad. Sci. USA* **1989**, *86*, 6959-6962.
13. Drain, C. M. *Proc. Natl. Acad. Sci., USA* **2002**, *99*, 5178-5182.
14. Drain, C. M.; Smeureanu, G.; Batteas, J. D.; Patel, S. in *Encyclopedia of Nanoscience and Nanotechnology*; Marcel Dekker: New York, 2004.
15. Drain, C. M.; Chen, X. in *Encyclopedia of Nanoscience and Nanotechnology*; American Scientific Publishers: New York, 2004.
16. Milic, T.; Garno, J. C.; Batteas, J. D.; Smeureanu, G.; Drain, C. M. *Langmuir* **2004**, *20*, 3974-3983.
17. Milic, T. N.; Chi, N.; Yablon, D. G.; Flynn, G. W.; Batteas, J. D.; Drain, C. M. *Angew. Chem. Int. Ed.* **2002**, *41*, 2117-2119.
18. Bruce, D. W.; Wali, M. A.; Wang, Q. M. *Chem. Commun.* **1994**, 2089-2090.
19. Wang, Q. M.; Bruce, D. W. *Angew. Chem. Int. Ed. Engl.* **1997**, *36*, 150-152.
20. (a) Drain, C. M.; Lehn, J.-M. *J. Chem. Soc., Chem. Commun.* **1994**, 2313-2315(correction 1995, p503). (b) Drain, C.M.; Russell, K.C.; Lehn, J.-M. *Chem. Commun.* **1996**, 337-337. (c) Drain, C. M.; Fischer, R.; Nolen, E.; Lehn, J.-M. *Chem. Commun.* **1993**, 243-245.
21. (a) Drain, C. M.; Hupp, J. T.; Suslick, K. S.; Wasielewski, M. R.; Chen, X. *J. Porph. Phthal.* **2002**, *6*, 241-256. (b) Shmilovits, M.; Diskin-Posner, Y.; Vinodu, M.; Goldberg, I. *Crystal Growth & Design* **2003**, *3*, 855-863.
22. Latterini, L.; Blossey, R.; Hofkens, J.; Vanoppen, P.; De Schryver, F. C.; Rowan, A. E.; Nolte, R. J. M. *Langmuir* **1999**, *15*, 3582-3588.
23. Drain, C. M.; Nifiatis, F.; Vasenko, A.; Batteas, J. D. *Angew. Chem., Int. Ed.* **1998**, *37*, 2344-2347.
24. Slone, R. V.; Hupp, J. T. *Inorg. Chem.* **1997**, *36*, 5422-5437.

25. Merlau, M. L.; Mejia, M. D. P.; Nguyen, S. T.; Hupp, J. T. *Angew. Chem., Int. Ed.* **2001**, *40*, 4239-4242.

26. Drain, C. M.; Batteas, J. D.; Flynn, G. W.; Milic, T.; Chi, N.; Yablon, D. G.; Sommers, H. *Proc. Natl. Acad. Sci. U.S.A.* **2002**, *99*, 6498-6502.

27. Liu, G.; Xu, S.; Qian, Y. *Acc. Chem. Res.* **2000**, *33*, 457-466.

28. Gong, X.; Milic, T.; Xu, C.; Batteas, J. D.; Drain, C. M. *J. Am. Chem. Soc.* **2002**, *124*, 14290-14291.

29. Fujita, M.; Kwon, Y. J.; S.Washizu; Ogura, K. *J. Am. Chem. Soc.* **1994**, *116*, 1151-1152.

30. Slone, R. V.; Hupp, J. T.; Stern, C. L.; Albrecht-Schmitt, T. E. *Inorg. Chem.* **1996**, *35*, 4096-4097.

31. Linton, B.; Hamilton, A. D. *Chem. Rev.* **1997**, *97*, 1669-1680.

32. Yuan, H.; Thomas, L.; Woo, L. K. *Inorg. Chem.* **1996**, *35*, 2808-2817.

33. Stang, P. J.; Fan, J.; Olenyuk, B. *Chem. Commun.* **1997**, 1453-1454.

34. Hunter, C. A.; Sanders, J. K. M. *J. Am. Chem. Soc.* **1990**, *112*, 5525-5534.

35. Chou, J.-H.; Kosal, M. E.; Nalwa, H. S.; Rakow, N. A.; Suslick, K. S. in *The Porphyrin Handbook*; Academic Press: New York, 2000; Vol. 6.

36. Chambron, J.-C.; Heitz, V.; Sauvage, J.-P. in *The Porphyrin Handbook*; Academic Press: New York, 2000; Vol. 6.

37. Dagani, R. *Chem. Eng. News* **1998**, *76*, 1-32.

38. Qi, L.; Colfen, H.; Antonietti, M. *Nano Lett.* **2001**, *1*, 61-65.

39. Maiti, N. C.; Mazumdar, S.; Periasamy, N. *J. Phys. Chem. B* **1998**, *102*, 1528-1538.

40. Kano, K.; Minamizono, H.; Kitae, T.; Negi, S. *J. Phys. Chem. A* **1997**, *101*, 6118-6124.

41. Purrello, R.; Scolaro, L. M.; Bellacchio, E.; Gurrieri, S.; Romeo, A. *Inorg. Chem.* **1998**, *37*, 3647-3648.

42. Delamarche, E.; Bernard, A.; Schmid, H.; Bietsch, A.; Michel, B.; Biebuyck, H. *J. Am. Chem. Soc.* **1998**, *120*, 500-508.

43. Helt, J. M.; Drain, C. M.; Batteas, J. D. *J. Am. Chem. Soc. 136*, **2004**, 628-634.

44. Groves, J. T.; Nemo, T. E. *J. Am. Chem. Soc.* **1983**, *105*, 5786-5791..

45. Lindsey, J. S. in *The Porphyrin Handbook*; Academic Press: New York, 2000; Vol. 1.

46. Drain, C. M.; Gong, X. *Chem. Commun.* **1997**, 2117-2118.

Metallodendrimers

Chapter 14

Metallodendrimers: Fractals and Photonics

Tae Joon Cho, Charles N. Moorefield, Pingshan Wang,
and George R. Newkome[*]

Departments of Polymer Science and Chemistry, The University of Akron,
Akron, OH 44325

Incorporation of 2,2':6',2"-terpyridinyl moieties into
monomeric, or building block, units has facilitated nanoscale
dendritic construction whereby metal centers are precisely
juxtaposed relative to each other and the framework. The
terpyridine-metal-terpyridine connectivity combined with the
iterative dendritic protocol has led to the creation of new
fractal constructs that can be accessed by step-wise and/or
self-assembly methods. Employment of these polyterpyridinyl
ligands for the production of novel cyclic supramolecules has
led to the creation of new molecular fractal motifs. These
highly ordered metallomacrocyclic architectures have been
created using *bis*terpyridine ligands with Ru and/or Fe. An
alkyl-modified cationic complex is counter-balanced with an
anionic dendrimer to produce an ordered hexamer-dendrimer
composite.

Introduction

Since the initiation of dendrimer chemistry,[1,2] these molecules have been demonstrated to be well-defined and highly branched tree-like structures, which exhibit unique molecular properties and provide uniform architectural foundations for creative modification.[3] In addition, potential intra- and inter-dendrimer interactions provide entry to the supramolecular regime.[4] This is exemplified in the metallodendrimer arena whereby metal complex formation has been extensively employed internally and peripherally for chemicophysical modifications. The use of polypyridyl-based ligands, such as 2,2':6',6''-terpyridine[5,6] and 2,2'-bipyridine,[7,8] has allowed access to many new macromolecular materials. Notably, for a number of years, the 2,2':6',6''-terpyridine ligand[9,10] has been of interest in the assembly of metallomacro-molecules and supramolecules,[11] owing to its metal-coordinating ability and the subsequent application in areas such as magnetic, electronic, electrochemical, photooptical, and catalytic potential. We herein report the construction and electrochemical properties of several unique metallodendrimers possessing terpyridine-metal-terpyridine connections ([–<M>–]).

Eloquent work in the area of self-assembly by Stang,[12] Lehn,[8] and many others,[13,14] has prompted our investigation of the potential to spontaneously construct metal-based (macro)molecules. More specifically, our strategy involved the preparation of a *bis*terpyridine monomer possessing a 120° angle with respect to the two ligating moieties. This would facilitate the assembly of 6 building blocks with 6 connecting metals in the ubiquitous benzenoid architecture, which would be the basis of a "modular building block set"[15] capable of being used to access "higher order" (fractal) architectures. The potential to synthesize such constructs, with little equilibration (metal – ligand exchange) under mild physicochemical condition, is predicated on the unique strength of the terpyridine-metal coordination.[16] It was also envisioned that these rigid structures, which possess an overall 12^+ charge, would be an ideal counter ion to a low generation dendrimer possessing 12 carboxylate surface groups;[17] this complementary interaction would form a hexamer-dendrimer composite, as a *suprasupermolecule*.

Results & Discussion

During initial efforts, Newkome et al.[6] first reported the preparation of metallodendrimers (Scheme 1) incorporating convenient terpyridine-metal-terpyridine (denoted as [–<M>–]) connectivity for building block attachment.

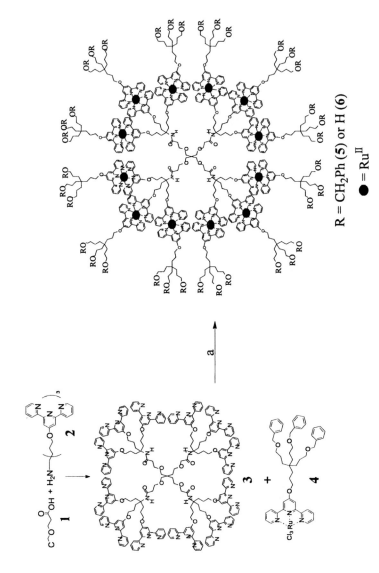

Scheme 1: Synthesis of dodecaRu^{II} metallodedrimer; a) 4-ethylmorpholine, MeOH, reflux.

R = CH$_2$Ph (**5**) or H (**6**)

● = Ru^{II}

Such a mode of connectivity permits the step-wise construction of macromolecular components and the assembly process can be evaluated by the quantification of the new metal diamagnetic center(s). The resulting metallodendrimer **5** possessed twelve pseudo-octahedral Ru[II] centers,[18] of which each metal center is coordinated by two orthogonal 4'-substituted 2,2':6',2"-terpyridines. Dendrons **2** and **4** were synthesized *via* the facile alkoxylation of 4'-chloroterpyridine. Dendritic assembly was then readily accomplished by coupling of tetracarboxylic acid **1** with amine **2** to afford the uncomplexed 12-terpyridinyl dendritic core **3**, which generated the metallodendrimer **5** when treated with 12 equivalents of the paramagnetic Ru[III] adducted building block **4** under reducing conditions.

Later, this strategy of [–<M>–] connectivity was applied to the development of metal complexes that were designed to explore the flexibility of this mode of connectivity toward the construction of precisely connected multiple dendritic assemblies or networks.[19] Metallo*bis*dendrimers were subsequently prepared (Figure 1) *via* a single [–<Ru>–] connection of two independently prepared dendrons. These two coupled dendrons loosely mimic a *"key"* A and a *"lock"* B, due to the proximity of the incorporated terpyridine units to the core branching centers.

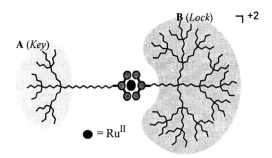

Figure 1: Newkome's "Key and Lock" system, 2nd G-<Ru>-3rd G.

Both the 1st generations of keys and locks were synthesized in a similar method starting from the alkoxylation of 4'-chloroterpyridine by hydroxy-carboxylic acid to provide the requisite cores. For the lower generations, of engaged keys and locks ([1st–<Ru>–1st], [2nd–<Ru>–2nd], and [2nd–<Ru>–3rd]), the cationic and anionic scans in cyclic voltammogram exhibit electrochemically and chemically reversible processes. When the generation increases, the Ru[II]/Ru[III] couple exhibits a larger ΔE_p value, indicating a slower electron transfer as the steric hindrance increases, however, at a very small difference. Completely irreversible behavior is clearly measured for the higher generations ([1st–<Ru>–4th] and [2nd–<Ru>–4th]) of this series of engaged keys and locks. The oxidation of the Ru[II] center is much more positive than that of all other complexes, confirming the steric effect on the redox centers, but the peak reduction potentials remain remarkably constant. From the electrochemical results of the dendritic assemblies, it was concluded that as the bulky surroundings around [−<Ru>−] redox center increase, the reversible redox behavior becomes more irreversible.

Newkome and He[20] have expanded dendritic chemistry into nanoscopic modular chemistry (Figure 2). For the first time, the tailored synthetic strategy made it possible to mimic the nanoscale architecture of simple organic molecules, such as CR_4, which can be envisioned as 'dendritic tetrahedrons'. The first two macroscopic isomeric 'dendritic methane's 7-1 and 7-2 were synthesized by a combination of divergent and convergent method. The 1st generation core 8 was refluxed with the 2nd generation metalloappendage 9 in the presence of 4-ethylmorpholine affording isomer 7-1. Similarly, the 2nd generation of core 10 was treated with 1st generation of counterpart 11 to give isomer 7-2. Both metallo-CR_4 isomers (Figure 2) possess identical molecular weights and display identical spectroscopic data (i.e., ^1H and ^{13}C NMR) as well as other physical properties such as decomposition temperature, solubility, and color. However, the electrochemical study revealed that the internal density and void region of these isomers are different.

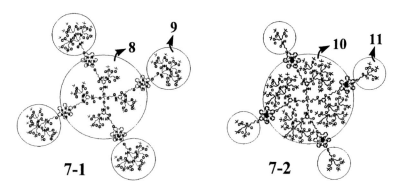

Figure 2: First two isomeric metallodendrimers mimicking CR_4.

The voltammograms showed one quasi-reversible pattern for the four Ru atoms of 7-1 and two waves for the metallic centers of 7-2. This suggested that, as opposed to 7-1, in which the bulky hyperbranched cluster is located on the periphery of the molecule, the internal dendritic structure of 7-2 may provide enough rigidity so that electrochemical communication between the Ru atom is possible.[21] In general, positively charged [−<M>−] assemblies are counter-balanced with ions, such as Cl^-, BF_4^-, PF_6^-; however, to date, there has been a derth of study relating to the zwitterionic forms of these types of complexes and their effects on macromolecular architecture. Newkome et al.[22] reported the construction of overall charge neutral dendritic metallomacromolecules (Scheme 2) without external counterions that incorporate [–<Ru^{II}>–] complexes with internally off-setting charges. Core 13 possessing eight easily hydrolyzed t-butyl ester moieties was refluxed with 4 equivalents of paramagnetic Ru^{III} adduct 12 in EtOH in the presence of 4-ethylmorpholine, as reducing agent, to assemble metallodendrimer 14, (Figure 3) which is still positively charged. The t-butyl groups on 14 were removed by treatment with formic acid affording acidic complex 15 (Figure 3). Finally, the neutral metallodendrimer 16 was formed by addition of a slightly excess of KOH into a H$_2$O-MeOH solution of 15. After dialysis, the desired neutral metallodendrimer 16 was analyzed (0.0 % of Na or Cl). All the intermediates and final complex exhibit correct mass numbers, respectively by MALDI-TOF mass spectra.

Electrochemical experiments with these metallodendrimers gave further insights to their electrocatalytic potential.

Figure 3 (a) shows the two reversible waves that characterize the cathodic CV response of the two terpyridine ligands of 14. After internal deprotection of the t-butyl groups, the presence of the carboxylic acid moieties in 15 results in a merging of the two redox waves and the virtual disappearance of the corresponding anodic signal [Figure 3 (b)]. The observed irreversibility is due to an electrochemical-chemical reaction in which a proton from the vicinal carboxylic acid group quenches the aromatic anion radical resulting in a probable 1,4-reduction of one of the pyridine rings of each terpyridine ligand. This explanation is further supported by CV experiments with neutral dendrimer 16; as seen in Figure 3 (c), the lack of neighboring acidic protons, readily available in 15, resulted in the recovery of the typical 'two-wave' reversible response of the terpyridine ligands.

As a *dendritic supramolecule*, the first example of metallodendritic spiranes has been obtained[23] *via* incorporation of single terpyridine units within each dendritic quadrant (Scheme 3).

By an NMR study, the absence of any free terpyridine moiety, the observed downfield and upfield shifts of 3', 5', and 6,6" terpyridine carbons, respectively, as well as the assignable, symmetric ^{13}C NMR spectra and furthermore, MALDI-TOF-MS (molecular peaks at m/z 3691 [M - PF$_6$]$^+$ for 18 and m/z 3186 [M - 2PF$_6$]$^{2+}$ for 20) supported the assigned intramolecular-based structures. The cyclic voltammograms of both 18 and 20 were similar and exhibited two quasi-reversible waves at negative potentials corresponding to redox process on two electroactive terpyridine groups.

Employment of polyterpyridinyl ligands has allowed access to other novel macrocyclic supramolecules as well as afforded the chemical approaches to new fractal motifs. Highly ordered metallomacrocyclic architectures have been created using *bis*terpyridine ligands with Ru and/or Fe. Newkome and Cho subsequently reported[24] the creation of these ligands and their step-wise and self-assembled supramolecules as hexagonal macrocyclic Ru^{II} and/or Fe^{II} complexes (Scheme 4).

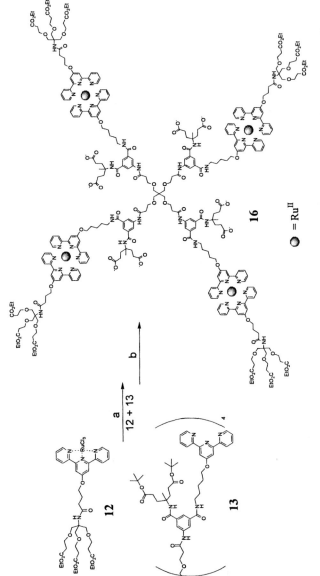

Scheme 2: Synthesis of neutral dendritic metallomacromolecule; a) 4- ethylmorpholine, MeOH, reflux; b) formic acid, r.t.

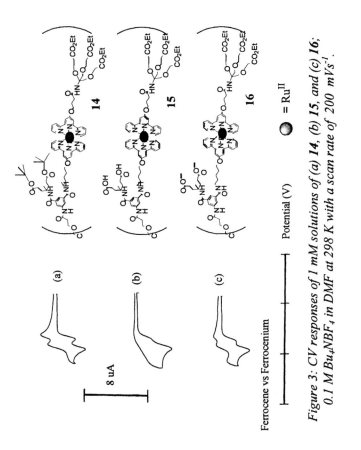

*Figure 3: CV responses of 1 mM solutions of (a) **14**, (b) **15**, and (c) **16**; 0.1 M Bu₄NBF₄ in DMF at 298 K with a scan rate of 200 mVs⁻¹.*

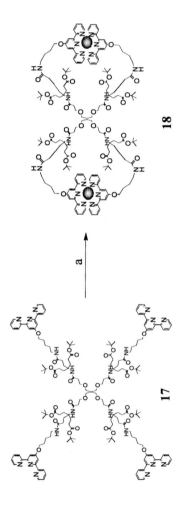

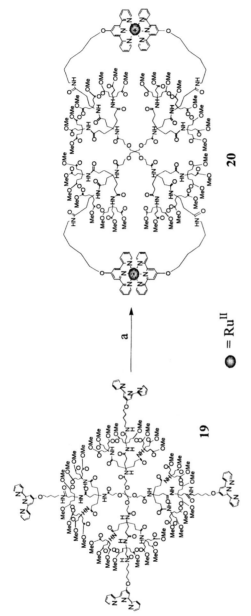

Scheme 3: Syntheses of spirometallodendrimers; a) 4-ethylmorpholine, EtOH, reflux.

19

20

⬤ = Ru^{II}

a

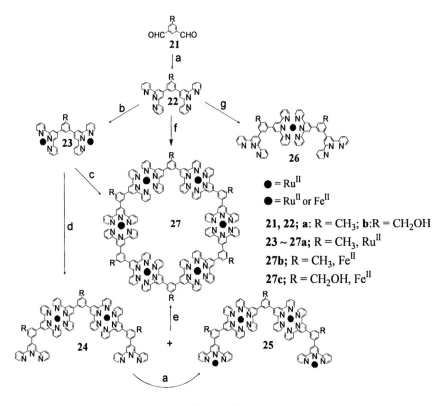

Scheme 4: Syntheses of hexa-Ru^{II} or -Fe^{II} macrocycles via self- and directed-assembly; a) i: acetylpyridine, NaOH, EtOH, r.t.; ii:NH$_4$OAc/HOAc, reflux; b) 2 eq. of RuCl$_3$, EtOH, reflux; c) 1 eq. Of 22, 4-ethylmorpholine, MeOH, reflux; d) 2 eq. of 22, 4-ethyl-morpholine, MeOH, reflux; e) 4-ethylmorpholine, MeOH, reflux; f) 1eq. of FeCl$_2$,MeOH, reflux; g) 0.5 eq. of RuCl$_3$, MeOH, reflux.

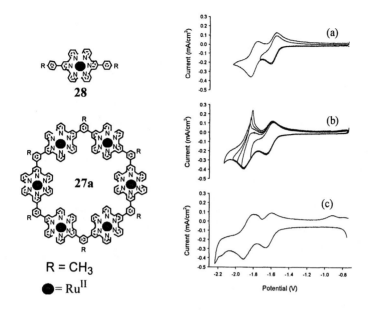

Figure 4: CV responses of 1mM solutions of monoRu^{II} complex **28** (a), and *macrocycle* **27a** *(b,c); 0.1 M Bu$_4$NBF$_4$ in DMF at 298 K with a scan rate of 100 mVs^{-1}.*

Synthesis of requisite building block began with treatment of the known dialdehyde **21** with excess 2-acetylpyridine and base (1N NaOH followed by NH$_4$OAc and acetic acid) to afford the desired angular bisterpyridine **22**. Reaction of hexagon precursor **22a** with 2 eq. of RuCl$_3$ produced paramagnetic bisRu^{III} adduct **23**, which was treated without further purificaton with 1 eq. of ligand **22a** under reducing conditions (4-ethylmorpholine) to yield the self-assembled, diamagnetic, Ru^{II} hexagonal complex **27a**. The structure of this supramolecule **27a** was verified by ^{1}H, ^{13}C, COSY, and HETCOR NMR studies and UV, MALDI-TOF mass spectra. In order to ensure structural verification of macrocycle **27a**, a stepwise, directed route to the material was devised. Subsequently, the diamagnetic trimer **24** was prepared from paramagnetic bisRu^{III} adduct **23** by sequential treatment with 2 eq. building block **22a** under reducing conditions. Finally, reaction of trimer **24** with 1 eq. of its paramagnetic bisRu^{III} adduct **25** yielded a material possessing identical spectral and physical characteristics to that of the self-assembled hexamer **27a**. CV experiments with **27a** further supported its proposed structure (Figure 4).

The electrochemical response of monoRu^{II} complex **28** considered as a monomeric unit of **27a**, showed two reversible waves that correspond to the monoelectronic reduction of each one of the two terpyridine ligands surrounding the Ru atom. The electrochemical response of **27a** in the same potential region revealed that the most positive wave was quasi-reversible and that the most negative one was characterized by a sharp oxidation peak that increased its size as the switching potential became more negative. CV response of **27a** in a slightly wider potential window in the cathodic scan (Figure 4-c) is characterized by an irreversible reduction at very negative potentials that resulted in the absence of the sharp oxidation peak observed in Figure 4-b during the anodic scan. This suggested that there is an irreversible reaction at about 2.1 V *vs.* ferrocenium/ferrocene that was either disconnected some high percentage of metal complex or formed a chemically different species that did not adsorb on the electrode surface.

Successful construction of macrocycle suggested the creation of easily modifiable functionality, as well as different metals, such as Fe^{II} (Scheme 4).[25] The related diamagnetic, hexameric Fe^{II} metallomacrocycles (**27b**, **27c**) were prepared, through self-assembly, (> 85 % yields) by reacting 1 eq. of the *bis*terpyridine ligand **22a** or **22b** with 1 eq. of $FeCl_2$. ^{1}H, ^{13}C, COSY, NOESY, and HETCOR NMR studies were performed on ligand **22a**, and the self-assembled hexagon **27b** verifying the peak assignments and coupling patterns. Transmission electron microscopy (TEM) images of macrocycle **27c** revealed single, hexagonal-shaped particles with the predicted diameter of about 37 Å, based on molecular modeling studies.

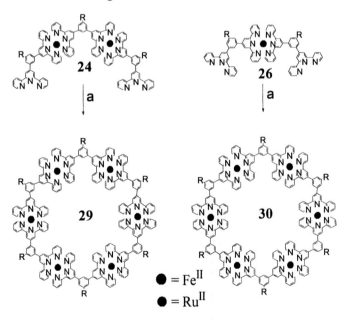

Scheme 5: Syntheses of heteronuclear (mixed Ru^{II} and Fe^{II}) macrocycles via semi-self-assembly; a) 1 eq. of $FeCl_2$, MeOH, reflux.

Access to more complicated building blocks prompted the construction of heteronuclear (mixed Fe^{II} and Ru^{II} complexes) hexagonal macrocycles (Scheme 5) through a semi-self-assembly approach.[26] Reaction of 1 eq. of diamagnetic trimer **24** with 1 eq. of $FeCl_2$ gave macrocycle **29**, which exhibited (^{1}H NMR) two distinct singlets in a 2:1 ratio for the methyl groups flanked by either Ru/Ru or Ru/Fe, respectively. The 6,6"-tpy$H$ and 3',5'-tpy$H$ absorptions were also observed as two sets of resolved peaks due to the coordination spheres. Construction of the alternating heteronuclear metallocycle **30** was accomplished by preparation of diamagnetic Ru^{II} dimer **26** and reacting it with 1 eq. of $FeCl_2$. The alternating architecture of **30** exhibited (^{1}H NMR, Figure 5) signals for all protons were divided into two sets of symmetric peaks with same integration value, respectively, only except the 2-ArH signal.

Further support for the proposed structures of heteronuclear metallomacrocycles **29** and **30** was provided by observation of their UV absorption spectra (MeCN at 25 °C). The macrocycles exhibited two MLCT bands attributed to the $-<Ru^{II}>-$ and $-<Fe^{II}>-$ complexes at λ_{max} 496 nm and 576 nm, respectively. Extinction coefficients for the Ru-tpy MLCT bands of **29** and **30** showed 3.7, and 3-fold increase for λ_{max} at 496 nm, respectively, related to the analogous coefficient for the Ar$-<Ru^{II}>-$Ar complex. Similarly, extinction coefficient for the $-<Fe^{II}>-$ MLCT bands of **29** and **30** showed 2.2, and 2.7-fold increase for λ_{max} at 576 nm, respectively, related to the analogous coefficient for the Ar$-<Fe^{II}>-$Ar complex.

Recently, Newkome and Wang reported the synthesis of functionalized *bis*terpyridine ligands,[27] which have useful groups at 5 position on benzene ring (Scheme 6). In a preliminary experiment, reaction of hexyloxy-*bis*terpyridine **33a** with 1 eq. of $RuCl_3$ under reducing condition afforded *via* spontaneous self-assembly, the desired hemeric, Ru^{II} metallomacrocycle **34**.

The structure of **34** was confirmed by NMR experiments and MALDI-TOF MS, which exhibited peaks at the same amu accordant with the calculated mass. Notably, this 12^+ charged hexamer possessed significantly enhanced solubilities in common organic solvents in contrast to that of nonalkylated metallohexamers. The analogous Fe^{II} hexamer was also prepared by an one-pot, self-assembly method *via* reaction of ligand **33a** with $FeCl_2$. To provide an organizational superstructure for the formation of a noncovalently-bonded network, the counter ions (PF_6^-) in hexamer **34** were changed with an dodecacarboxylate-terminated dendrimer[17] to give a [34^{12+}(dendrimer^{12-})], as a *suprasupermolecular* network,[28] (Figure 6) which is an ordered-hexamer-dendrimer composite.

Conclusions

The preceding studies on bisterpyridine-based metallodendrimers and metallomacrocycles have presented a case for the structural novelities, which enter the nanomolecular regime and open the door to new suprasupermolecular

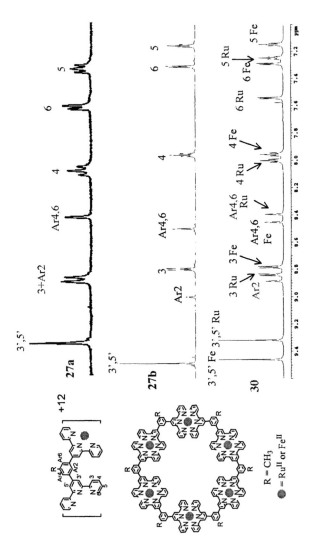

Figure 5: ^1H *NMR spectra of hexa-Ru^{II} -Fe^{II}, and alternating Ru^{II}-Fe^{II} metallomacrocycles 27a, 27b, and 30.*

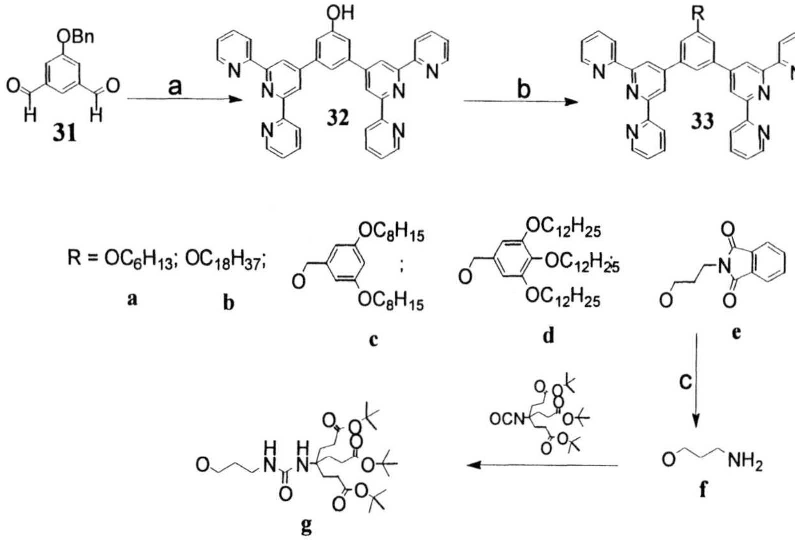

Scheme 6: Syntheses of O-*alkyl-bis(terpyridinyl)phenol derivatives;*
a) i: 2-pyCOMe, NaOH, EtOH; ii: NH₄OH/AcOH, reflux; iii: Pd/C,
H₂, EtOH/THF; b) Br or Cl alkylating agents, K₂CO₃, DMF;
c) N₂H₄

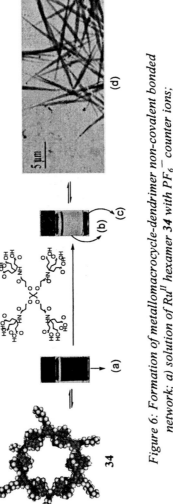

Figure 6: Formation of metallomacrocycle-dendrimer non-covalent bonded network; a) solution of Ru^{II} hexamer 34 with PF$_6^-$ counter ions; b) solution of containing only PF$_6^-$; c) precipitate comprised of a hexamer-dendrimer composite; d) TEM image of hexamer-dendrimer composite

constructs capable of energy storage and photon capture. Experiments are ongoing for access to more complicate dendritic hexagonal metallomacrocycles and shape-persistent metallodendrimers as well as larger fractal architectures.

Acknowledgments

We gratefully acknowledge financial support from the National Science Foundation (DMR-041780 and CHE-0135786), Korean Research Foundation (KRF-2003-042-C0006), the Air Force Office of Scientific Research (F49620-02-1-0428,02) and the Ohio Board of Regents.

References

1. Newkome, G. R.; Yao, Z.; Baker, G. R.; Gupta, V. K. *J. Org. Chem.* **1985**, *50*, 2003-2004.
2. Tomalia, D. A.; Baker, H.; Dewald, J.; Hall, M.; Kallos, G.; Martin, S.; Roeck, J.; Ryder, J.; Smith, P. *Polym. J. (Tokyo)* **1985**, *17*, 117-132.
3. Newkome, G. R.; Moorefield, C. N.; Vögtle, F. In *Dendrimers and Dendrons: Concepts, Syntheses, Applications*; Wiley - VCH: Weinheim, Germany, 2001.
4. Newkome, G. R.; He, E.; Moorefield, C. N. *Chem. Rev.* **1999**, *99*, 1689-1746.
5. Ziessel, R. *Synthesis* **1999**, *11*, 1839-1865.
6. Newkome, G. R.; Cardullo, F.; Constable, E. C.; Moorefield, C. N.; Thompson, A. M. W. C. *J. Chem. Soc., Chem. Commun.* **1993**, 925-927.
7. Eisenbach, C. D.; Schubert, U. S.; Baker, G. R.; Newkome, G. R. *J. Chem. Soc. , Chem. Commun.* **1995**, 69-70.
8. Baxter, P. N. W.; Lehn, J.-M.; Kneisel, B. O.; Baum, G.; Fenske, D. *Chem. Eur. J.* **1999**, *5*, 113-120.
9. Heller, M.; Schubert, U. S. *Eur. J. Org. Chem.* **2003**, 947-961.
10. Thompson, A. M. W. C. *Coord. Chem. Rev.* **1997**, *160*, 1-52.
11. Schubert, U. S.; Eschbaumer, C. *Angew. Chem. Int. Ed.* **2002**, *41*, 2892-2926.; Hofmeier, H.; Schubert, U. S. *Chem. Soc. Rev.* **2004**, *33*, 373-399.
12. Stang, P. J. *Chem. Eur. J.* **1998**, *4*, 10-27.
13. Zimmerman, S. C. *Curr. Opin. Colloid Interface Sci.* **1997**, *2*, 89-99.
14. Rebek, J., Jr. *Acc. Chem. Res.* **1999**, *32*, 278-286.
15. Young, J. K.; Baker, G. R.; Newkome, G. R.; Morris, K. F.; Johnson, C. S., Jr. *Macromolecules* **1994**, *27*, 3464-3471.
16. Constable, E. C.; Thompson, A. M. W. C.; Tocher, D. A.; Daniels, M. A. M. *New J. Chem.* **1992**, *16*, 855-867.

17. Newkome, G. R.; Young, J. K.; Baker, G. R.; Potter, R. L.; Audoly, L.; Cooper, D.; Weis, C. D.; Morris, K. F.; Johnson, C. S., Jr. *Macromolecules* **1993**, *26*, 2394-2396.

18. Sauvage, J.-P.; Collin, J.-P.; Chambron, C.; Guillerez, S.; Coudret, C.; Balzani, V.; Barigelletti, F.; De Cola, L.; Flamigni, L. *Chem. Rev.* **1994**, *94*, 993-1019.

19. Newkome, G. R.; Güther, R.; Moorefield, C. N.; Cardullo, F.;Echegoyen, L.; Pérez-Cordero, E.; Luftmann, H. *Angew. Chem., Int.Ed. Engl.* **1995**, *34*, 2023-2026.

20. Newkome, G. R.; He, E.; Godínez, L. A.; Baker, G. R. *J. Am. Chem.Soc.* **2000**, *122*, 9993-10006.

21. Cuadrado, I.; Casado, C. M.; Alonso, B.; Morán, M.; Losada, J.;Belsky, V. *J. Am. Chem. Soc.* **1997**, *119*, 7613-7614.

22. Newkome, G. R.; He, E.; Godínez, L. A.; Baker, G. R. *Chem.Commun.***1999**, 27-28.

23. Newkome, G. R.; Yoo, K. S.; Moorefield, C. N. *Chem. Commun.* **2002**,2164-2165.

24. Newkome, G. R.; Cho, T. J.; Moorefield, C. N.; Baker, G. R.;Saunders, M. J.; Cush, R.; Russo, P. S. *Angew. Chem. Int. Ed.* **1999**,*38*, 3717-3721.

25. Newkome, G. R.; Cho, T. J.; Moorefield, C. N.; Cush, R.; Russo, P. S.;Godínez, L. A.; Saunders, M. J. *Chem. Eur. J.* **2002**, *8*, 2946-2954.

26. Newkome, G. R.; Cho, T. J.; Moorefield, C. N.; Mohapatra, P. P.; Godínez, L. A. *Chem. Eur. J.* **2004**, *10*, 1493-1500.

27. Wang, P.; Moorefield, C. N.; Newkome, G. R. *Org. Lett.* **2004**, *6*, 1197-1200.

28. Unpublished data.

Chapter 15

Encapsulation Effects on Homogeneous Electron Self-Exchange Dynamics in Tris(Bipyridine) Iron-Core Dendrimers

Young-Rae Hong and Christopher B. Gorman[*]

Department of Chemistry, North Carolina State University, Box 8204, Raleigh, NC 27695–8204
[*]Corresponding author: chris_gorman@ncsu.edu

A series of tris(bipyridine) iron core dendrimers was synthesized up to the second generation. Their homogeneous electron self-exchange dynamics were studied by NMR line-broadening. While the first generation dendrimers showed dramatically attenuated electron self-exchange rate with respect to dimethyl-bipyridine iron complex, little rate change between the first and second generation dendrimers was observed. These results suggest that the effective distance of electron transfer in this homogeneous self-exchange case is not a simple function of dendrimer generation.

Background

Understanding electron transfer is important because it can provide us fundamental knowledge associated with many biological and chemical processes such as those found in metalloproteins, luminescent molecules, and the emerging area of molecular electronics.[1-6] The factors that govern the rate and driving force for electron transfer can be investigated by systematically varying the structure of model compounds. To this end, redox-active core dendrimers are attractive molecules because each generation of dendrimer nominally produces a different degree-of-encapsulation of the redox-active core resulting in changes of electron transfer dynamics. We and others have prepared various redox-active core dendrimers and studied their electron transfer dynamics in various scenarios including heterogeneous electron transfer at a metal-solution interface and electron hopping through the dendrimer film.[7-9] By observing changes in the rate and/or driving force for electron transfer, several effects of dendrimer generation and primary structure can be rationalized to affect the macromolecular conformation and thus the degree of encapsulation of the redox-active core.

One scenario for electron transfer that, to our knowledge has not yet been explored with redox-active core dendrimers is homogeneous, bimolecular electron transfer in solution. Is the rate of this reaction attenuated in the same manner observed previously for heterogeneous electron transfer? Moreover, can this approach allow us to measure a greater degree of rate attenuation? In heterogeneous electron transfer reactions, only a four to five order of magnitude rate decrease can be observed before voltammetry becomes non-quantitative. In contrast, in solution, a variety of spectroscopic techniques are available to measure electron transfer rates, and these can span a much greater range. Although one cannot directly compare heterogeneous and homogeneous electron transfer rates as they represent different kinetic orders, wider detection limits for homogeneous electron transfer rate determination should help us to probe encapsulation effects especially for the slow electron transfer behavior found in redox-active core dendrimers.

We recently reported an isostructural series of redox-active, metal tris(bipyridine) core dendrimers that are amenable to study of homogeneous electron transfer kinetics.[10] The central redox units of these dendrimers can be chemically oxidized. Here, we report our first results on homogeneous electron self-exchange kinetics between oxidized and reduced dendrimer species. As will be shown, the attenuation of the rate of this reaction with dendrimer generation is not the same as the behavior found in heterogeneous, electrochemical rate determinations.

Objectives

In this study, our goals were to understand how the encapsulation of a redox-active, iron tris(bipyridine) core using dendrons affects the homogeneous electron self-exchange rate. This rate was measured by NMR line broadening technique in partially oxidized dendrimer solutions of generations "0" through "2" as defined below.

Experimental

Materials

All synthetic efforts were reported previously[10] except the second generation dendron and dendrimer which were prepared in a similar fashion and characterized as below.

G2-Bpy. Yield: 92% (1.79 g); ^{1}H NMR (CDCl$_3$) δ (ppm) 1.2 (m, 4H), 1.5-1.8 (br, 30H), 2.1-2.4 (m, 12H), 2.6 (t, 4H), 3.9 (t, 8H), 5.0 (s,16H), 6.8 (d, 8H), 6.9 (d, 16H), 7.0-7.2 (m, 24H), 7.3-7.6 (m, 42H), 8.2 (s, 2H), 8.6 (d, 2H). Anal. Calcd for C$_{170}$H$_{168}$N$_2$O$_{12}$: C, 83.99; H, 6.97; N, 1.15; found: C, 83.75; H, 6.83; N, 1.32.

[Fe(G2-Bpy)$_3$](PF$_6$)$_2$. Yield: 91% (0.48 g); ^{1}H NMR (CD$_2$Cl$_2$) δ (ppm) 1.2 (m, 4H), 1.4-1.6 (br, 30H), 2.0-2.4 (m, 12H), 2.6 (t, 4H), 3.7 (t, 8H), 5.0 (s,16H), 6.7 (d, 8H), 6.8 (d, 16H), 6.9-7.1 (m, 26H), 7.2-7.4 (m, 42H), 8.2 (s, 2H). Anal. Calcd for C$_{510}$H$_{504}$F$_{12}$FeN$_6$O$_{36}$P$_2$: C, 80.18; H, 6.65; N, 1.10; found: C, 80.13; H, 6.67; N, 1.08.

NMR kinetic measurements

All deuterated solvents were purchased from Cambridge Isotope Laboratories, Inc, dried following standard procedures, and stored in a dry box under nitrogen atmosphere. NMR spectra were obtained using Varian 300 MHz NMR spectrometer at 25 °C unless otherwise noted. Sample solutions for NMR measurement were prepared freshly in the dry box prior to use. A solution of Iron(II) core dendrimer in CD$_2$Cl$_2$:CD$_3$CN (5:1 v/v) were prepared and transferred to 5 mm Kontes brand threaded NMR tube in the dry box. Aliquots (typically 10 μL) of the oxidants (NOPF$_6$ or Fe(bpy)$_3$$^{3+}$) in CDCN$_3$ were added to the NMR tube between each NMR measurement. Line widths were determined by fitting the experimental signals to a Lorenzian function using commercially available NUTS software. Temperatures were maintained by temperature controller available in Varian spectrometers.

The second order electron self-exchange rate constants were measured by NMR line broadening techniques. The proton NMR spectra of the partially oxidized dendrimer solutions were interpreted by use of the two-site exchange model between diamagnetic iron(II) and paramagnetic iron(III) cores. The approximate Bloch-McConnell equations for the electron self-exchange rate constants in the fast exchange and slow exchange limits were used to calculate the rate constants.[11-13]

$$k_{ex} = \frac{4\pi f_p f_d (\delta v)^2}{(W_{DP} - f_d W_D - f_p W_P) C_{tot}} \qquad \text{for the fast exchange limit}$$

$$k_{ex} = \frac{\pi (W_{DP} - W_D)}{[P]} \qquad \text{for the slow exchange limit}$$

In the equation, δv is the contact shift in Hz (chemical shift movements caused by paramagnetic electrons, that is the chemical shift difference between pure diamagnetic and pure paramagnetic species), f_p and f_d are the mole fractions of paramagnetic and diamagnetic species respectively, W_P, W_D, and W_{DP} are the peak width (full line width at half-maximum) for paramagnetic species only, diamagnetic species only, and the mixture of two species respectively, C_{tot} is total molar concentration and [P] is molar concentration of paramagnetic species. Values for f_p for the fast exchange system were more precisely determined using the relationship

$$f_p = \frac{\left| v_{dp} - v_d \right|}{\delta v}$$

assuming that the chemical shifts vary linearly with mole fraction. v_d and v_{dp} are the resonance frequency of the diamagnetic species and the mixture respectively.

Results & Discussion

Synthesis

We previously reported the synthesis and preliminary electrochemical characterization for a series of metal tris(bipyridine) core dendrimers.[10] Second generation of the dendron (G2-Bpy) was prepared by similar procedures with high yields. This dendron ligand was then successfully attached to the iron(II) metal core and characterized by ^{1}H-NMR spectroscopy. Figure 1 shows the dendrimer structures employed in this study and defines, in this case, what is regarded as the zeroth, first and second generation of this dendrimer.

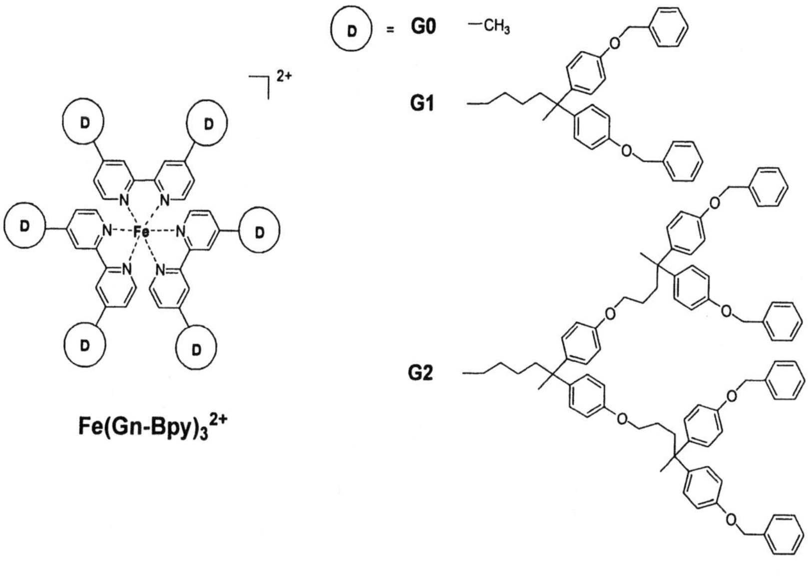

Figure 1. Structures of the iron tris(bipyridine) core dendrimers employed in this study. The counter anion was PF_6^-.

Determination of homogeneous electron self-exchange rate constants

The homogeneous electron self-exchange rate of dimethyl-bipyridine iron complex [Fe(G0-Bpy)$_3$] in CD$_2$Cl$_2$/CD$_3$CN (5:1 v/v) was measured by NMR line broadening techniques. A solvent mixture was used in this study to ensure the solubility of the oxidants and dendrimers. The rate of electron self-exchange for $Fe(G0\text{-}Bpy)_3^{2+/3+}$ was found to be in the fast exchange limit, indicated by the linear relationship between the mole fraction, f_p and chemical shift.[12, 13] The singlet corresponding to the methyl protons at 2.5 ppm on the methyl-bipyridine ligand was used in the calculation of the rate constant as it showed sufficient chemical shift movements and line broadening during the experiments (Figure 2).

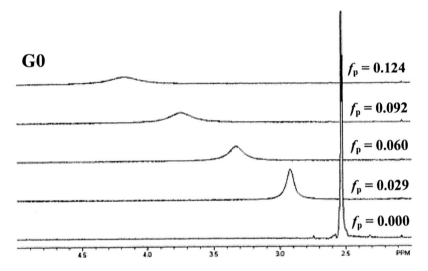

G0

$f_p = 0.124$

$f_p = 0.092$

$f_p = 0.060$

$f_p = 0.029$

$f_p = 0.000$

Figure 2. ^1H NMR spectra of $Fe(G0\text{-}Bpy)_3^{2+/3+}$ systems at 20 °C in CD_2Cl_2/CD_3CN (5:1 v/v) mixture. The counter anion was PF_6^-.

Although $NOPF_6$ was an acceptable oxidant for $Fe(G0\text{-}bpy)_3^{2+}$, oxidation of the core of $Fe(G1\text{-}bpy)_3^{2+}$ by $NOPF_6$ was unsuccessful because the dendron itself was oxidized to some extent by $NOPF_6$. As the dendron is intended only as an encapsulating moiety, an alternate, milder oxidant was required. Fortunately, it was determined that unsubstitued iron tris(bipyridine) [$Fe(bpy)_3^{2+/3+}$] had slightly higher redox potential compared to the methyl substituted iron complex [$Fe(G0\text{-}bpy)_3^{2+/3+}$] and, upon addition of $Fe(bpy)_3^{3+}$ to $Fe(Gn\text{-}bpy)_3^{2+}$ (n = 0, 1 2) the former oxidized the latter. The redox potential difference between these two species (ca. 160 mV in CH_3CN) provided a very mild route to prepare the oxidized dendrimers. Moreover, the resulting $Fe(bpy)_3^{2+}$ did not interfere with the spectroscopic observation of the desired, line broadened signals.

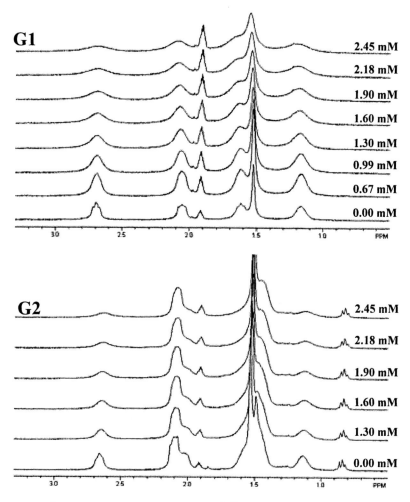

Figure 3. ^1H NMR spectra of $Fe(G1-Bpy)_3{}^{2+/3+}$ and $Fe(G2-Bpy)_3{}^{2+/3}$ systems at 25 °C in CD_2Cl_2/CD_3CN (5/1 v/v) mixture. The counter anion was PF_6^-.

The electron self-exchange rate for the first and second generation dendrimers were found to be in the slow exchange limit as determined by a lack of noticeable change in chemical shift over the concentration range examined (Figure 3). Data were fit to the slow exchange equation and average k_{ex} values were determined from slopes of plots $\pi(W_{DP}-W_D)$ vs $[Fe(Gn-Bpy)_3{}^{3+}]$ (Figure 4). The spin-spin coupling of the observed proton peaks around 2.7 ppm with the methylene protons next to it hampered the determination of the line width. Thus, triplet peaks in pure $Fe(Gn-Bpy)_3{}^{2+}$ (n = 1 and 2) were deconvoluted to determine W_D values. W_{DP} values of the sample mixture were measured after line broadening was corrected for this spin-spin coupling.

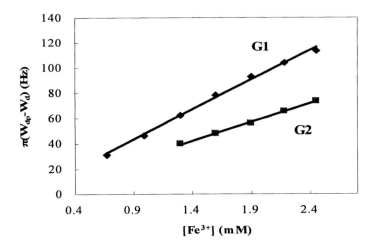

Figure 4. A plot of line broadening versus the concentration of Fe(Gn-Bpy)$_3^{3+}$ (n = 1, 2).

Table 1. Electron self-exchange rate constants

Systems	C_{tot} (mM)	k_{ex} (M^{-1}s^{-1})[a]	
Fe(G0-Bpy)$_3^{2+/3+}$	11.13	2.59 (0.19) x 10^7	fast exchange[b]
Fe(G1-Bpy)$_3^{2+/3+}$	8.56	4.71 (0.27) x 10^4	slow exchange[c]
Fe(G2-Bpy)$_3^{2+/3+}$	8.51	3.00 (0.21) x 10^4	slow exchange[c]

a. values in parentheses represent the magnitude of the 90% confidence interval
b. fast exchange equation was applied.
c. slow exchange equation was applied.

This treatment of data gave more acceptable results for both molecules, giving an intercept much closer to the expected value of zero (Figure 4). Observed electron self-exchange rate constants for all generations were tabulated in Table 1.

As can be seen from the data in Table 1, the electron self-exchange rate for the first generation dendrimer was retarded dramatically relative to Fe(G0-bpy)$_3^{2+/3+}$ system. In contrast, the attenuation of the electron self-exchange rate between the first and the second generation dendrimer was insignificant compared to that between Fe(G0-Bpy)$_3^{2+/3+}$ and Fe(G1-Bpy)$_3^{2+/3+}$. This result was unexpected as it does not correlate with the "encapsulation" behaviors illustrated in heterogeneous electron transfers in dendrimers previously.[7] This result is perhaps even more surprising if one regards homogeneous electron transfers between the redox centers in Fe(G2-Bpy)$_3^{2+}$ and Fe(G2-Bpy)$_3^{3+}$ must occur effectively through four "layers" of repeat units (e.g. two for each molecule).

We rationalize these results by considering the possibility of relatively rapid core motion within the dendrimer architecture. Previously, the effects of the core motion were argued to be important in governing the rate of heterogeneous electron transfer iron-sulfur cluster core dendrimers containing alternately flexible and rigid repeat units and in a series of iron-sulfur cluster core dendrimer constitutional isomers.[7-9] Most notably, in films of these molecules, the rate of the electron hopping through these films was observed to be in the slow exchange realm and no significant variation in the electron self-exchange rate constants was observed with generation.[8] Similar explanations used to rationalize the behavior in these systems may be applicable here. Over the time-scale of the slow exchange limit, the redox core unit within the dendrimer can move so as to achieve a relatively close approach with another redox core in a neighboring dendrimer. Thus, the effective distance for electron self-exchange appears to be not strongly affected by dendrimer size in the slow electron exchange limit. While these results do not span the range of dendrimer size (generation), core and repeat unit required to make this conclusion in a general sense, these preliminary data do represent a result that is difficult to interpret without core mobility playing an important role in the kinetics of this process.

Conclusions

As dendrimers are studied in a greater variety of ways, the idea of dendritic encapsulation is less and less well described by a static dendrimer model. Here, in exploring electron transfer between dendrimers in solution, the simple generational dependence of encapsulation on rate attenuation does not suffice. To understand the effects of encapsulation for the electron transfer kinetics, the time-scale of the electron transfer as well as the structure and conformational mobility of the dendrimers are important factors. In this case, a simple rationale for the behaviors observed is that the redox core within these dendrimers in the slow electron self-exchange limit can move in a non-rate limiting fashion toward a neighboring redox core with the result that the structural effect of the dendrimer is reduced and electron transfer is facilitated in larger dendrimers.

Acknowledgements

The National Science Foundation (CHE-0315311) is gratefully acknowledged for the support of this research. We thank Dr. Sabapathy Sankar and Dr. Hanna Gracz for assistance with NMR measurements.

References

1   Campagna, S.; Denti, G.; Serroni, S.; Juris, A.; Venturi, M.; Ricevuto, V.; Balzani, V. *Chem.-Eur. J.* **1995**, *1*, (4), 211-221.
2   Gorman, C. B. *Adv. Mater.* **1997**, *9*, (14), 1117-1119.
3   Vögtle, F.; Plevoets, M.; Nieger, M.; Azzellini, G. C.; Credi, A.; De Cola, L.; De Marchis, V.; Venturi, M.; Balzani, V. *J. Am. Chem. Soc.* **1999**, *121*, (26), 6290-6298.
4   Hecht, S.; Fréchet, J. M. J. *Angew. Chem. Int. Ed.* **2001**, *40*, (1), 74-91.
5   Cameron, C. S.; Gorman, C. B. *Adv. Funct. Mater.* **2002**, *12*, (1), 17-20.
6   Gorman, C. B.; Smith, J. C. *Acc. Chem. Res.* **2001**, *34*, (1), 60-71.
7   Gorman, C. B.; Smith, J. C.; Hager, M. W.; Parkhurst, B. L.; Sierzputowska-Gracz, H.; Haney, C. A. *J. Am. Chem. Soc.* **1999**, *121*, (43), 9958-9966.
8   Gorman, C. B.; Smith, J. C. *J. Am. Chem. Soc.* **2000**, *122*, (38), 9342-9343.
9   Chasse, T. L.; Sachdeva, R.; Li, C.; Li, Z. M.; Petrie, R. J.; Gorman, C. B. *J. Am. Chem. Soc.* **2003**, *125*, (27), 8250-8254.
10 Hong, Y. R.; Gorman, C. B. *J. Org. Chem.* **2003**, *68*, (23), 9019-9025.
11 Sandström, J., *Dynamic NMR Spectroscopy*. ed.; Academic Press: New York, 1982.
12 Chan, M. S.; Wahl, A. C. *J. Phys. Chem.* **1978**, *82*, (24), 2542-2549.
13 Triegaardt, D. M.; Wahl, A. C. *J. Phys. Chem.* **1986**, *90*, (9), 1957-1963.

Chapter 16

Extraction of Metal Nanoparticles from within Dendrimer Templates

Joaquin C. Garcia-Martinez, Orla M. Wilson, Robert W. J. Scott, and Richard M. Crooks[*]

Department of Chemistry, Texas A&M University, P.O. Box 30012, College Station, TX 77842–3012

The synthesis and characterization of Pd, Au, Ag and bimetallic AuAg dendrimer-encapsulated nanoparticles (DENs), and their subsequent extraction from their dendrimer templates with alkanethiol ligands to yield near-monodisperse metal monolayer-protected clusters (MPCs) are reported. In particular, the factors affecting the extraction experiment - ionic strength, thiol concentration, dendrimer generation and DEN size - are examined, and it is conclusively shown that under optimal conditions, the particles remain intact throughout the extraction process. UV-vis spectroscopy and HRTEM are used to characterize the nanoparticles before and after extraction, and FTIR confirms that the dendrimer remains in the aqueous phase after the extraction. Furthermore, the affinity of specific ligands for different metal surfaces can be used to preferentially extract nanoparticles of one metal from a mixture of Au and Ag DENs in a process referred to as selective extraction. An application of the selective extraction process to the chemical characterization of bimetallic AuAg DENs is briefly discussed.

Background

Here we demonstrate that Pd, Au, Ag and bimetallic AuAg nanoparticles can be extracted from within dendrimer templates and transferred to an organic phase using alkanethiol and alkanoic acid surfactants while leaving the dendrimer intact in the aqueous phase (Scheme 1).[1-5] The extraction of metal nanoparticles from dendrimers is a facile route for preparing monodisperse monolayer-protected clusters (MPCs),[6,7] and it avoids the need for multi-step purification processes. The results very strongly suggest that individual nanoparticles are extracted from the dendrimer without significant loss of metal or aggregation. FT-IR spectroscopy was used to ascertain the whereabouts of the dendrimer after extraction; the amide bands, characteristic of PAMAM dendrimers, were only present in the aqueous phase. Thus, the dendrimer can be recycled and used for the preparation of another batch of dendrimer-encapsulated nanoparticles (DENs). One further benefit of the extraction experiment is that the resulting MPCs can be easily characterized by techniques such as mass spectroscopy or electrochemistry, and thus the extraction experiment can be used to provide information about the original size and shape distributions of the DENs.[3] We will also examine recent successful extractions of Au, Ag, and bimetallic AuAg DENs using alkanethiol and alkanoic acid surfactants,[4,5] and show how the extraction method can be used to gain chemical information of the surface structure of nanoparticles.

In 1998 we reported the synthesis of dendrimer-encapsulated Cu nanoparticles,[8] and shortly thereafter found that it was also possible to prepare a number of other types of near-monodisperse dendrimer-encapsulated nanoparticles (DENs),[9-11] including metals such as Pd,[12-14] Au,[15] and Ag.[16] We further showed that the resulting organic/inorganic composites were catalytically active for certain simple types of reactions.[9-11] DENs are prepared by a two-step process. First, metal ions are sequestered within the dendrimer, followed by the addition of a reducing agent such as $NaBH_4$. This results in the reduction of the metal ions to form a zerovalent metal nanoparticle. Because the synthesis takes place in a well-defined dendrimeric template, the resulting metal nanoparticle (the replica) can be quite monodisperse in size. Among the desirable characteristics of DENs are that they can be solubilized in nearly any solvent,[9-11] the nanoparticle surface is unpassivated and therefore catalytically active,[11] and the dendrimer branches can be used as selective gates to control access of small molecules to the encapsulated nanoparticles.[13]

This extraction method is important for several reasons. First, it demonstrates that nanometer-scale materials prepared within a molecular template can be removed leaving both the replica and template undamaged. Second, it provides a straightforward approach for preparing highly monodisperse metallic and bimetallic MPCs without the need for subsequent purification. Third, it demonstrates that multiple, fairly complex operations, including formation of covalent bonds, electron-transfer, molecular transport, heterogeneous self assembly, and nanoparticle transport, can all be executed within the interior of a dendrimer.

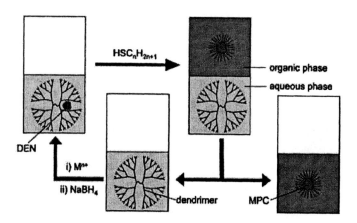

Scheme 1. Reprinted with permission from J. Am. Chem. Soc. 2004, 126, 16170-16178. Copyright 2004 American Chemical Society.

Objective

Here, we present a detailed study of the parameters that control the extraction of nanoparticles from dendrimer templates. Some of the fundamental questions to be answered include: how can such large nanoparticles escape the confinement of the dendrimer interior, and what are the driving forces for the extraction? To begin, we have examined the impact of the following parameters on the extraction process: (1) different metals such as Pd, Au, and Ag; (2) the nanoparticle size; (3) the concentration of the *n*-alkanethiol extractant; and (4) the generation and the peripheral functionalization of the dendrimer. Furthermore, we will show how the extraction process can be used to separate Au and Ag nanoparticles from an aqueous mixture of Au and Ag DENs. This selective extraction approach takes advantage of the affinity of alkanethiol and alkanoic acids for Au and Ag oxide surfaces, respectively. As an application for this process, we will briefly show how the extraction technique can allow the separation and structural determination of AuAg core/shell nanoparticles.

Experimental

Fourth-, sixth- and eighth-generation hydroxyl-terminated PAMAM dendrimers (G4-OH, G6-OH, and G8-OH, respectively) having an ethylenediamine core were obtained as 10-25% aqueous solutions from Dendritech, Inc. (Midland, MI). All n-alkanethiols, ascorbic acid, n-undecanoic acid, NaBH$_4$, HAuCl$_4$, and AgNO$_3$ were used as received from the Aldrich Chemical Co. (Milwaukee, WI). HPLC grade ethanol, toluene, and hexane were purchased from EMD Chemicals Inc., and 18 MΩ·cm Milli-Q water (Millipore, Bedford, MA) was used to prepare aqueous solutions. Cellulose dialysis sacks having a molecular weight cutoff of 12,000 were purchased from Sigma Diagnostics, Inc. (St. Louis, MO).

Synthesis and extraction of metal nanoparticles from the interior of PAMAM dendrimers. The method used to prepare and extract DENs has previously been described in the literature,[1-5] but small deviations from this procedure were necessary in some cases. A short summary of the basic procedure follows, with specific reference to the synthesis of fourth-generation, hydroxyl-terminated PAMAM dendrimers containing 55-atom Au nanoparticles (G4-OH(Au$_{55}$)). To prepare the Au DENs, a 10 mL aqueous solution containing 2.0 μM G4-OH and 110 μM HAuCl$_4$ was vigorously stirred for 2 min, followed by the addition of a 5-fold molar excess of NaBH$_4$ (150 mM in 0.3 M NaOH). The reduction of the intradendrimer Au^{3+} complex to zerovalent Au can be easily followed as the color changes from yellow to brown upon addition of NaBH$_4$. After the synthesis of the Au DENs was complete, a 150-fold molar excess of NaBH$_4$ was added to the aqueous Au DEN solution, followed by the addition of 10 mL of a 20 mM n-dodecanethiol (HSC$_{12}$) solution in toluene. The vial was shaken for 5 min, and the resulting emulsion was allowed to settle for 5 min, after which phase separation was complete. The toluene layer containing the Au MPC product was purified by first concentrating the solution to 1 mL on a rotary evaporator, and then adding 15 mL of ethanol to precipitate the MPCs. Centrifugation resulted in separation of the Au MPCs from excess free n-alkanethiol and others impurities. The MPCs were washed and centrifuged twice with ethanol to ensure complete purification. Similar procedures were followed for the synthesis and extraction of G4-OH(Pd$_{40}$) DENs. G6-OH(Ag$_{110}$) DENs were synthesized using an intradendrimer displacement reaction, which converts Cu DENs to Ag DENs.[16] Specifically, a 0.55 mM solution of G6-OH(Cu$_{55}$) DENs was synthesized as follows: 0.28 mL of a 20.0 mM solution of Cu(NO$_3$)$_2$ was added to 9.06 mL of a 11.0 μM solution of G6-OH. The solution was purged with N$_2$ for 15 min prior to reduction with a 3-fold excess of BH$_4^-$. Excess BH$_4^-$ was removed by addition of acid, and then 1.1 mL of 0.01 M AgNO$_3$ was added to the Cu DEN solution. As a consequence of the enhanced nobility of Ag and the stronger oxidizing power of Ag$^+$ compared to Cu, the Ag$^+$

ions oxidize the Cu^0 nanoparticles to Cu^{2+} and in doing so are themselves reduced to Ag^0.[16]

Characterization. UV-vis absorbance spectra were recorded with a Hewlett-Packard HP 8453 UV-vis spectrometer using quartz cells. UV-vis spectra of DENs were collected using deionized water as the reference. UV-vis spectra of metal MPCs were obtained immediately after extraction and without purification, using either toluene or hexane as the reference. HRTEM was performed using a JEOL 2010 transmission electron microscope (JEOL USA Inc., Peabody, MA). Samples were prepared by placing one drop of a solution on a holey-carbon-coated Cu grid (EM science, Gibbstown, NJ) and allowing the solvent to evaporate in air.

Results & Discussion

Extraction of Palladium and Gold Nanoparticles

Pd or Au nanoparticles are extracted from their dendrimer templates by first adding a large excess of $NaBH_4$ to the aqueous solution and then introducing a toluene solution containing an appropriate n-alkanethiol (Scheme 1). After shaking for 5 min, the aqueous phase turned from brown to colorless and the toluene phase turned from colorless to brown, indicating extraction of the nanoparticles from the dendrimer and transport of the resulting MPCs into the toluene phase. Note that Esumi and coworkers have previously reported that n-alkanethiols adsorb onto the surface of Au DENs,[17,18] but they did not observe extraction of the nanoparticles from the dendrimers. This is likely a consequence of our finding that the presence of both an alkanethiol surfactant and a sufficiently high ionic strength is required for the extraction of DENs.[1-3] Here, we rely on an excess of $NaBH_4$ to increase the ionic strength, but identical results were obtained when $NaCl$, $Mg(NO_3)_2$, or Na_2SO_4 were used for this purpose.

Figure 1 shows UV-vis absorption spectra of aqueous-phase G4-OH(Pd_{40}) and G8-OH(Au_m) (m = 55, 147, 1022) prior to extraction (black lines) and the corresponding toluene-phase, n-alkanethiol-stabilized MPCs after extraction (red lines). In the case of Pd DENs (Figure 1a), the aqueous G4-OH(Pd_{40}) solution displays one intense band at 200 nm, corresponding to absorption by the dendrimer and BH_4^-, and a gradually increasing absorption towards higher energy, which is characteristic of Pd colloids. All the Au DEN aqueous solutions also display an increasing absorbance toward higher energy, which results from interband electronic transitions of the encapsulated zerovalent Au nanoparticles.[12-15,19] In addition, a plasmon band at 500-550 nm appears in the G8-OH(Au_{1022}) solution, which is not present for the smaller gold DENs.

Plasmon bands arising from Au nanoparticles larger than 2 nm are typically observed in this range.[20,21] Therefore the broad width and weak absorbance of the plasmon band appearing for the G8-OH(Au$_{1022}$) solution suggests that the particles are in the 2-3 nm range.[22-24] In addition, all the UV-vis spectra of the Au DENs (Figure 1b-d) contain a small peak at 290 nm superimposed on the rising background. Although there is controversy in the literature regarding the origin of this band,[25,26] it is reproducibly observed during Au DEN syntheses[15,25] and does not seem to affect nanoparticle properties.

The UV-vis spectra of the MPCs in the toluene phase after extraction of the nanoparticles are very similar to those of the aqueous-phase DENs in both form and intensity (the spectra are cut off below 285 nm due to toluene absorption). It is especially interesting to point out the absence of a plasmon band in the UV-vis spectra of the Au$_{55}$ and Au$_{147}$ nanoparticles after extraction, which indicates that no significant change in particles size occurs during the extraction process. The small variations in the UV-vis spectra of the aqueous DEN solutions and the extracted MPC solutions probably result from changes in the environment of the metal surface of the nanoparticles after extraction, and it is known that changes in solvent and adsorbed ligands have a profound influence over the UV-vis spectra of metal nanoparticles.[22,23] Regardless of these small differences, however, the main point is that Pd and Au DENs can be extracted from dendrimers with no significant increase in the average nanoparticle size.

To demonstrate that the dendrimer template remains in the aqueous phase, we examined the toluene phase and remaining aqueous phase after extraction. After extraction of G4-OH(Pd$_{40}$) DENs, the aqueous phase was dialyzed for 48 h to remove toluene and BH$_4^-$, and the resulting absorbance spectrum of this solution (blue line, Figure 1a) indicated that only the G4-OH dendrimer was present. That is, the characteristic interband absorption signature of Pd DENs was absent. FT-IR spectroscopy of both the toluene and aqueous phases after extraction showed that the dendrimer is only present in the aqueous phase.[1,3] This means that the dendrimer can be recovered and recycled to prepare additional DENs. This principle has been demonstrated and the results indicate that more than 80% of the original dendrimer can be recovered after two cycles of DEN preparation and subsequent nanoparticle extraction.[3]

HRTEM micrographs were obtained to evaluate the average particle size and size distributions of Pd and Au nanoparticles before and after the extraction from the dendrimer templates. The average diameters were obtained by measuring the size of 150 randomly selected particles. In all cases, the average size of the nanoparticles was identical before and after extraction within experimental error. Specifically, before extraction, the G4-OH(Pd$_{40}$), G8-OH(Au$_{55}$), G8-OH(Au$_{147}$), and G8-OH(Au$_{1022}$) DENs, had average nanoparticle diameters of 1.7 ± 0.4 nm, 1.2 ± 0.2 nm, 1.5 ± 0.3 nm , and 2.1 ± 0.7 nm, respectively, which agree with previous results obtained for similarly-prepared

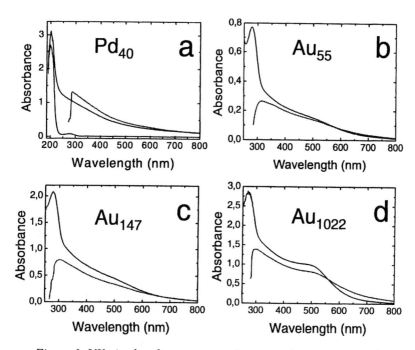

Figure 1. UV-vis absorbance spectra demonstrating extraction of (a) Pd nanoparticles from a 6.2 μM G4-OH(Pd$_{40}$) DEN solution, (b) Au$_{55}$ and (c) Au$_{147}$ nanoparticles from 2.0 μM G8-OH(Au$_{55}$) and G8-OH(Au$_{147}$) DEN solutions, and (d) Au$_{1022}$ nanoparticles from a 0.50 μM G8-OH(Au$_{1022}$) DEN solution. The black line corresponds to the aqueous phase before extraction, and the red line was obtained from the organic phase after extraction with a toluene solution containing 20 mM of n-hexanethiol for the extraction of Pd$_{40}$ and 20 mM of n-dodecanethiol for the extraction of Au$_m$ nanoparticles. The blue line in spectrum (a) corresponds to the aqueous phase after extraction. Reprinted with permission from J. Am. Chem. Soc. 2003, 125, 11190-11191 and J. Am. Chem. Soc. 2004, 126, 16170-16178. Copyright 2003 and 2004 American Chemical Society.

materials in our group.[12-15] After extraction the average nanoparticle sizes were
1.5 ± 0.3 nm, 1.3 ± 0.4 nm, 1.5 ± 0.3 nm, 2.2 ± 0.5 nm for MPC(Pd_{40}),
MPC(Au_{55}), MPC(Au_{147}) and MPC(Au_{1022}), respectively. The data strongly
suggest that the individual nanoparticles are extracted from the dendrimer
without significant loss of metal or aggregation.

The extraction of the nanoparticles is also nearly independent of the
peripheral functionality or the generation of the dendrimer. Au DENs containing
55 Au atoms were prepared using three different generations of PAMAM
dendrimers: G4-OH(Au_{55}), G6-OH(Au_{55}), and G8-OH(Au_{55}). We have
previously shown that the particle size only depends on the dendrimer-to-metal-
ion molar ratio and is independent of the dendrimer generation.[15] Indeed, these
three samples have very similar nanoparticle sizes and distributions; specifically,
the average nanoparticle diameters were 1.2 ± 0.3 nm, 1.3 ± 0.2 nm, and $1.2 \pm$
0.2 nm for the G4-OH(Au_{55}), G6-OH(Au_{55}), and G8-OH(Au_{55}) DENs,
respectively. These results are very close to the calculated values assuming that
these nanoparticles are spherical in shape.[27] After the extraction, the average
particle size was 1.2 ± 0.3 nm, 1.3 ± 0.4 nm, and 1.3 ± 0.4 nm for the Au_{55}
nanoparticles extracted from the interior of the G4-OH, G6-OH, and G8-OH
dendrimers, respectively. This indicates that the particle size of the final MPCs
depends only on the molar ratio of dendrimer to metal ions used to form the
nanoparticles. In addition, similar-sized Au nanoparticles have been prepared
and successfully extracted using amine-terminated PAMAM dendrimers instead
of hydroxyl-terminated dendrimers. The UV-vis spectra and HRTEM
micrographs of the resulting Au MPCs also show that are essentially identical to
those of the Au DEN precursors.[3]

Figure 2 shows the effect of concentration of the n-alkanethiol ligand during
the extraction of the Au nanoparticles from the dendrimers. Specifically, four 5.0
mL aliquots of G4-OH(Au_{55}) ([Au]= 110 μM) were extracted using 5.0 mL of
dodecanethiol solutions of 110, 330, 550, and 1100 μM in toluene. These
solutions correspond to HSC_{12}/Au molar ratios of 1, 3, 5, and 10, respectively.
Note that only 76% of the Au atoms in an ideal Au_{55} cluster are on the surface,
so in all cases a stoichiometric excess of HSC_{12} is present in solution. UV-vis
spectra of the toluene phases after extraction are essentially identical to that
shown in Figure 1b (prepared with a HSC_{12}/Au molar ratio = 182), except for the
case where the HSC_{12}/Au molar ratio = 1, for which a plasmon band is observed
near 500 nm.[3] This suggests that at low n-alkanethiol ligand concentrations the
extraction process results in particle aggregation during extraction. HRTEM data
confirms this hypothesis, as shown in Figure 2. Figure 2a-d shows micrographs
and particle-size distributions for extractions of Au DENs using HSC_{12}/Au molar
ratios of 1, 3, 5, and 10, respectively. The results from these data are
summarized in Figure 2e, which shows that both the average particle size and the
particle-size distribution decrease as the HSC_{12}/Au molar ratio increases from 1

to 10. We rationalize this finding in terms of the following proposed extraction mechanism: when low concentrations of *n*-alkanethiols are used, the nanoparticles undergo relatively slow extraction and the nascent MPCs are only partially passivated for some time following extraction. During this period, the average particle size increases via metal aggregation. It is possible to decrease the period of partial thiol passivation, and thus the likelihood of aggregation, by increasing the concentration of the *n*-alkanethiol.

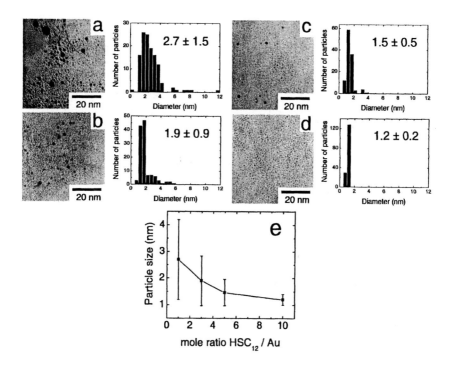

*Figure 2. HRTEM micrographs and particle-size distributions for Au$_{55}$ nanoparticles extracted from G4-OH dendrimers using the following HSC$_{12}$/Au molar ratios: (a) 1, (b) 3, (c) 5, and (d) 10. The horizontal axis of the histograms is the same for each panel to facilitate comparison of the size distributions. Figure 2(e) shows the average particle size (from HRTEM) vs. the HSC$_{12}$/Au molar ratio. The vertical bars indicate the standard deviation of the particle-size distribution. Reprinted with permission from J. Am. Chem. Soc. **2004**, 126, 16170-16178. Copyright 2004 American Chemical Society.*

Selective Extraction of Silver and Gold Nanoparticles

Since reporting the extraction of Pd and Au nanoparticles from within dendrimer templates,[1-3] we have been investigating the potential of using the extraction process as a characterization tool.[4,5] In particular, we were interested in determining whether the extraction method could be used to characterize the surface structure of monometallic and bimetallic DENs. As a proof-of-concept experiment, we reasoned that if we could find surfactants that had highly selective affinities for different metals, then it would be possible to separate mixtures of nanoparticles. This idea is based on the concept of "orthogonal assembly", which has its roots in a 1989 paper by Wrighton and Whitesides.[28] The authors demonstrated that two different surfactants could selectively interact with each of two different metal or metal-oxide surfaces. More specifically, several groups have shown that n-alkanoic acids bind to metal-oxide surfaces such as Ag oxides, but not pure metallic surfaces such as Au.[29,30] Here, we report the separation of Au and Ag dendrimer-encapsulated nanoparticles from an aqueous mixture of the two using a selective extraction approach, Figure 3.[4]

Figure 3. Demonstration of the selective extraction of Ag MPCs and simultaneous extraction of both Au and Ag MPCs from an aqueous mixture of similarly-sized Au and Ag DENs. Reprinted with permission from Chem. Mater. 2004, 16, 4202-4204. Copyright 2004 American Chemical Society.

In this method, an n-alkanoic acid present in hexane and having a high affinity for Ag oxide surfaces (but not Au) is added to an aqueous mixture of similarly-sized Ag[16] and Au[15] DENs. Through strong ligand-nanoparticle interactions, the Ag nanoparticles are selectively extracted into the organic phase resulting in a solution of Ag MPCs.[4] Subsequent addition of an organic phase containing an n-alkanethiol leads to extraction of the remaining Au DENs. It is also possible to extract both metals simultaneously by employing the strong affinity of n-alkanethiol molecules for both Au and Ag surfaces (by addition of a strong reducing agent to remove the Ag oxide layer), or selectively extract just Au nanoparticles from a mixture of DENs. These results are important, because they demonstrate a simple chemical approach for separating nanoparticles having different chemical compositions.

Figure 4 shows UV-vis absorption spectra for an aqueous mixture of G4-OH(Au$_{147}$) and G6-OH(Ag$_{110}$) DENs prior to extraction (black line), the hexane phase after the first extraction with n-decanoic acid (red line), and the second hexane phase after subsequent extraction with n-dodecanethiol (green line). After the first extraction, the n-decanoic acid/hexane phase reveals a peak at 420 nm, which corresponds to the plasmon absorption of Ag.[31] The position and magnitude of this band are nearly identical to those of the original G6-OH(Ag$_{110}$) solution, indicating a quantitative extraction of the Ag nanoparticles. Likewise, after the second extraction, the spectrum of Au$_{147}$ MPCs is almost identical to that of the G6-OH(Au$_{147}$) DENs prior to extraction. The presence of a much larger plasmon band for Ag DENs compared to Au DENs is in accordance with their fundamental optical properties and previous literature reports.[24,32]

HRTEM was used in combination with large-area energy dispersive X-ray spectroscopy (EDS) to analyze the particle size and atomic composition of the samples before and after extraction. Prior to extraction the mixture of DENs have a particle size of 1.6 ± 0.3 nm and large-area EDS analysis of 10 areas on the TEM grid indicates an average composition of 43% Ag and 57% Au. Analogous analysis of the hexane phase after the first extraction with n-decanoic acid results in particles with a diameter of 1.7 ± 0.4 nm and an average composition of 95 ± 6% Ag and 5 ± 6% Au. This is convincing evidence that n-alkanoic acids preferentially extract the Ag$_{110}$ nanoparticles from the DEN mixture. HRTEM and EDS analysis of the particles resulting from the second extraction with n-dodecanethiol indicate a particle-size distribution of 1.4 ± 0.3 nm and a metal composition of 8 ± 6% Ag and 92 ± 8% Au. From these results we conclude that similarly-sized Au and Ag DENs can be separated.

In the presence of a reducing agent no oxide is present on the Ag surface, and therefore it is possible to simultaneously extract both Ag and Au DENs using n-dodecanethiol. Further confirmation of the need for a Ag oxide layer for selective extraction with n-decanoic acid was obtained by adding NaBH$_4$ to the mixture of Ag and Au DENs. Under these reducing conditions, neither of the DENs extract with n-decanoic acid, presumably due to the absence of the Ag oxide layer. NMR spectroscopy confirmed that BH$_4^-$ did not reduce or otherwise react with the acid.

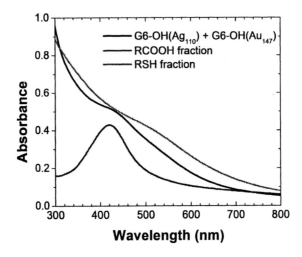

Figure 4. UV-vis absorbance spectra demonstrating the separation of G6-OH(Au₁₄₇) and G6-OH(Ag₁₁₀) DENs by selective extraction. The black line corresponds to the aqueous phase mixture of G6-OH(Au₁₄₇) and G6-OH(Ag₁₁₀) DENs before extraction, the red line was obtained from the hexane phase after extraction with a solution containing n-decanoic acid, the green line was obtained from the second hexane phase after extraction with a solution containing n-dodecanethiol. Reprinted with permission from Chem. Mater. 2004, 16, 4202-4204. Copyright 2004 American Chemical Society.

Extraction of AuAg Bimetallic Nanoparticles

Since the selective extraction experiment indicated that orthogonal assembly of surfactants onto metal nanoparticles having different surface compositions provides a basis for separation of Au and Ag DENs, we sought to further test its adaptability as a characterization tool of AuAg bimetallic DENs.[5] Three structurally unique bimetallic DENs were synthesized and characterized: AuAg alloys, core/shell [Au](Ag) and core/shell [AuAg alloy](Ag) (for structured materials, brackets indicate the core metal and parentheses indicate the shell metal). The bimetallic DENs were characterized using UV-vis spectroscopy, HRTEM, and single-particle X-ray EDS. We have shown that depending on the surface metal and its oxidation state, these nanoparticles can be extracted from the dendrimer into an organic phase with either *n*-alkanethiol or *n*-alkanoic acid molecules. While *n*-dodecanethiol in the presence of NaBH₄ will quantitatively extract all AuAg nanoparticles into the organic phase regardless of structure, *n*-undecanoic acids will extract only AuAg alloy particles having significant Ag loadings, and Ag shell nanoparticles. Thus, this selective extraction strategy provides an important new tool for chemical analysis of the structure of bimetallic nanoparticles in the 1-3 nm diameter size range. We envision that under certain conditions alloy nanoparticles could be separated from core/shell particles or from monometallic nanoparticles that may have formed during the reduction step. We are currently exploring the extension of this selective extraction strategy to other metals, and our aim is to explore new ligands, such as phosphines, *n*-alkylamines, and *n*-alkaneisocyanides, that can interact either with Au, Ag, or other metals, such as Pd, Pt, Cu, or Fe. This will make it possible to lay out a wet-chemical roadmap for the qualitative analysis of the surface structure and composition of monometallic and bimetallic DENs.

Conclusions

To summarize, we have shown that both Pd and Au DENs with diameters of less than 2.2 nm can be extracted intact from within the interior of dendrimer templates using alkanethiol ligands. Extraction proceeds quickly, regardless of the size of the DEN, the dendrimer generation, and the peripheral functionalization of the dendrimer. The success of the extraction experiment depends only on the ionic strength of the dendrimer solution and the thiol concentration. The average particle size and optical properties of the extracted nanoparticles are the same as the precursor DENs as evidenced by HRTEM and UV-vis spectroscopy. The mechanism proposed for the extraction involves penetration of the dendrimer by the alkanethiol, adsorption of the thiol to the nanoparticle surface, and extraction of the resulting MPC into the organic phase. We have further shown that selective extraction, based on the affinity of ligands for different metal surfaces, can be used to separate a mixture of Au and Ag DENs. The selective extraction technique can be used as a chemical characterization tool which, we believe, will greatly aid in the structure elucidation of bimetallic nanoparticles with diameters in the 1 - 3 nm range. We are further exploring other metal and surfactant combinations to fully realize the power of this technique for purifying and characterizing the smallest of nanoparticles.

Acknowledgments

We gratefully acknowledge the U. S. Department of Energy, DOE-BES Catalysis Science grant no. DE-FG02-03ER15471, the Robert A. Welch Foundation, and the U. S. National Science Foundation (Grant No. 0211068) for financial support of this work. Dr. Joaquin C. Garcia-Martinez thanks the

Ministerio de Educacion, Cultura y Deporte of Spain for postdoctoral fellowship support. We also thank Dr. Zhiping Luo of the TAMU Microscopy and Imaging Center for assistance with HRTEM and EDS measurements.

References

(1) Garcia-Martinez, J. C.; Scott, R. W. J.; Crooks, R. M. *J. Am. Chem. Soc.* **2003**, *125*, 11190-11191.

(2) Garcia-Martinez, J. C.; Crooks, R. M. *Polym. Preprint* **2004**, *45*, 462-463.

(3) Garcia-Martinez, J. C.; Crooks, R. M. *J. Am. Chem. Soc.* **2004**, *126*, 16170-16178.

(4) Wilson, O. M.; Scott, R. W. J.; Garcia-Martinez, J. C.; Crooks, R. M. *Chem. Mater.* **2004**, *16*, 4202-4204.

(5) Wilson, O. M.; Scott, R. W. J.; Garcia-Martinez, J. C.; Crooks, R. M. *J. Am. Chem. Soc.* **2005**, *127*, 1015-1024.

(6) Templeton, A. C.; Wuelfing, W. P.; Murray, R. W. *Acc. Chem. Res.* **2000**, *33*, 27-36.

(7) Whetten, R. L.; Shafigullin, M. N.; Khoury, J. T.; Schaaff, T. G.; Vezmar, I.; Alvarez, M. M.; Wilkinson, A. *Acc. Chem. Res.* **1999**, *32*, 397-406.

(8) Zhao, M.; Sun, L.; Crooks, R. M. *J. Am. Chem. Soc.* **1998**, *120*, 4877-4878.

(9) Crooks, R. M.; Zhao, M.; Sun, L.; Chechik, V.; Yeung, L. K. *Acc. Chem. Res.* **2001**, *34*, 181-190.

(10) Crooks, R. M.; Lemon, B. I.; Sun, L.; Yeung, L. K.; Zhao, M. *Top. Curr. Chem.* **2001**, *212*, 81-135.

(11) Niu, Y.; Crooks, R. M. *C. R. Chim.* **2003**, *6*, 1049-1059.

(12) Zhao, M.; Crooks, R. M. *Angew. Chem. Int. Ed.* **1999**, *38*, 364-366.

(13) Niu, Y.; Yeung, L. K.; Crooks, R. M. *J. Am. Chem. Soc.* **2001**, *123*, 6840-6846.

(14) Scott, R. W. J.; Ye, H.; Henriquez, R. R.; Crooks, R. M. *Chem. Mater.* **2003**, *15*, 3873-3878.

(15) Kim, Y.-G.; Oh, S.-K.; Crooks, R. M. *Chem. Mater.* **2004**, *16*, 167-172.

(16) Zhao, M.; Crooks, R. M. *Chem. Mater.* **1999**, *11*, 3379-3385.

(17) Satoh, K.; Yoshimura, T.; Esumi, K. *J. Colloid. Interface. Sci.* **2002**, *255*, 312-322.

(18) Esumi, K.; Satoh, K.; Torigoe, K. *Langmuir* **2001**, *17*, 6860 - 6864.

(19) Gröhn, F.; Bauer, B. J.; Akpalu, Y. A.; Jackson, C. L.; Amis, E. J. *Macromolecules* **2000**, *33*, 6042-6050.

(20) Rao, C. N. R.; Kulkarni, G. U.; Thomas, P. J.; Edwards, P. P. *Chem. Eur. J.* **2002**, *8*, 28-35.

(21) Alvarez, M. M.; Khoury, J. T.; Schaaff, T. G.; Shafigullin, M. N.; Marat, N.; Vezmar, I.; Whetten, R. L. *J. Phys. Chem. B* **1997**, *101*, 3706-3712.

(22) Templeton, A. C.; Pietron, J. J.; Murray, R. W.; Mulvaney, P. *J. Phys. Chem., B* **2000**, *104*, 564-570.

(23) Underwood, S.; Mulvaney, P. *Langmuir* **1994**, *10*, 3427-3430.

(24) Mulvaney, P. *Langmuir* **1996**, *12*, 788-800.

(25) Lee, W. I.; Bae, Y.; Bard, A. J. *J. Am. Chem. Soc.* **2004**, *126*, 8358-8359.

(26) Esumi, K.; Suzuki, A.; Aihara, N.; Usui, K.; Torigoe, K. *Langmuir* **1998**, *14*, 3157-3159.

(27) Leff, D. V.; Ohara, P. C.; Heath, J. R.; Gelbart, W. M. *J. Phys. Chem.* **1995**, *99*, 7036-7041.

(28) Laibinis, P. E.; Hickman, J. J.; Wrighton, M. S.; Whitesides, G. M. *Science* **1989**, *245*, 845-847.

(29) Schlotter, N. E.; Porter, M. D.; Bright, T. B.; Allara, D. L. *Chem. Phys. Lett.* **1986**, *132*, 93-98.

(30) Tao, Y.-T. *J. Am. Chem. Soc.* **1993**, *115*, 4350-4358.

(31) Creighton, J. A.; Eadon, D. G. *J. Chem. Soc. Faraday Trans.* **1991**, *87*, 3881-3891.

(32) Kreibig, U.; Vollmer, M. *Optical Properties of Metal Clusters*; Springer, 1995; Vol. 25.

Chapter 17

Metallo Groups Linked to the Surface of Phosphorus-Containing Dendrimers

Jean-Pierre Majoral[*], Anne-Marie Caminade[*], and Regis Laurent

Laboratoire de Chimie de Coordination du CNRS, 205 route de Narbonne, 31077 Toulouse Cedex 4, France

This paper is a review concerning metallic derivatives linked to the surface of phosphorus-containing dendrimers, that are dendrimers possessing one phosphorus at each branching point. The synthetic aspects will be described first, concerning in particular the complexation of transition metals (Fe, W, Rh, Pt, Pd, Ru) by monophosphine or diphosphine end groups, the grafting of metallocene derivatives (zirconocene and ferrocene), and the interaction with various clusters (Au-Fe, Au, Ti). Some applications of these metallo phosphorus dendrimers will be described in the field of materials science (modification of the surface of existing materials or creation of new materials), and in the field of catalysis (Stille couplings, Knoevenagel condensations, Michael additions).

Introduction

The grafting of metallic derivatives to the surface of dendrimers[1,2] is an area of current and constant interest, mainly driven by the catalytic properties[3,4] of some of the resulting metallo-dendrimers.[5-8] In many cases, the complexation of metals has been conducted with organic dendrimers; however, many examples exist also concerning metallic derivatives linked to the surface of phosphorus-containing dendrimers, that are dendrimers having one phosphorus atom at each branching point.[9-10] This paper is a review concerning phosphorus dendrimers in which metallic derivatives are linked to their periphery, mostly but not exclusively through phosphine end groups. In most cases, the internal structure of these dendrimers is constituted of $OC_6H_4CH=NNMeP(S)$ linkages, emanating from a trifunctional (P=S) or hexafunctional (N_3P_3) core. In most cases, the complexing end groups are constituted either of phosphines or diphosphines. The structure of the first generation of three of the most widely used phosphorus dendrimers is shown in Figures 1 and 2; these compounds were synthesized up to the tenth generation for dendrimers 1-G_n (3072 phosphines),[11] up to the fourth generation for dendrimers 2-G_n (48 diphosphines),[12] and up to the fifth generation for dendrimers 3-G_n (192 phosphines).[11]

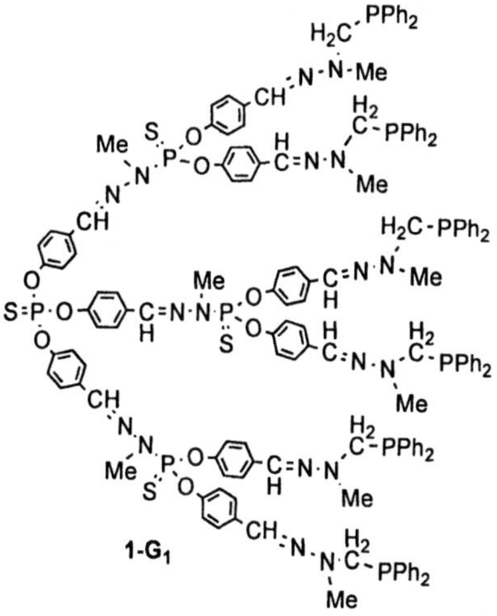

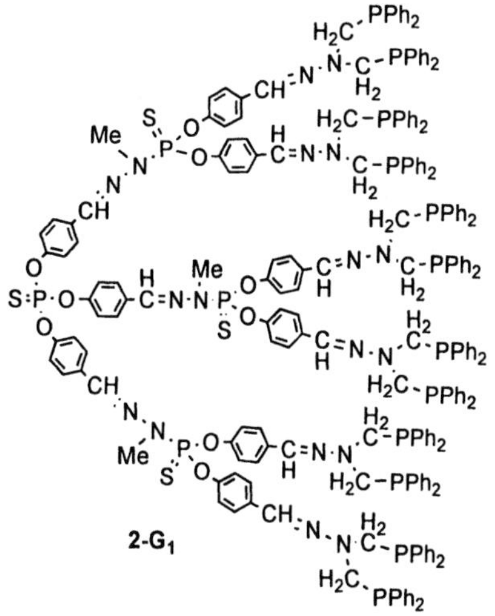

Figure 1. Monophosphine and diphosphine end groups linked to the surface of first generation of phosphorus dendrimers built from a trifunctional core.

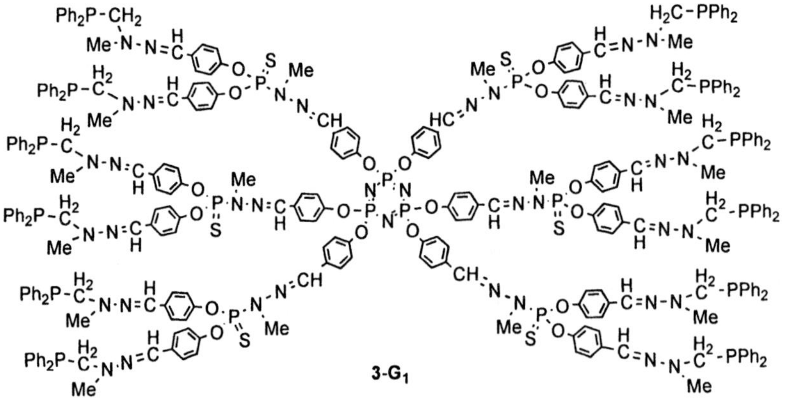

Figure 2. Monophosphine end groups linked to the surface of the first generation of a phosphorus dendrimer built from a hexafunctional core.

Compounds **1-G$_n$** and **3-G$_n$** afforded the first examples of large phosphine complexes linked to the surface of any type of dendrimers.[11] In the first part of this review, the grafting of transition metals, metallocenes, and organometallic clusters using various types of end groups will be described; then, some properties of these metallo dendrimers will be discussed, particularly in the fields of new materials and of catalysis.

Complexation of transition metals

The advantage of phosphines as end groups is that a single type of dendrimer may serve as complexing agent for different types of metals, as illustrated in Scheme 1 for metals of groups 6, 8, and 9. In this Scheme, as well as in several other schemes throughout this review, **Dendri** indicates that the internal structure is constituted of $OC_6H_4CH=NNMeP(S)$ linkages. First experiments were conducted with common M(0) derivatives [$Fe(CO)_4$ and $W(CO)_5$], then, with two types of Rh(I) derivatives. In most cases, experiments were carried out at least up to the fourth generation, and in some cases up to the fifth or sixth generation. For the largest compounds, 192 Rh(acac)(CO) groups have been linked to the surface of dendrimer **1d-G$_6$** for the **1-G$_n$** series,[13] whereas 96 RhCl(COD) have been complexed by dendrimer **3c-G$_4$**, which is the largest compound of series **3-G$_n$**.[14] In all cases, the complexation goes easily to completion within a few hours at room temperature, as show by ^{31}P NMR.

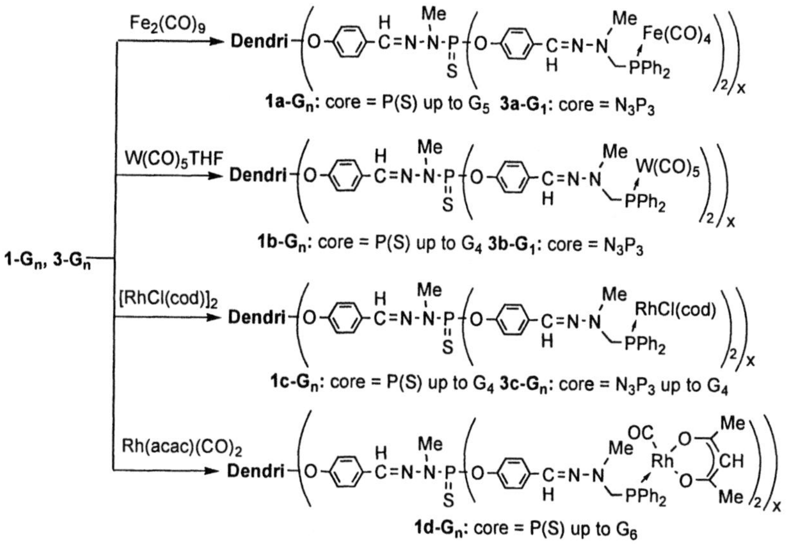

Scheme 1. Complexation of transition metals by monophosphine end groups. Only the most external layers are shown in all cases.

The complexation properties of phosphino groups linked to another type of end groups were also tested with $Fe(CO)_4$ and $W(CO)_5$ (Scheme 2).[15] This series of compounds possesses several types of functions at each end groups, besides the phosphine complexes: allyl groups, secondary amines, and tertiary amines, illustrating the concept of "multiplurifunctionalization".[16]

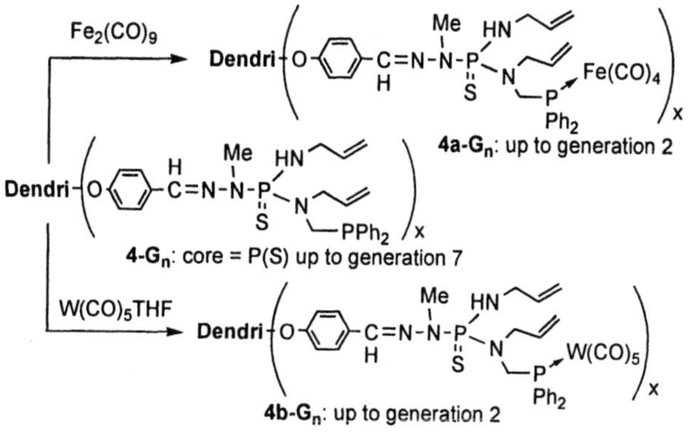

Scheme 2. Complexation of Fe and W derivatives by another type of monophosphine end groups.

In the previous examples, all end groups of the dendrimers are identical, whereever their locations are on the surface. However, some special dendrimers exist, in which part of the surface possesses one type of end groups, and another part possesses another type of end groups. These special dendrimers, called "surface block",[17] are generally built by association of two different dendrons (dendritic wedges) by their core or on a core. However, the diblock compound 5-G_1G_0 was synthesized by the step-by-step growth on one side to graft ammonium groups, then on the other side to graft phosphino groups. Finally, iron is complexed by the phosphines, on one third of the end groups.[18]

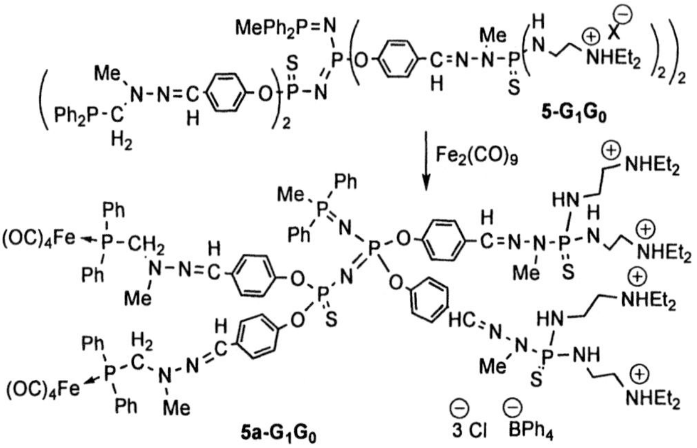

Scheme 3. Complexation of iron by 1/3 of the end groups of a small diblock dendrimer.

The grafting of diphosphine end groups instead of monophosphines allows the expansion of the scope of the complexing properties of dendrimers, as shown by the complexation of $PtCl_2$ (Scheme 4).[12] It is important to note that the complexation of each metal cleanly occurs between two phosphorus linked to the same nitrogen, and not randomly.

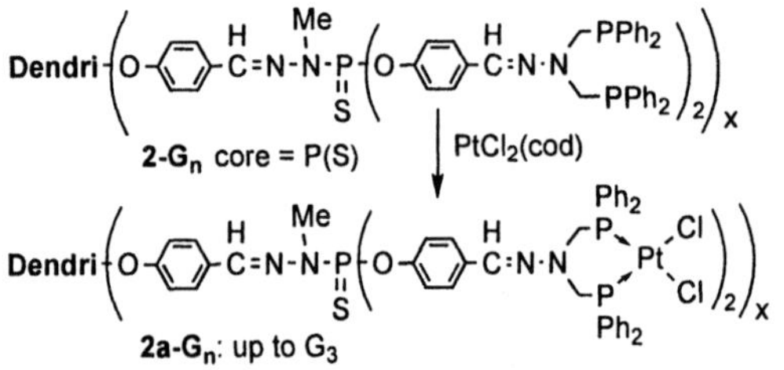

Scheme 4. Complexation of platinum(II) by diphosphine end groups.

The same type of experiment was carried out with $PdCl_2$. In this case, a real organometallic chemistry was developed on the surface of the dendrimer, as illustrated in Scheme 5[12]. An easy halogen exchange occurs when **2b-G$_1$** is reacted with KBr, whereas halogen exchange and alkylation of palladium occur simultaneously with the Grignard reagent MeMgBr. A monomethylation is also observed with Cp_2ZrCl_2, leading to complexes **2e-G$_n$** (n ≤ 3). The presence of Pd-Me bonds allowed to carry out CO insertion; a second insertion was observed using norbornene, which took place readily in the Pd-acetyl bond, leading to **2g-G$_n$**.

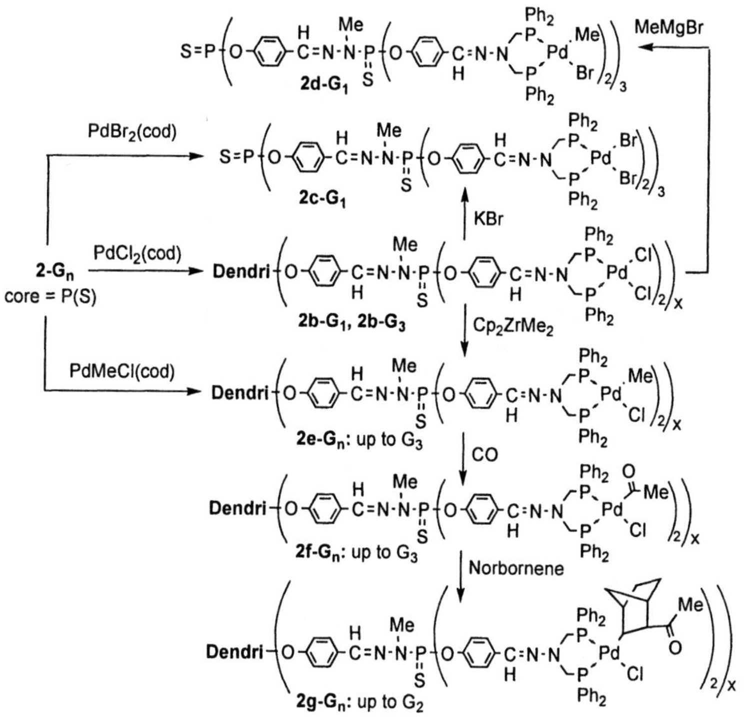

Scheme 5. *Complexation of palladium derivatives by diphosphine end groups.*

Dendrimers **2-G$_n$** were also used to complex $Rh(acac)$[12] and ruthenium hydride derivatives[19] (Scheme 6). In the latter case, two different behaviors were observed for the resulting complexes, depending on the nature of the hydrido complex. In the case of **2i-G$_n$**, the complex is stable, whereas dendrimers **2k-G$_n$** exist as a mixture of isomers, depending on the relative position of PCy_3, both hydrides and H-H ligands. Furthermore, a fluxional behaviour is observed. However, in both cases addition of CO gives the analogous complexes **2j-G$_n$** and **2l-G$_n$** in which either one PPh_3 or the dihydrogen is replaced by CO.

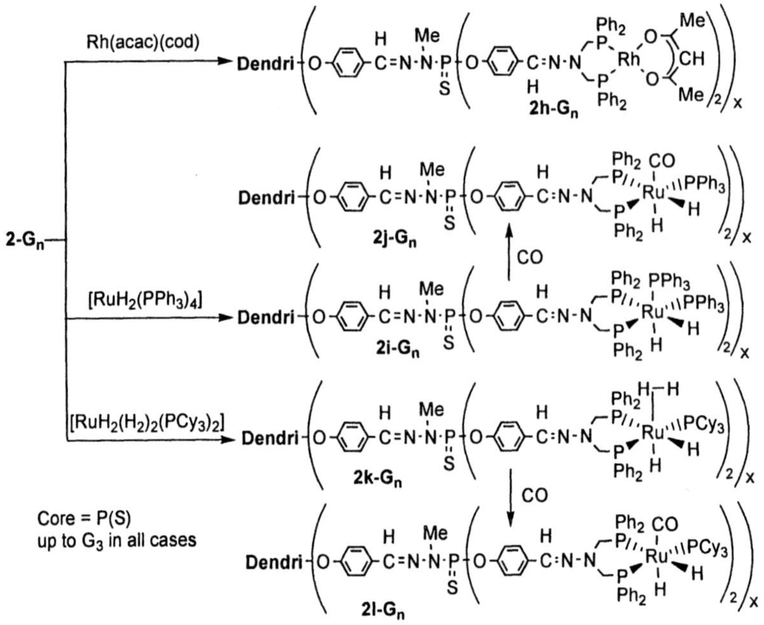

Scheme 6. Complexation of rhodium and ruthenium derivatives by diphosphine end groups.

Other bidentate ligands have been grafted to the surface of phosphorus dendrimers, in particular P,N ligands, which are able to complex PdCl$_2$. These reactions have been applied both to a dendrimer leading to 6a-G$_1$[20], and a dendron leading to 7a-G$_2$[21] (Scheme 7).

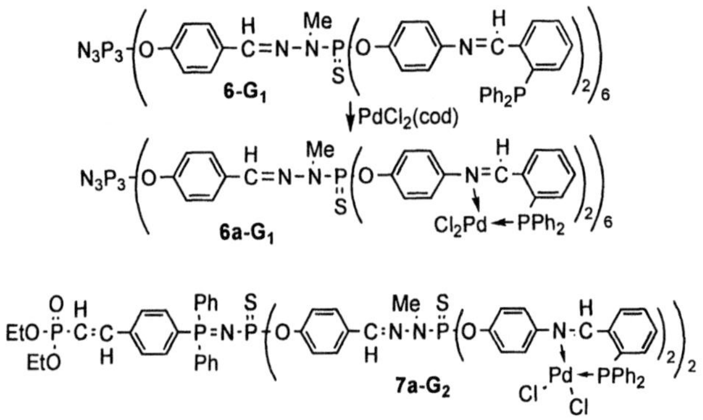

Scheme 7. Complexation of PdCl$_2$ by bidentate end groups.

Metallocene derivatives

Besides phosphino complexes, metallocene derivatives constitute another important way to graft organometallic derivatives on the surface of dendrimers.[22,23] Concerning particularly phosphorus-containing dendrimers, two types of metallocenes have been used: zirconocenes and ferrocenes. The first example concerns the formal [3+2] cycloaddition of 2-phosphino-1-zirconaindene **9** with various aldehydes, for example as when applied to the aldehyde-terminated dendrimers **8-G$_4$** and **8-G$_8$** (Scheme 8). These reactions lead to unusually stable anionic zirconocene complexes on the surface of dendrimers **8a-G$_4$** and **8a-G$_8$**,[24] which constitute the first polyzwitterionic metallo dendrimers.

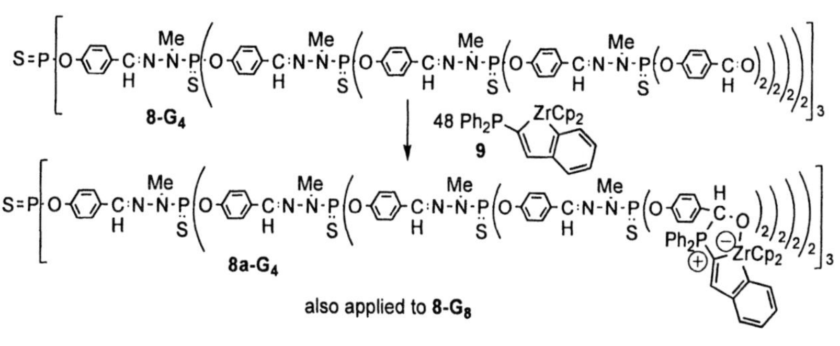

Scheme 8. *Grafting of zirconocene derivatives leading to polyzwitterions.*

Ferrocene derivatives are the most popular metallocenes, mainly due to their high thermal stability and their well-behaved electrochemical properties. They are linked to the surface of phosphorus-containing dendrimers through a phenol linkage, which reacts in basic conditions with P(S)Cl$_2$ end groups. First experiments were carried out with nonchiral ferrocenes, leading to dendrimers **10a-G$_n$** (n ≤ 9).[25] The largest compound in this series (**10a-G$_9$**) possesses theoretically 1536 ferrocenyl end groups (Scheme 9). Among numerous types of end groups, these ferrocenes afford one of the highest thermal stability to phosphorus-containing dendrimers (376 °C for **10a-G$_5$**).[26]

Several types of enantiomerically pure ferrocenyl compounds having a planar chirality were also grafted to the surface of phosphorus-containing dendrimers.[27] In some cases, these ferrocenes bear another function, such as an aldehyde (**10f-G$_n$**) or a phosphine protected by borane (**10c-G$_n$**), which can be deprotected by DABCO to afford **10d-G$_n$** (Scheme 9). All these reactions have been carried out at least up to the ninth generation, and even up to the eleventh generation for **10e-G$_{11}$**. Study of the chiroptical properties within each series shows that the value of the molar rotation divided by the number of chiral end groups versus generation is a constant.

Concerning electrochemical properties, in all cases dendrimers $10a\text{-}f\text{-}G_n$ exhibit a single oxidation wave, corresponding to a multielectronic transfer of equivalent and electrochemically independent ferrocenyl end groups; this fact confirms the absence of steric hindrance, even for high generations. Furthermore, changes in solubility are observed in the oxidation state, leading to stable modified electrodes. Indeed, the multiferroceniums obtained upon exhaustive electrolysis at controlled potential deposit onto the Pt electrode surface, forming a blue conducting film.

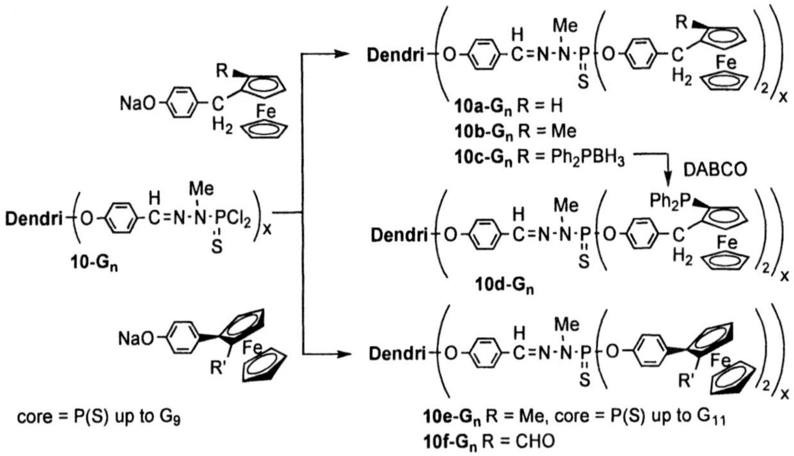

Scheme 9. Grafting of ferrocene derivatives on the surface of dendrimers.

Clusters as end groups. Toward materials.

Dendrimers, and in particular phosphorus-containing dendrimers, have been used in many cases in the field of materials science,[28] either to modify the surface of existing materials, as illustrated above by the modified electrodes, or to create new materials. This latter field can be illustrated by the interaction of organometallic clusters with dendrimers. In a first attempt, the clusters were created on the surface of phosphorus dendrimers, starting from compounds $1e\text{-}G_n$ and $3e\text{-}G_n$ terminated by Au-Cl end groups. These end groups were particularly useful for imaging dendrimers by transmission electron microscopy (TEM),[11] and were able to undergo a reaction of the Au-Cl linkages, as shown by the reaction with Cp_2ZrMe_2 (Scheme 10). Thus, the reaction with the monoanionic complex **11**, and the dianionic complex **12** leads to dendrimers **1g-G_n** and **1h-G_n**, possessing neutral $AuFe_2$ clusters or monoanionic $AuFe_3$ clusters as end groups.[29]

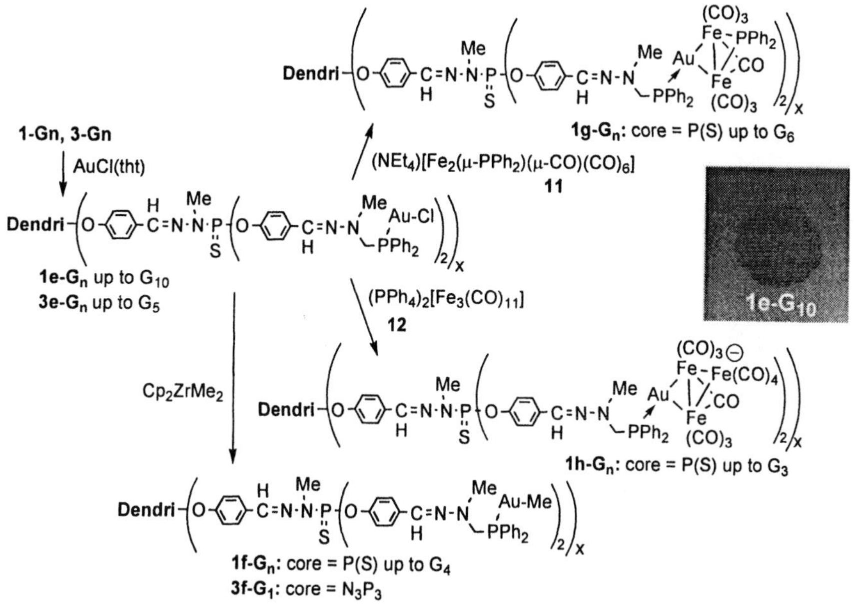

Scheme 10. Reactivity of Au-Cl end groups, and characterization by electron microscopy. Synthesis of Gold-Iron clusters from these Au-Cl end groups.

The second type of interaction between clusters and phosphorus-containing dendrimers was carried out with the thiol terminated dendrimer **11-G_n**, in order to take profit of the known propensity of thiols to form strong bonds with gold. The result of the interaction gives microcrystals (Scheme 11), which were shown by TEM, small- and wide-angle X-ray diffraction (SAXRD, WAXRD) IR spectroscopy, and energy-dispersive X-ray spectroscopy (EDX) analyses to be constituted for the first time of naked Au_{55} clusters, which have lost both all the PPh_3 and Cl ligands. A thin amorphous shell, presumably made of dendrimers, protects the microcrystals and induces their growing, as shown in Scheme 11. These crystals of naked Au_{55} clusters might be candidates for future nanoelectronic devices working with quantum dots.[30]

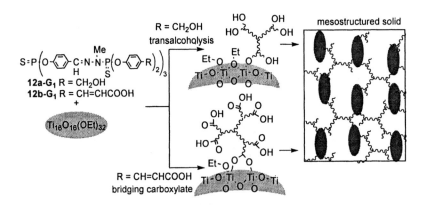

$N_3P_3\left(O-\langle\bigcirc\rangle-\overset{\overset{\displaystyle H}{|}}{C}\overset{\overset{\displaystyle Me}{|}}{\underset{\displaystyle S}{\parallel}}N\cdot N\cdot P\left(O-\langle\bigcirc\rangle-\overset{\overset{\displaystyle H}{|}}{C}\overset{\overset{\displaystyle Me}{|}}{\underset{\displaystyle S}{\parallel}}N\cdot N\cdot P\left(O-\langle\bigcirc\rangle-\overset{\overset{\displaystyle Me}{|}}{C}=N\cdot N\cdot P\left(O-\langle\bigcirc\rangle-\overset{\overset{\displaystyle Me}{|}}{C}=N\cdot N\overset{\overset{\displaystyle H}{|}}{\underset{O}{}}SH\right)_2\right)_2\right)_2\right)_6$

11-G₃

+ Au₅₅(PPh₃)₁₂Cl₆

● = Au₅₅(PPh₃)₁₂Cl₆

= 11-Gn (n = 3, 4, 5)

200 nm

Scheme 11. Synthesis of nanocrystals of naked Au₅₅. TEM image of a monocrystal of Au₅₅, enveloped by an amorphous layer of 11-G₄.

The third experiment concerning clusters led to hybrid organic-inorganic materials, obtained by the assembly of two nanobuilding blocks with well-defined structures, that are dendrimers **12a,b-G₁** and the titanium oxo cluster Ti₁₆O₁₆(OEt)₃₂ (Scheme 12). The hybrid interface is obtained by transalcoholysis with **12a-G₁**, and nucleophilic substitution affording bridging carboxylates with **12b-G₁**. In both cases, ^{17}O MASNMR, FTIR, and X-ray diffraction show that the integrity of the titanium oxo bricks is preserved, and that the dendrimers act like spacers in the array of clusters, to form a long range ordered structure.[31,32]

R = CH₂OH
transalcoholysis

mesostructured solid

$S:P\left(O-\langle\bigcirc\rangle-\overset{\overset{\displaystyle Me}{|}}{C}\overset{\overset{\displaystyle H}{|}}{\underset{\displaystyle S}{\parallel}}N\cdot N\cdot P\left(O-\langle\bigcirc\rangle-R\right)_2\right)_3$

12a-G₁ R = CH₂OH
12b-G₁ R = CH=CHCOOH
+

Ti₁₆O₁₆(OEt)₃₂

R = CH=CHCOOH
bridging carboxylate

Scheme 12. Synthesis of mesostructured solids via the interaction of dendrimers with titanium clusters.

Larger dendrimers **12b-G$_5$** and **12b-G$_7$** were used in sol-gel processes with titanium alkoxydes (Ti(OEt)$_4$, Ti(OiPr)$_4$, Ti(OBu)$_4$), and Ce(OiPr)$_4$. The carboxylic acid functions of the dendrimers act as anchoring sites for the development of the inorganic network. Thermolysis of the hybrid solids at 450°C induces the decomposition of the organic constituent (the dendrimer), and affords sponge-like mesostructured materials.[33]

Catalytic properties.

Metallo dendrimers might be ideal catalysts that combine the advantages of both homogeneous and heterogeneous catalysis, because they are generally easily soluble, and their large size allows an easy recovery. Various phosphorus-containing dendrimers have been used as catalysts,[34] concerning metals linked to the surface of these dendrimers, mainly three types of catalytic reactions have been tested: i) Stille couplings using the palladium derivatives **2b-G$_n$**[35] and **6a-G$_1$**,[20] ii) Knoevenagel condensations using the ruthenium derivatives **2i-G$_n$**,[35] and iii) Michael additions using also **2i-G$_n$**.[35] In most cases the catalysts can be reused at least twice. Despite numerous data, no rule can be deduced concerning the efficiency of the catalysis using dendrimers compared to monomers. As shown in Scheme 13 for various Stille couplings, the efficiency of the conversion using dendrimer **6a-G$_1$** compared to the corresponding monomer is either worse (case a), identical (case b), or better (case c).[20]

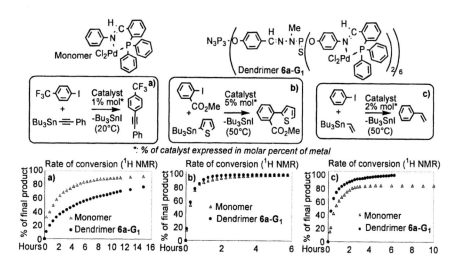

Scheme 13. Various Stille couplings catalyzed by a monomer or dendrimer
***6a-G$_1$**. Rate of conversion measured by ^1H NMR*

Conclusion

Numerous examples concerning the presence of metallic derivatives as end groups of phosphorus-containing dendrimers have already been described. Depending on the type of these end groups, the metal induces particular properties such as an increased thermal stability, various uses in the field of materials science, or in catalysis. Concerning this latter point, work is in progress to develop enantioselective catalyses using metallo phosphorus dendrimers decorated with chiral ligands.[36]

References

1   Fréchet, J. M. J.; Tomalia, D. A. Eds, *Dendrimers and other dendritic polymers*, John Wiley and Sons, Chichester, **2001**.

2   Newkome, G. R.; Moorefield, C. N.; Vögtle, F. *Dendrimers and dendrons. Concepts, syntheses, applications*, Wiley VCH, Weinheim, **2001**.

3   Kreiter, R.; Kleij, A. W.; Gebbink, R. J. M. K.; van Koten, G. *Topics Cur. Chem.* **2001**, *217*, 163.

4   Astruc, D.; Chardac, F. *Chem. Rev.* **2001**, *101*, 2991.

5   Cuadrado, I.; Moran, M.; Casado, C. M.; Alonso, B.; Losada, J. *Coord. Chem. Rev.* **1999**, 193-195, 395.

6   Newkome, G. R.; He, E.; Moorefield, C. N. *Chem. Rev.* **1999**, 99, 1689.

7   Rossell, O.; Seco, M.; Caminade, A. M.; Majoral, J. P. *Gold Bull.* **2001**, 34, 88.

8   van Manen, H. J.; van Veggel, F. C. J. M.; Reinhoudt, D. N. *Topics Cur. Chem.* **2001**, 217, 121.

9   Majoral, J. P.; Caminade, A. M. *Chem. Rev.* **1999**, 99, 845.

10 Caminade, A. M.; Laurent, R.; Chaudret, B.; Majoral, J. P. *Coord. Chem. Rev.* **1998**, 178-180, 793.

11 Slany, M.; Bardají, M.; Casanove, M. J.; Caminade, A. M.; Majoral, J. P.; Chaudret, B. *J. Am. Chem. Soc.* **1995**, 117, 9764.

12 Bardají, M.; Kustos, M.; Caminade, A. M.; Majoral, J. P.; Chaudret, B. *Organometallics* **1997**, 16, 403.

13 Slany, M.; Bardaji, M.; Caminade, A. M.; Chaudret, B.; Majoral, J. P. *Inorg. Chem.* **1997**, 36, 1939.

14 Bardají, M.; Slany, M.; Lartigue, M. L.; Caminade, A. M.; Chaudret, B.; Majoral, J. P. *Main Group Chemistry* **1997**, 2, 133.

15 Slany, M.; Caminade, A. M.; Majoral, J. P. *Tetrahedron Lett.* **1996**, 37, 9053.

16 Lartigue, M. L.; Slany, M.; Caminade, A. M.; Majoral, J. P. *Chem. Eur. J.* **1996**, 2, 1417.

17 Hawker, C. J.; Fréchet, J. M. J. *J. Am. Chem. Soc.* **1992**, *114*, 8405.

18 Maraval, V.; Sebastian, R. M.; Ben, F.; Laurent, R.; Caminade, A. M.; Majoral, J. P. *Eur. J. Inorg. Chem.* **2001**, 1681.

19 Bardají, M.; Caminade, A. M.; Majoral, J. P.; Chaudret, B. *Organometallics* **1997**, 16, 3489.

20 Koprowski, M.; Sebastian, R. M.; Maraval, V.; Zablocka, M.; Cadierno, V.; Donnadieu, B.; Igau, A.; Caminade, A. M.; Majoral, J. P. *Organometallics* **2002**, 21, 4680.

21 Sebastian, R. M.; Griffe, L.; Turrin, C. O.; Donnadieu, B.; Caminade, A. M.; Majoral, J. P. *Eur. J. Inorg. Chem.* **2004**, 2459.

22 Casado, C. M.; Cuadrado, I.; Moran, M.; Alonso, B.; Alonso, B.; Garcia, B.; Gonzales, B.; Losada, J. *Coord. Chem. Rev.* **1999**, 53, 185-186,.

23 Astruc, D.; Blais, J. C.; Cloutet, E.; Djakovitch, L.; Rigaut, S.; Ruiz, J.; Sartor, V.; Valério, C. *Topics Curr. Chem.* **2000**, 210, 229.

24 Cadierno, V.; Igau, A.; Donnadieu, B.; Caminade, A. M.; Majoral, J. P. *Organometallics* **1999**, 18, 1580.

25 Turrin, C. O.; Chiffre, J.; de Montauzon, D.; Daran, J. C.; Caminade, A. M.; Manoury, E.; Balavoine, G.; Majoral, J. P. *Macromolecules* **2000**, 33, 7328.

26 Turrin, C. O.; Maraval, V.; Leclaire, J.; Dantras, E.; Lacabanne, C.; Caminade, A. M.; Majoral, J. P. *Tetrahedron* **2003**, 59, 3965.

27 Turrin, C. O.; Chiffre, J.; Daran, J. C.; de Montauzon, D.; Caminade, A. M.; Manoury, E.; Balavoine, G.; Majoral, J. P. *Tetrahedron* **2001**, 57, 2521.

28 Caminade, A. M.; Majoral, J. P. *Acc. Chem. Res.* **2004**, 37, 341.

29 Benito, M.; Rossell, O.; Seco, M.; Segalés, G.; Maraval, V.; Laurent, R.; Caminade, A. M.; Majoral, J. P. *J. Organomet. Chem.* **2001**, 622, 33.

30 Schmid, G.; Meyer-Zaika, W.; Pugin, R.; Sawitowski, T.; Majoral, J. P.; Caminade, A. M.; Turrin, C. O. *Chem. Eur. J.* **2000**, 6, 1693.

31 Boggiano, M. K.; Soler-Illia, G. J. A. A.; Rozes, L.; Sanchez, C.; Turrin, C. O.; Caminade, A. M.; Majoral, J. P. *Angew. Chem. Int. Ed.* **2000**, 39, 4249.

32 Sanchez, C.; Soler-Illia, G. J. A. A.; Rozes, L.; Caminade, A. M.; Turrin, C. O.; Majoral, J. P. *Mat. Res. Soc. Symp. Proc.* **2000**, 628, CC6.2.1.

33 Bouchara, A,.; Rozes, L.; Soler-Illia, G. J. A. A.; Sanchez, C.; Turrin, C. O.; Caminade, A. M.; Majoral, J. P. *J. Sol-Gel Sci. Tech.* **2003**, 26, 629.

34 Caminade, A. M.; Maraval, V.; Laurent, R.; Majoral, J. P. *Curr. Org. Chem.* **2002**, 6, 739.

35 Maraval, V.; Laurent, R.; Caminade, A. M.; Majoral, J. P. *Organometallics* **2000**, 19, 4025.

36 Majoral, J. P.; Caminade, A. M.; Laurent, R. *Polym. Preprint* **2004**, 45, 364.

Chapter 18

Hyperbranched Polyynes Containing Cobaltcarbonyls as Precursors to Nanostructured Magnetoceramics

Matthias Häußler, Jacky W. Y. Lam, Hongchen Dong, Hui Tong, and Ben Zhong Tang[*]

Department of Chemistry, Institute of Nano Materials and Technology, and Center for Display Research, The Hong Kong University of Science and Technology, Clear Water Bay, Kowloon, Hong Kong, China

Abstract: Facile complexation of the acetylenic triple bonds in the hyperbranched polyynes (hb–PYs) with cobalt carbonyls readily furnishes cobalt-containing organometallic polymers at room temperature. The cobalt-polyyne complexes are highly resistant to thermolysis. Pyrolyses of the polyyne complexes under nitrogen afford soft ferromagnetic ceramics with high magnetizability (M_s up to ~118 emu/g) and low coercivity (H_c down to ~0.045 kOe).

Background

Organometallic polymers are hybrid macromolecules consisting of organic and inorganic species, which often show unique magnetic, electronic, catalytic, sensory, and optical properties. They are also potential candidates as precursors for the fabrication of nanostructured materials and advanced ceramics.[1]

Although a vast variety of linear organometallic macromolecules have been prepared, we are interested in the development of their nonlinear congeners with novel structures and unique properties. Our group has used desalt polycoupling for the synthesis of hyperbranched organometallic poly(ferrocenylenesilyne)s.[2] Upon pyrolyses, the polymers are transformed to mesoporous ceramics, whose iron contents are much higher than those of the ceramics prepared from their linear counterparts of poly(ferrocenylenesilene)s. We studied their properties and found that the ceramics showed excellent soft ferromagnetic behaviors with extremely low hysteresis losses.[2,3]

Acetylenic triple bond is a valuable ligand in organometallic chemistry and readily forms complexes with different types of cobalt carbonyls (Scheme 1).[4] We have recently developed a new protocol for the synthesis of hyperbranched polyynes (*hb*-PY)s through oxidative polycoupling initiated by CuCl (Scheme 2).[5] We envision that the acetylenic moieties of the *hb*-PYs will readily form complexes with transition metals and that pyrolyses of the resultant polymer complexes will transform them into nanostructured magnetic ceramics.[5a,6] In this paper, we report the recent results of our ongoing studies on the syntheses of hyperbranched organometallic polymers and their utilizations as precursors to advanced magnetoceramics.

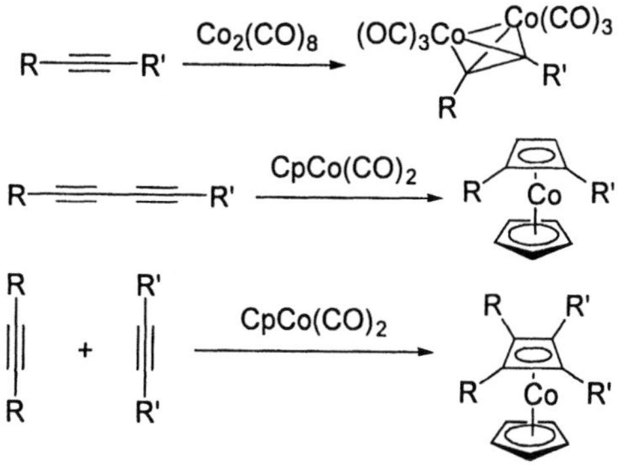

Scheme 1. Complexation Motifs of Cobalt Carbonyls with Acetylenic Triple Bonds.

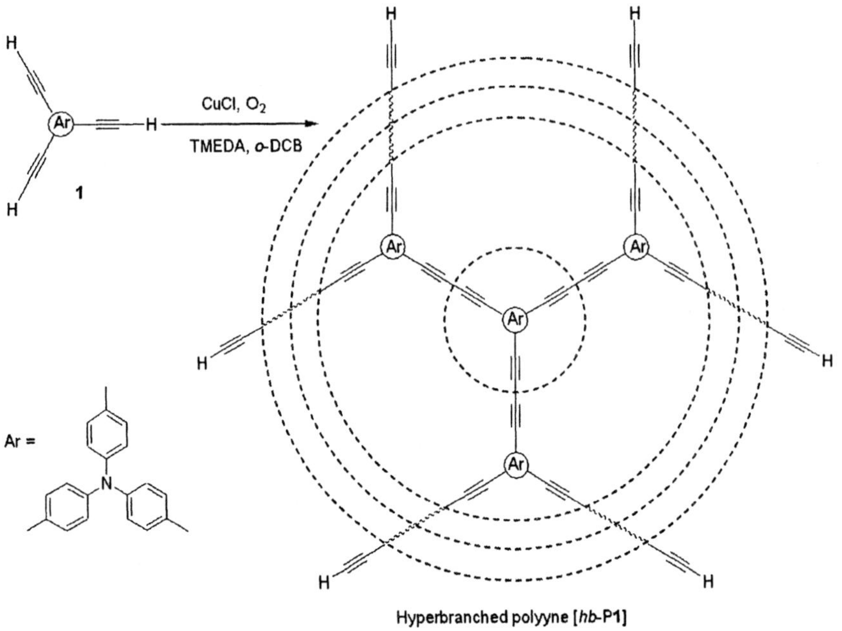

CuCl, O$_2$

TMEDA, o-DCB

1

Ar =

Hyperbranched polyyne [hb-P1]

Scheme 2. Oxidative Polycoupling of Triyne.

Objective

The objective of this investigation is to develop a synthetic route towards high metal-loaded magnetoceramics by using metal-containing hyperbranched polyynes as precursor materials.

Experimental

N,N,N',N'–tetramethylethylenediamine (TMEDA), *o*-dichlorobenzene (*o*-DCB) and copper(I) chloride (CuCl) were purchased from Aldrich and used as received. Octacarbonyldicobalt [Co$_2$(CO)$_8$] and cyclopentadienyldicarbonyl-cobalt [CpCo(CO)$_2$] were purchased from Fluka and Strem, respectively, and stored in a dark, cold, dry place. All other reagents and solvents were purchased from Aldrich and used without further purification.

The oxidative polycoupling of triyne **1** was effected by a mixture of CuCl and TMEDA in *o*-DCB, furnishing hyperbranched polyyne (*hb*–**P1**) in 51% yield with a high molecular weight (M_w = 24100, M_w/M_n = 1.6). Typical procedures for polycoupling of **1** can be found in our previous publications.[5] For the metal complexation reaction, the cobalt complex was dissolved in dry THF and added dropwise into a THF solution of *hb*–**P1**. After 1-h stirring, the THF solvent was evaporated to about half of its original volume under reduced pressure and polymer-metal complex **2** or **3** was obtained by precipitation of the reaction mixture to dry hexane through a cotton filter. The dark yellowish-brown precipitate was washed three times with hexane and dried under vacuum.

Ceramics **4** and **5** were fabricated from polymer precursors **2** and **3**, respectively, by high temperature pyrolysis in a Lindberg tube furnace with a maximum temperature of 1700 °C. In a typical ceramization experiment, 31 mg of **2a** was placed in a porcelain crucible, which was heated to 1000 °C at a heating rate of 10 °C/min under a steam of nitrogen (0.2 L/min). The sample was sintered for 1 h at 1000 °C and black ceramic **4a** was obtained in 52.6% yield (16.3 mg) after cooling. Ceramics **4b** and **5** were prepared by similar fabrication procedures. Yield: 49.9% (**4b**); 64.8% (**5**). Pyrolysis of **2b** for 1 h at 1200 °C under nitrogen yielded a ceramic material in 42.4%.

The X-ray photoelectron spectroscopy (XPS) experiments were conducted on a PHI 5600 spectrometer (Physical Electronics) and the core level spectra were measured using a monochromatic Al K_α X-ray source ($h\nu$ = 1486.6 eV). The analyzer was operated at 23.5 eV pass energy and the analyzed area was 0.8 mm in diameter. The binding energies were referenced to the adventitious hydrocarbon C1*s* line at 285.0 eV and the curve fitting of the XPS spectra was performed using the least square method. The energy-dispersion X-ray (EDX) microanalyses were performed on a JEOL-6300 SEM system with quantitative elemental mapping and line scan capacities operating at an accelerating voltage of 15 kV. The X-ray diffraction (XRD) diagrams were recorded on a Philips PW 1830 powder diffractometer using the monochromatized X-ray beam from a nickel-filtered Cu K_α radiation (λ = 1.5406 Å). Magnetization curves were recorded on a Lake Shore 7037/9509-P vibrating sample magnetometer (VSM) at room temperature.

Results & Discussion

Metallic Complexation

Acetylene triple bond is a versatile ligand frequently used in organometallic chemistry.[7] Examples of acetylene-metal reactions include facile complexations of one triple bond with $Co_2(CO)_8$[8] and of two triple bonds with $CpCo(CO)_2$[9] (cf., Scheme 1). Hyperbranched polymer *hb*-**P1** contains many triple bonds and should be readily metallified through their complexations with cobalt carbonyls. When a mixture of *hb*-P1 and octacarbonyldicobalt with a $[Co_2(CO)_8]/[C≡C]$ ratio of 1:1 is stirred in THF at room temperature, the solution color changes from yellow to brown accompanying CO gas evolution. The mixture remains homogenous towards the end of reaction, and the organometallic hyperbranched polymer **2a** (Scheme 3) is obtained by pouring the THF solution into hexane in 88.6% yield (Table 1).[10] Although **2a** is completely soluble in the reaction mixture, it becomes insoluble after purification, possibly due to the formation of supramolecular aggregates in the precipitation and drying processes or due to slow decarbonylation of the cobalt clusters.[16] Analogous result is obtained when a higher $[Co_2(CO)_8]/[C≡C]$ feed ratio (1.5:1) is used: the reaction solution is homogeneous but the purified product (**2b**) is insoluble. The yield of **2b** is higher (~94%), suggesting a more complete complexation of the metallic species with the triple bond and thus a higher cobalt loading.

Complexes **2** are stable in air,[16] whose IR spectra exhibit strong vibration bands typical of cobalt carbonyl absorptions at 2090, 2055, and 2025 cm^{-1},[8c] verifying the integration of the metallic species into the polyyne structure at the molecular level. Alternatively, *hb*-P1 is metallized through intra- and inter-molecular complexations of its triple bonds with $CpCo(CO)_2$, giving a hyper-branched polyyne complex carrying cyclopentadienylcyclobutadienylcobalt moieties (**3**) in a satisfactory yield (59.5%; Table 1, no. 3).

Elemental analysis is used to estimate the compositions of the metal-polyyne complexes. Complex **2a** contains ~28 wt% of cobalt and its congener **2b** shows a cobalt content as high as ~37 wt% (Table 1). The cobalt content of **3** is similar (35 wt%). This seems odd at the first sight because according to Scheme 1, *one* Co atom of $[CpCo(CO)_2]$ reacts with *two* acetylenic triple bonds, whereas *two* Co atoms of $[Co_2(CO)_8]$ reacts with only *one* triple bond. However, simultaneous to the incorporation of the two cobalt moieties, six CO ligands are introduced into the complex. This is why the apparent Co content of complex **2** is comparable to that of **3**.

Pyrolytic Ceramization

We have previously found that hyperbranched organometallic polymers, in comparison to their linear counterparts, are better precursors to magnetoceramics in terms of ceramization yield and magnetic susceptibility, because the three

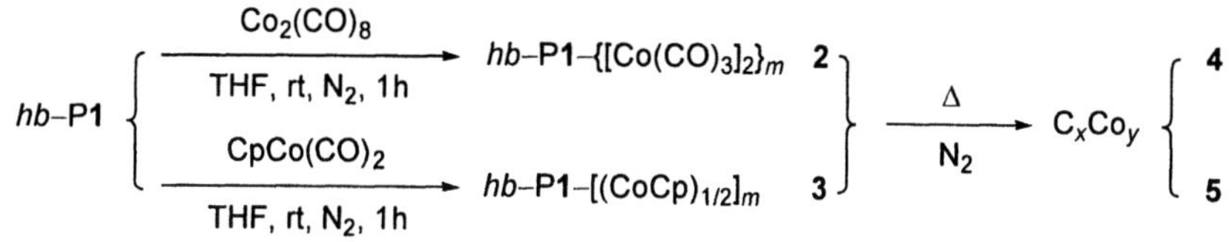

Scheme 3. Complexation with Cobalt Carbonyls and Ceramization to Magnetic Ceramics.

Table 1. Syntheses of Cobalt-Containing Polyynes[a] and Ceramics

no	cobalt carbonyl	[Co]: [C≡C][b]	complex yield[c] (%)	complex composition[d] (%)				ceramic yield[f] (%)
				C	H	N	Co[e]	
1	$Co_2(CO)_8$	1.0:1.0	88.6 (2a)	43.8	3.1	2.5	27.8[g]	52.6 (4a)
2	$Co_2(CO)_8$	1.5:1.0	93.8 (2b)	29.6	2.0	1.7	36.7[g]	49.9[h] (4b)
3	$CpCo(CO)_2$	1.5:1.0	59.5 (3)	58.1	4.2	2.7	35.0	64.8 (5)

[a] Carried out under nitrogen; [*hb*-P1] = 6 mg/mL. [b] Ratio of cobalt complex to acetylene bond in the feed mixture. [c] Calculation based on 100% substitution. [d] Determined by elemental analysis (EA). [e] Calculated by subtracting the contents of C, H and N from 100 wt%. [f] Weight of the residue left after pyrolysis at 1000 °C for 1 h under a steam of nitrogen (flow rate: ~0.2 L/min). [g] Assuming that Co is bonded as $Co_2(CO)_6$ in the complex with a [Co]:[O] ratio of 1:3 (cf., Scheme 1). [h] Weight of the residue after pyrolysis at 1200 °C for 1 h under nitrogen was 42.4%.

dimensional cages of hyperbranched polymers enable better retention of the pyrolytic species and steadier growth of the magnetic crystallites.[2,3] Will the hyperbranched polyyne-cobalt complexes (2 and 3) also serve as precursors to magnetic ceramics? The answer to this question is a firm yes. Ceramization of the complexes in a tube furnace (with a heating rate of 10 °C/min) at 1000 or 1200 °C for 1 h under a steam of nitrogen gives ceramic products 4 and 5 (Scheme 3 and Table 1) in ~42–65% yields. All the ceramics are magnetizable and can be readily attracted to a bar magnet, which prompts us to further investigate their structures and properties.

Structure and Composition

To get a first impression of the resultant metalloceramics, we used scanning electron microscopy (SEM) to examine their surface morphologies. Figure 1 shows a typical SEM image. The ceramic produced at 1000 °C under nitrogen is a compact but rough material, with its surface decorated by numerous nanoparticles or clusters.

Pyrolysis of organic materials under nitrogen leads to not only the formation of carbon but also the evolution of volatiles such as CH_4 and H_2.[11] We thus used elemental and energy-dispersion X-ray (EDX) analyses to determine the atomic compositions. Neglecting traces of oxygen, the cobalt contents of the ceramics 4a, 4b and 5 determined by EA are 100% minus the percentages of C, H and N, which are 15.6, 27.0 and 8.1 atom% (or 47.5, 64.5 and 29.9 wt%), respectively (Table 2). The effect of the CO ligands found by the EA measurements of the organometallic polymer precursors (Table 1) becomes apparent. All of the CO moieties in 2a and 2b are easily removed during the pyrolysis processes, thus yielding ceramics with higher metal loadings. The compositions obtained from EDX are in fair agreements with those from EA.

In order to study the surface and bulk distributions of cobalt in the ceramics, XPS studies are carried out. The XPS analyses reveal cobalt contents of 3.3, 3.8 and 1.3 atom% (or 14.2, 15.9 and 5.7 wt%) for 4a, 4b and 5, respectively, which are much lower compared to the values found indirectly from the EA measurements (Table 2). On the contrary, the carbon compositions rise to values as high as 93.1%. This composition gradient from the bulk to the surface suggests that the ceramization process starts from the formation of the cobalt nanocluster cores, which are covered by the carboneous shells.

The XPS analyses offer further information on the chemical structures of the surfaces of the ceramic products. Figure 2 shows the magnified Co2p and O1s

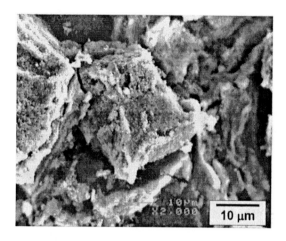

Figure 1. SEM micrographs of ceramic 5 prepared by pyrolytic ceramization of hyperbranched organometallic polymer 3 at 1000 °C under nitrogen.

Table 2. Syntheses of Cobalt-Containing Ceramics

	EA (atom %)				EDX (atom %)				XPS (atom %)			
	C	H	N	Co[a]	C	N	O	Co	C	N	O	Co
4a	84.0	0	0.4	15.6	73.7	4.4	6.8	15.1	89.5	0.1	7.1	3.3
4b	73.0	0	0	27.0	50.2	7.8	17.0	25.0	88.3	0.5	7.4	3.8
5	90.8	0	1.1	8.1	84.8	0	8.1	7.1	93.1	0.8	4.8	1.3

[a] Calculated cobalt content by subtracting the percentages of C, H, and N from 100%.

photoelectron spectra. All ceramics exhibit two major peaks at binding energies of 796.5 and 780.2 eV corresponding to different types of Co_xO_y. Additionally, for **4b** and more pronounced for **5**, a small shoulder is located at 778.2 eV, which is associated with the absorption of metallic cobalt. The appearance of cobalt in the surface layer mainly as a mixed oxide can be further verified by inspecting the O1s photoelectron spectra. All ceramics show a strong peak at a binding energy of 529.9 eV corresponding to cobalt oxides. The variation in the height reflects their cobalt abundances on the ceramic surface.

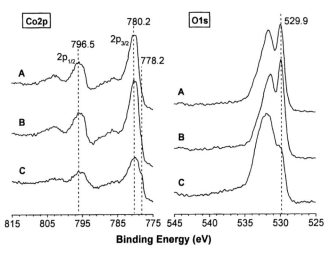

Figure 2. Co2p and O1s photoelectron spectra of ceramics (A) **4a**, *(B)* **4b** *and (C)* **5** *prepared by ceramization of hyperbranched cobalt-polyyne complexes* **2** *and* **3** *at 1000 °C under nitrogen.*

To gain more insights into the bulk structures of the ceramics, their powder X-ray diffraction (XRD) patterns are measured. Figure 3 shows the XRD pattern of polymer **2** and ceramics **4** and **5**. Polymer **2** shows no Bragg reflections and is thus amorphous in nature. Ceramic **4b**, prepared at 1000 °C, however, exhibit three peaks located at 2θ angles of 44.2°, 51.5°, and 75.9°. The reflection peaks can be identified, according to the database files of JCPDS–International Centre for Diffraction Data (ICDD), as reflections of the cobalt metal (ICDD-data file #15-0806). No peaks associated with the presence of Co_xO_y are found, suggesting that those oxygenic species only exist in amorphous state. Using the full width at half-maximum (fwhm) of the reflection peaks, crystallite sizes of ~33.8 nm can be calculated from the Scherrer equation.[2,12,13] It is well known that changes in the pyrolysis conditions directly effect the crystallite growth.[2,3,17] Indeed, raising the sintering temperature from 1000 to 1200 °C intensifies and sharpens the Bragg reflections, suggesting that the crystallites are better packed and bigger in size. The calculated average crystallite size increases to ~43.5 nm. Ceramic **5** exhibits similar Bragg reflections, associated with the formation of cobalt metal. The reflection peaks are, however, broader, indicating that the crystallites in the ceramic are less perfect in packing and smaller in size. This is understandable, considering the lower cobalt-loading of the precursor polymers. This finding is further supported by calculated crystallite size of ~26.3 nm from their fwhm values.

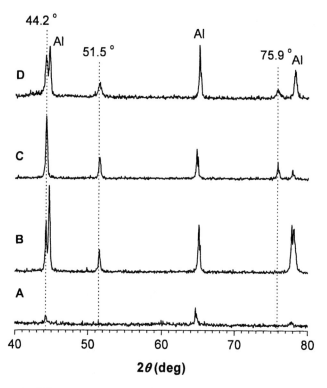

*Figure 3. XRD diffractograms of ceramics (A) **2b**, (B) **4b** (prepared at 1000 °C), (C) **4b** (prepared at 1200 °C) and **5**. The peaks labeled with Al are caused by the aluminum sample holder.*

Magnetic Susceptibility

It now becomes clear that all of the ceramic materials contain nanoscopic cobalt species, which are expected to be magnetically susceptible. Figure 4 shows magnetization curves of the ceramics. With an increase in the strength of externally applied magnetic field, the magnetization of **4b**, swiftly increases and ultimately levels off at a saturation magnetization (M_s) of ~118 emu/g. This value is impressively high, taking into account that the M_s value of maghemite (γ-Fe_2O_3) is 74 emu/g. It is known that cobalt itself is ferromagnetic but its oxides (Co_3O_4 and CoO) are paramagnetic at room temperature.[14] The high M_s value of **4b** thus reveals that only thin layers of paramagnetic cobalt oxides exist on the surfaces of the ceramics and that wrapping by carboneous shells prevents oxidation of the ferromagnetic cobalt nanoclusters in the cores. The hyperbranched architecture, furthermore, endows the materials with a high metal retention ability upon pyrolysis, thus leading to better precursors for the preparation of magnetoceramics. The M_s values of **4a** (~57.8 emu/g) and **5** (~26 emu/g) are lower, which is understandable, because their cobalt contents are lower (Table 2).

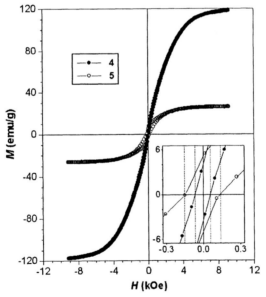

*Figure 4. Plots of magnetization (M) versus applied magnetic field (H) at 300 K for magnetoceramics **4b** (prepared at 1200 °C) and **5**. Insets: enlarged portions of the M–H plots in the low strength region of the applied magnetic field.*

The hysteresis loops of our magnetoceramics are small. From the enlarged $H-M$ plots shown in the insets of Figure 4, the coercivities (H_c) of **4b** and **5** are found to be 0.058 and 0.142 kOe, respectively. An H_c value as low as 0.045 kOe is observed in the magnetization of a ceramic (**4a**) made from the complex of hb–**P1** and $Co_2(CO)_8$ with a $[Co_2(CO)_8]/[C≡C]$ feed ratio of 1:1,[6a] suggesting that the low coercivity is a general property of this family of magnetic ceramics. The high magnetizability (M_s ~118 emu/g) and low coercivity (H_c <0.06 kOe) of **4** make it an excellent soft ferromagnetic material,[14,15] which may find an array of high-tech applications in various electromagnetic systems.

Conclusions

In this work, a group of hyperbranched cobalt-polyyne complexes has been successfully synthesized through facile complexation of acetylenic triple bonds with cobalt carbonyls at room temperature. The hyperbranched organometallic polymers are found to strongly resistant to thermolysis and efficiently carbonized upon pyrolysis (up to ~65 wt%). The resultant ceramic materials consist of bulk metallic cobalt nanoparticles (with sizes up to ~34 nm) wrapped by amorphous carbon shells. Pyrolytic ceramizations of the cobalt-polyyne complexes yield soft ferromagnetic materials with high magnetic susceptibilities (M_s up to ~118 emu/g) and low hysteresis losses (H_c down to ~0.045 kOe).

Acknowledgments

This work was partly supported by the Hong Kong Research Grants Council (603304, 604903, N_HKUST606_03, HKUST6085/02P and 6121/01P) and the University Grants Committee of Hong Kong through an Area of Excellence (AoE) Scheme (AoE/P-10/01-1-A).

References

1   (a) Bill, J.; Aldinger, F. *Adv. Mater.* **1995**, *7*, 775. (b) Interrante, L. V.; Liu, Q.; Rushkin, I.; Shen, Q. *J. Organomet. Chem.* **1996**, *521*, 1. (c) Corriu, R. J. P. *Angew. Chem. Int. Ed.* **2000**, *39*, 1376. (d) Manners, I. *Science* **2001**, *294*, 1664.

2   (a) Sun, Q.; Lam, J. W. Y.; Xu, K.; Xu, H.; Cha, J. A. K.; Zhang, X.; Jing, X.; Wang, F.; Tang, B. Z. *Chem. Mater.* **2000**, *12*, 2617. (b) Sun, Q.; Xu, K.; Peng, H.; Zheng, R.; Häussler, M.; Tang, B. Z. *Macromolecules* **2003**, *36*, 2309.

3   (a) Tang, B. Z.; Xu, K.; Peng, H.; Luo, J.; Zhang, X.; Sun, Q.; Lam, W.
     Y.; Cha, J. A. K. US Patent 6,759,502, 2004. (b) Sun, Q.; Peng, H.; Xu,
     K.; Tang, B. Z. In *Macromolecules Containing Metal- and Metal-like
     Elements*; Abd-El-Aziz, A., Carraher, C., Pittman, C., Sheats, J., Zeldin,
     M., Eds.; Wiley: New York, 2004; Vol. 2, Chapter 2, pp 29–59. (c)
     Häußler, M.; Sun, Q.; Xu, K.; Lam, J. W. Y.; Dong, H.; Tang, B. Z. *J.
     Inorg. Organomet. Polym.* **2005**, *15*, in press.

4   (a) Fritch, J. R.; Vollhardt, K. P. C. *Organometallics* **1982**, *1*, 590. (b)
     Corriu, R. J. P.; Devykler, N.; Guerin, C.; Henner, B.; Jean, A. *J.
     Organomet. Chem.* **1996**, *509*, 249.

5   (a) Häussler, M.; Zheng, R.; Lam, W. Y. J.; Tong, H.; Dong, H.; Tang, B.
     Z. *J. Phys. Chem. B* **2004**, *108*, 10645. (b) Häußler, M.; Lam, J. W. Y.;
     Tong, H.; Zheng R.; Tang, B. Z. *Polym. Mater. Sci. Eng.* **2004**, *90*, 412.
     (c) Häußler, M.; Lam, J. W. Y.; Tong, H.; Zheng R.; Tang, B. Z. *Polym.
     Prepr.* **2004**, *45 (1)*, 895.

6   (a) Häußler, M.; Lam, J. W. Y.; Tang, B. Z. *Polym. Prepr.* **2004**, *45 (1)*,
     448. (b) Häußler, M.; Lam, J. W. Y.; Dong, H.; Tong, H.; Tang, B. Z.
     Polym. Prepr. **2004**, *45 (2)*, 827.

7   For recent reviews, see: (a) Babudri, F.; Farinola, G. M.; Naso, F. *J.
     Mater. Chem.* **2004**, *14*, 11. (b) Long, N. J.; Williams, C. K. *Angew.
     Chem. Int. Ed.* **2003**, *42*, 2586.

8   (a) Newkome, G. R.; He, E. F.; Moorefield, C. N. *Chem. Rev.* **1999**, *99*,
     1689. (b) Chauhan, B. P. S.; Corriu, R. J. P.; Lanneau, G. F.; Priou, C.;
     Auner, N.; Handwerker, H.; Herdtweck, E. *Organometallics* **1995**, *14*,
     1657. (c) Chan, W. Y.; Berenbaum, A.; Clendenning, S. B.; Lough, A. J.;
     Manners, I. *Organometallics* **2003**, *22*, 3796.

9   (a) Nishihara, H.; Kurashina, M.; Murata, M. *Macromol. Symp.* **2003**, *196*,
     27. (b) Altmann, M.; Bunz, U. H. F. *Angew. Chem. Int. Ed.* **1995**, *34*, 569.

10   Hexane is a good solvent of the cobalt complexes. Using hexane as a
     precipitation medium here is to make sure that the unreacted cobalt
     carbonyl, if any, will be completely removed, although the isolation yield
     suffers because some fractions of the reaction product may also be
     dissolved into the hexane solvent.

11   Corriu, R. J. P. *Angew. Chem. Int. Ed.* **2000**, *39*, 1376.

12   (a) Tang, B. Z.; Geng, Y.; Lam, J. W. Y.; Li, B.; Jing, X.; Wang, F.;
     Pakhomov, A. B.; Zhang, X. X. *Chem. Mater.* **1999**, *11*, 1581. (b) Tang,
     B. Z.; Geng, Y.; Sun, Q.; Zhang, X.; Jing, X. *Pure Appl. Chem.* **2000**, *72*,
     157.

13   Rasburn, J.; Petersen, R.; Jahr, T.; Rulkens, R.; Manners, I.; Vancso, G. J.
     Chem. Mater. **1995**, *7*, 871.

14   Goldmann, A. *Handbook of Ferromagnetic Materials*; Kluwer: Boston,
     MA, 1999.

15 A ferromagnetic material with a coercivity smaller than 0.126 kOe (or 10^4 A/m) is termed a soft magnet: Askeland, D. R. *The Science and Engineering of Materials*, 3rd ed.; PWS: Boston, MA, 1994.

16 Since the metal complexation reactions were not performed in complete darkness, decarbonylation reactions due to exposure of the materials to room light cannot be completely ruled out. Further investigations are underway to clarify this issue.

17 Ginzburg, M.; MacLachlan, M. J.; Yang, S. M.; Coombs, N.; Coyle, T. W.; Raju, N. P.; Greedan, J. E.; Herber, R. H.; Ozin, G. A.; Manners, I. *J. Am. Chem. Soc.* **2002**, *124*, 2625.

Chapter 19

Syntheses and Nonlinear Optical Properties of Alkynylruthenium Dendrimers

Mark G. Humphrey[1], Clem E. Powell[1], Marie P. Cifuentes[1], Joseph P. Morrall[1,2], and Marek Samoc[2]

[1]Department of Chemistry and [2]Research School of Physical Sciences and Engineering, Australian National University, Canberra ACT 0200, Australia

Low-generation alkynylruthenium dendrimers with arylethynyl branching and spacer units have been synthesized. A "steric control" methodology to rapidly afford the necessary organometallic dendrons has been developed. The dendrimer complexes possess interesting nonlinear optical properties, including very large two-photon absorption cross-sections, the wavelength dependence of which has been examined in one case. They also undergo reversible oxidation in solution, which results in both linear and nonlinear electrochromism.

Background

The nonlinear optical (NLO) properties of materials have come under increasing scrutiny. This is because interactions of light with materials possessing NLO properties can result in new electromagnetic field components (e.g. with differing phase, frequency, amplitude, polarization, path, etc). NLO materials have potential applications in optical signal processing, switching and frequency generation and may also contribute to optical data storage, optical communication, and image processing.[1] Inorganic salts were the initial focus of attention as second-order NLO materials, after the invention of the laser made it possible to observe these effects, but the potential of organic molecules was soon realized and this became a major area of study. More recently, organometallic complexes have attracted attention, as they possess the advantages of organic molecules (a purely electronic nonlinearity with a time domain of fs, ready processability into films, and ready access to systematically-modified compounds which permits structure-property studies and optimization of responses), together with greater structural diversity and the accessibility of more than one oxidation state for the metal.

The NLO effects from organometallic complexes can be explained by considering the effect of a local electric field E_{loc} acting on a molecule. This can distort the molecular electron density distribution $\rho(r)$, the distortion being described in terms of changes in the dipole moment μ. Changes in the dipole moment induced by a weak field are linear with the magnitude of the field, but when E_{loc} is comparable in strength to the internal electric fields within the molecule, the distortion and the induced dipole moment should be treated as nonlinear functions of the field strength:

$$\mu = \mu_0 + \alpha E_{loc} + \beta E_{loc}E_{loc} + \gamma E_{loc}E_{loc}E_{loc} + \dots \qquad (1)$$

The tensors α, β and γ defined by the above equation are the linear polarizability, the second-order or quadratic polarizability (the first hyperpolarizability) and the third-order or cubic polarizability (the second hyperpolarizability), respectively. Both μ and E_{loc} are vectors, α is a second-rank tensor, β is a third-rank tensor, and γ is a fourth-rank tensor. Many tensor components of α, β, and γ are equivalent by symmetry rules or equal to zero. Polarizabilities are invariant with respect to all point group symmetry operations, so all the components of β vanish in centrosymmetric point groups.

The electric field of a light wave can be expressed as:

$$E(t) = E_0\cos(\omega t)$$

Equation (1) can therefore be written as:

$$\mu(t) = \mu_0 + \alpha E_0 \cos(\omega t) + \beta E_0^2 \cos^2(\omega t) + \gamma E_0^3 \cos^3(\omega t) + \ldots$$

Trigonometric relations such as $\cos^2(\omega t) = 1/2 + 1/2\cos(2\omega t)$ reveal that the nonlinear terms in the dipole moment expansion introduce contributions at different frequencies, the second-order (β) term introducing a time-independent (d.c.) contribution and a term oscillating at 2ω (second-harmonic generation), and providing a frequency mixing phenomenon if the input field is a sum of two components with different frequencies. A constant (d.c.) field may influence an oscillating field if the two are combined in a medium containing second-order nonlinear molecules (the electro-optic effect). The cubic term in Equation (1) leads to several NLO effects, one being oscillation of the induced dipoles at 3ω (third-harmonic generation). The magnitude of these effects depends on the magnitude of the NLO coefficients β and γ, so a major focus of studies has been to prepare materials with large NLO coefficients that are stable to processing and subsequent device operating conditions.

Second-order effects have been the subject of an enormous number of studies, and consequently structure-NLO activity relationships have been developed that enable one to design compounds with optimized quadratic NLO response. While less is known of molecular structure-NLO activity relationships for third-order properties than is known for second-order properties, it has been established with organic compounds that increase in π-delocalization possibilities (e.g. progressing from small molecules to π-conjugated polymers), the introduction of strong donor and acceptor functional groups, controlling chain orientation, packing density, and conformation, and increasing dimensionality can all result in increased cubic nonlinearity. However, with the exception of several studies of organometallic polymers, few large organometallic molecules have been examined, and so structure-cubic NLO activity studies are sparse.

Dendrimers are monodisperse hyperbranched molecules that have attracted significant interest recently as novel materials with possible uses in medical diagnostics, molecular recognition, catalysis, and photoactive device engineering. Although organic dendrimers dominate the field, organometallic dendrimers have been the focus of considerable interest because the metal may imbue the dendritic material with specific optical, electronic, magnetic, catalytic, and other properties. The great majority of organometallic dendrimers are peripherally-metalated organic dendrimers. In contrast, considerably fewer core-metalated and other-shell-metalated dendrimers have been reported.

Organometallic dendrimers with transition metals in every generation are comparatively rare, and rigid, π-delocalizable examples even more so. Examples incorporating 16-electron group 10 metals within an arylalkynyl structure have been recently reported by several groups.[2] Dendrimers of this type are also of interest because of possible applications in nonlinear optics: NLO materials with a dendritic construction may have enhanced nonlinearities coupled to favorable transparency and processing characteristics, because the 1,3,5-trisubstituted benzene branching points in arylalkynyl dendrimers may permit extensive π-delocalization without appreciable red-shift of the important linear optical absorption band(s).

Objective

The discussion above suggests that metal-containing π-delocalizable dendrimers may have enhanced nonlinearities. This study involved development of synthetic procedures to electron-rich (18-valence-electron) bis(diphosphine)ruthenium-containing arylalkynyl dendrimers and investigation of their cubic NLO properties; the latter has revealed very large two-photon absorption cross-sections, the wavelength dependence of which has been examined. As these compounds have fully reversible redox processes, the possibility of "switching" the NLO properties of these complexes using electrochemical stimuli has also been probed.

Experimental

Experimental procedures to and spectroscopic characterization of the complexes $1,3,5-C_6H_3\{4-C\equiv CC_6H_4C\equiv C-trans-[RuX(dppe)_2]\}_3$ (X = C≡CPh (**1**), $4-C\equiv CC_6H_4NO_2$ (**2**), $4-C\equiv CC_6H_4NEt_2$ (**3**)), $1,3,5-C_6H_3(4-C\equiv CC_6H_4C\equiv C-trans-[Ru(dppe)_2]C\equiv C-3,5-C_6H_3\{4-C\equiv CC_6H_4C\equiv C-trans-[RuX(dppe)_2]\}_2)_3$ (X = C≡CPh (**4**), $4-C\equiv CC_6H_4NO_2$ (**5**)),[3] $1,3,5-C_6H_3(4-C\equiv CC_6H_4C\equiv C-3,5-C_6H_3\{C\equiv C-trans-[RuX(dppe)_2]\}_2)_3$ (X = C≡CPh (**6**), $4-C\equiv CC_6H_4NO_2$ (**7**)) (*4,5*) and $1,3,5-C_6H_3\{4-C\equiv CC_6H_4C\equiv C-trans-[RuCl(dppe)_2]\}_3$ (**8**)[3] have been reported in detail elsewhere.

UV/vis spectra were recorded as THF solutions in 1 cm cells using a Cary 5 spectrophotometer. The cyclic voltammetric measurement was recorded using a MacLab 400 interface and MacLab potentiostat from ADInstruments (using Pt disc working, Pt auxiliary and Ag-AgCl reference mini-electrodes from Cypress Systems). The scan rate was 100 mV s⁻¹. The electrochemical solution contained

0.1 M (NnBu$_4$)PF$_6$ and ca. 10^{-3} M complex in CH$_2$Cl$_2$. The solution was purged and maintained under an atmosphere of argon, and referenced to an internal ferrocene/ferrocenium couple (E^0 at 0.56 V). Spectroelectrochemical data were recorded on a Cary 5 spectrophotometer (45 000 - 4 000 cm^{-1}) in CH$_2$Cl$_2$. The solution spectra of the oxidized species at 253 K were obtained by electrogeneration (Thompson 401E potentiostat) at a Pt gauze working electrode within a cryostatted optically transparent thin-layer electrochemical (OTTLE) cell, path-length 0.5 mm, mounted within the spectrophotometer. The electrogeneration potential was ca. 300 mV beyond $E_{1/2}$, to ensure complete electrolysis. The efficiency and reversibility was tested by applying a sufficiently negative potential to regenerate the starting compound; stable isosbestic points were observed in the spectral progression.

Z-scan measurements at 800 nm (12 500 cm^{-1}) were performed using 100 fs pulses from a system consisting of a Coherent Mira Ti-sapphire laser pumped with a Coherent Verdi cw pump and a Ti-sapphire regenerative amplifier pumped with a frequency-doubled Q-switched pulsed Nd:YAG laser (Spectra Physics GCR) at 30 Hz and employing chirped pulse amplification. Tetrahydrofuran solutions were examined in a 0.1 cm path length cell. The closed-aperture and open-aperture Z-scans were recorded at a few concentrations of each compound and the real and imaginary part of the nonlinear phase shift determined by numerical fitting using equations given in reference.[6] The real and imaginary part of the hyperpolarizability of the solute was then calculated by linear regression from the concentration dependencies. The nonlinearities and light intensities were calibrated using measurements of a 1 mm thick silica plate for which the nonlinear refractive index $n_2 = 3 \times 10^{-16}$ cm^2 W^{-1} was assumed.

"Switching" the cubic nonlinearity at 800 nm was performed as follows. An argon-saturated dichloromethane solution containing ca. 0.3 M (NnBu$_4$)PF$_6$ supporting electrolyte was examined in an OTTLE cell (with Pt auxiliary-, Pt working- and Ag-AgCl reference electrodes), path length 0.5 mm, with the 800 nm laser beam passing through a focussing lens and directed along the axis passing through a 1.5 mm diameter hole in the Pt sheet working electrode. The electrochemical cell was mounted on a computer driven translation stage, as usual in Z-scan measurements.[6] The w_0 parameter of the beam (the radius at the $1/e^2$ intensity point) was chosen to be in the range 35 - 45 μm. The Rayleigh length $z_R = \dfrac{\pi w_0^2}{\lambda}$, where w_0 is the Gaussian beam waist and λ is the wavelength, was therefore taken to be $z_R > 3$ mm. A "thin sample" assumption was therefore considered to be justified. In effect, one can then treat the total effect of the third-order nonlinearity of all the components of the system, the solution (solvent and dissolved materials) and the glass walls of the cell, as being an additive quantity. The beam "cropping" by the aperture was also negligible over the range of travel of the cell (z = -3 cm to +3 cm from the focal plane), the beam radius growing by roughly a factor of ten (i.e. to about 350-450 μm) over the distance of ten Rayleigh lengths, but still providing for almost complete transmission through the 1.5 mm aperture. The beam transmitted through the electrochemical cell was split in two, one part being focussed on an "open aperture" detector, the other part being transmitted through a 1 mm aperture to provide the "closed aperture" signal. Z-scans were collected with the electrochemical cell. The electrogeneration potential was 0.8 V to ensure complete electrolysis; this required approx. 5 min. The Z-scan measurements were carried out during the electrolysis and were continued while the electrode potential was cycled from zero to 0.8 V and back to zero again. The real and imaginary parts of the nonlinear phase change were determined for the resting state and electrochemicaly modified molecules in the same way as for our standard Z-scan measurements. For absorbing solutions it was assumed that the absorption saturation process can be modelled by a linear dependence of the absorption coefficient on the light intensity.[7] The nonlinearities and light intensities were again calibrated against silica.

Wavelength dependence of the cubic nonlinearity was assessed by Z-scan measurements performed using two amplified femtosecond laser systems. The first system was based on a Coherent Mira-900D Ti-sapphire oscillator and included a chirped pulse Ti-sapphire amplifier operating at a repetition rate of 30 Hz. This system was used at wavelength 800 nm and provided ca 150 fs FWHM pulses. The second system was a Clark-MXR CPA-2001 regenerative amplifier pumping a Light Conversion TOPAS optical parametric amplifier. This system was operated at a repetition rate of 250 Hz (reduced from the usual default rate of 1 kHz to minimize potential problems with thermal effects and sample photodecomposition). The output of the optical parametric amplifier was tuned in the range 650 nm to 1300 nm using the second harmonic of the signal, the second harmonic of the idler, or the signal, respectively, in three wavelength ranges for tuning the system. The pulse duration was ca 150 fs. The Z-scan set-ups used lenses with the focal lengths suitable for creating focal spots with the $1/e^2$ radius, w_0, being in the range 40 – 65 μm. This resulted in the Rayleigh lengths being greater than 3 mm throughout the wavelength range employed; the measurements of solutions in 1 mm thick glass cells with ca. 1 mm thick glass walls could therefore be always treated in the thin sample approximation. Due to the deviations from Gaussian character of the beam from the OPA, it was necessary to perform some spatial filtering of the beam. This resulted in the beam approximating the truncated Airy pattern case discussed by Rhee et al.[8] All measurements were calibrated against Z-scans taken on the pure solvent and on silica and glass plates of thicknesses in the range 1 - 2.5 mm. It was assumed that the dispersion of the nonlinear refractive index of silica can be neglected in the range of wavelengths investigated, and so the value $n_2 = 3 \times 10^{-16}$ cm^2 W^{-1} was adopted throughout the range. The light intensities used in the three different wavelength ranges differed somewhat, but as a rule the intensities were adjusted to obtain nonlinear phase shifts for the measured samples in the range 0.5 – 1.0 rd, which corresponded to peak intensities of the order of 100 GW cm^{-1}.

Results & Discussion

Our initial attempt to prepare a zero-generation dendrimer utilized 1,3,5-triethynylbenzene, but while chloro(triphenylphosphine)gold(I) reacted cleanly to afford [1,3,5-{(Ph$_3$P)Au}$_3$C$_6$H$_3$], excess *cis*-[RuCl$_2$(dppm)$_2$] (dppm = bis(diphenylphosphino)methane) afforded the bis-adduct [1,3-*trans*-[RuCl(dppm)$_2$]$_2$–5-HC$_2$C$_6$H$_3$] only, structural studies confirming that steric constraints had restricted the extent of reaction.[9] We therefore inserted arylethynyl "spacer" units between the 1,3,5-triethynylbenzene core and the ligated metal units and were successful in obtaining the desired dendrimers,[3] the convergent procedure that we employed being shown in Figure 1.

The lengthy syntheses in Figure 1 are not ideal when the intent is to promulgate physical property investigations, so we developed a more rapid methodology for dendron synthesis. Although an undesired product from an attempt to trimetalate triethynylbenzene, [1,3-*trans*-[RuCl(dppm)$_2$]$_2$–5-HC$_2$C$_6$H$_3$] can also be considered as an archetypal organometallic dendron, with the AB$_2$ 1,3,5-trisubstituted benzene composition required for alkynylruthenium dendrimer construction. Replacing the bidentate phosphine dppm with 1,2-bis(diphenylphosphino)ethane (dppe) has permitted the reaction sequence shown in Figure 2, and thereby rapid access to nanometer-sized π-delocalized peripherally-metalated complexes.[4]

Cubic NLO studies at 800 nm

Linear optical and cubic NLO data for **1-5** are collected in Table 1. Increasing the size of the complexes in proceeding from **1** and **2** to **4** and **5**, respectively, does not reduce optical transparency significantly; the small blue shift observed in proceeding from **1** to **4** may indicate a lack of co-planarity through the dendritic structure of **4**. The significant γ_{imag} values for all complexes are indicative of two-photon absorption, which becomes important as λ_{max} approaches the wavelength corresponding to 2ω (ω = frequency of incident radiation). The γ values and two-photon absorption cross-sections σ_2 of these dendritic complexes are amongst the largest for organometallic complexes thus far (*10,11*).

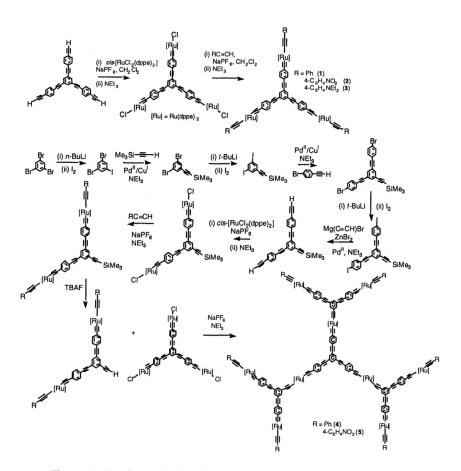

Figure 1. Synthesis of alkynylruthenium dendrimers by a convergent procedure.

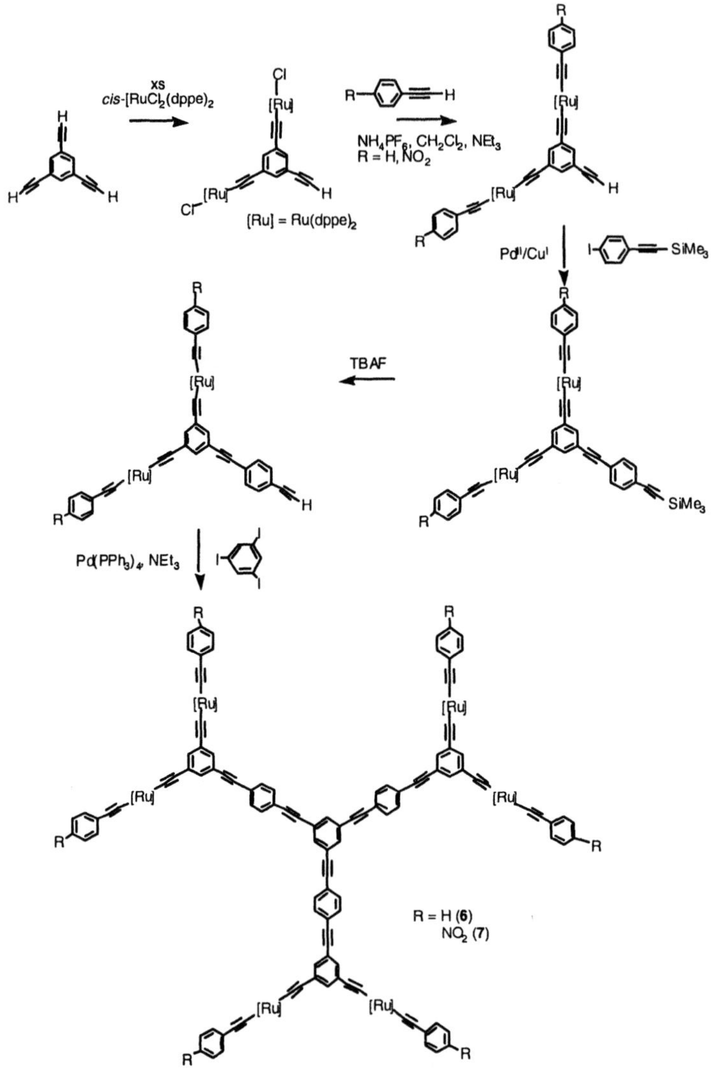

Figure 2. Synthesis of alkynylruthenium dendrimers by a convergent procedure, and employing "steric control" for dendron synthesis.

Table 1. Linear Optical and Cubic Nonlinear Optical Data

Cmpd	λ_{max}, ε [a]	γ_{real} [b]	γ_{imag} [b]	$|\gamma|$ [b]	σ_2 [c]
1	412, 11.6	-600 ± 200	2900 ± 500	3000 ± 500	700 ± 120
2	459, 8.9	-5000 ± 1000	5600 ± 1000	7500 ± 1400	1300 ± 200
3	410, 15.3	-8300 ± 3000	4600 ± 900	9500 ± 3100	1000 ± 200
4	402, 42.1	-5050 ± 500	20100 ± 2000	20700 ± 2100	4800 ± 500
5	467, 16.0	-14900 ± 3000	18200 ± 3000	23500 ± 4200	4400 ± 700

All measurements in THF. All complexes are optically transparent at 800 nm. [a] (nm), (10^4 M^{-1} cm^{-1}). [b] (10^{-36} esu), Z-scan at 800 nm, 100 fs pulses, repetition rate 30 Hz, referenced to the nonlinear refractive index of silica ($n_2 = 3 \times 10^{-16}$ cm^2 W^{-1}). [c] (10^{-50} cm^4 s), calculated using the equation $\sigma_2 = \hbar\omega\beta/2\pi N$, where β is the two-photon absorption coefficient.

Wavelength dependence of cubic NLO properties

The studies of cubic NLO properties summarized above were undertaken at 800 nm. Examination of the literature revealed no extant studies of the dispersion of γ_{real} and γ_{imag} of organometallic (or indeed inorganic) complexes. We therefore extended the scope of our studies by determining the wavelength dependence of the nonlinearity of 7,[5] preliminary results from which are summarized in Figure 3. A striking feature is that the sign of the imaginary part of γ and therefore that of the nonlinear absorption coefficient β is negative in the range ca 700 – 850 nm, corresponding to absorption saturation. Elsewhere, γ_{imag} is positive, corresponding to two-photon absorption. The real part of γ is negative across most of the measurement range, positive values being obtained around 1200 nm. The overall shape of the dispersion is not unexpected, given that the real and imaginary parts of the hyperpolarizability are related through Kramers-Kronig relations. However, the existence of a region of negative β suggested to us that the obtained results should be treated as arising from a superposition of two competing processes: two-photon absorption, involving the two excited states seen in the one-photon absorption, and absorption bleaching within the absorption tail of the first excited state.

We attempted to interpret the dispersion using a sum-over-states approach, assuming that the dispersion of the complex cubic optical nonlinearity can be expressed by equations similar to those presented infor example, Ref. 12, with for the incorporation of damping factors. We limited the dispersion relation to a small number of terms that are close to a resonance under our experimental conditions. For two-photon resonances with two excited states at ω_a and ω_b, the leading dispersion terms are those containing denominators with frequency

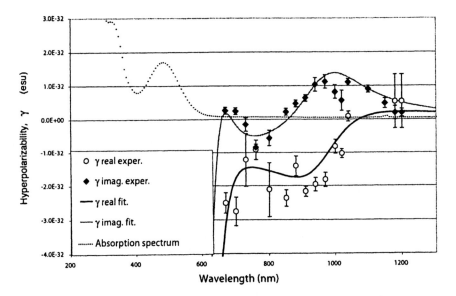

Figure 3. *Experimental and calculated values of γ_real and γ_imag for 7. The absorption spectrum (against air, in arbitrary units) of a dilute solution of the dendrimer in dichloromethane is also shown (the small peak at ca 1150 nm is due to overtone absorption of the solvent). Experimental values of γ were obtained using Z-scan with 150 fs pulses at repetition rate of 250 Hz. (Adapted with permission from Powell, C.E.; Morrall, J.P.; Ward, S.A.; Cifuentes, M.P.; Notaras, E.G.A.; Samoc, M.; Humphrey, M.G. J. Am. Chem. Soc. 2004, 126, 12234).*

differences ω_a-2ω and ω_b-2ω. The wavelength dependence of the absorption bleaching process can be modelled with terms containing single-photon resonances, and therefore ω_a-ω in the denominator. The line-fitting in Figure 3 was derived from the following simplified formula:

$$\gamma = \left[\frac{A}{\nu_a - 2\nu - i\Gamma_1} + \frac{B}{\nu_b - 2\nu - i\Gamma_2} - \frac{C}{\nu_a - \nu - i\Gamma_1} \right] \times \frac{1}{(\nu_a - \nu - i\Gamma_1)^2}$$

where A , B and C were adjustable constants and wave numbers ν were used instead of frequencies. Optimizing the expression to fit the experimental data for the real and imaginary parts of γ resulted in (i) the damping constants Γ_1 and Γ_2 being 2660 and 2211 cm^{-1}, respectively, (ii) the resonance frequencies ν_a and ν_b being 20462 and 31250 cm^{-1}, respectively, corresponding to wavelengths of 489 nm and 320 nm (values close to those for the absorption spectra maxima), and (iii) the constants A, B and C being 5.17 x 10^{-21}, 4.05 x 10^{-21} and 6.78 x 10^{-21} cm^2 statVolt^{-2}, respectively. There is satisfactory agreement of the experimentally observed values of γ_{imag} and γ_{real} with those predicted by the above equation. The γ_{imag} dispersion curve shows the expected two-photon absorption peak at twice the wavelength of the lowest one-photon transition, the second peak arising from resonance at $\nu = \nu_b/2$ being distorted by a negative nonlinear absorption contribution from the absorption bleaching due to the nearby one-photon resonance at $\nu = \nu_a$.

This component of our study has demonstrated that dispersion of cubic nonlinearity in complex molecules such as organometallic dendrimers can be understood in terms of an interplay of two-photon absorption and absorption saturation, and that simple dispersion relations can reproduce the behavior of both the real and imaginary components of the hyperpolarizability.

Switching the NLO properties of alkynylruthenium complexes

Attention has recently moved to examining ways to reversibly modulate ("switch") optical nonlinearities. Procedures that modify λ_{max} or ε may also modify NLO coefficients, possibilities including protonation/deprotonation, oxidation/reduction and photoisomerization sequences. We have successfully examined this idea by protonation/deprotonation,[13] because these alkynyl complexes can reversibly afford vinylidene complexes upon protonation – optical nonlinearities for the alkynyl/vinylidene complex pairs varied by up to a factor of five. However, such chemical transformation remote from the optical bench is a less-than-ideal approach to switching nonlinearity. We recently demonstrated *in situ* switching utilizing an optically-transparent thin-layer electrochemical (OTTLE) cell.[14] Compound **8** (the precursor to **1-3**: see Figure 4) undergoes reversible oxidation in solution, assigned to metal-centered Ru[II/III] processes. The UV-vis-NIR spectroelectrochemistry of **8** reveals the resting state to be essentially transparent below 20 000 cm^{-1} (wavelengths greater than 500 nm), whereas the oxidized form **8^{3+}** has a strong absorption band at 11 200 cm^{-1} (893 nm); oxidation therefore "switches on" a transition with appreciable intensity at the frequency of a Ti-sapphire laser (800 nm). As was summarized above, dendritic compounds such as **8** have significant γ_{imag} values at 800 nm. The sign of the imaginary part of the third-order nonlinearity is positive for two-photon absorption but, under strong one-photon absorption conditions in the oxidized form, absorption saturation is possible, reversing the sign of the nonlinearity, as displayed in Figure 4. At the same time, the sign of the real part of the nonlinearity changes as well.

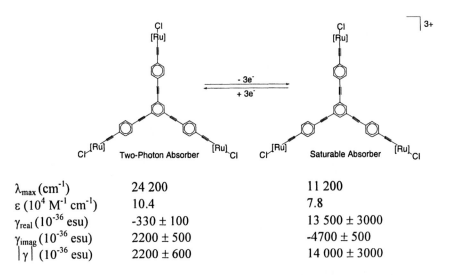

	Two-Photon Absorber	Saturable Absorber
λ_{max} (cm^{-1})	24 200	11 200
ε (10^4 M^{-1} cm^{-1})	10.4	7.8
γ_{real} (10^{-36} esu)	-330 ± 100	13 500 ± 3000
γ_{imag} (10^{-36} esu)	2200 ± 500	-4700 ± 500
$|\gamma|$ (10^{-36} esu)	2200 ± 600	14 000 ± 3000

Figure 4. Reversible electrochemical switching of 8 and its effect on the cubic nonlinear optical properties determined by the Z-scan technique. (Adapted with permission from Powell, C.E.; Cifuentes, M.P.; Humphrey, M.G.; Morrall, J.P.; Samoc, M.; Luther-Davies, B. J. Phys. Chem. A 2003, 107, 11264).

Oxidation therefore results in changes in both imaginary (absorptive) and real (refractive) parts of the nonlinearity. This phenomenon has also been probed by a combination of fs degenerate four-wave mixing (DFWM) and pump-probe experiments on **8** at 12 500 cm^{-1}, which have identified fs time-scale processes as being responsible for this nonlinear electrochromism.[15]

Conclusions

The studies summarized above have embraced the synthesis of low-generation alkynylruthenium dendrimers that incorporate arylethynyl branching and spacer units – these syntheses involved the development of "steric control" to rapidly afford the organometallic dendrons. The resultant dendrimer complexes possess interesting nonlinear optical properties, including very large imaginary components of the cubic nonlinearity and two-photon absorption cross-sections, the wavelength dependence of which has been examined. These complexes also undergo reversible oxidation in solution, which results in both linear and nonlinear electrochromism.

Acknowledgments

MGH thanks the Australian Research Council (ARC) for financial support and an ARC Australian Professorial Fellowship, and Johnson-Matthey Technology Centre for a generous loan of ruthenium salts. MPC thanks the ARC for an ARC Australian Research Fellowship.

References

1. Humphrey, M.G.; Powell, C.E. *Coord. Chem. Rev.* **2004**, *248*, 725.
2. See, for example: (a) Onitsuka, K.; Fujimoto, M.; Ohshiro, N.; Takahashi, S. *Angew. Chem. Int. Ed.* **1999**, *38*, 689. (b) Leininger, S.; Stang, P.J.; Huang, S. *Organometallics* **1998**, *17*, 3981.
3. McDonagh, A.M.; Powell, C.E.; Morrall, J.P.; Cifuentes, M.P.; Humphrey, M.G. *Organometallics* **2003**, *22*, 1402.
4. Hurst, S.K.; Cifuentes, M.P.; Humphrey, M.G. *Organometallics* **2002**, *21*, 2353.
5. Powell, C.E.; Morrall, J.P.; Ward, S.A.; Cifuentes, M.P.; Notaras, E.G.A.; Samoc, M.; Humphrey, M.G. *J. Am. Chem. Soc.* **2004**, *126*, 12234.
6. Sheikh-bahae, M.; Said, A.A.; Wei, T.; Hagan, D.J.; van Stryland, E.W. *IEEE J. Quantum Electr.* **1990**, *26*, 760.
7. Sutherland, R. L. *Handbook of Nonlinear Optics*; Marcel Dekker: New York, 1996; Vol. 52.
8. Rhee, B. K.; Byun, J. S.; van Stryland, E. W. *J. Opt. Soc. Am. B* **1996**, *13*, 2720.
9. Whittall, I.R.; Humphrey, M.G.; Houbrechts, S.; Maes, J.; Persoons, A.; Schmid, S.; Hockless, D.C.R. *J. Organomet. Chem.* **1997**, *544*, 277.

10. McDonagh, A.M.; Humphrey, M.G.; Samoc, M.; Luther-Davies, B.; Houbrechts, S.; Wada, T.; Sasabe, H.; Persoons, A. *J. Am. Chem. Soc.* **1999**, *121*, 1405.

11. McDonagh, A.M.; Humphrey, M.G.; Samoc, M.; Luther-Davies, B. *Organometallics* **1999**, *18*, 5195.

12. Willetts, A.; Rice, J. E.; Burland, D. M.; Shelton, D. P. *J. Chem. Phys.* **1992**, *97*, 7590.

13. Hurst, S.; Cifuentes, M.P.; Morrall, J.P.L.; Lucas, N.T.; Whittall, I.R.; Humphrey, M.G.; Samoc, M.; Luther-Davies, B.; Asselberghs, I.; Persoons, A.; Willis, A.C. *Organometallics* **2001**, *20*, 4664.

14. (a) Cifuentes, M.P.; Powell, C.E.; Humphrey, M.G.; Heath, G.A.; Samoc, M.; Luther-Davies, B. *J. Phys. Chem. A* **2001**, *105*, 9625. (b) Powell, C.E.; Cifuentes, M.P.; Morrall, J.P.; Stranger, R.; Humphrey, M.G.; Samoc, M.; Luther-Davies, B.; Heath, G.A. *J. Am. Chem. Soc.* **2003**, *125*, 602.

15. Powell, C.E.; Cifuentes, M.P.; Humphrey, M.G.; Morrall, J.P.; Samoc, M.; Luther-Davies, B. *J. Phys. Chem. A* **2003**, *107*, 11264.

Organometallic Polymers, Materials, and Nanoparticles

Ferrocene-Containing Systems

Chapter 20

Synthesis, Self-Assembly, and Applications of Polyferrocenylsilane Block Copolymers

Xiaosong Wang, Mitchell A. Winnik[*], and Ian Manners[*]

Department of Chemistry, University of Toronto, 80 St. George Street,
Toronto, Ontario M5S 3H6, Canada

This article reviews recent developments in the synthesis and self-assembly of iron-containing PFS block copolymers with well-defined architectures as well as the exploration of PFS self-assembled nanostructures in material science. An emphasis is placed on the work in our research groups.

Introduction

High molecular weight polyferrocenylsilane (PFS) discovered in the early 1990's represents an interesting class of metal-containing macromolecules.[1] Depending on the substituted groups on skeletal silicon elements, PFS could be semicrystalline or amorphous.[2] In this paper, we refer PFS to semicrystalline dimethyl substituted polyferrocenylsilane unless otherwise indicated. Due to the presence of iron, PFS exhibits a range of intriguing properties that traditional organic polymers do not possess, including redox-activity (Fe(II)/Fe(III)), charge-dissipativity, and an ability to act as a magnetic ceramic precursor.[3]

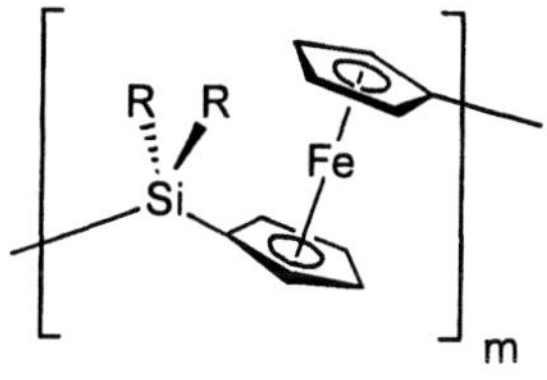

PFS

The incorporation of PFS segments into self-organizing motifs, such as block copolymers, provides further possibilities for supramolecular chemistry and the development of functional nanomaterials.[4]

This article summarizes recent developments in the synthesis and self-assembly of PFS block copolymers as well as their applications in material science.

Synthesis of PFS Block Copolymers

E = H or SiMe$_3$

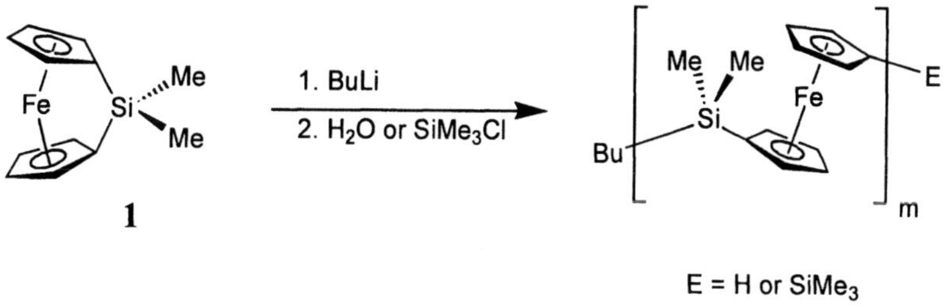

Scheme 1

We reported that silicon-bridged [1]ferrocenophanes (**1**) could undergo living anionic ring-opening polymerization (ROP) at room temperature using initiators such as BuLi (see Scheme 1),[5] permitting the synthesis of PFS with controlled molecular weights and narrow molecular weight distributions (polydispersities < 1.2). The comparative reactivity of PFS anionic end group follows the sequence of PS > PFS > PDMS (PS: polystyrene, PDMS: polydimethylsiloxane).[5] Therefore, by sequential addition of monomers in an order of relative activity, we successfully produced block copolymers with skeletal transition metal atoms. The first "prototypical" block copolymers that we synthesized by consecutive addition of monomers in a one-pot anionic polymerization were organic-organometallic (PS-*b*-PFS) (see Scheme 2) and organometallic-inorganic (PFS-*b*-PDMS) block copolymers (see Scheme 3). Simply extending this method to other monomers allows the synthesis of PI-*b*-PFS[6] (PI: polyisoprene) and PFS-*b*-PMVS[7] (polymethylvinylsiloxane). In principle, this sequential anionic polymerization technique can be applied to integrate a PFS block with any polymers compatible with the anionic mechanism. One exception is a methacrylate type of polymers, because side reactions such as nucleophilic attack of the carbanionic chain ends to the carbonyl groups of methacrylates may occur if the monomers are added directly at the anionic chain end derived from **1**.

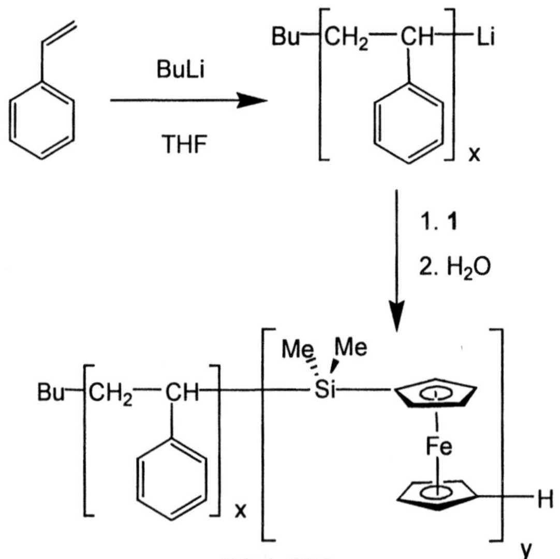

PS-*b*-PFS

Scheme 2

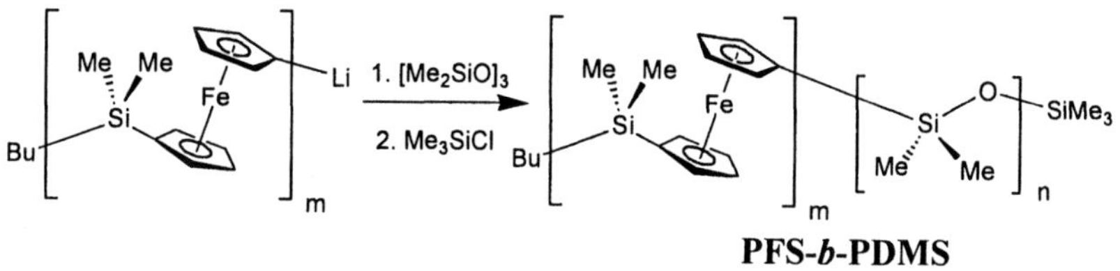

PFS-*b*-PDMS

Scheme 3

Recently, we synthesized the first PFS-methacrylate block copolymer by a two-step anionic polymerization (see Scheme 4).[8] Hydroxy-terminated PFS chains (PFS-OH) were first synthesized using *tert*-butyldimethylsilyloxy-1-propyllithium as an OH-protected initiator. Deprotonation of PFS-OH using potassium hydride gives an oxanionic chain end which initiates the growth of dimethylaminoethylmethacrylate (DMAEMA) as a second block.

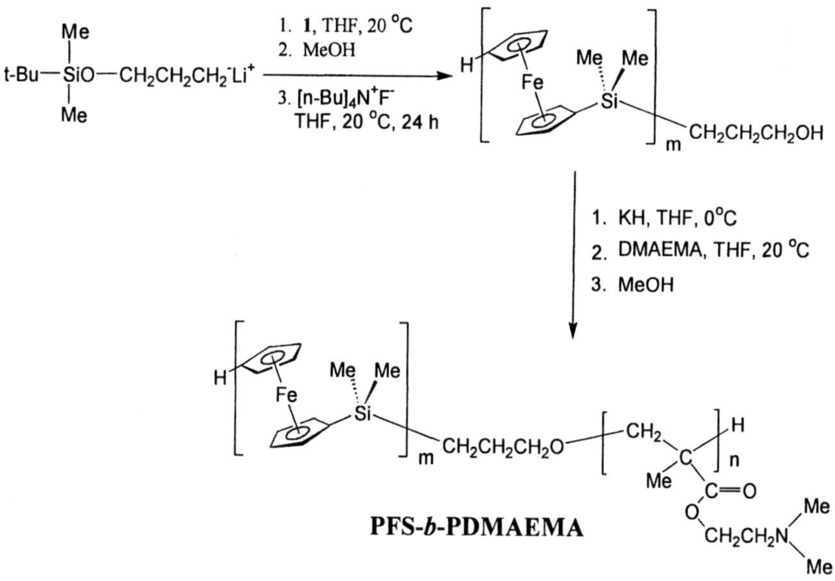

PFS-*b*-PDMAEMA

Scheme 4

Rehahn and coworker also reported a method to synthesize PFS-*b*-PMMA block copolymers (see Scheme 5).[9] They transferred living PFS chain ends to DPE anionic species (DPE: 1,1-diphenylethylene) by sequentially adding 1,1-dimethylsilacyclobutane (DMSB) and DPE at the end of the living anionic chain end derived from **1**, followed by the polymerization of MMA to subsequently grow the second block of PMMA. Under optimum conditions, PFS-*b*-PMMA block copolymers can be obtained in >90% yield (in addition to <10% PFS homopolymer) with polydispersity indices below 1.1.

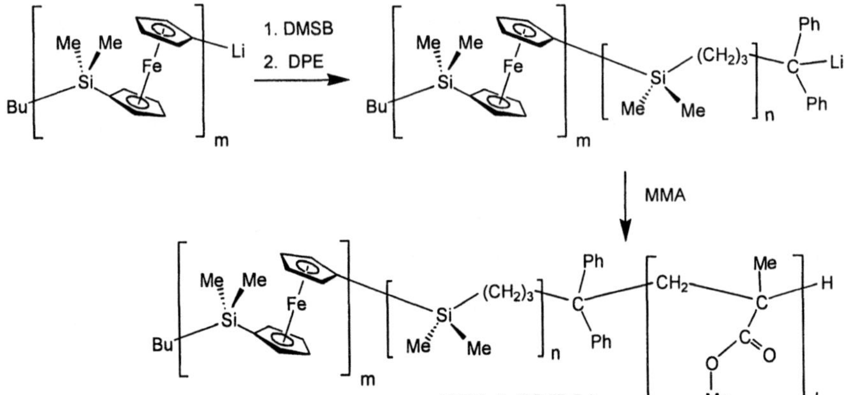

PFS-*b*-PMMA

Scheme 5

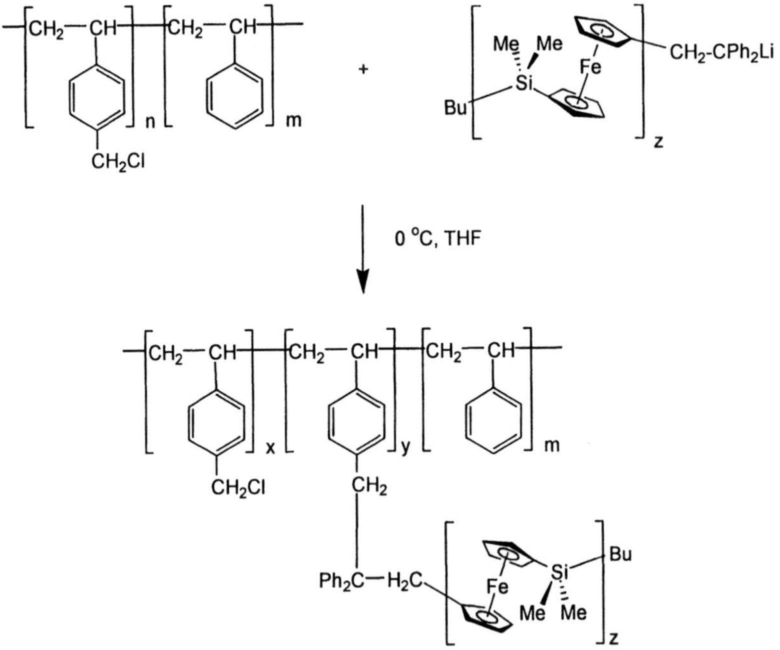

PS-g-PFS

Scheme 6

We observed that, at elevated temperature (50 °C), DPE could relatively effectively cap the living anionic polyferrocenylsilane directly.[10] Therefore, the addition of DMSB to "pump up" the activity of PFS anionic chain ends can be ommitted but the yield of the isolated PFS-*b*-PMMA is lower.[11] The living DPE-capped PFS polymers can also react with chloromethyl functionalities of poly(styrene-co-chloromethylstyrene) (PS-co-PCMS), leading to the first PS-*g*-PFS graft copolymers (see Scheme 6).

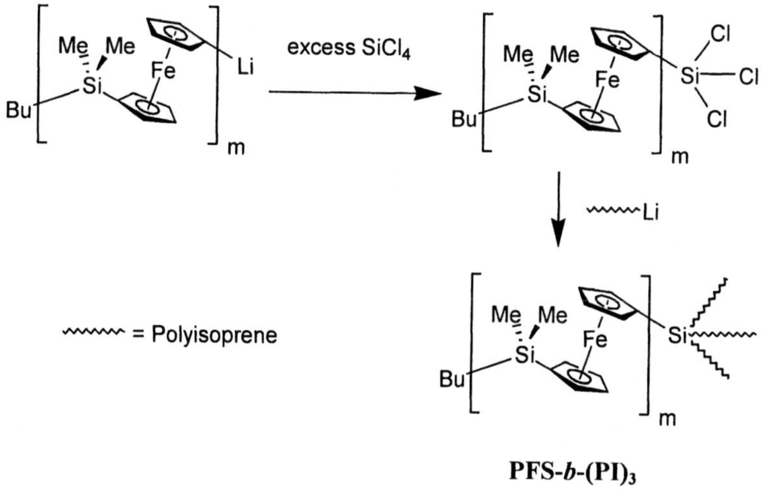

PFS-*b*-(PI)₃

Scheme 7

The first organometallic miktoarm star copolymer, PFS(PI)₃, with PDI of 1.04 was synthesized through an anionic polymerization by using SiCl₄ as the coupling agent as shown in Scheme 7.[12] the well-defined structure was confirmed by the characterizations of GPC and ¹H NMR spectroscopy.

The synthesis of PFS block copolymers by the combination of the ROP of **1** and a living free radical polymerization has also been attempted. PFS-*b*-PMMA was synthesized as shown in Scheme 8 that involved a step of Ru-catalyzed living free radical polymerization.[13] The molecular weights of the block copolymers are well controlled with narrow polydispersities, although the conversion of MMA is incomplete. In view of the versatility of living free radical polymerization in terms of reaction conditions, the choice of monomers and the compatibility with functional groups, further studies in this methodology may provide a route to access a range of novel PFS block copolymers under mild conditions.

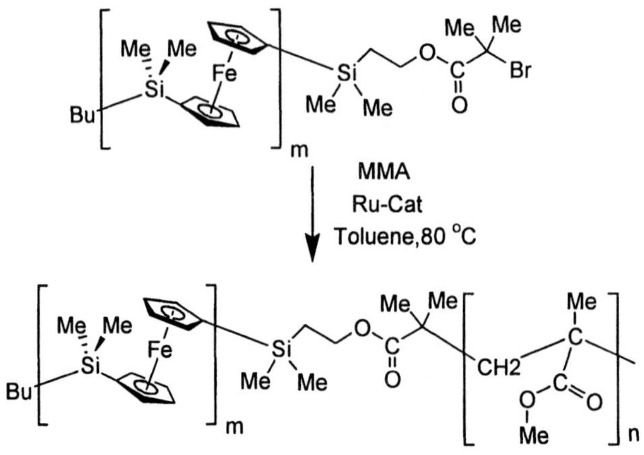

PFS-*b*-PMMA

Scheme 8

In this context, it is worthwhile to note that we recently reported a new photolytic living anionic ROP of **1** which proceeds in the presence of anionic promoter Na[C$_5$H$_5$] via the cleavage of the Fe-Cp bonds in the monomer (see Scheme 9).[14] This method differs fundamentally from the previously reported living anionic ROP in the presence of organolithium initiators as illustrated in Scheme 1 which proceeds in the absence of UV irradiation and involves Si-Cp bond cleavage. The propagating centers for the new photolytic methodology are silyl-substituted cyclopentadienyl anions, which are less basic than iron coordinated Cp anions. This method not only provides a new concept for living anionic polymerization but also offers opportunities to synthesize PFS block copolymers at mild conditions.

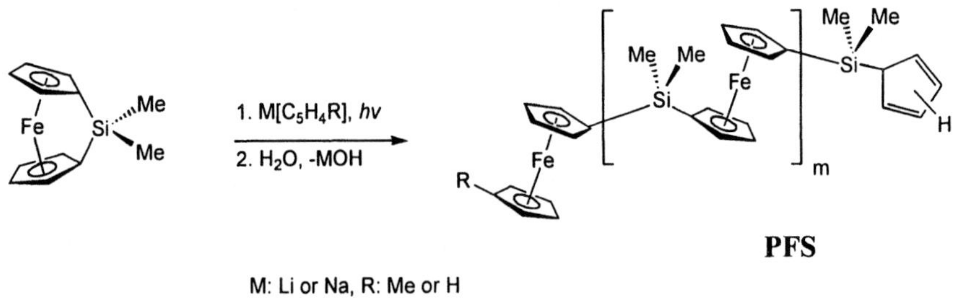

M: Li or Na, R: Me or H

Scheme 9

Solution Self-Assembly of PFS Block Copolymers

Solution self-assembly of PFS block copolymers allows the generation of discrete supramolecular organometallic nanomaterials with a range of one-dimensional morphologies including cylinders,[15] tubes,[16] fibers,[17] tapes.[6] When a bulk polymer sample of PFS_{50}-b-$PDMS_{300}$ (block ratio of PFS:PDMS = 1 : 6) was heated with n-hexane, a selective solvent for PDMS, in a sealed vial at 80°C, the cylinders were solubilized and could be visualized by TEM (see Figure 1a), in which the high electron density of iron-rich PFS blocks provides the contrast in the image.[15] As schematically illustrated in Figure 1b, the cylinders possess an iron-rich, organometallic core of PFS surrounded by an insulating sheath of PDMS.

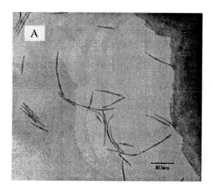
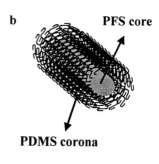

Figure 1. (a) TEM image and (b) scheme for PFS nanocylinders self-assembled from PFS$_{50}$-b-PDMS$_{300}$ (PFS : PDMS = 1 : 6) in hexane. (Reproduced with permission from reference 4.)

Apart from cylinders, PFS-b-PDMS block copolymers can also self-assemble into tubular structures when the block ratios of PFS : PDMS reaches 1 : 12.[16] As shown in Figure 2, PFS_{40}-b-$PDMS_{480}$ (PFS : PDMS = 1 : 12) were self-organized in hexane into nanotubes, in which the PFS blocks aggregate to form a shell with a cavity in the middle of the tube, while the PDMS blocks form the corona (see Figure 2b).[16] Several additional TEM techniques including dark field TEM and energy-filtered TEM also supported the nanotube structures.[18] The presence of the hollow cavities was further confirmed by trapping tetrabutyllead in the voids and performing energy-dispersive X-ray measurements on the resulting structure.[16]

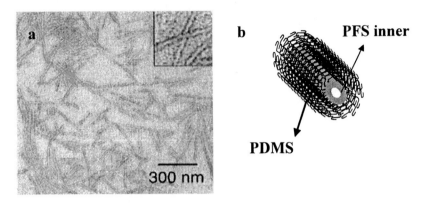

Figure 2. (a) TEM image and (b) scheme for PFS nanotubes self-assembled from PFS$_{40}$-b-PDMS$_{480}$ (PFS : PDMS = 1: 12) in hexane. (Reproduced with permission from reference 16.)

The essential role of block ratio in determining the micellar morphologies of PFS block copolymers was also discerned in our studies on PFS-PI system.[6] PI$_{30}$-b-PFS$_{60}$ and PI$_{320}$-b-PFS$_{53}$ have similar PFS length but differ significantly in PI length, with block ratio (PI : PFS) of 1:2 and 6:1 respectively. As shown in Figure 3, the block copolymers of PI$_{320}$-b-PFS$_{53}$ with longer PI chains, cylinders were observed (see Figure 3a), while the tape-like micelles as shown in Figure 3b self-assembled from PI$_{30}$-b-PFS$_{60}$.

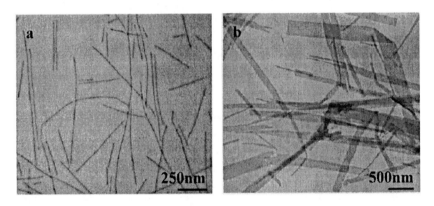

Figure 3. TEM images for the micelles self-assembled from (a) PI$_{320}$-b-PFS$_{53}$ in THF/Hexane (2/8, v/v) and (b) PI$_{30}$-b-PFS$_{60}$ THF/Hexane (3/7, v/v). (Reproduced with permission from reference 6.)

Under certain circumstances, spherical aggregates were also found. For example, when PFS tube-like micelles were heated above the T$_m$ of PFS blocks (ca. 120-145°C)[16] or an amorphous poly(ferrocenylmethylphenylsilane) block was used to replace semicrystalline PFS in the block copolymers,[19] spheres became the only morphology observed in TEM characterizations. These experiments suggested that the crystallization of PFS chains could be a possible reason for the formation of observed one-dimensional supramolecular assemblies. PFS crystallization diffraction was indeed detected in either PFS nanocylinders[19] or nanotubes[16] by WAXD experiments on the micelles solid samples dried from the corresponding solutions.

One potential application of these well-defined aggregates may be of use as etching resists for semiconducting substrates, such as GaAs or Si, and offer potential access to magnetic or semiconducting nanoscopic patterns on various substrates.[20] In a collaboration with J. Spatz and M. Moller at the University of Ulm, Germany, the PFS cylinders have been positioned on the surface of a GaAs resist by capillary forces along grooves, which were previously formed from electron beam etching of the surface. Subsequent reactive plasma etching generated connected ceramic lines of reduced size (see Figure 4).[21]

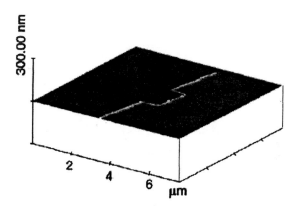

Figure 4. Scanning force micrograph for the ceramic line dervied from aligned PFS-b-PDMS cylindrical micelles by H_2 plasma eatching. (Reproduced with permission from reference 21.)

Despite the success in the fabricatin of nanoceramic lines by plasma etching of aligned micellar precursors, our further exploitations of PFS nanostructures were often frustrated by the labile nature of the self-assembled micelles that easily undergo dissociation in a common solvent or shape deformation at an elevated temperature.

Shell-Crosslinked Nanocylinders and Nanotubes

Several groups have demonstrated that the stability of self-assembled micelles can be improved by performing a shell-crosslinking reaction.[22] The synthesis of PFS-*b*-PMVS and PI-*b*-PFS block copolymeris through the sequential anionic polymerization, followed by the self-assembly in hexane allowed us to access either nanocylinders or nanotubes with vinyl groups dangling from the corona chains. By taking advantage of pendent vinyl groups, we performed a Pt-catalyzed hydrosilylation crosslinking reaction with tetramethyldisiloxane as a crosslinker, leading to shell-crosslinked nanocylinders and nanotubes.[23, 7]

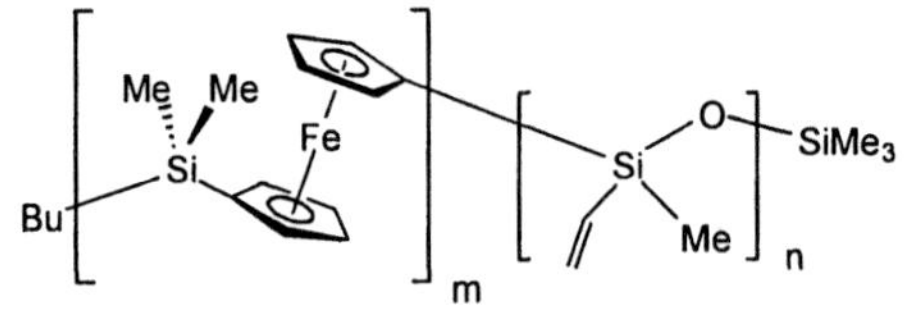

PFS-*b*-PMVS

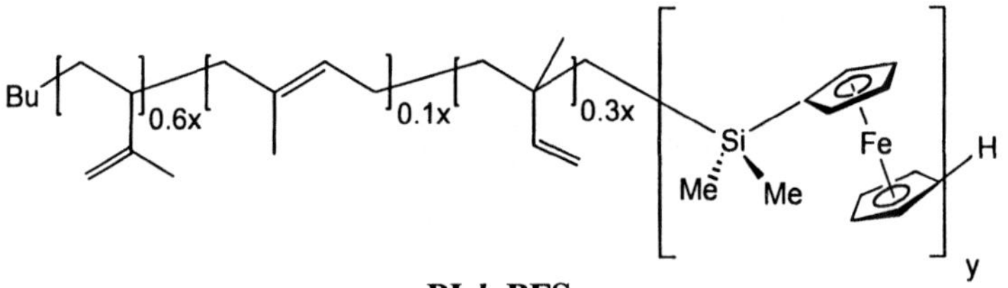

PI-*b*-PFS

As a result of shell-crosslinking, the one-dimensional micellar structures were locked-in and preserved even if transferred from hexane to a common solvent.[23, 7] Figure 5 illustrates a TEM image for PI_{320}-*b*-PFS_{53} shell-crosslinked micelles. The TEM sample was prepared from a micellar solution in THF, a common solvent for both PFS and PI blocks. We have further proved that the shell-crosslinking reaction is an important step in making use of PFS nanostrucutres.

We pyrolyzed shell-crosslinked nanocylinders of PI_{320}-*b*-PFS_{53} upon heating up to 600 °C with a temperature ramp of 1 °C / min under N_2. The resulting ceramics with excellent shape retention were characterized by TEM (see Figure 6). In a control experiment, pyrolysis of uncrosslinked PI_{320}-*b*-PFS_{53} cylindrical micelles was found to lead to the destruction of the structure. This comparison indicated that the shell-crosslinking plays an essential role in the shape retention

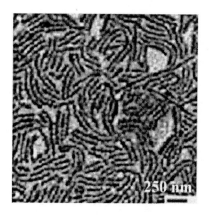

Figure 5. TEM image for shell-crosslinked PI₃₂₀-b-PFS₅₃ cylindrical micelle solid samples prepared by drying a drop of THF solution on a carbon-coated copper grid. (Reproduced with permission from reference 23.)

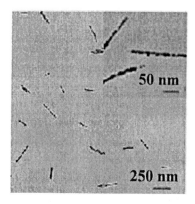

Figure 6. TEM image for cylinderical nanoceramic replica derivied from PI₃₂₀-b-PFS₅₃ shell-crosslinked micelles through a pyrolysis process under N₂ with a temperature ramp of 1°C / min up to 600 °C. (Reproduced with permission from reference 23.)

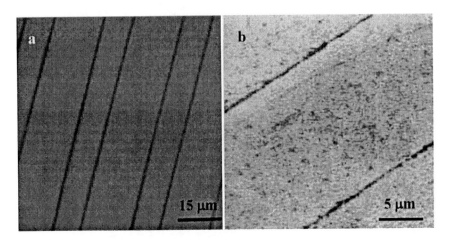

Figure 7. (a) Optical and (b) SEM images for microfluidically aligned shell-crosslinked PI$_{320}$-b-PFS$_{53}$ nanocylinders. (Reproduced with permission from reference 23.)

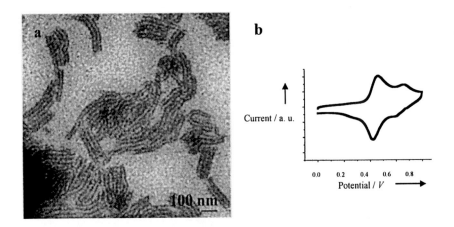

Figure 8. Shell-crosslinked nanotubes of PFS$_{48}$-b-PMVS$_{300}$ block copolymers (a) TEM image for the sample prepared for THF solution. (b) cyclic voltammetery in dichloromethane : benzonitrile (2:1) with 0.1 M [Bu$_4$N][PF$_6$] as supporting electrolyte. (Reproduced with permission from reference 7.)

and permits the formation of a ceramic replica, presumably due to high ceramic yield resulted from crosslinked structure.

We also aligned and patterned shell-crosslinked micelles on a flat silicon substrate by a microfluidic technique (see Figure 7). Ordered magnetic nanoceramic arrays could be subsequently generated by a pyrolysis process, which may be of great interesting as magnetic memory materials or as catalysts for the growth of carbon nanotubes.

As for shell-crosslinked nanotubes (see Figure 8a), we proved that a redox-activity due to PFS chains is associated with the structures by performing cyclic voltammetry experiments (see Figure 8b).[7] This new type of nanotubes provides a chance to encapsulate guest compounds or particles in the cavities through an in-situ redox reaction using PFS as a reductant.

Self-Assembly of PFS Block Copolymers in Solid State

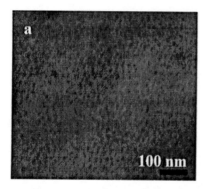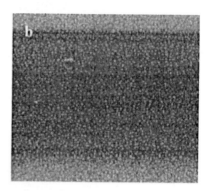

Figure 9. (a) TEM image for PS_{374}-b-PFS_{45} thin film microtomed parallel to the substrate, (b) AFM image (2 μm × 2 μm) of pyrolyzed sample (600 °C, 2 h, N_2) of a PS_{374}-b-PFS_{45} thin film in which the PS matrix has been crosslinked by UV radiation. (Reproduced with permission from reference 24.)

Polyferrocenylsilane block copolymers also phase separate in solid state to generate periodic, nanoscopic iron-rich domains that can be observed by TEM without resorting to staining techniques.[5] Studies in this area are of growing interest.[4] As shown in Figure 9a, perpendicular ordering of PFS microdomains traversing the film thickness was observed throughout the sectioned sample from a PS_{374}-b-PFS_{45} thin film. This thin film was pyrolyzed above 600 °C, leading to iron-rich ceramic nanodomain replicas as shown in Figure 9b.[24] In a collaboration with Ajayan and Ryu at Rensselaer Polytechnic Institue, we have used the iron-rich ceramics as catalysts for the growth of single-walled carbon nanotubes (SWCN) on the substrate by a simple one step chemical vapor deposition (CVD). The resulting SWCN lying on the surface was characterized by SEM (Figure 10).[25] Multiwalled CN have also been reported by Hinderling et al using a similar procedure but involving prior plasma treatment.[26]

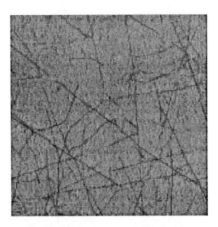

Figure 10. AFM image for single-walled carbon nanotubes (SWCN) synthesized by chemical vapor deposition (CVD) with iron-rich ceramic particles derived from a PS-b-PFS thin film as catalyst. (Reproduced with permission from reference 25.)

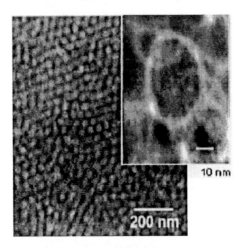

Figure 11. TEM image for a cryomicrotomed PFS_{90}-b-$PDMS_{900}$ film cast from toluene. Inset: High-resolution STEM image of this sample obtained in the dark field mode. (Reproduced with permission from reference 28.)

Most studies[27] indicated that the self-assembly of PFS block copolymers in bulk state follows the theory that the longer block forms the continuous phase, whereas the shorter block forms the imbedded structures. In a recent study, we discovered an astonishing exception.[28]

As illustrated in Figure 11, the morphology of PFS_{90}-b-$PDMS_{900}$ film casting from toluene shows somewhat a disordered hexagonal packed objects that are formed by the PDMS blocks (white areas), while the continuous phase appears gray (Figure 11). Scanning TEM (STEM) in the dark field mode to view an image of the electrons that are elastically scattered by the PFS domains also confirmed this morphology. As shown in the inset of Figure 11, we observed the presence of PFS rings (white) in the midst of the relatively electron-poor PDMS domains (black and gray).

The small-angle X-ray scattering (SAXS) pattern of the film using a Cu Kα source shows broad peaks with relative positions of $3^{1/2}$, $7^{1/2}$, and $21^{1/2}$, suggesting a hexagonally packed cylinders.

Therefore, it was confirmed that an asymmetric PFS_{90}-b-$PDMS_{900}$ copolymer undergoes phase segregation where the longer block forms isolated cylinders surrounded by a shell of PFS, with the remaining PDMS filling the interstitial spaces.

Summary

Living anionic ROP of **1** represents an unprecedented breakthrough in the controlled synthesis of metal-containing polymers, which allows the access to a range of well-defined block copolymers with predicted molecular weights, narrow polydispersities and designed architectures.

Self-assembly of PFS block copolymers either in solution or in bulk generated a number of interesting nanostructures with a iron-rich, redox-active PFS domains. These nanomaterials have been proved useful for various applications including the fabrication of nanoceramics and precessable precursors of catalysts for carbon nanotube growth.

References

1. Foucher, D. A.; Tang, B. Z.; Manners, I. *J. Am. Chem. Soc.* **1992**, *114*, 6246-6248.
2. Rulkens, R.; Lough, A. J.; Manners, I.; Lovelace, S. R.; Grant, C.; Geiger, W. E. *J. Am. Chem. Soc.* **1996**, *118*, 12683-12695.
3. a) Manners, I. *Science* **2001**, *294*, 1664-1666. b) Kulbaba, K.; Manners, I. *Macromol. Rapid Commun.* **2001**, *22*, 711-724.
4. a) Massey, J.; Power, K.N.; Winnik, M.A.; Manners, I. *Adv. Mater.* **1998**, *10*, 1559-1562. b) Lammertink, R. G. H.; Hempenius, M. A.; Thomas, E. L.; Vancso, G. J. *J. Poly. Sci. B-Poly. Phys.* **1999**, 37, 1009-1021. c) Li, W.; Sheller, N.; Foster, M. D. Balaishis, D.; Manners, I.; Annis, B.; Lin, J. S. *Polymer* **2000**, *41*, 719-724. d) Eitouni, H. B.; Balsara, N. P. *J. Am. Chem. Soc.* **2004**, 126, 7446-7447.
5. Ni, Y.; Rulkens, R.; Manners, I. *J. Am. Chem. Soc.* **1996**, *118*, 4102-4114.
6. Cao, L.; Manners, I; Winnik, M. A. *Macromolecules* **2002**, 35, 8258-8260.
7. Wang, X. S.; Winnik, M. A.; Manners, I. *Angew. Chem. Int. Ed.* **2004**, *43*, 3703-3707.
8. Wang, X. S.; Winnik, M. A.; Manners, I. *Macromol. Rapid Commun.* **2002**, *23*, 210-213.
9. Kloninger, C; Rehahn M. *Macromolecules* **2004**, 37, 1720-1727.
10. Power-Billard, K. N.; Wieland, P.; Schafer, M.; Nuyken, O.; Manners, I. *Macromolecules* **2004**, 37, 2090-2095.
11. Vanderark, L.; Manners, I. unpublished results.
12. Wang, X. S.; Winnik, M. A.; Manners, I. *Macromol. Rapid Commun.* **2003**, *24*, 403-407.
13. Korczagin, I.; Hempenius, M. A.; Vancso, G. J. *Macromolecules* **2004**, 37, 1686-1690.
14. Tanabe, M.; Manners, I. *J. Am. Chem. Soc.* **2004**, *126*, 11434-11435.
15. Massey, J. A.; Power, K. N.; Manners, I.; Winnik, M. A. *J. Am. Chem. Soc.* **1998**, *120*, 9533-9540.
16. Raez, J.; Manners, I.; Winnik, M. A. *J. Am. Chem. Soc.* **2002**, *124*, 10381-10395.
17. Raez, J.; Manners, I.; Winnik, M. A. *Langmuir* **2002**, *18* 7229-7239.
18. Frankowski, D. J.; Raez, J.; Manners, I.; Winnik, M. A.; Khan, S. A.; Spontak, R. J. *Langmuir* **2004**, *20*, 9304-9314.
19. Massey, J. A.; Temple, K.; Cao, L.; Rharbi, Y.; Raez, J.; Winnik, M. A.; Manners, I. *J. Am. Chem. Soc.* **2000**, 122, 11577-11584.
20. Cao L.; Massey, J. A.; Winnik, M. A.; Manners, I.; Riethmuller, S.; Banhart F.; Spatz, J. P.; Möller, M. *Adv. Funct. Mater.* **2003**, *13*, 271-276.
21. Massey, J. A.; Winnik, M. A.; Manners, I.; Chan, V. Z-H.; Ostermann, J. M.; Enchelmaier, R.; Spatz, J. P.; Möller, M. *J. Am. Chem. Soc.* **2001**, *123*, 3147-3148.

22. (a) Thurmond, K. B.; Kowalewski, T.; Wooley, K. L. *J. Am. Chem. Soc.* **1996**, 118, 7239-7240. (b) Thurmond, K. B.; Kowalewski, T.; Wooley, K. L. *J. Am. Chem. Soc.* **1997**, 119, 6656-6665. (c) Ding, J.; Liu, G. *Macromolecules* **1998**, 31, 6554-6558. (d) Bütün, V.; Lowe, A. B.; Billingham, N. C.; Armes, S. P. *J. Am. Chem. Soc.* **1999**, 121, 4288-4289.

23. Wang, X. S.; Arsenault, A.; Ozin, G. A.; Winnik, M. A.; Manners, I. *J. Am. Chem. Soc.* **2003**, *125,* 12686-12687.

24. Temple, K.; Kulbaba, K.; Power-Billard, K. N.; Manners, I.; Leach, K. A.; Xu, T.; Russell, T. P.; Hawker, C. J. *Adv. Mater.* **2003**, *15,* 297-300.

25. Lastella, S.; Jung, Y. J.; Yang, H., Vajtai, R.; Ajayan, P. M.; Ryu, C. Y.; Rider, D. A.; Manners, I. *J. Mater. Chem.* **2004**, 14, 1791-1794.

26. Hinderling, C.; Keles, Y.; Stockli, T.; Knapp, H. E.; de los Arcos, T.; Oelhafen, P.; Korczagin, I.; Hempenius, M. A.; Vancso, G. J.; Pugin, R. L.; Heinzelmann, H. *Adv. Mater.* **2004**, 16, 876-879.

27. Lammertink, R. G. H.; Hempenius, M. A.; Vancso, G. J.; Shin, K.; Rafailovich, M. H.; Sokolov, J. *Macromolecules* **2001**, 34, 942-950.

28. Raez, J.; Zhang, Y.; Cao, L.; Petrov, S.; Görgl, R.; Wiesner, U.; Manners, I.; Winnik, M. A. *J. Am. Chem. Soc.* **2003**, *125,* 6010-6010.

Chapter 21

Carbanion-Pump Mediated Synthesis and Bulk Morphologies of Ferrocenyldimethylsilane–Methyl Methacrylate Diblock Copolymers

Christian Kloninger and Matthias Rehahn[*]

Ernst-Berl-Institute for Chemical Engineering and Macromolecular Science, Darmstadt University of Technology, Petersenstrasse 22, D–64287 Darmstadt, Germany

Block copolymers from [1]dimethylsilaferrocenophane (FS) and methyl methacrylate (MMA) were synthesized via sequential anionic polymerization. The key step in this synthesis is the 1,1-dimethylsilacyclobutane-mediated endcapping of the living poly(1,1'-ferrocenyldimethylsilane) precursor. It allows efficient grafting of MMA and thus the formation of well-defined diblock copolymers having molecular weights of around 100.000 $g \cdot mol^{-1}$ and polydispersities below 1.1. Investigation of the bulk microphase behavior using transmission electron microscopy shows strong phase-segregation of the copolymer segments. Depending on their overall composition, various microphase morphologies are observed, including a bicontinuous gyroidic.

Background

Block copolymers can be used as thermoplastic elastomers,[1,2] to turn brittle polymers into high-impact materials[3] or as phase compatibilizers,[4-6] turning incompatible polymer mixtures into blends with improved properties.[7] In addition to such well-established applications, there might be further fields where multiphasic block copolymers will play a key role in the future. For example, they could allow combination of the excellent mechanical properties of conventional thermoplastics with the electrical conductivity of functional polymers in one single material. In order to examine the potential of this concept, synthetic strategies have to be developed first which make such novel materials accessible.

Poly(methyl methacrylate) (PMMA) is a transparent thermoplastic with excellent mechanical, thermal and chemical properties. For various purposes, it could be very useful to combine it with an electrically conducting counterpart. In this concern, polyferrocenylsilane (PFS) is a promising candidate:[8,9] It is thermally very stable, readily soluble in a variety of solvents, and – since the iron(II) centers are integral parts of the polymer main chains – has interesting magnetic and redox properties.[10,11] Also, it can be converted into semiconducting materials by exposure to an oxidizing agent.[12,13] Consequently, PFS-based block copolymers might be useful as organic semiconductors, electrochromic compounds or antistatic coatings. Moreover, the utilization of PFS-based block copolymers as auxiliars in nanolithography is currently a topic of special interest.[14-18] Therefore, it was a milestone when I. Manners et al. could show that PFS-based polymers are not only accessible via thermal or transition-metal catalyzed polymerization but also via living anionic polymerization of strained [1]silaferrocenophane (FS) monomers.[19,20] The anionic polymerization technique allows control of the PFS molar masses and molar mass distribution (PDI) as well as the formation of block copolymers. Consequently, various diblock and triblock block copolymers containing this organometallic system have been prepared during recent years.[21-26] Block copolymers from ferrocenylsilanes and MMA, however, have not yet been described to the best of our knowledge. Hence, we decided to develop a synthetic access to these block copolymers. The general synthetic strategy selected for this purpose is shown in Scheme 1. Since the direct block-copolymerization procedure may have some risks to failure because of side reactions of the carbanionic chain ends with the MMA's carbonyl groups, appropriate measures might be required to ensure high block efficiency and narrow block-length distribution. This is important because the optional benefits and perspectives of PFS-b-PMMA diblock copolymers are strongly correlated with their phase morphology: for well-defined diblock copolymers one can expect – in addition to the three classic micromorphologies, i.e. spheres, cylinders and lamellae – a bicontinuous gyroidic morphology. This gyroidic phase in particular might allow combination of the PMMA's mechanical properties with the benefits of the functional PFS block in an ideal manner.

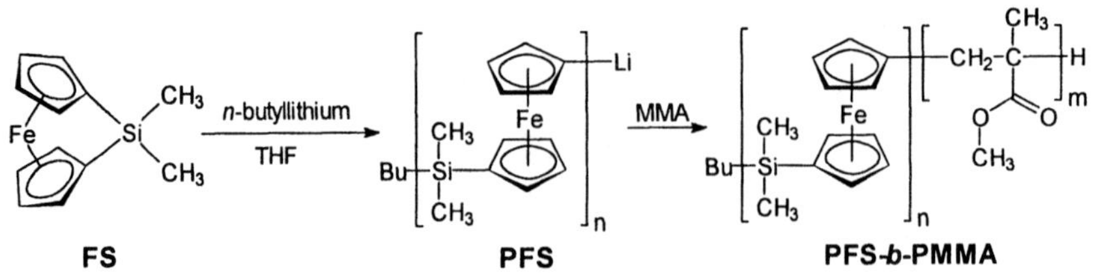

Scheme 1. General synthetic strategy applied for the preparation of PFS-b-PMMA diblock copolymers.

Objective

In this communication we report on an efficient synthetic access to well-defined PFS-*b*-PMMA diblock copolymers via living anionic polymerization. Also, we describe some first results obtained by transmission electron microscopy (TEM) concerning the bulk micromorphologies formed by these new materials.

Experimental

Synthesis of [1]dimethylsilaferrocenophane (FS) and of PFS homopolymers was carried out according to the literature.[27] For the synthesis of PFS-*b*-PMMA diblock copolymers, a representative entry is the following: FS (1.5 g, 6.19 mmol) was polymerized in 20 mL THF by addition of n-BuLi (50 μL, 0.08 mmol, 1.6 M in hexane) to give a living PFS precursor with M_n = 18,000 g/mol and PDI = 1.08. After complete conversion of the monomer DPE (56 μL, 0.32 mmol) was added, followed by DMSB (21 μL, 0.16 mmol; [n-BuLi] / [DMSB] / [DPE] = 1/2/4). In a second ampoule dry LiCl (34 mg, 0.8 mmol) was dissolved in THF (30 mL). Both ampoules were attached to the polymerization reactor, and after combining the solutions the reactor was immersed into a liquid nitrogen / isopropanol bath at –78 °C. For the polymerization of the second block, MMA (1.8 g, 18 mmol) was rapidly introduced from a syringe into the reactor through a Teflon-coated silicon septum under vigorous stirring. The polymerization was terminated after stirring the solution for 75 min at –78 °C by adding degassed methanol. After precipitating in methanol and drying under vacuum, 3.3 g (98.5%) of crude PFS-*b*-PMMA was isolated. It was dissolved in THF (200 mL) and hexane (200 mL) was added dropwise until the pure diblock started to precipitate. The pure diblock copolymer was isolated by ultracentrifuging (2.5 g, 75% yield).

Quantitative SEC analysis was performed with THF as the mobile phase on a modular setup consisting of a Waters model 515 pump, a Waters model 410 refractive-index (RI) detector and a three column set (PSS-SDV gel; 5μ; 10^6, 10^5, 10^4 Å). Molecular weight determination was carried out with respect to PMMA standards (purchased from PSS Polymer Standards Service GmbH, Mainz, Germany). TEM experiments were carried out using a Zeiss CEM 902 electron microscope operating at 80 kV. All images were recorded with a Slow-Scan CCD camera obtained from ProScan Inc. in the bright-field mode. Ultrathin polymer slices (50 nm) were cut with a diamond knife at room temperature using a Reichert Ultracut E or a Leica Ultracut UCT and collected on copper grids (400 mesh). No staining was required due to the high electron density of the organometallic phase.

Results & Discussion

Profound knowledge of the kinetics of FS homopolymerization is required for efficient block copolymerization processes. Therefore, first, kinetic investigations were carried out in THF at room temperature at constant initial monomer concentration. The concentration of n-BuLi as the initiator was varied from 1.68 × 10^{-5} to 4.80 × 10^{-5} mol·L^{-1}. After addition of the initiator to the monomer solution, samples were taken from the reaction mixture at regular intervals, terminated with degassed and water-free methanol, and analyzed using GC in order to determine the FS conversion as a function of time. It was shown that the reaction follows first order kinetics: at higher initiator concentrations, the kinetics shows a linear dependency and therefore the polymerization seems to be truly living. Only at lower initiator concentrations, the plots get slightly curved. This indicates a certain extent of chain termination as the chain growth proceeds to higher conversions. In full agreement with this conclusion, the obtained polymers exhibit narrow molar mass distributions and molecular weights in excellent agreement with the values expected from the reaction stoichiometry.

Block copolymerization of FS and MMA

First we tried to check whether or not the living PFS chain ends can be used directly to initiate the MMA block copolymerization. Hence homopolymerization of FS was carried out at room temperature. When this conversion was complete, the reaction mixture was cooled down to −78 °C and MMA was added. Unfortunately, the reactivity of the living PFS chain end proved to be too high even at low temperatures, resulting in quantitative chain termination. Consequently, measures had to be taken to lower the reactivity of the living PFS chain ends towards the MMA carbonyl groups. A well established method is to end-cap the living chain ends of the first block using 1,1-diphenylethylene (DPE): DPE living chain termini do not attack carbonyl groups but readily start MMA polymerization even at −78 °C. In practice, after polymerization of FS under the above conditions, a small quantity of DPE was added to the reaction mixture at room temperature (1.2 equiv. DPE were typically applied per PFS chain end equivalent). Again, the results were very disappointing: the conversion of the living PFS chain ends with DPE proved to be very incomplete (~80%) even after reaction at room temperature for 1.5 days.

The idea which was followed next was to take advantage of a highly reactive auxiliary for improving the efficiency of the DPE attachment to the PFS chain ends. This auxiliary should guarantee faster conversion of the PFS chain termini as well as accelerated subsequent addition of DPE. A potential candidate for solving this problem was reported by Sheikh et al. who used 1,1-dimethylsila-cylobutane as a "carbanion pump".[28] The resulting carbanion could be trapped efficiently with 1,1-diphenylethylene (88%).[29] This activation method was tested to achieve more efficient end-capping of the PFS chains with DPE. Again, FS was polymerized until complete conversion was achieved, followed by simultaneous addition of DMSB and DPE in a 2/4 ratio related to the living chains (Scheme 2). A change in color to deep red was observed within only a few minutes, indicating a very fast end-capping reaction. Then, a solution of MMA and LiCl in THF was added to the resulting deep red solution at −78 °C. After 75 min reaction time, the mixture was quenched, and the polymer was precipitated in methanol and subsequently analyzed using NMR and SEC.

It was evident that 90-95% block efficiency could be achieved. The remaining 5-10% are either due to spontaneous chain termination or homocoupling. Therefore, last traces of PFS homopolymer have to be removed by selective precipitation of the block polymers. Advantage can be taken here of the fact that the solubility of PMMA is quite different from that of PFS: while PMMA readily dissolves in polar solvents like THF, acetone or ethyl acetate, PFS is additionally soluble in less polar solvents like cyclohexane or mixtures of THF and hexane. Hence, selective precipitation of the block copolymers was achieved by redissolving the material in THF, followed by slow addition of n-hexane which is not a solvent for PMMA. Applying this method, a series of very pure and really high-molecular-weight PFS-b-PMMA-block copolymers could be obtained. In the following, these block copolymers are named $F_kM_l^x$ where F and M are the abbreviations of the respective blocks, PFS and PMMA. The subscripts k and l represent the volume fractions (in %) while the superscript x represents the over-all number-average molecular weight (in kg·mol^{-1}).

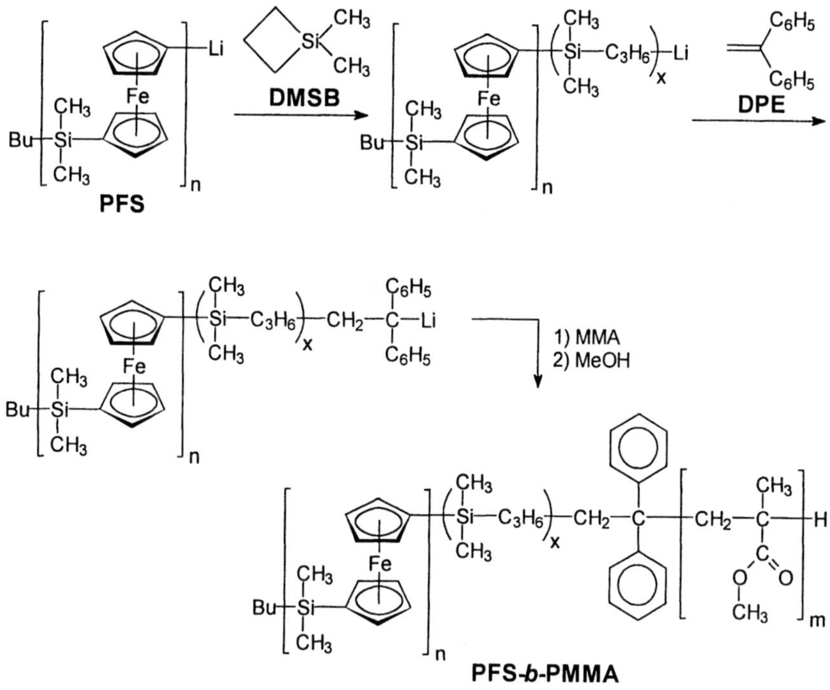

Scheme 2. 1,1-Dimethylsilacyclobutane-mediated synthesis of poly(ferrocenyl-silane-b-methylmethacrylate) diblock copolymers; x ranges from 1 to 3.

Bulk Morphology

In order to characterize the bulk morphology of the PFS-*b*-PMMA diblock copolymers using TEM, thin films had to be prepared from a non-selective sol-vent. Based on Hildebrand's solubility parameters, δ, appropriate solvents might be THF, methylene chloride and chloroform. Hence, 5 - 10 wt.-% solutions of the PFS-*b*-PMMA diblock copolymers were prepared, films were cast onto glass plates, and the solvent was allowed to evaporate slowly, over a period of 1 week, at room temperature in a desiccator containing a saturated solvent atmosphere. In order to remove last traces of solvent from the obtained orange, transparent films, and to perfectionize the morphologies, the films had to be annealed under reduced pressure (ca. 1 mbar) at temperatures clearly above the glass-transition temperatures of both blocks, and above the melting point of PFS (T_m = 143 °C) as well. At these temperatures, degradation might occur. In order to make sure that there is no degradation, the thermal stability of the block copolymers was analyzed using thermogravimetrical analysis (TGA). Four steps of weight loss can be detected, starting at 128, 275, 340 and 435 °C, respectively. The first step can be assigned to the loss of remaining solvent. Comparison of the TGA curve with those of PFS and PMMA homopolymers allows the conclusion that the weight loss at 340 °C is due to thermal degradation of PMMA while at 435 °C thermal degradation of PFS sets in. Only the origin of the degradation starting at 275 °C remained unclear. Maybe there is a further degradation caused by cooperative effects in the block copolymer structure. A very similar degradation behavior was found for samples obtained from THF solution while the degradation of films prepared from chloroform decomposed already at approx. 180 °C. Impurities in the solvent such as hydrochloric acid might force this early decomposition. Consequently, THF and methylene chloride were used as the solvents for film casting. The solution-cast films were first heated gradually, over a period of 2 h, from room temperature to above the PMMA's glass transition temperature (ca. 130 °C). After 1 h, the final annealing temperature was adjusted in one single temperature-rising step. Finally, the films were quenched in ice water to prevent crystallization of the PFS blocks. Since for PFS homopolymers it is known that the time required for crystallization is approx. 20 min at temperatures of 80 – 120 °C,[30,31] amorphous materials were obtained by the above procedure. Ultrathin samples of all PFS-b-PMMA diblock copolymers were prepared using an ultramicrotome, placed onto copper grids and transferred into the TEM without further pre-treatment.

One of the first results of the TEM investigations was that the method of film preparation is an important parameter for the long-range order within the samples. Figure 1 shows bright-field pictures of the classic morphologies, i.e., spheres, cylinders and lamellae, of pure PFS-b-PMMA diblock copolymers. The PMMA domains appear in white while the PFS is black. For $F_{25}M_{75}^{68}$ (Figure 1a), which is the sample with the lowest PFS content ($\phi_{PFS} = 0.25$), sphere-like PFS particles can be seen in a matrix of PMMA. Further raise of the PFS volume fraction results in cylindric morphologies. Expressed in PFS volume fractions, ϕ_{PFS}, the region of stability of the cylindric morphology ranges from $\phi_{PFS} \approx 0.25$ to $\phi_{PFS} \approx 0.37$. The right hand side of Figure 1b represents a projection along the cylinder axis while on the left hand side the cylinders are cut parallel to their axis. In this perspective, the hexagonal symmetry can be seen which is characteristic for cylindric phases. Moreover, in particular for samples of rather low molecular weight a long-range order was found of – in some cases – up to more than 3-4 μm. Above $\phi_{PFS} = 0.40$ the PFS-b-PMMA diblock copolymers form lamellar phases. Figure 1c displays the TEM picture of $F_{48}M_{52}^{49}$. Here again, regions of different orientation are evident: while some lamellae are cut parallel to the lamella plane, tilted orientations are shown at other places.

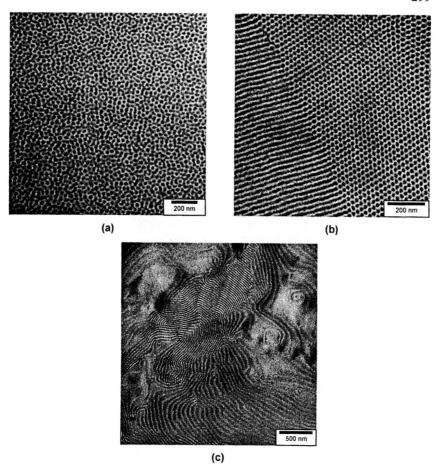

(a) (b)

(c)

Figure 1. TEM pictures of (a) $F_{25}M_{75}{}^{68}$ (annealed at 190 °C), (b) $F_{37}M_{63}{}^{38}$ (annealed at 165 °C) and (c) $F_{48}M_{52}{}^{49}$ (annealed at 190 °C); all samples were cast from methylene chloride solution. White: PMMA phase, black: PFS phase.

In summary, most of the prepared samples show the classic morphologies, irrespective of the way of sample preparation. Only in a very few cases the film preparation procedure proved to be significant. Quite special projections were obtained from $F_{40}M_{60}{}^{41}$, for example: when films were prepared from methylene chloride solution followed by annealing at 165 or 190 °C, TEM shows coexistence of hexagonally arranged cylinders and lamellae. When annealing was done at 220°C, on the other hand, only the lamellar morphology was found. This observation seems to be the result of a small amount of PFS homopolymer still present in the block copolymer: approx. 1 wt.-% of PFS homopolymer could be detected in the $F_{40}M_{60}$[41] sample by careful analysis of the SEC trace. During film formation, the homopolymer seems to accumulate in some sub-volumes of the sample, causing a slightly higher PFS volume fraction there. Consequently, since the overall composition of the material is very close to the phase boundary between cylindric and lamellar phases, the morphology switches into some sub-volumes. The disappearence of the cylindric morphology after annealing at 220 °C, on the other hand, is the result of the lower melt viscosity at elevated annealing temperature: at 220 °C efficient interdiffusion of homo- and block copolymers seems to be possible which equilibrates the PFS volume fraction over the whole sample and thus leads to the formation of only one single morphology. A quite similar behavior was found for $F_{43}M_{57}$[51]: again, the TEM pictures show coexisting cylindric and lamellar phases, separated from each other by sharp grain boundaries. An implicit conclusion of these latter TEM pictures is that in the PFS-b-PMMA diblock copolymers the lamellar phase follows directly the cylindrical one. Hence, at least for high-molecular-weight samples no complex (gyroidic) phases can be expected at a ϕ_{PFS} of around 0.40. When we take into account that a gyroidic phase is stable only for weakly segregated materials, i.e. for moderate values of $\chi \cdot N$, the product of the Flory-Huggins interaction parameter χ and the overall degree of polymerization N, it is obvious that PFS-b-PMMA diblock copolymers represent a system in the strong segregation limit.

In order to investigate the coexistence of cylindrical and lamellar morphologies in PFS-b-PMMA diblock copolymers at $\phi_{PFS} \approx 0.40$ further, a series of pure block copolymers was blended with PFS homopolymers having the same molecular weight as the corresponding PFS block in the diblock copolymer. For example, blend films of $F_{36}M_{64}$[48] + 4% PFS (overall PFS content ϕ_{PFS} = 0.40), and of $F_{37}M_{63}$[38] + 2 % PFS (overall PFS content ϕ_{PFS} = 0.39) were prepared and annealed at 190 °C. In all cases pure hexagonal phases developed. Additional regions with a stable lamellar phase were not observed. Quite similar results were obtained when the annealing was carried out at 165 °C. These investigations reconfirmed that the transition between cylindrical and lamellar morphology in PFS-b-PMMA diblock copolymers is localized at ϕ_{PFS} = 0.40. A hint towards the existence of a gyroidic phase could not be found in the samples discussed so far. However, the conclusions concerning the microphase separation behavior of PFS-b-PMMA diblock copolymers drawn so far base upon the morphological characterization of rather high-molecular-weight samples. In the following, additional blends were prepared from sample $F_{45}M_{55}$[27]. This copolymer has a rather moderate molecular weight and a nearly symmetric composition. Thus it gives a lamellar morphology when studied as pure material (Figure 2a).

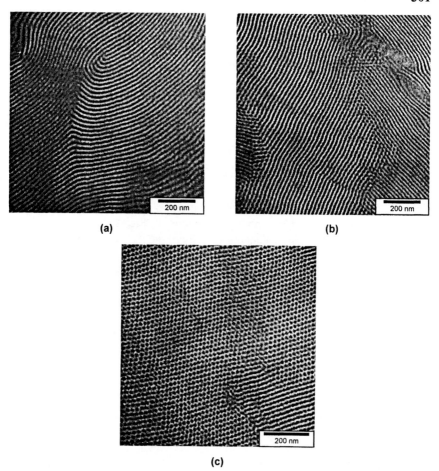

Figure 2. TEM pictures of (a) $F_{45}M_{55}{}^{27}$ diblock copolymer, and for blends of $F_{45}M_{55}{}^{27}$ with (b) 5 vol.-% and (c) 6 vol.-% PMMA homopolymer (annealed at 165 °C).

In order to reach the already mentioned region of composition next to the transition from the lamellar to the cylindrical phase, the volume fraction of PFS must be lowered. Therefore, low-molecular-weight PMMA homopolymer ($M_n = 9600$ g·mol⁻¹; PDI = 1,06) was synthesized and used as blend partner. First, annealing was done at 165 °C. Starting from the lamellar structure of the pure diblock copolymer (Figure 2a), 5 vol.-% of PMMA homopolymer were added (Figure 2b). A region is reached where cylinders and lamellae coexist.

This coexistence, however, is found in particular at the boundaries between lamellar grains of different orientation. Upon further increase of the PMMA content (Figure 2c), a sharp transition occurs into the cylindrical regime. This result is in full agreement with the results obtained for the high-molecular-weight samples. A completely different situation was found when the annealing temperature was raised to 190 °C (Figure 3). This annealing at higher temperatures is equivalent – due to the temperature-dependence of χ – to a decrease of $\chi \cdot N$ and thus to an attenuation of the incompatibility between PFS and PMMA. Upon addition of 5 vol.-% PMMA homopolymer the original lamellar phase of the pure diblock copolymer (Figure 2a) switches to a more complex superstructure (Figure 3a) in which, in addition to the clearly visible lamellar regions, a further phase arises. However, exact interpretation of this phase is difficult based only on the TEM pictures. The structural assignment of the morphology found for the second blend with 6 vol.-% PMMA homopolymer shown in Figure 3b is much easier. This projection can be assigned unambiguously to the well-known wagon-wheel projection of the bicontinuous gyroidic phase. The expression "wagon-wheel" was chosen because of the characteristic pattern which is observed in TEM pictures when the cubic gyroid structure is viewed in the 111 direction. For better illustration of the structure found for $F_{45}M_{55}{}^{27}$ + 6 vol.-% PMMA, Figure 3c shows additionally a Fourier Transform filtered section of the picture given in Figure 3b.

Conclusions

Very pure FS monomers and DMSB-mediated DPE end-capping allow the production of living PFS precursors readily activated for block copolymerization with MMA. High-molecular-weight PFS-*b*-PMMA block copolymers are thus available with PDIs < 1.1 and different PFS contents. The thermal properties and the phase behavior of these polymers were investigated with the aid of TGA, DSC and TEM. In order to obtain equilibrium structures, the samples for the TEM investigations were prepared from homogeneous solution and subsequently annealed at temperatures between 165 and 220 °C. The analysis of the obtained films did not show any evidence of a thermal degradation. Investigation of the phase structures using TEM showed the diblock copolymers to be systems in the strong segregation limit and hence the expected microphase structures spheres, cylinders and lamellae. According to these studies, the range of stability of the cylindrical phase is between $\phi_{PFS} \approx 0.25$ and $\phi_{PFS} \approx 0.40$ for the PFS-*b*-PMMA diblock copolymers. Below this volume fraction, the spheric morphology is found, above the lamellar one. In samples having a PFS volume fraction of $\phi_{PFS} = 0.40$, moreover, biphasic morphologies from lamellae and cylinders were found to be coexistent. This shows that at least for molecular weights higher than $M_n = 40,000$ g·mol^{-1} no complex morphologies like the bicontinuous gyroidic

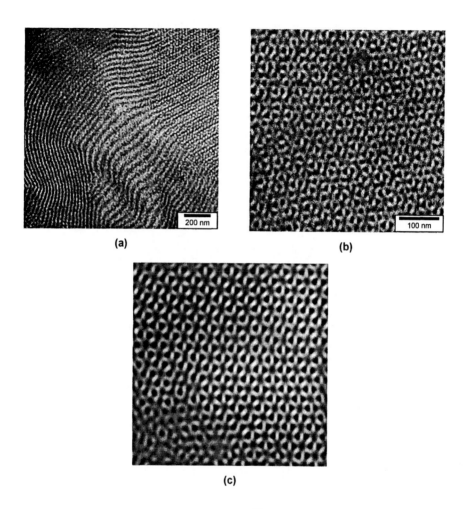

Figure 3: TEM pictures of blends of $F_{45}M_{55}{}^{27}$ with (a) 5 vol.-% and (b) 6 vol.-% PMMA homopolymer. Annealing was done at 190 °C. For (b) the fast-fourier transformation (FFT) filtered image of the considered section is shown in (c).

one can be expected. In continuation of this work we investigated blends of rather low-molecular-weight PFS-*b*-PMMA diblock copolymers with either PFS or PMMA homopolymers. The results of this work illustrate that even in the system PFS / PMMA there is a region of stability for the bicontinuous gyroidic phase – at least when annealing is done at higher temperatures. This is an important result when we consider potential applications of these materials in which the accessibility of a bicontinous phase may be of great significance for the macroscopic behavior.

However this morphological behavior is observed only in blends of PFS-*b*-PMMA diblock copolymer and PMMA homopolymer so far. Therefore it remains a challenge to prove the existence of a gyroidic phase in the pure diblock copolymer as well. A further point of interest is the analysis of the long-range order in the phases using techniques such as small angle x-ray scattering. For a more substantial understanding of the PFS/PMMA-phase behavior, finally, determination of the Flory-Huggins interaction parameters is necessary. These studies are subject of current investigations.

Acknowledgments

We thank the Fonds der Chemischen Industrie e.V. (FCI), the Otto Röhm Stiftung, and the Vereinigung der Freunde der TU Darmstadt e.V. for financial support of this work.

References

1   Holden, G., Legge, N. R., Quirk, R., Schroeder, H. E. (Eds.) *Thermoplastic Elastomers*, 2nd ed.; Hanser: Munich, Vienna, New York, 1996.
2   Spontak, R. J.; Patel, N. P. *Curr. Opin. Colloid Interface Sci.* **2000**, *5*, 334.
3   Knoll, K.; Nießner, N. *Macromol. Symp.* **1998**, *132*, 231.
4   Roe, R.-J.; Rigby, D. *Adv. Polym. Sci.* **1987**, *82*, 103.
5   Fayt, R.; Jérome, R.; Teyssié, P. *Am. Chem.Soc.; Symp.Ser.* **1989**, *27*, 775.
6   Jiang, M.; Xie, H. K. *Prog. Polym. Sci.* **1991**, *16*, 977.
7   Hamley, I. W. *The Physics of Block Copolymers*, Oxford University Press: Oxford, 1998.
8   Manners, I. *Angew. Chem.* **1996**, *108*, 1712.
9   Manners, I. *Can. J. Chem.* **1998**, *76*, 371.
10 Manners, I. *Chem. Commun.* **1999**, 857.
11 Manners. I. *Macromol. Rapid Commun.* **2001**, *22*, 711.
12 Rulkens, R.; Resendes, R.; Verma, A.; Manners, I.; Murti, K.; Fossum, E.; Miller, P.; Matyjaszewski, K. *Macromolecules*, **1997**, *30*, 8165.

13 Bakueva, L.; Sargent, E. H.; Resendes, R.; Bartole, A.; Manners, I. *J. Mater. Sci.: Mater. Electr.* **2001**, *12*, 21.

14 Lammertink, R. G. H.; Hempenius, M. A.; van den Enk, J. E.; Chan, V. Z.-H.; Thomas, E. L.; Vancso, G. J. *Adv. Mater.* **2000**, *12*, 98.

15 Cheng, J. Y.; Ross, C. A., Chan, V. Z.-H.; Thomas, E. L.; Lammertink, R. G. H.; Vancso, G. J. *Adv. Mater.* **2001**, *13*, 1174.

16 Lammertink, R. G. H.; Hempenius, M. A.; Vancso, G. J.; Shin, K.; Rafailovich, M. H.; Sokolov, J. *Macromolecules* **2001**, *34*, 942.

17 Lammertink, R. G. H.; Hempenius, M. A.; Chan, V. Z.-H.; Thomas, E. L.; Vancso, G. J. *Chem. Mater.* **2001**, *13*, 429.

18 Cao, L.; Massey, J. A.; Winnik, M. A.; Manners, I.; Riethmüller, S.; Banhart, F.; Spatz, J. P.; Möller, M. *Adv. Funct. Mater.* **2003**, *13*, 271.

17 Rulkens, R.; Ni, Y; Manners, I. *J. Am. Chem. Soc.* **1994**, *116*, 12121.

20 Ni, Y.; Rulkens, R.; Manners, I. *J. Am. Chem. Soc.* **1996**, *118*, 4102.

21 Resendes, R.; Massey, J.; Dorn, H.; Winnik, M. A.; Manners, I. *Macromolecules* **2000**, *33*, 8.

22 Massey, J.; Power, K. N.; Manners, I.; Winnik, M. A. *J. Am. Chem. Soc.* **1998**, *120*, 9533.

23 Massey, J. A.; Power, K. N.; Winnik, M. A.; Manners, I. *Adv. Mater.* **1998**, *10*, 1559.

24 Lammertink, R. G. H.; Versteeg, D. J.; Hempenius, M. A.; Vancso, G. J. *J. Polym. Sci. A: Polym. Chem.* **1998**, *36*, 2147.

25 Lammertink, R. G. H.; Hempenius, M. A.; Thomas, E. L.; Vancso, G. J. *J. Polym. Sci. B: Polym. Phys.* **1999**, *37*, 1009.

26 Li, W.; Sheller, N.; Foster, M. D.; Balaishis, D.; Manners, I.; Annis, B.; Lin, J.-S. *Polymer* **2000**, *41*, 719.

27 Kloninger, C.; Rehahn, M. *Macromolecules* **2004**, *37*, 1720.

28 Sheikh, Md. R. K.; Imae, I.; Tharanikkarasu, K.; LeStrat, V. M.-J.; Kawakami, Y. *Polym. J.* **2000**, *32*, 527.

29 Sheikh, Md. R. K.; Tharanikkarasu, K.; Imae, I.; Kawakami, Y. *Macromolecules*, **2001**, *34*, 4384.

30 Lammertink, R.G.H.; Hempenius, M.A; Vancso, G.J., *Macromol. Chem. Phys.* **1998**, *199*, 2141.

31 Lammertink, R.G.H.; Hempenius, M.A.; Manners, I.; Vancso, G.J. *Macromolecules* **1998**, *31*, 795.

Chapter 22

Lithographic Patterning and Reactive Ion Etching of a Highly Metallized Polyferrocenylsilane

Scott B. Clendenning, Alison Y. Cheng, and Ian Manners[*]

Department of Chemistry, University of Toronto, 80 St. George Street, Toronto, Ontario M5S 3H6, Canada
*Corresponding author: imanners@chem.utoronto.ca

A highly metallized cobalt-clusterized polyferrocenylsilane (Co-PFS) has been patterned using electron-beam lithography, UV-photolithography and a selective dewetting soft lithographic process. Subsequent pyrolysis and/or plasma reactive ion etching to fabricate metal-containing magnetic ceramic films has been demonstrated.

Introduction

New approaches to the patterning of metallic structures using a minimum number of processing steps are of intense current interest. Direct patterning of processible, high resolution lithographic resists with high metal content is one very promising avenue.[1] To date the majority of direct-write resists for metallic structures consist of surfactant-stabilized metal colloids,[2] molecular organometallic species,[3] metal-containing composites[4] or purely inorganic materials.[5] These resists operate in negative-tone mode such that exposed regions undergo a chemical change rendering them insoluble in the developing medium. Studies of transition metal-containing polymers are extremely rare. One recent example is the use of thin films of the organometallic cluster polymer $[Ru_6C(CO)_{15}Ph_2PC_2PPh_2]_n$ as a negative-tone electron-beam resist to direct-write conducting wires of metal nanoparticles as small as 100 nm in width.[6] As a metal source, metallopolymers offer the advantages of ease of processibility, atomic level mixing and stoichiometric control over composition.[7] The use of a metallopolymer precursor to form metal containing ceramics has also been shown.[8] Plasma reactive ion etching (RIE) provides a rapid and convenient means to chemically and physically modify surfaces in a uniform manner over square centimeter areas. In the case of metallopolymers, their generally higher etch resistance[9] in a variety of plasmas in comparison to organic polymers, coupled with the possibility of depositing metal-containing ceramic materials with interesting physical properties, is highly intriguing. In this paper we review the patterning of thin films of a highly metallized polyferrocene polymer by electron-beam lithography (EBL), UV-photolithography and a selective dewetting soft lithographic process as well as its processing by pyrolysis and RIE to fabricate metal-containing magnetic ceramics.

The ring-opening polymerization of sila[1]ferrocenophanes (1) by thermal,[10] anionic[11] and transition metal-catalyzed[12] routes yields high molecular weight, soluble polyferrocenylsilanes (2, PFS) containing coordinatively bonded iron atoms in the main chain (Scheme 1). The incorporation of PFS into patterned surfaces has already yielded materials with tunable magnetic properties that may find applications as protective coatings, magnetic recording media and in anti-static shielding.[13] Furthermore, the low plasma etch rates of polymers containing organometallic moieties in comparison to their purely organic counterparts[9a] suggest their use as etch masks which can also deposit interesting materials.

The introduction of additional metals into the PFS chain can increase metal loadings and allow access to binary or higher metallic alloy species. We have

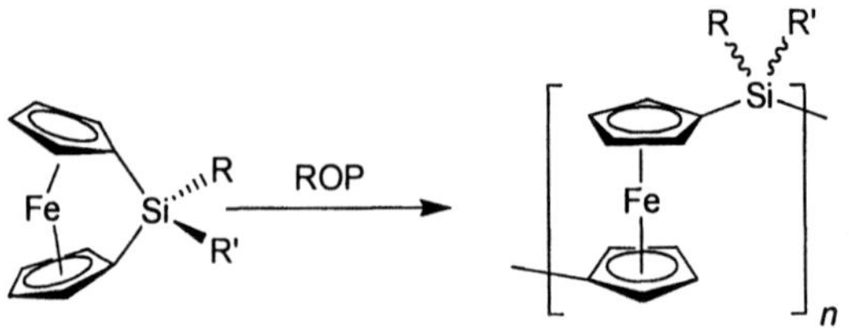

Scheme 1. Ring-opening polymerization of a sila[1]ferrocenophane (1) to afford a polyferrocenylsilane (2).

found in our research that PFS with acetylenic substituents at silicon[14] (Scheme 2, **2a**) can be clusterized by treatment with dicobalt octacarbonyl to yield the highly metallized, soluble, air-stable cobalt-clusterized polyferrocenylsilane (Co-PFS) (**3**) which contains three metal atoms per repeat unit.[15]

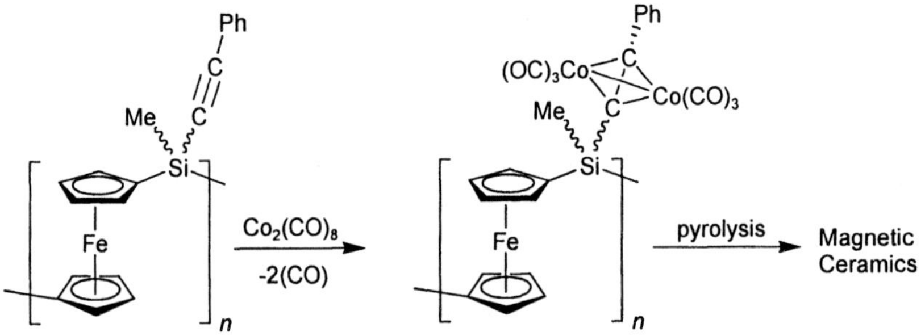

Scheme 2. Cobalt clusterization of the acetylenic substituents of a polyferrocenylsilane (2a) to yield the highly metallized Co-PFS (3), a precursor to magnetic ceramics.

Pyrolysis of bulk **3** under a N_2 atmosphere at either 600°C or 900°C affords black magnetic ceramics in relatively high yield (72% and 59%, respectively). Powder X-ray diffraction (PXRD) studies revealed the ceramic residual was composed of Fe/Co alloy particles embedded in an amorphous C/SiC matrix. Figure 1 shows the cross-sectional transmission electron micrographs of the ceramics prepared at 600 °C and 900 °C. The presence of electron-rich metal nanoparticles can clearly be seen. The chemical compositions of these metal particles were determined by element mapping electron spectroscopic imaging (ESI) experiments. The results indicated that both iron and cobalt were localized in the same nanoparticles. Superconducting quantum interference device (SQUID) magnetometry showed that ceramics formed at 900°C were ferromagnetic with no blocking temperature up to 355 K.

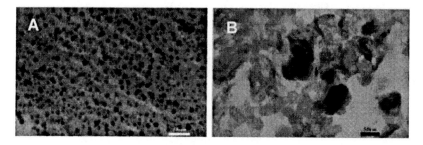

Figure 1. Cross-sectional TEM images of ceramics resulting from pyrolysis of Co-PFS (3) at (a) 600°C and (b) 900°C. The scale bars are 50 nm. (Reproduced with permission from Reference 15.)

One of the advantages of polymers is the ease with which they can form films. We have recently demonstrated the patterning of thin films of Co-PFS (3) by electron beam lithography (EBL), [16] UV-photolithography, [17] and a soft lithography/selective dewetting technique[18] as well as the formation of magnetic ceramic films via pyrolysis and/or reactive ion etching (RIE) in a secondary magnetic field.[19]

Cobalt-Clusterized Polyferrocenylsilane as an Electron-Beam Lithography Resist

Electron-beam lithography (EBL) is a maskless lithography in which a resist material is exposed to an electron-beam with controlled dose in a predetermined pattern.[20] Areas exposed to the beam in positive-tone resists become more soluble in a developing medium whereas exposed areas in a negative-tone resist become less soluble. In order to determine whether Co-PFS could function as an electron beam resist, uniform thin films (*ca.* 200 nm thick) of 3 were spin-coated on to silicon substrates and electron beam lithography was carried out at varying doses and current in a modified scanning electron microscope. The treated films were then developed in THF prior to characterization. Resists of 3 were found to operate in a negative-tone fashion.[16] Shapes including dots and bars were successfully fabricated (Figure 2).

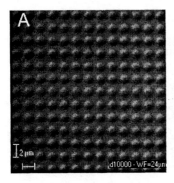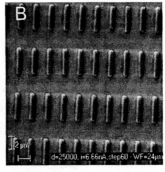

Figure 2. SEM images of (a) dots and (b) bars fashioned by EBL using a Co-PFS (3) resist. (Reproduced with permission from Reference 16)

The elemental composition of these patterns was investigated using time-of-flight secondary ion mass spectrometry (TOF-SIMS) and X-ray photoelectron spectroscopy (XPS). Iron and cobalt elemental maps of the arrays of bars on the Si substrate were obtained using TOF-SIMS. The mapping clearly revealed that Fe and Co were concentrated within the bars. Information regarding the chemical environment and distribution of Fe and Co throughout the bars was obtained via XPS and compared to data obtained from an untreated film of Co-PFS. The atomic ratio of Fe:Co for the bars was in agreement with the theoretical value of 1:2. Detailed scans for iron and cobalt revealed no change in binding energy for the elements at 3 nm and 12 nm depths indicating uniformity in average chemical environment. Overall, XPS revealed little change in the chemical composition of the polymer resist after EBL. Finally, magnetic force microscopy (MFM) indicated no appreciable magnetic field gradient above the bars following EBL.

In order to enhance the magnetism of the patterned bars, an array of bars on a Si substrate was pyrolyzed at 900 °C under a N_2 atmosphere to promote the formation metallic nanoclusters[15,21] The same array of bars was characterized before and after pyrolysis by tapping mode atomic force microscopy (AFM). Comparison of the images and cross-sectional profiles suggests there is excellent shape retention accompanying a decrease in the dimensions of the bars. MFM studies of the bars indicated that they consist of heterogeneous ferromagnetic clusters which magnetic dipoles appear to be randomly oriented. The magneto-optic Kerr effect (MOKE) was also used to independently confirm the ferromagnetic behavior of the ceramic bars.

Although the aforementioned proof-of-concept experiments dealt with the formation of micron-scale objects, EBL can be used to routinely fabricate structures down to 30-50 nm.[22] Indeed, this process has already been extended to the patterning of sub 500 nm bars and dots from Co-PFS.

Cobalt-Clusterized Polyferrocenylsilane (Co-PFS) as a UV-Photoresist

The serial nature of EBL can result in long processing times especially if large doses are required. In UV-photolithography, square centimeter areas of the resist can be exposed through a mask at the same time normally leading to much faster processing. Organic polymers incorporating acetylenic moieties have been shown to be thermally crosslinkable. Upon heating, the acetylene groups from adjacent chains undergo cyclotrimerization and coupling reactions, creating crosslinks in the polymer.[23] In addition, photo-induced polymerization of alkyl- and aryl-substituted acetylenes are known to be catalyzed by metal carbonyls such as $Cr(CO)_6$, $Mo(CO)_6$ and $W(CO)_6$.[24] Recently, Bardarau *et al.* demonstrated that polyacrylates with pendent acetylenic side groups could be photocrosslinked in the presence of $W(CO)_6$ as catalyst.[25] In the case of Co-PFS (**3**), which inherently contains both the acetylenic unit and the metal carbonyl catalyst, is a promising candidate as a resist material for UV-photolithography.[17]

To study this possibility, a thin film (*ca.* 200 nm) of **3** on Si substrate was exposed to near UV radiation (λ=350-400 nm, 450 W) for 5 minutes. The exposed film was developed in THF before characterization. Co-PFS was found to be a negative-tone photoresist. This appears to be consistent with photo-initiated crosslinking mechanism of acetylenes in the presence of metal carbonyls. However, it is also possible to have crosslinking in Co-PFS as a result of decarbonylation of the Co-cluster. The thickness of the film before and after UV treatment was determined by ellipsometry. A 200 nm thick film of Co-PFS had a thickness of *ca.* 170 nm after exposure to UV radiation and solvent development. The decrease in thickness is probably a reflection of the decreased volume of the polymer upon crosslinking.

Patterning of Co-PFS films was accomplished using a metal foil shadow mask fabricated by micromachining with features *ca.* 50-300 μm. Figure 3a is an optical micrograph of a straight line of Co-PFS patterned with the shadow mask. Smaller features (*ca.* 10-20 μm) were obtained using a chrome contact mask (Figure 3b). In both cases the unexposed polymer was completely removed during development with THF, leaving behind patterns with well-defined edges.

Patterned Co-PFS was pyrolyzed at 900 °C under a N_2 atmosphere in an attempt to fabricate magnetic ceramic lines. The resulting ceramics lines have the same dimensions as the polymer precursor showing excellent shape retention in the lateral directions. Inspection of the ceramic line at higher magnifications revealed the formation of what appeared to be Co/ Fe nanoparticles throughout the line.

UV-photolithography using Co-PFS (**3**) as a resist provides a convenient route to deposit patterned polymer and magnetic ceramic onto flat substrates. Due to the excellent RIE resistance of the polymer, Co-PFS patterned by UV-photolithography could also potentially be utilized for pattern transfer onto the underlying substrate using conventional plasma etching techniques. Investigations are underway to improve the resolution of this resist.

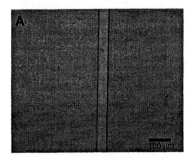
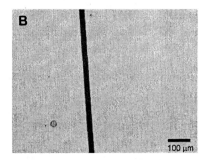

Figure 3. Optical micrographs of Co-PFS (3) lines fabricated by UV-photolithography using (a) a shadow mask and (b) a chrome contact mask. (Reproduced with permission from Reference 17.)

Ordered 2D Ring Arrays of Cobalt-Clusterized Polyferrocenylsilane

Arrays of nanoparticle rings are a desirable target in terms of expanding nanofabrication techniques as well as potential applications in high density magnetic data storage devices.[26] Recently, long range order and excellent size control in 2D arrays of polymer rings were achieved by Yang *et al.* using selective dewetting of a solution of a polymer-containing film on a substrate patterned with micron-sized hydrophilic and hydrophobic areas using soft lithography.[27] In collaboration with the group of Prof. G.A. Ozin at the University of Toronto, we used a modified version of this technique to pattern arrays of rings of Co-PFS (3) with diameters of *ca.* 2, 5 and 12 μm onto a thin gold film (*ca.* 100 nm) sputtered onto a silicon wafer that had been primed with a 5 nm layer of titanium.[18] Figure 4a shows a scanning electron micrograph of an array of *ca.* 5 μm diameter rings. The rings had a volcano-like morphology with the light interior corresponding to the hydrophilic area while the area outside the rings corresponded to the hydrophobic area. The rings were then treated with a hydrogen plasma (20 mTorr, 100 W, 12 min) in order to remove carbonaceous material, thereby consolidating the structure, along with the excess polymer around the rings (Figure 4b). Traces of gold were removed using a KI_3 etch prior to pyrolysis under a dinitrogen atmosphere at 900 °C for 5 h. Visualization of the pyrolyzed rings using SEM (Figure 4c) revealed a highly ordered array of nanoparticle rings with inner and outer diameters of 5.4 ± 0.2 μm and 6.3 ± 0.2 μm, respectively. The presence of Fe and Co in the bright nanoparticles was confirmed by EDX. Interestingly, the nanoparticles appeared to self-assemble into concentric rings to give an overall annular width of *ca.* 800 nm. The particles, which likely have a metal oxide shell,[21] exhibited a bimodal size distribution with the larger particles 111 ± 20 nm and the smaller ones less than 70 nm. The majority of particles were larger than the estimated magnetic single-domain size for spherical particles of iron (14 nm) and Co (70 nm).[28] Magnetic force microscopy confirmed that the nanoparticles in the rings were magnetic. While it is also possible to image superparamagnetic metallic nanoparticles using MFM,[29] the patterned Fe/Co alloy nanoparticles resulting from the pyrolysis of microbars of **3** (vide supra) under similar conditions proved to be ferromagnetic by magneto-optic Kerr effect (MOKE) measurements[16] thus supporting the presence of ferromagnetic nanoparticles in the case of the patterned rings. Therefore, by employing a consolidating hydrogen plasma treatment followed by pyrolysis, we have demonstrated the transformation of metallopolymer rings into a highly ordered array of ferromagnetic nanoparticles. Through judicious choice of the metallopolymer precursor and its concentration in solution, ordered arrays of many different metallic nanoparticle rings with a range of dimensions should be accessible.

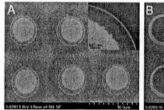 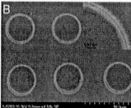 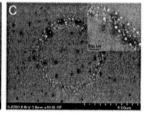

Figure 4. Field Emission SEM images of 5 μm Co-PFS (3) rings (a) as prepared, (b) following H₂-RIE (20 mTorr, 100 W, 12 min) and (c) after subsequent pyrolysis (N₂, 900 °C, 5 h). Insets show a representative section of rim at 40k magnification. (Reproduced with permission from Reference 18.)

Polyferrocenylsilanes as Reactive Ion Etch Resists

Pyrolytic routes to ceramics generally involve the input of large amounts of thermal energy for prolonged periods of time in order to cause the decomposition of the ceramic precursor, removal of volatiles and the formation of the desired ceramic material. We wished to explore the possibility of carrying out such a transformation using plasma reactive ion etching (RIE).[20] It has been demonstrated that polyferrocenylsilanes possess low plasma etch rates which can be attributed to the formation of a protective layer of involatile iron and silicon compounds. This property has been exploited with PFS block copolymers self-assembled into micelles[30] or phase-separated thin films[31] to deposit patterned ceramics as well as for pattern transfer to the substrate. In our research, we are interested in the direct formation of high metal content magnetic ceramic films by RIE treatment of Co-PFS in the presence of a secondary magnetic field.[19]

Thin films (*ca.* 100-200 nm thick) of Co-PFS (**3**) on a Si substrate were treated with either H_2 or O_2 plasma (10-20 mTorr, 100 W). The effects of the plasmas on the chemical composition of the treated films were studied by TOF-SIMS depth profiling. In both cases, by plotting the intensity of the Si^+, Fe^+, FeO^+, Co^+ and CoO^+ signals as a function of depth, it could be seen that the plasma affected only the top 10 nm of the films, leaving the underlying polymer untouched. The chemical composition of the modified films was analyzed by XPS. In both cases a Co: Fe ratio of approximately 1:1 was found at 3 nm depth. In comparison to the 2:1 ratio of Co: Fe ratio found in the untreated polymer, this deficiency of Co in the film surface is most likely a reflection of the volatility of the cobalt carbonyl clusters during the high vacuum processing required for RIE. Formation of iron oxides (FeO and Fe_2O_3) as well as cobalt oxides (CoO and Co_3O_4) was observed at 3 nm depth in the O_2-plasma treated films. In the case of H_2-RIE, small amounts of oxides were also observed at 3 nm depth, presumably due to the oxidation of reduced metal upon exposure to atmospheric oxygen.

Morphological changes in the surface of Co-PFS (**3**) films following RIE were investigated by transmission electron microscopy (TEM) and AFM. Thin films of the polymer (*ca.* 50 nm thick) on a carbon-coated copper TEM grid were exposed to oxygen and hydrogen plasma. In both cases, analysis by TEM revealed the presence of electron rich nanoworms with widths ranging from 4-12 nm which remained on the supporting carbon film. Untreated samples were featureless, supporting plasma-induced nanoworm formation. Electron energy-loss spectroscopic (EELS) elemental mapping and energy-dispersive X-ray (EDX) indicated that there was a high concentration of Fe, Co and Si in the nanoworms. In contrast to the essentially flat and featureless surface of the untreated film, AFM images of both plasma treated films exhibited a pervasive interconnected reticulated structure which would appear as nanoworms in projection normal to the plane of the sample. These features are reminiscent of surface reticulations observed by Thomas *et al.* by AFM in organosilicon polymers following ambient temperature O_2-RIE, which were hypothesized to originate due to a polarity difference between the overlying inorganic layer and the native polymer causing dewetting or spinodal decomposition of the strained inorganic layer.[32]

In order to access films with useful magnetic properties, plasma-induced crystallization of metallic nanoclusters was attempted by carrying out H_2-RIE on a thin film of Co-PFS in a secondary magnetic field. The Co-PFS films (*ca.* 200 nm thick) on Si substrates were placed between two samarium-cobalt (SmCo) magnets aligned with opposite magnetic poles facing each other during H_2- or O_2-RIE. The magnets caused the formation of an intense plasma plume around the sample which in turn resulted in intense etching conditions. Nanostructures obtained under these conditions were proven to be magnetic by MFM. A tapping mode AFM micrograph and the corresponding magnetic force micrograph of a H_2 plasma-treated film are shown in Figure 5. The reticulations were much larger than those found for samples treated under similar plasma conditions in the absence of the samarium-cobalt magnets. We postulate that the secondary magnetic field concentrated the plasma thereby accelerating nanoworm formation through more efficient removal of carbonaceous material and silicon while the additional thermal energy present in the plasma plume promoted metal crystallization. To the best of the authors' knowledge this is the first example of the formation of a ferromagnetic film directly from the plasma treatment of a metallopolymer.

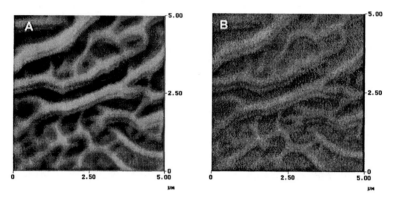

Figure 5. Tapping mode AFM images of thin film of Co-PFS (3) following H₂-RIE in a secondary magnetic field (a) and the corresponding MFM image (b). (Reproduced with permission from Reference 19.)

Conclusions

High molecular weight Co-PFS (**3**) is readily accessible via transition metal–catalyzed ROP of sila[1]ferrocenophane (**2a**) followed by clusterization of the acetylenic substituents with $Co_2(CO)_8$. The intrinsic high metal content, air-stability and solution processability make Co-PFS an excellent candidate for lithographic patterning. Fabrication of patterned arrays of polymer and magnetic ceramics on the sub-micron scale via EBL was demonstrated while UV-photolithography gave micron scale features. Furthermore, the use of a template-assisted selective dewetting technique involving soft lithography led to large 2D arrays of Co-PFS rings which could be transformed into magnetic nanoparticle ring arrays. We have also demonstrated the use of this highly metallized polymer as a RIE resist. Treatment of the polymer films with either H_2 or O_2 plasma in the presence of a secondary magnetic field afforded magnetic ceramics. We are currently studying the use of these patterned ceramics in spintronics as an isolating, magnetic layer in a nano-granular in-gap structure[33] as well as catalysis.[34] Concurrently, we are developing new highly metallized polymers containing Fe, Co, Mo and/or Ni[35] while working on improving resist sensitivity and resolution.

Acknowledgments

I.M. would like to thank the Canadian Government for a Canadian Research Chair. S.B.C. is grateful to NSERC for a PDF. The authors would also like to acknowledge the excellent work of our collaborators: *EBL*: Prof. Harry E. Ruda; Dr. Stephane Aouba (Center of Advanced Nanotechnology, University of Toronto); *AFM/ MFM*: Prof. Christopher M. Yip, Mandeep S. Rayat, Patrick Yang (Department of Chemical Engineering and Applied Chemistry, University of Toronto); *TEM:* Dr. Neil Coombs, *Ellipsometry:* Chantal Paquet (Department of Chemistry, University of Toronto); *RIE, UV Photolithography, XPS:* Prof. Zheng-Hong Lu, Dr. Sijin Han, Dr. Dan Grozea (Department of Materials Science and Engineering, University of Toronto); *TOF-SIMS*: Dr. Rana N. S. Sodhi, Dr. Peter M. Brodersen, (Surface Interface Ontario, Department of Chemical Engineering and Applied Chemistry, University of Toronto); *MOKE*: Prof. Mark R. Freeman, Jason B. Sorge (Department of Physics, University of Alberta).

References

1. Gonsalves, K. E.; Merhari, L.; Wu, H.; Hu, Y. *Adv. Mater.* **2001**, *13*, 703-714.

2. For recent examples see a) Hamann, H. F.; Woods, S. I.; Sun, S. *Nano Lett.* **2003**, *3*, 1643-1645. b) Werts, M. H. V.; Lambert, M.; Bourgoin, J.-P.; Brust, M. *Nano. Lett.* **2002**, *2*, 43-47. c) Bedson, T. R.; Palmer, R. E.; Wilcoxon, J. P. *Microelectronic Eng.* **2001**, *57-58*, 837-841. d) Lohau, J.; Friedrichowski, S.; Dumpich, G.; Wassermann, E. F.; Winter, M.; Reetz, M. T. *J. Vac. Sci. Technol. B* **1998**, *16*, 77-79 and references cited therein.

3. For recent examples see a) Saifullah, M. S. M.; Subramanian, K. R. V.; Tapley, E.; Kang, D.-J.; Welland, M. E.; Butler, M. *Nano Lett.* **2003**, *3*, 1587-1591. b) Alexe, M.; Harnagea, C.; Erfurth, W.; Hesse, D.; Gösele, U. *Appl. Phys. A* **2000**, *70*, 247-251. c) Koops, H. W. P.; Kaya, A.; Weber, M. *J. Vac. Sci. Technol. B* **1995**, *13*, 2400-2403 and references cited therein.

317

4. Stellacci, F.; Bauer, C. A.; Meyer-Friedrichsen, T.; Wenseleers, W.; Alain, V.; Kuebler, S. M.; Pond, S. J. K.; Zhang, Y.; Marder, S. R.; Perry, J. W. *Adv. Mater.* **2002**, *14*, 194-198.

5. a) Fujita, J.; Watanabe, H.; Ochiai, Y.; Manako, S.; Tsai, J. S.; Matsui, S. *Appl. Phys. Lett.* **1995**, *66*, 3064-3066. b) Watanabe, H.; Fujita, J.; Ochiai, Y.; Matsui, S.; Ichikawa, M. *Jpn. J. Appl. Phys.* **1995**, *34*, 6950-6955.

6. Johnson, B. F. G.; Sanderson, K. M.; Shephard, D. S.; Ozkaya, D.; Zhou, W.; Ahmed, H.; Thomas, M. D. R.; Gladden, L.; Mantle, M. *Chem. Commun.* **2000**, 1317-1318.

7. Manners, I. *Synthetic Metal-Containing Polymers*; Wiley-VCH: Weinheim, 2004.

8. a) MacLachlan, M. J.; Ginzburg, M.; Coombs, N.; Coyle, T. W.; Raju, N. P.; Greedan, J. E.; Ozin, G. A.; Manners, I. *Science* **2000**, *287*, 1460-1463. b) Häussler, M.; Zheng, R.; Lam, J. W. Y.; Tong, H.; Dong, H.; Tang, B. Z. *J. Phys. Chem. B.* **2004**, *108*, 10645-10650.

9. a) Taylor, G. N.; Wolf, T. M.; Stillwagon, L. E. *Solid State Technol.* **1984**, *27*, 145-155. b) Brodie, I.; Muray, J. J. *The Physics of Micro/Nano-Fabrication*; Plenum Press: New York, 1992. c) Deng, T.; Ha, Y.-H.; Cheng, J. Y.; Ross, C. A.; Thomas, E. L. *Langmuir* **2002**, *18*, 6719-6722. d) Taylor, G. N.; Hellman, M. Y.; Wolf, T. M.; Zeigler, J. M. *Advances in Resist Technology and Processing V;* SPIE Vol. 920; SPIE: Bellingham, WA, 1988, p 274-290.

10. Foucher, D. A.; Tang, B. Z.; Manners, I. *J. Am. Chem. Soc.* **1992**, *114*, 6246-6248.

11. Ni, Y. Z.; Rulkens, R.; Manners, I. *J. Am. Chem. Soc.* **1996** *118*, 4102-4114.

12. Ni, Y. Z. ; Rulkens, R.; Pudelski, J. K.; Manners, I. *Macromol. Rapid Comm.* **1995**, *16*, 637-641.

13. For a review on polyferrocenylsilane chemistry, see: Kulbaba, K.; Manners, I. *Macromol. Rapid Comm.* **2001**, *22*, 711-724.

14. Berenbaum, A.; Lough, A. J.; Manners, I. *Organometallics* **2002**, *21*, 4415-4424.

15. Berenbaum, A.; Ginzburg-Margau, M.; Coombs, N.; Lough, A. J.; Safa-Safat, A.; Greedan, J. E.; Ozin, G. A.; Manners, I. *Adv. Mater.* **2003**, *15*, 51-55.

16. Clendenning, S. B.; Aouba, S.; Rayat, M. S.; Grozea, D.; Sorge, J. B.; Brodersen, P. M.; Sodhi, R. N. S.; Lu, Z.-H.; Yip, C. M.; Freeman, M. R.; Ruda, H. E.; Manners, I. *Adv. Mater.*, **2004**, *16*, 215-219.

17. Cheng, A. Y.; Clendenning, S. B.; Yang, G.; Lu, Z.-H.; Yip, C. M.; Manners, I. *Chem. Commun.* **2004**, 780-781.

18. Clendenning, S. B.; Fournier-Bidoz, S.; Pietrangelo, A.; Yang, G.; Han, S.; Brodersen, P. M.; Yip, C. M.; Lu, Z.-H.; Ozin, G. A.; Manners, I. *J. Mater. Chem.* **2004**, *14*, 1686-1690.

19. Clendenning, S. B.; Han, S.; Coombs, N.; Paquet, C.; Rayat, M. S.; Grozea, D.; Broderson, P. M.; Sodhi, R. N. S.; Yip, C. M.; Lu, Z.-H.; Manners, I. *Adv. Mater.* **2004**, *16*, 291-296.

20. Madou, M. J. *Fundamentals of Microfabrication*, 2nd ed.; CRC Press: Boca Raton, 2002.

21. Ginzburg, M.; MacLachlan, M. J.; Yang, S. M.; Coombs, N.; Coyle, T. W.; Raju, N. P.; Greedan, J. E.; Herber, R. H.; Ozin, G. A.; Manners, I. *J. Am. Chem. Soc.* **2002**, *124*, 2625-2639.

22. Sheats, J. R.; Smith, B. W.; *Microlithography*; Marcel Dekker Inc.: New York 1998.

23. Swanson, S. A.; Fleming, W. W.; Hofer, D. C. *Macromolecules* **1992**, *25*, 582-588.

24. a) Masuda, T.; Yamamoto, K.; Higashimura, T. *Polymer* **1982**, *23*, 1663-1666. b) Masuda, T.; Kuwane, Y.; Yamamoto, K.; Higashimura, T. *Polym. Bull.* **1980**, *2*. 823-827. c) Landon, S. J.; Shulman, P. M.; Geoffroy, G. L. *J. Am. Chem. Soc.* **1985**, *107*, 6739-6740.

25. Bardarau, C.; Wang, Z. Y. *Macromolecules* **2003**, *36*, 6959-6961.

26. Castaño, F. J.; Ross, C. A.; Frandsen, C.; Eilez, A.; Gil, D.; Smith, H. I.; Redjdal, M.; Humphrey, F. B. *Phys. Rev. B* **2003**, *67*, 184425.

27. Lu, G.; Li, W.; Yao, J.; Zhang, G.; Yang, B.; Shen, J. *Adv. Mater.* **2002**, *14*, 1049-1053.

28. Kittel, C. *Phys. Rev.*, **1946**, *70*, 965-971.

29. Pedreschi, F.; Sturm, J. M.; O'Mahony, J. D.; Flipse, C. F. J. *J. Appl. Phys.*, **2003**, *94*, 3446-3450.

30. a) Massey, J. A.; Winnik, M. A.; Manners, I.; Chan, V. Z. H.; Ostermann, J. M.; Enchlmaier, R.; Spatz, J. P.; Möller, M. *J. Am. Chem. Soc.* **2001**, *123*, 3147-3148. b) Cao, L.; Massey, J. A.; Winnik, M. A.; Manners, I.; Riethmuller, S.; Banhart, F.; Spatz, J. P.; Möller, M. *Adv. Funct. Mater.* **2003**, *13*, 271-276.

31. a) Lammertink, R. G. H.; Hempenius, M. A.; van der Enk, J. E.; Chan, V. Z. H.; Thomas, E. L.; Vancso, G. J. *Adv. Mater.* **2000**, *12*, 98-103. b) Lammertink, R. G. H.; Hempenius, M. A.; van der Enk, J. E.; Chan, V. Z. H.; Thomas, E. L.; Vancso, G. J. *Chem. Mater.* **2001**, *13*, 429-433.

32. Chan, V. Z. H.; Thomas, E. L.; Frommer, J.; Sampson, D.; Campbell, R.; Miller, D.; Hawker, C.; Lee, V.; Miller, R. D. *Chem. Mater.* **1998**, *10*, 3895-3901.

33. Jalil, M. B. A. *IEEE Tran. Magn.* **2002**, *38*, 2613-2615.

34. a) Lastella, S.; Jung, Y. J.; Yang, H.; Vajtai, R.; Ajayan, P. M.; Ryu, C. Y.; Rider, D. A.; Manners, I. *J. Mater. Chem.* **2004**, *14*, 1791-1794. b) Hinderling, C.; Keles, Y.; Stoeckli, T.; Knapp, H. F.; De Los Arcos, T.; Oelhafen, P.; Korczagin, I.; Hempenius, M. A.; Vancso, G. J.; Pugin, R.; Heinzelmann, H. *Adv. Mater.* **2004**, *16*, 876-879.
35. Chan, W. Y.; Berenbaum, A.; Clendenning, S. B.; Lough, A. J.; Manners, I. *Organometallics* **2003**, *22*, 3796-3808.

Chapter 23

Poly(Ferrocenylsilane-*block*-methacrylate)s via Sequential Anionic and Atom Transfer Radical Polymerization

Mark A. Hempenius[*], Igor Korczagin, and G. Julius Vancso[*]

MESA[+] Institute for Nanotechnology, University of Twente, P.O. Box 217, 7500 AE Enschede, The Netherlands

The synthesis and characterization of PFS-*b*-PMMA block copolymers, obtained by a sequential polymerization approach, is described. The poly(ferrocenylsilane) (PFS) block was prepared by anionic polymerization of [1]dimethylsilaferrocenophane, and end-capped using 3-(*tert*-butyldimethylsiloxy)propyldimethylchlorosilane. Deprotection yielded a 3-hydroxypropyl-terminated PFS which was acylated with 2-bromoisobutyric anhydride. The 2-bromoisobutyryl end-functionalized PFS was used as a macroinitiator for controlled ATRP. Well-defined PFS-*b*-PMMA block copolymers with low polydispersities ($M_w/M_n < 1.1$) and controlled compositions were obtained.

Background

The self-organization of block copolymers constitutes a versatile means of producing ordered periodic structures with phase-separated microdomain sizes in the order of tens of nanometers.[1] By varying the relative molar masses of the blocks and the total molar mass, control over morphology, microdomain size and -spacing is easily achieved in the bulk state. In thin films, asymmetric block copolymers can self assemble to ordered monolayers, provided that next to block copolymer composition, film thickness is controlled. Block copolymer thin films[2] are of significant interest as templates for the fabrication of a variety of nanoscale structures such as periodic dot arrays,[3,4] nanopores[5] and nanowires.[6,7] In a number of applications, the utility of the produced patterns, such as etch resistance or conductivity relies on the introduction of inorganic elements in separate loading steps. Poly(ferrocenylsilane)s[8] owe a number of useful characteristics such as redox activity and a very high resistance to reactive ion etching[9] to the presence of alkylsilane and ferrocene units in the main chain, and combine these characteristics with the properties and processability of polymers. Asymmetric organic-organometallic block copolymers featuring isoprene[10] or styrene[11] units as organic, and ferrocenyldimethylsilane units as organometallic blocks, form monolayer thin films of densely packed organometallic spheres in an organic matrix. Due to the very high resistance of poly(ferrocenylsilane)s to reactive ion etching,[9] these thin films can be used as nanolithographic templates, enabling a direct transfer of the nanoscale patterns into underlying silicon or silicon nitride substrates with high aspect ratios.[12,13] The hybrid block copolymers also allow the patterning of a variety of thin metal films into dot arrays for e.g. high density magnetic data storage,[14] and are catalyst precursors for templated carbon nanotube growth.[15]

Objective

In block copolymer thin films, the presence of a substrate and surface can induce orientation of the microphase structure,[16] and can result in changes in domain dimensions or phase transitions[17,18] due to substrate or surface preferences of one of the blocks. After having established the thin film morphology of PI-b-PFS and PS-b-PFS diblock copolymers, we were interested to have access to PFS block copolymers with a larger variety of polar organic blocks, in particular poly(methyl methacrylate), which has a higher affinity for silicon substrates. In addition to organic-organometallic diblock copolymers featuring polar blocks, amphiphilic organometallic block copolymers are accessible when e.g. trimethylsilyl- or aminoalkyl methacrylate repeat units, which are readily transformed into ionic blocks by hydrolysis or quaternization, are employed. Until recently, the only reported example of a poly(ferrocenylsilane-*block*-methacrylate) block copolymer, featuring a 2-(dimethylaminoethyl) methacrylate block,[19] was obtained in a route applicable for this specific monomer. Thus, poly(ferrocenylsilane)-poly(methacrylate) block copolymers have remained largely unexplored. In this paper, we report the synthesis and characterization of poly(ferrocenylsilane-*block*-methyl methacrylate) block copolymers,[20] in which the organometallic block is formed by anionic polymerization, and the organic block by Atom Transfer Radical Polymerization (ATRP).[21] The latter technique allows the controlled polymerization of a wider variety of vinyl monomers than anionic polymerization schemes and imposes less strict requirements on monomer purity.

Simultaneous with this work, an elegant route to PFS-*b*-PMMA block copolymers was reported by Kloninger and Rehahn, who used anionic polymerization to form both the PFS and the PMMA block.[22] In this approach, living PFS chains react with 1,1-dimethylsilacyclobutane which acts as a "carbanion pump" to increase the reactivity of the living end, and are subsequently capped with 1,1-diphenylethylene. The resulting carbanion is a useful initiator for methyl methacrylate polymerization. In an alternative approach by Manners, living PFS chains in THF were activated to react with 1,1-diphenylethylene by heating the solution to 50 °C.[23]

Experimental

Materials. Ferrocene (98%), N,N,N',N'-tetramethylethylenediamine (TMEDA, 99.5+%), *n*-butyllithium (1.6 M in hexanes), chlorodimethylsilane (98%), dichlorodimethylsilane (99%), allyloxy-*tert*-butyldimethylsilane (97%), tetrabutylammonium fluoride (1.0 M in THF), diisobutylaluminium hydride (1.0 M in toluene), pyridine (anhydrous), methyl methacrylate (MMA, 99%), 2-bromo-2-methylpropionic acid (98%), 2-bromoisobutyryl bromide (98%), dichloro(*p*-cymene)ruthenium(II) dimer and tricyclohexylphosphine were obtained from Aldrich. Allyloxytrimethylsilane (>97%) and triethylamine (>99.5%) were obtained from Fluka. Platinum cyclovinylmethylsiloxane complex in cyclic methylvinylsiloxanes (3-3.5% Pt) was obtained from ABCR, Karlsruhe. *n*-Heptane (for synthesis) and 4-(dimethylamino)pyridine (for synthesis) were purchased from Merck. Tetrahydrofuran (THF) for anionic polymerizations was distilled from sodium-benzophenone under argon, degassed in three freeze-pump-thaw cycles and distilled by vacuum condensation from *n*-butyllithium. Toluene for ATRP was degassed on a high-vacuum line and distilled by condensation. 2-Bromoisobutyric anhydride was prepared according to a literature procedure.[24] ^{1}H NMR (CDCl$_3$): δ 2.00 (CH$_3$, s). The ATRP catalyst, (*p*-cymene)ruthenium(II) chloride-tricyclohexylphosphine, was synthesized as described earlier.[25]

Techniques. ^{1}H NMR spectra were recorded in CDCl$_3$ on a Varian Unity Inova (300 MHz) instrument at 300.3 MHz and on a Varian Unity 400 spectrometer at 399.9 MHz. A solvent chemical shift of δ = 7.26 ppm was used as a reference. GPC measurements were carried out in THF at 25 °C, using microstyragel columns (bead size 10 mm) with pore sizes of 10^5, 10^4, 10^3 and 10^6 Å (Waters) and a dual detection system consisting of a differential refractometer (Waters model 410) and a differential viscometer (Viscotek model H502). Molar masses were determined relative to polystyrene standards.

Monomer purification. [1]Dimethylsilaferrocenophane was prepared and purified as described earlier.[26,27] Methacrylate monomers were degassed on a high-vacuum line in three freeze-pump-thaw cycles and distilled under vacuum from calcium hydride immediately before use.

Anionic polymerization. [1]Dimethylsilaferrocenophane polymerizations were carried out in THF in an MBraun Labmaster 130 glove box under an atmosphere of prepurified nitrogen (<0.1 ppm H$_2$O), using *n*-butyllithium as initiator.

End-capping reagents.

3-(Trimethylsiloxy)propyldimethylchlorosilane (**1**). A two-necked 100 mL round bottom flask fitted with a septum and connected to a Schlenk line was evacuated and filled with argon. Toluene (10 mL), allyloxytrimethylsilane (8.2 g, 63 mmol), chlorodimethylsilane (11.9 g, 126 mmol) and platinum catalyst (2-3 droplets, ~ 5·10^{-6} mol Pt) were added. After stirring the mixture under argon at room temperature for four days, complete conversion was achieved. The reaction mixture was degassed on a vacuum line in three freeze-pump-thaw cycles and toluene and excess (CH$_3$)$_2$SiHCl were removed by vacuum condensation. The product was purified by vacuum distillation (b.p. 37-38 °C, 0.1 mm) and obtained as a colorless oil (isolated yield 12.7 g, 90%). ^{1}H NMR (CDCl$_3$): δ 3.56 (CH$_2$O, t, 6.6 Hz, 2H); 1.62 (CH$_2$, m, 2H); 0.81 (CH$_2$Si, m, 2H); 0.42 (ClSi(CH$_3$)$_2$, s, 6H); 0.10 (OSi(CH$_3$)$_3$, s, 9H).

3-(*Tert*-butyldimethylsiloxy)propyldimethylchlorosilane (**2**) was prepared by a Pt-catalyzed hydrosilylation reaction of allyloxy-*tert*-butyldimethylsilane and chlorodimethylsilane in toluene, similar as described for **1**. The product was purified by vacuum distillation (b.p. 50-52 °C, 0.1 mm) and obtained as a colorless oil (isolated yield 13.5 g, 86%). ^{1}H NMR (CDCl$_3$): δ 3.60 (CH$_2$O, t, 6.6 Hz, 2H); 1.62 (CH$_2$, m, 2H); 0.90 (SiC(CH$_3$)$_3$, s, 9H); 0.82 (CH$_2$Si, m, 2H); 0.42 (ClSi(CH$_3$)$_2$, s, 6H); 0.06 (s, 6H, OSi(CH$_3$)$_2$).

Macroinitiators. In a typical experiment, [1]dimethylsilaferrocenophane (600 mg, 2.48 mmol) in THF (5 mL) was polymerized by adding *n*-BuLi (0.25 mL of a 0.2 M solution in *n*-heptane, 5·10^{-5} mol) at room temperature. After 15 min, the solution was cooled to –70 °C and end-capper **1** (56 mg, 2.5·10^{-4} mol) was added. After stirring for 2 h at -70 °C, the mixture was allowed to come to room temperature and added to MeOH (50 mL) to precipitate **3a**, which was dried under vacuum. The trimethylsilyl end group was cleaved in a mixture of THF (5 mL), H_2O (0.4 mL) and AcOH (2 droplets) by stirring at room temperature for 24 hours, yielding hydroxypropyl-terminated PFS **4**, which was precipitated in *n*-heptane and dried under vacuum. Alternatively, living PFS was end-capped with **2** at 20 °C to produce **3b**. Cleavage of the TBDMS ether, by stirring with *i*-Bu_2AlH (0.5 mL, $5 \cdot 10^{-4}$ mol) in CH_2Cl_2 (10 mL) at 20 °C for 24 h under argon, followed by precipitation in MeOH, stirring in CH_2Cl_2/H_2O (pH = 5) and precipitation in *n*-heptane gave **4**. Acylation of **4** ($5 \cdot 10^{-5}$ mol OH) was carried out with 2-bromoisobutyric anhydride (0.16 g, $5 \cdot 10^{-4}$ mol) in dry pyridine (5 mL) in the presence of 4-(dimethylamino)pyridine (20 mg, $1.6 \cdot 10^{-4}$ mol). The mixture was stirred at room temperature under dry N_2 for 48 h. 2-Bromoisobutyryl end-functionalized PFS **5b** was precipitated in MeOH and dried under vacuum. $M_n = 1.20 \cdot 10^4$ g/mol, $M_w = 1.26 \cdot 10^4$ g/mol, $M_w/M_n = 1.05$.

Block copolymer synthesis. A glass tube containing a magnetic stirring bar was charged with a PFS macroinitiator (60-210 mg), (*p*-cymene)ruthenium(II) chloride-tricyclohexylphosphine (5 mg), methyl methacrylate (0.80 g, 8.0 mmol) and degassed toluene (0.8-1.5 mL) in the glove box and sealed. The mixture was stirred in a thermostated oil bath (80 °C) for 14 h. After cooling, the mixture was diluted with THF and dropwise added to MeOH to precipitate the product. The block copolymers obtained in this process were dried under vacuum.

Results & Discussion

Our synthetic approach to PFS-*b*-PMMA block copolymers combines the living anionic ring opening polymerization of [1]dimethylsilaferrocenophane[28] with a controlled/"living" radical polymerization of methyl methacrylate by means of ATRP.[29] Anionic polymerization of [1]dimethylsilaferrocenophanes allows one to form well-defined organometallic blocks with controlled block lengths and low polydispersities.[26,28]

In ATRP, free radicals are generated through a reversible redox process catalyzed by a transition metal complex. Uniform growth of chains is accomplished through fast initiation and a rapid reversible deactivation of free radicals.[21] An ATRP system consists of an initiator, a catalyst (transition metal complex) and monomer. For the synthesis of the methacrylate blocks, a ruthenium-based catalyst [$RuCl_2$(*p*-cymene)(PR_3)], with R=cyclohexyl, was chosen. Due to its high catalytic activity and control over the polymerization process, poly(methyl methacrylate) (PMMA) with polydispersities as low as $M_w/M_n < 1.1$ can be obtained using this catalyst complex.[30]

Various α-haloesters have been successfully employed as ATRP initiators. Among these, 2-bromoisobutyryl groups are particularly useful as initiator for the ATRP of methyl methacrylate,[31] as this initiator produces a radical which is structurally nearly identical to the propagating radical. Acrylates are polymerized successfully using 2-halopropionyl- and 2-haloisobutyryl groups as initiator.[21b,29]

Here, PFS homopolymers end-capped with a 2-bromoisobutyryl moiety serve as macroinitiators for methacrylate polymerization. This group is easily introduced by reacting hydroxyl-terminated PFS with e.g. 2-bromoisobutyryl bromide or 2-bromoisobutyric anhydride. We attempted to prepare hydroxyalkyl-terminated PFS by treating living PFS anions with styrene oxide, but incomplete end-functionalizations were found. Similar results were reported for the end-capping of living polystyrene with this reagent.[29]

We then explored the use of 3-(trimethylsiloxy)propyldimethylchlorosilane 1 and 3-(*tert*-butyldimethylsiloxy)propyldimethylchlorosilane 2 (Scheme 1) as end-capping reagents for living anionic PFS. Chlorosilanes in general are particularly successful end-capping reagents in anionic polymerization due to their high reactivity and lack of side reactions[32] and their utility can be increased further by incorporating protected functional groups. Nevertheless, only a few accounts have appeared in the literature on the use of such reagents in the end-functionalization of living polymer anions.[33,34] Chlorosilanes 1 and 2 were synthesised by the hydrosilylation of allyloxyalkylsilanes with chlorodimethylsilane in toluene (Scheme 1).

$$CH_2=CHCH_2OSiMe_2R \xrightarrow[\text{Toluene}]{\substack{HSiMe_2Cl \\ Pt\text{-}cat}} ClSiMe_2CH_2CH_2CH_2OSiMe_2R$$

1 R = Me, **2** R = *t*-Bu

Scheme 1. Synthesis of end-capping reagents with a protected hydroxyl functionality. Reprinted with permission from ref. 20. Copyright 2004 ACS.

The hydroxyl-protecting group should be stable in the presence of the PFS carbanion and cleave quantitatively under conditions that do not deteriorate the PFS chain. Trimethylsilyl ethers are cleaved under very mild conditions but are also prone to nucleophilic attack by organolithium compounds.[35] To prevent reaction of living PFS anions with the trimethylsilyl ether moiety of **1**, end-capping reactions were carried out at −70 °C. Hydrolysis of the trimethylsilyl ether in a mixture of THF, water and acetic acid produced the hydroxypropyl-terminated PFS **4**, which was then acylated with 2-bromoisobutyric anhydride in pyridine with 4-(dimethylamino)pyridine as a catalyst[24,36] to macroinitiator **5** (Scheme 2).

Tert-butyldimethylsilyl (TBDMS) ethers are stable towards organometallic reagents, allowing one to end-cap living polymers with **2** at room temperature. TBDMS ethers are usually cleaved either under acidic conditions or by treatment with tetra-*n*-butylammonium fluoride (TBAF) in THF.[37] Acidic hydrolysis is a less favorable option for cleaving a *tert*-butyldimethylsilyl protecting group, as PFS chains may not tolerate strongly acidic conditions. TBAF cleaves *tert*-butyldimethylsilyl ethers quantitatively but it also is a powerful desilylation reagent.[38] We observed a significant broadening and tailing to lower molar mass values of PFS in GPC traces upon deprotection of **3b** by TBAF. Clearly, TBAF causes chain scission in PFS. By treating **3b** with diisobutylaluminium hydride,[39] however, a clean deprotection of the TBDMS ether was achieved, yielding **4** without any molar mass decline (Scheme 2).

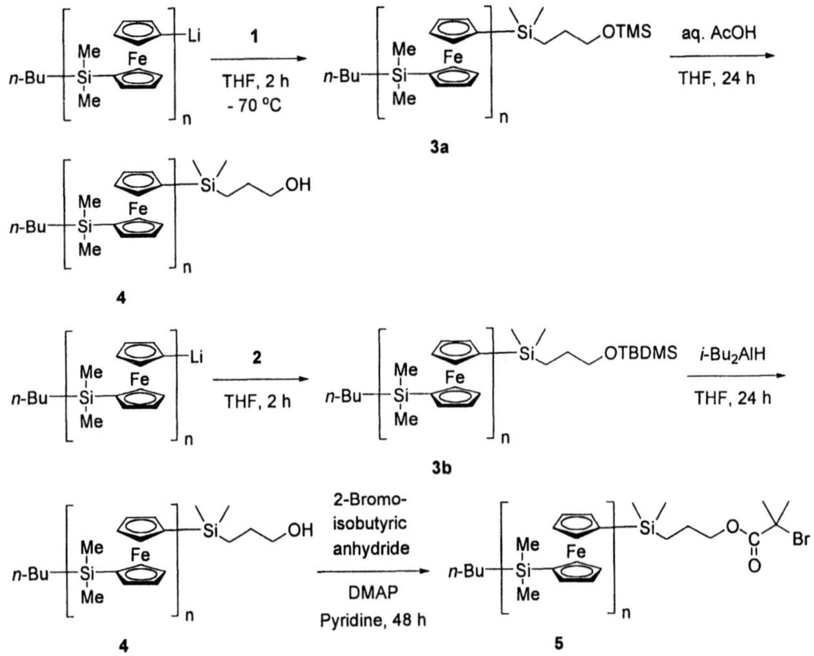

Scheme 2. *Synthesis of a poly(ferrocenylsilane) macroinitiator for ATRP. Reprinted with permission from ref. 20. Copyright 2004 ACS.*

Hydroxypropyl-terminated PFS **4** and corresponding macroinitiators **5** of varying molar mass were obtained and characterized using GPC and ^{1}H NMR. GPC showed **4** and **5** to be well-defined polymers, with polydispersities $M_w/M_n < 1.1$, indicating that the end-capping and deprotection steps did not cause any molar mass decline of the PFS chains. A diagnostic end group signal clearly observed in the ^{1}H NMR spectra of **4** was a triplet at $\delta = 3.56$ ppm, belonging to the methylene protons adjacent to the hydroxyl group (Figure 1). On several occasions, the signal of the CH$_2$ next to the OH appeared as a quartet rather than as the triplet shown in Figure 1. This is typical of such moieties and supports the assignment made for this signal. This methylene signal was no longer visible in the ^{1}H NMR spectra of **5**, nor was the signal of the acyloxymethylene group due to overlap with the ferrocenyl signals of the PFS chain. Clearly present, however, was the characteristic singlet at δ = 1.93 ppm due to the 2-bromoisobutyryl moiety (Figure 2). Also, the signal due to the CH$_3$ group of the *n*-butyl initiator moiety was observed at δ = 0.90 ppm. End group analysis of **5** was performed, based on the ratio of the initiator and end-capper integrals to the ferrocenyl integral at δ = 4.01 and 4.22 ppm.

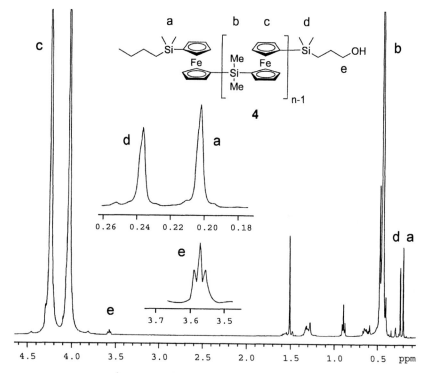

*Figure 1. 400 MHz ^{1}H NMR spectrum of the hydroxypropyl-terminated PFS **4**.*

Similarly, the dimethylsilyl signal of the end-capper moiety was compared with the dimethylsilyl signal of the polymer backbone. Table 1 summarizes the number-average degrees of polymerisation DP$_n$ obtained by GPC and by ^{1}H NMR. Comparison of the dimethylsilyl signals of the initiator (s, δ = 0.21 ppm) and end-capper (s, δ = 0.24 ppm) moieties of **4**, and of the corresponding signals of **5** constitutes another way to gauge the degree of end-functionalization. Based on the ^{1}H NMR integrals one can conclude that end-capping living PFS with 3-(*tert*-butyldimethylsiloxy)propyldimethylchlorosilane **2** (Table 1, **5c**) leads to higher degrees of functionalization compared to 3-(trimethylsiloxy)propyl-dimethylchlorosilane **1**. GPC measurements following the formation of the second block should provide further information on end-capping efficiency.

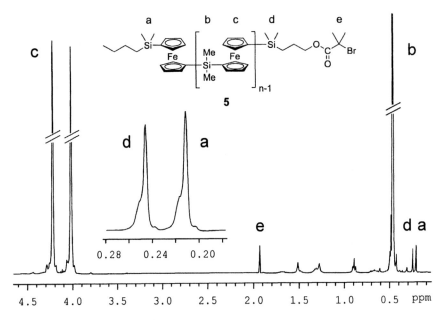

Figure 2. 400 MHz ^{1}H NMR spectrum of the PFS macroinitiator 5.

The 2-bromoisobutyryl end-capped poly(ferrocenylsilane) **5** was then used as initiator in the ruthenium-mediated living polymerization of methyl methacrylate (Scheme 3). In a typical experiment, macroinitiator **5a** (60 mg), ruthenium complex (5 mg), methyl methacrylate (0.80 g, 8.0 mmol) and toluene (0.8 mL) were placed in a glass tube under a nitrogen atmosphere. The tube was sealed and the reaction mixture was stirred at 80 °C for 14 h. After precipitation in methanol and drying under vacuum, PFS-*b*-PMMA block copolymer **6a** was obtained.

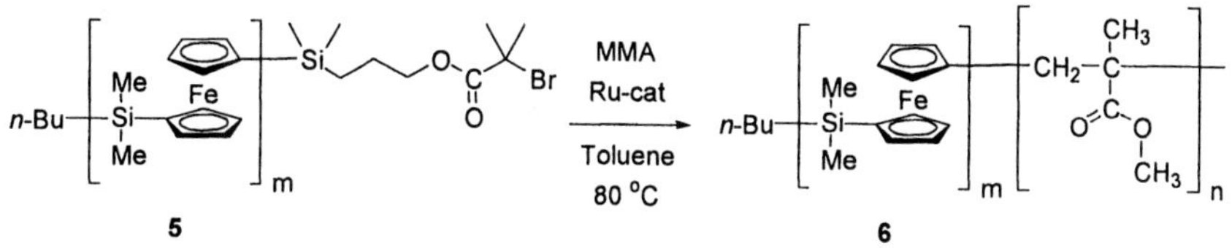

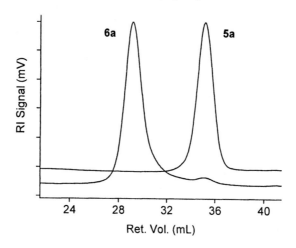

Scheme 3. Synthesis of poly(ferrocenyldimethylsilane-block-methyl methacrylate) block copolymers. Reprinted with permission from ref. 20. Copyright 2004 American Chemical Society.

Figure 3 shows the GPC traces of macroinitiator **5a** and PFS-*b*-PMMA block copolymer **6a**. The main part of the PFS macroinitiator was converted into block copolymer, indicating that the end-capping strategy (Scheme 2) enabled an efficient incorporation of 2-bromoisobutyryl groups.

*Figure 3. GPC traces of PFS macroinitiator **5a** and its corresponding PFS-b-PMMA block copolymer **6a**. Reprinted with permission from ref. 20. Copyright 2004 American Chemical Society.*

Control over block copolymer composition and molar mass was achieved by varying the initiator concentration $[I]_0$ with respect to monomer concentration $[M]_0$. Alternatively, the molar mass of the PFS macroinitiator can be tuned, while keeping the monomer to initiator ratio $[M]_0/[I]_0$ constant. Characteristic GPC traces of obtained PFS-*b*-PMMA block copolymers are shown in Figure 4. The traces of block copolymers **6a**, **6b$_1$** and **6b$_2$** show small amounts of the PFS block, which was end-capped using **1**. In the GPC trace of block copolymer **6c**, however, where end-capper **2** was used, the initiator block was not detected. Based on the ^{1}H NMR and the GPC results, we estimate that end-capping with **2**, followed by reductive cleavage of the TBDMS ether by i-Bu$_2$AlH and acylation gives a degree of end-functionalization of 90-95%.

For block copolymers **6a**, **6b$_1$**, **6b$_2$** and **6c**, MMA conversions of 45%, 60%, 58% and 60% were reached, respectively, in 14 h. Block copolymer compositions were determined using ^{1}H NMR, by relating the integral of the CH$_3$O signal at $\delta = 3.59$ ppm belonging to the methacrylate repeat units, to the ferrocenyl signal integral at $\delta = 4.01$ and 4.22 ppm. Combined with M_n values of the PFS block obtained by GPC, total block copolymer molar masses M_n were calculated and compared with values obtained by GPC. The results, summarized in Table 2, are in good agreement. The block copolymers described here are asymmetric, with PFS weight fractions ranging from 7 to 26%.

Table 1. Degrees of Polymerization DP$_n$ of Various PFS Macroinitiators 5 Based on GPC and ^{1}H NMR Analysis.

	$M_{n,GPC}$ [a] (g/mol)	M_w/M_n	$DP_{n,GPC}$ [b]	$DP_{n,NMR}$ [c] $-CH_3$ $(n$-$Bu)$	$DP_{n,NMR}$ [d] $-C(CH_3)_2Br$	$DP_{n,NMR}$ [e] $-Si(CH_3)_2$
5a	8.000	1.02	33	34	49	46
5b	12.000	1.05	50	44	64	72
5c [f]	12.000	1.07	50	46	60	52

[a] Measured by GPC, relative to polystyrene standards. [b] Calculated from M_n determined by GPC. [c] Calculated from ferrocenyl and initiator integrals (CH$_3$). [d] Calculated from ferrocenyl and 2-bromoisobutyryl integrals. [e] Calculated from backbone Si(CH$_3$)$_2$ and end-capper Si(CH$_3$)$_2$ integrals. [f] Obtained via **3b**, deprotected using i-Bu$_2$AlH. Reprinted with permission from ref. 20. Copyright 2004 American Chemical Society.

Conclusions

In summary, a route to 2-bromoisobutyryl end-functionalized poly(ferrocenyldimethylsilane), allowing essentially quantitative end-capping, is described. Following the formation of the organometallic block by anionic polymerization, a ruthenium-mediated controlled radical polymerization of methyl methacrylate was employed to grow the second block. Well-defined PFS-b-PMMA block copolymers with low polydispersities ($M_w/M_n < 1.1$) and controlled compositions were obtained. The use of ATRP to form the second block implies that a wide variety of (meth)acrylic and other vinyl monomers can be used, thus opening up the way to novel hybrid organic-organometallic block copolymers and amphiphilic organometallic block copolymers.

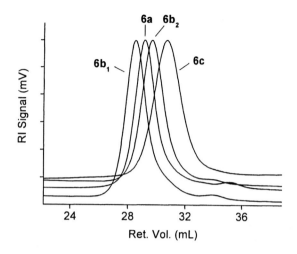

Figure 4. GPC traces of representative PFS-b-PMMA block copolymers. Reprinted with permission from ref. 20. Copyright 2004 ACS.

Table 2. Molar Mass Characteristics of PFS-*b*-PMMA Block Copolymers 6.

	$M_{n,GPC}$ [a] (g/mol) PFS-Br	$M_{n,NMR}$ [b] (g/mol) PMMA	$M_{n,NMR}$ (g/mol) PFS-b-PMMA	$M_{n,GPC}$ [a] (g/mol) PFS-b-PMMA	M_w/M_n PFS-b-PMMA	Wt. % PFS [c]
6a [d]	8.000 (5a)	114.000	122.000	95.700	1.06	7
6b$_1$ [d]	12.000 (5b)	116.300	128.300	101.700	1.10	9
6b$_2$ [d]	12.000 (5b)	53.200	65.200	66.900	1.06	18
6c [d]	12.000 (5c)	33.600	45.600	40.500	1.18	26

[a] Measured by GPC, relative to polystyrene standards. [b] Calculated from ferrocenyl and CH_3O integrals, based on $M_{n,GPC}$ of the PFS block. [c] Calculated from ferrocenyl and CH_3O integrals. [d] **6a**: MMA = $8.0 \cdot 10^{-3}$ mol, PFS-Br = $7.5 \cdot 10^{-6}$ mol, toluene 0.8 mL, Ru catalyst 5 mg, MMA conversion 45%. **6b$_1$**: MMA = $7.3 \cdot 10^{-3}$ mol, PFS-Br = $8.5 \cdot 10^{-6}$ mol, toluene 1.0 mL, Ru catalyst 5 mg, MMA conversion 60%. **6b$_2$**: MMA = $4.0 \cdot 10^{-3}$ mol, PFS-Br = $8.8 \cdot 10^{-6}$ mol, toluene 1.0 mL, Ru catalyst 5 mg, MMA conversion 58%. **6c**: MMA = $7.2 \cdot 10^{-3}$ mol, PFS-Br = $1.8 \cdot 10^{-5}$ mol, toluene 1.5 mL, Ru catalyst 5 mg, MMA conversion 60%. Reprinted with permission from ref. 20. Copyright 2004 American Chemical Society.

Acknowledgements

The authors thank C. J. Padberg (University of Twente) for the GPC measurements, and gratefully acknowledge the University of Twente and the Masif Program of the MESA$^+$ Institute for Nanotechnology for financial support.

References

1   (a) Hamley, I. W. *Nanotechnology* **2003**, *14*, R39-R54. (b) Park, C.; Yoon, J.; Thomas, E. L. *Polymer* **2003**, *44*, 6725-6760.

2   Fasolka, M. J.; Mayes, A. M. *Annu. Rev. Mater. Res.* **2001**, *31*, 323-355.

3   Park, M.; Harrison, C.; Chaikin, P. M.; Register, R. A.; Adamson, D. H. *Science* **1997**, *276*, 1401-1404.

4   Spatz, J. P.; Herzog, T.; Mößmer, S.; Ziemann, P.; Möller, M. *Adv. Mater.* **1999**, *11*, 149-153.

5   Xu, T.; Stevens, J.; Villa, J. A.; Goldbach, J. T.; Guarini, K. W.; Black, C. T.; Hawker, C. J.; Russell, T. P. *Adv. Funct. Mater.* **2003**, *13*, 698-702.

6   Thurn-Albrecht, T.; Schotter, J.; Kästle, G. A.; Emley, N.; Shibauchi, T.; Krusin-Elbaum, L.; Guarini, K.; Black, C. T.; Tuominen, M. T.; Russell, T. P. *Science* **2000**, *290*, 2126-2129.

7   Lopes, W. A.; Jaeger, H. M. *Nature* **2001**, *414*, 735-738.

8   For reviews on poly(ferrocenylsilane)s see (a) Manners, I. *Macromol. Symp.* **2003**, *196*, 57-62. (b) Kulbaba, K.; Manners, I. *Macromol. Rapid Commun.* **2001**, *22*, 711-724. (c) Manners, I. *Chem. Commun.* **1999**, 857-865.

9   Lammertink, R. G. H.; Hempenius, M. A.; Chan, V. Z.-H.; Thomas, E. L.; Vancso, G. J. *Chem. Mater.* **2001**, *13*, 429-434.

10 Lammertink, R. G. H.; Hempenius, M. A.; Vancso, G. J. *Langmuir* **2000**, *16*, 6245-6252.

11 Lammertink, R. G. H.; Hempenius, M. A.; Vancso, G. J.; Shin, K.; Rafailovich, M. H.; Sokolov, J. *Macromolecules* **2001**, *34*, 942-950.

12 Lammertink, R. G. H.; Hempenius, M. A.; Van den Enk, J. E.; Chan, V. Z.-H.; Thomas, E. L.; Vancso, G. J. *Adv. Mater.* **2000**, *12*, 98-103.

13 Cheng, J. Y.; Ross, C. A.; Thomas, E. L.; Smith, H. I.; Vancso, G. J. *Appl. Phys. Lett.* **2002**, *81*, 3657-3659.

14 Cheng, J. Y.; Ross, C. A.; Chan, V. Z.-H.; Thomas, E. L.; Lammertink, R. G. H.; Vancso, G. J. *Adv. Mater.* **2001**, *13*, 1174-1178.

15 Hinderling, C.; Keles, Y.; Stöckli, T.; Knapp, H. F.; de los Arcos, T.; Oelhafen, P.; Korczagin, I.; Hempenius, M. A.; Vancso, G. J.; Pugin, R.; Heinzelmann, H. *Adv. Mater.* **2004**, *16*, 876-879.

16 Anastasiadis, S. H.; Russell, T. P.; Satija, S. K.; Majkrzak, C. F. *Phys. Rev. Lett.* **1989**, *62*, 1852-1855.

17 Spatz, J. P.; Sheiko, S.; Möller, M. *Adv. Mater.* **1996**, *8*, 513-517.

18 Spatz, J. P.; Möller, M.; Noeske, M.; Behm, R. J.; Pietralla, M. *Macromolecules* **1997**, *30*, 3874-3880.

19 Wang, X.-S.; Winnik, M. A.; Manners, I. *Macromol. Rapid Commun.* **2002**, *23*, 210-213.

20 Korczagin, I.; Hempenius, M. A.; Vancso, G. J. *Macromolecules* **2004**, *37*, 1686-1690.

21 (a) Wang, J.-S.; Matyjaszewski, K. *Macromolecules* **1995**, *28*, 7572-7573. (b) Matyjaszewski, K.; Xia, J. *Chem. Rev.* **2001**, *101*, 2921-2990.

22 Kloninger, C.; Rehahn, M. *Macromolecules* **2004**, *37*, 1720-1727.

23 Power-Billard, K. N.; Wieland, P.; Schäfer, M.; Nuyken, O.; Manners, I. *Macromolecules* **2004**, *37*, 2090-2095.

24 Ohno, K.; Wong, B.; Haddleton, D. M. *J. Polym. Sci. Part A: Pol. Chem.* **2001**, *39*, 2206-2214.

25 Demonceau, A.; Stumpf, A. W.; Saive, E.; Noels, A. F. *Macromolecules* **1997**, *30*, 3127-3136.

26 Ni, Y.; Rulkens, R.; Manners, I. *J. Am. Chem. Soc.* **1996**, *118*, 4102-4114.

27 Lammertink, R. G. H.; Hempenius, M. A.; Thomas, E. L.; Vancso, G. J. *J. Pol. Sci. Part B: Pol. Phys.* **1999**, *37*, 1009-1021.

28 Rulkens, R.; Ni, Y.; Manners, I. *J. Am. Chem. Soc.* **1994**, *116*, 12121-12122.

29 Acar, M. H.; Matyjaszewski, K. *Macromol. Chem. Phys.* **1999**, *200*, 1094-1100.

30 Simal, F.; Demonceau, A.; Noels, A. F. *Angew. Chem. Int. Edit.* **1999**, *38*, 538-540.

31 Ando, T.; Kamigaito, M.; Sawamoto, M. *Tetrahedron* **1997**, *53*, 15445-15457.

32 Hadjichristidis, N.; Fetters, L. J. *Macromolecules* **1980**, *13*, 191-193.

33 Peters, M. A.; Belu, A. M.; Linton, R. W.; Dupray, L.; Meyer, T. J.; DeSimone, J. M. *J. Am. Chem. Soc.* **1995**, *117*, 3380-3388.

34 Tezuka, Y.; Imai, H.; Shiomi, T. *Macromol. Chem. Phys.* **1997**, *198*, 627-641.

35 Greene, T. W.; Wuts, P. G. M. *Protective Groups in Organic Synthesis*, 3rd ed.; Wiley & Sons, Inc.: New York, 1999.

36 Höfle, G.; Steglich, W.; Vorbrüggen, H. *Angew. Chem.* **1978**, *90*, 602-615.

37 Corey, E. J.; Venkateswarlu, A. *J. Am. Chem. Soc.* **1972**, *94*, 6190-6191.

38 Oda, H.; Sato, M.; Morizawa, Y.; Oshima, K.; Nozaki, H. *Tetrahedron* **1985**, *41*, 3257-3268.

39 Corey, E. J.; Jones, G. B. *J. Org. Chem.* **1992**, *57*, 1028-1029.

Chapter 24

Synthesis and Layer-by-Layer Assembly of Water-Soluble Polyferrocenylsilane Polyelectrolytes

Zhuo Wang, Geoffrey A. Ozin[*], and Ian Manners[*]

Department of Chemistry, University of Toronto, 80 St. George Street, Toronto, Ontario M5S 3H6, Canada

Polyferrocenylsilanes (PFSs) are an interesting class of high molecular weight transition metal-containing polymers which exhibit attractive redox, semiconductive, preceramic and other physical properties. Convenient routes to a range of cationic and anionic water-soluble PFS polyelectrolytes have been achieved. These metal-containing materials are readily soluble in water and have been utilized in the self-assembly of multilayer superlattices. The electrostatic assembly of the water-soluble PFSs have also been used in tuning the optical properties of photonic crystals with nanometer scale precision.

Background

Polyelectrolytes are a class of macromolecules which possess ionic groups along the polymer chains. Most of these macromolecules are water soluble and are of considerable importance to many industrial and technological applications[1]. Polyelectrolytes have attracted intense interest since the early 1990's as the building blocks for electrostatic superlattices[2,3]. These are multilayer structures that are assembled by sequential adsorption of oppositely charged polyelectrolyte monolayers from aqueous solutions. The resulting multilayers are nanometer-scale ultra thin films which have found wide applications in many areas[3]. For example, the electrostatically assembled multilayers have been used successfully in surface property modifications and surface patterning, as well as in manufacturing of thin film devices and development of capsules for controlled drug releases[4]. Interestingly, despite extensive studies, the polyelectrolytes used in the layer-by-layer assembly process are generally limited to organic macromolecules. To date examples of inorganic polyelectrolytes are rare. This is especially the case for transition metal-containing polymers, which would be expected to bring many attractive features, such as conductive, optical, redox, and catalytic properties, to the ultra-thin multilayered films[5,6].

Polyferrocenylsilanes (PFSs) are a class of high molecular weight metal-containing polymers with many interesting properties[5]. These polymers are readily accessible via the thermal, anionic, transition metal-catalyzed, and most recently, photolytic anionic ring-opening polymerization (ROP) of strained silicon-bridged [1]ferrocenophanes[7,8].

The PFS materials have demonstrated reversible redox activity and semiconductivity after oxidation[7]. Their ability to dissipate charge makes them attractive materials as radiation protective coatings[9]. PFSs also provide access to molded networks which can be pyrolyzed to give nanostructured ceramic materials with tunable magnetic properties[10]. For example, ceramic films, macroscopic shapes, and micrometer-scale patterned structures have been successfully generated by this means[11,12]. The introduction of acetylene moieties to the PFS has made it possible to incorporate a variety of other transition metals to the polymers[13,14]. Highly metallized PFSs containing cobalt clusters have been synthesized and provide a route to both patterned and well-ordered 2-dimentional nanostructures with interesting magnetic properties[15-17]. In addition, PFS block copolymers have been generated via anionic and transition metal-catalyzed ROP and form phase-separated domains in the solid state and self-assembled micellar aggregates in solution[18,19]. Pyrolysis or plasma treatment of well-ordered PFS domains affords catalytically-active iron nanoclusters which can be utilized to generate carbon nanotubes[20]. PFS materials also offer potential applications in microsphere technology[21], as photonic band gap materials[22,23], as variable refractive index sensing materials[24] and as etch resists[25].

Objective

Organometallic superlattices formed by layer-by-layer electrostatic assembly of PFS polyelectrolytes are of great interest and expected to provide novel properties. The objectives of this study were to synthesize both cationically and anionically charged water-soluble PFSs, assemble organic-organometallic and all organometallic electrostatic superlattices on various substrates, and study the assembly behavior of the polyelectrolytes.

Experimental

All chemicals were purchased from Aldrich and used as received unless otherwise specified. (3-Aminopropyl)trimethoxysilane was purchased from Gelest Inc.. Polymer [fcSiMeCl]$_n$ (1)[26], fcSiMeCl (3), fcSiMe(2-C$_6$H$_4$CH$_2$NMe$_2$) (4)[27], and LiC≡CCH$_2$N(SiMe$_2$CH$_2$)$_2$ (6)[28] were synthesized according to literature procedures. NMR characterizations and molecular weight determinations were carried out as previously described[28-31].

The concentrations of the polyelectrolyte solutions were 10 mM PSS and 10 mM PFS in 0.1 M NaCl aqueous solutions. The preparation of Si and Au substrates and the layer-by-layer assembly of polyelectrolytes was carried out as previously described[32,33].

Quartz crystal microbalance (QCM) measurement data, ellipsometric data, UV-vis spectra of quartz-supported PSS-PFS films, X-ray photoelectron spectra, and advancing contact angles for water were obtained as previously described[32].

Results & Discussion

Synthesis of the Cationic PFS Polyelectrolytes

The first water-soluble PFS was prepared via nucleophilic substitution reactions on poly(ferrocenylmethylchlorosilane) **1** with alkoxide nucleophiles that possess dimethylamino functional groups (Scheme 1)[29]. Subsequent quaternization of the amino groups with methyl iodide afforded the cationic polymer **2**.

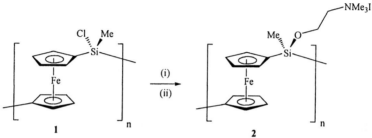

Scheme 1. (i) HOCH$_2$CH$_2$NMe$_2$, Et$_3$N, toluene, 25°C. (ii) MeI, CH$_2$Cl$_2$, reflux.

Another synthetic route to water-soluble cationic PFS is via thermal ring-opening polymerization of monomer **4** which was generated from fcSiMeCl (**3**) via substitution of the chloride with an aryl lithium reagent (Scheme 2)[27,34].

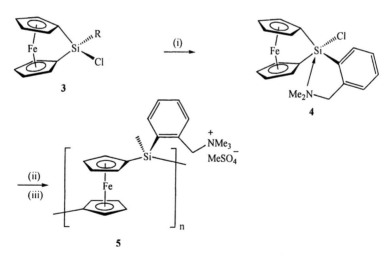

Scheme 2. (i) 2-C$_6$H$_4$CH$_2$NMe$_2$Li, THF, -78 °C. (ii) Xylene, 160 °C, 16 h. (iii) Me$_2$SO$_4$, CH$_2$Cl$_2$, 25 °C, 24 h.

Quaternization of the amino groups in the resulting polymer with dimethyl sulfate afforded the cationic polymer **5** (Scheme 2)[30]. Complete reaction was confirmed by ^1H, ^{13}C and ^{29}Si NMR in both D_2O and deuterated DMSO.

The cationic polyferrocenylsilanes were also envisioned to arise from the protected aminoalkynyl PFS **7**[28]. It was expected that, upon silyl group removal, polymer **7** would function as a precursor to PFS polyelectrolytes via quaternization and other derivatization methods. PFS **7** was readily accessible through nucleophilic substitution of the chlorine atoms on poly(ferrocenylmethylchlorosilane) **1**[26] with a lithium acetylide reagent (**6**) (Scheme 3)[28,35]. The lithium acetylide **6** was prepared in situ from $HC\equiv CCH_2N(SiMe_2CH_2)_2$[28,36] and *n*-BuLi. Complete substitution of the chlorine side groups of **1** was confirmed by ^1H, ^{13}C and ^{29}Si NMR. GPC analysis of polymer **7** showed that the material possessed a molecular weight of $M_w = 304,100$ and a PDI = 1.48. Subsequent removal of the cyclic disilyl protecting groups was achieved under mild conditions in a mixture of THF+MeOH solvents (3:1 v/v) to give the aminopropynyl PFS **8**. Further reduction of **9** with hydrazine under air afforded polymer **9**.

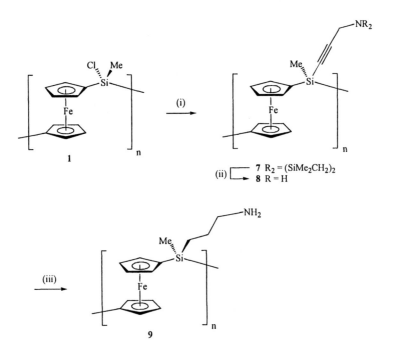

Scheme 3. (i) $LiC\equiv CCH_2N(SiMe_2CH_2)_2$ (6), THF, 0 to 25 °C. (ii) THF + MeOH, 25 °C, 4 days. (iii) Hydrazine, THF, 25 °C.

Alternatively, PFS **7** can be prepared via catalytic ROP of monomer **10**, which was derived from fcSiMeCl (**3**) via nucleophilic substitution reaction with LiC≡CCH₂N(SiMe₂CH₂)₂ (**6**) (Scheme 4)[28].

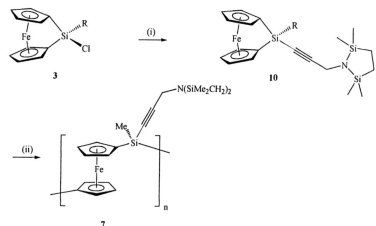

Scheme 4. (i) LiC≡CCH₂N(SiMe₂CH₂)₂ (6), THF, -78 to 25 °C. (ii) PtCl₂, Toluene, 25 °C.

PFS **11** and **12** with quaternary ammonium substituents were readily obtained through protonation of the amino polymers **8** and **9**, respectively, with 1 equiv of HCl (1 M solution in ether) (Figure 1). Both cationic polymers were

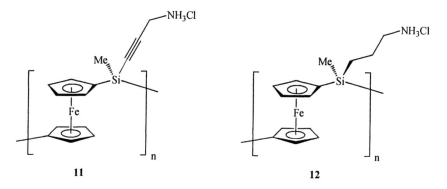

Figure 1. Cationic PFS polyelectrolytes 11 and 12.

very soluble in water and their solutions remained clear orange-yellow over a period of several months.

Synthesis of Anionic PFS Polyelectrolytes

The synthesis of anionic PFSs was achieved using the reaction of the polymer **7** with 1,3-propane sultone[34,35] in the presence of an excess amount of diisopropylethylamine in a mixture of solvents (THF+MeOH, 2:1 v/v) (Scheme 5). Precipitation of the crude product into an acetone solution containing sodium hexafluorophosphate or sodium tetraphenylborate for cation-exchange purposes afforded polymer **13** as an orange-yellow powder in high yield (≥90%). Subsequent reduction of **13** with hydrazine yielded PFS **14**. Both anionic polymers were yellow powders, highly soluble in water, and remained stable in water after a period of several months.

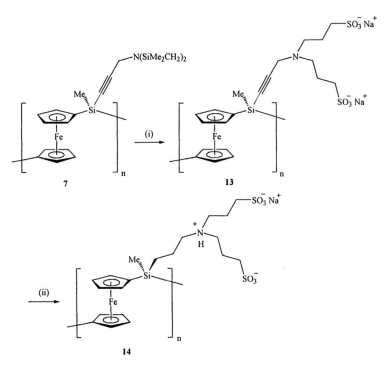

Scheme 5. (i) 1,3-propane sultone, (i-Pr)₂EtN, THF + MeOH, 25 °C. (ii) Hydrazine, H₂O, 25 °C..

Layer-by-Layer Assembly of Organic-Organometallic and All Organometallic Polymer Superlattices

The assembly of the electrostatic superlattices was accomplished by the alternate adsorption of the anionic and cationic polyelectrolytes on Si and Au substrates that have been chemically modified with (3-aminopropyl) trimethoxysilane (APMS) and 2-aminoethanethiol (AET), respectively[32]. In the following discussion, the organic-organometallic PSS-PFS multilayer films were assembled from the anionic poly(styrenesulfonate) (PSS) and cationic PFS **5**, while the all organometallic PFS-PFS multilayers were assembled from the anionic polyelectrolytes PFS **14** and cationic PFS **5** [32,33]. A schematic representation of the sequential adsorption process is shown in Figure 2.

The LbL assembly of the organic-organometallic polymer superlattices was monitored using a QCM derivatized with AET[32]. The measurement of the frequency change of a 10 MHz QCM was plotted against the number of bilayers deposited on the substrate (Figure 3a). A slight nonlinearity was observed in the initial three deposition steps. This is probably due to the initial substrate priming step or the LbL adsorption behavior of the polyelectrolytes in the early dipping steps. After the initial steps, a linear growth of the multilayers was observed.

The increase of film thickness on both Si and Au substrates during the assembly process was observed by using ellipsometry measurements (Figure 3b)[32]. The nearly superimposable curves for films grown on both substrates suggest that the growth pattern of the films is independent of the substrates. The average growth of thickness for each PSS and PFS layer is 7 ± 3 Å and 16 ± 5 Å, respectively. The thicker layer of the latter is attributed to its lower charge density, therefore a larger amount of the PFS polyelectrolyte is needed to compensate for the surface charges of the PSS layer.

The linear sequential buildup of the PSS-PFS superlattices on quartz substrates was followed by UV-vis spectroscopy (Figure 4)[32]. The characteristic absorptions of PFS are centered at 220 nm, an intense ligand-to-metal charge-transfer transition (LMCT), and at 450 nm, a much weaker d – d transition in the visible region. The PSS $\pi - \pi^*$ transition is centered at 260 nm. The absorbances at 220 nm, 260 nm and 450 nm increased linearly with the increase of the bilayer number. This indicates a regular increase of adsorbed polymer with each deposition step.

The linear relationship between the superlattice absorbance and the deposition layer number was again observed for the all organometallic PFS-PFS superlattice on quartz (Figure 5)[33]. These UV-vis results, in conjunction with the QCM and ellipsometry data, support the proposed LbL assembly of electrostatic superlattices as illustrated schematically in Figure 2.

The surface composition of the PSS-PFS films grown on Au substrates was measured by XPS[32]. For the film composed of 10 layers of polyelectrolytes, the photoelectron spectrum measured at take-off angle 90° was compared to the

342

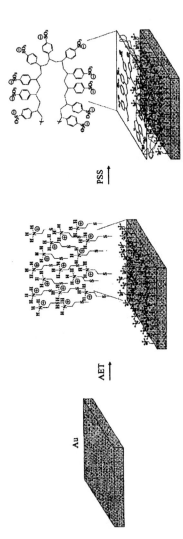

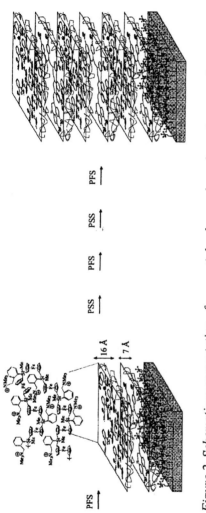

Figure 2. Schematic representation of sequential polymer electrolyte adsorption for growing PSS-PFS organic-organometallic multilayer thin films. (Reproduced with permission from reference 32.)

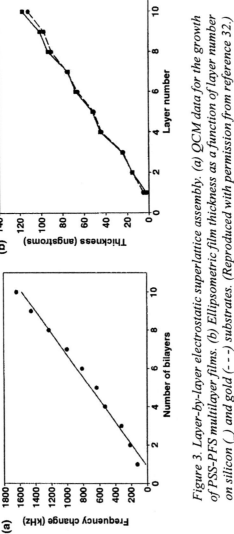

Figure 3. Layer-by-layer electrostatic superlattice assembly. (a) QCM data for the growth of PSS-PFS multilayer films. (b) Ellipsometric film thickness as a function of layer number on silicon (_) and gold (- - -) substrates. (Reproduced with permission from reference 32.)

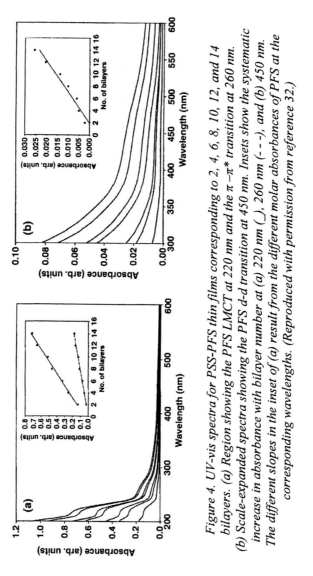

Figure 4. UV-vis spectra for PSS-PFS thin films corresponding to 2, 4, 6, 8, 10, 12, and 14 bilayers. (a) Region showing the PFS LMCT at 220 nm and the π–π* transition at 260 nm. (b) Scale-expanded spectra showing the PFS d-d transition at 450 nm. Insets show the systematic increase in absorbance with bilayer number at (a) 220 nm (—), 260 nm (- - -), and (b) 450 nm. The different slopes in the inset of (a) result from the different molar absorbances of PFS at the corresponding wavelengths. (Reproduced with permission from reference 32.)

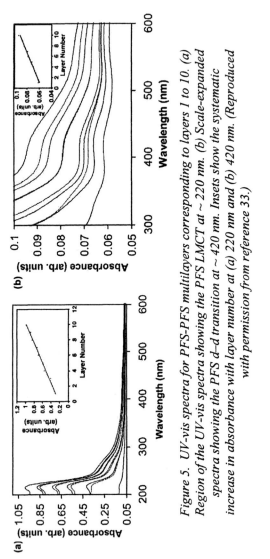

Figure 5. UV-vis spectra for PFS-PFS multilayers corresponding to layers 1 to 10. (a) Region of the UV-vis spectra showing the PFS LMCT at ~ 220 nm. (b) Scale-expanded spectra showing the PFS d–d transition at ~ 420 nm. Insets show the systematic increase in absorbance with layer number at (a) 220 nm and (b) 420 nm. (Reproduced with permission from reference 33.)

spectrum measured at take-off angle 20°. In the latter case, while peaks for C, N, O, S, and Fe, which are present at the surface, are clearly observed, the Au peaks are greatly reduced in comparison to the spectrum measured at take-off angle 90°. This result indicates that the Au surface is essentially completely covered by the 10 layer film. The magnified Fe (2p) and S (1s) regions for the films composed of 6 and 10 layers illustrate the expected increase in relative intensity with the increase in layer buildup.

The wettability properties of Au surfaces modified with the PSS-PFS layers were examined by using the advancing water contact angles measurements (Figure 6)[32]. With the exception of the first PSS monolayer adsorbed on an AET-primed gold surface which gave an advancing water contact angle of 55° ± 2°[37], the advancing contact angle systematically alternates between 75° ± 2° for PSS-terminated surfaces and 67° ± 5° for PFS-terminated surfaces. The consistent observation that the PFS-terminated surfaces have smaller advancing contact angle in comparison to the PSS-terminated surfaces indicates that the former are slightly more hydrophilic than the latter. These results also demonstrate that the wettability of surfaces can be simply tuned by altering the outermost surface layer of the multilayer assembly[37,38].

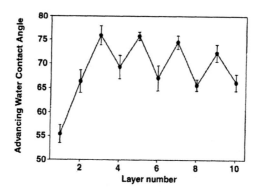

Figure 6. Advancing water contact angle on PSS (odd numbered layers) and PFS(even numbered layers)-terminated multilayer films grown on Au substrates. (Reproduced from reference 32 with permission.)

The topography of the PSS-PFS film composed of 10 layers of polyelectrolyte on a Au substrate was examined by AFM (Figure 7)[32]. The distribution of grain sizes in the gold-supported multilayers is $2000 - 10000$ nm^2, larger than the gold globules for the bare substrate surface (distribution of grain sizes: $450 - 2000$ nm^2). This indicates a smoothing of the Au substrate surface. In addition, the root-mean-square (rms) roughness values measured for the 10 layer film and the bare gold substrate were 6 and 9 Å, respectively, over a 500 nm^2 scanned area. The minor variation in the surface topography and the rms roughness between the bare substrate surface and the PSS-PFS-deposited surface indicates that the growth of the multilayers follows the substrate surface morphology.

The data from QCM, ellipsometry, UV-vis, XPS, contact angle and AFM measurements of the PSS-PFS and PFS-PFS multilayers demonstrate that LbL assembly of oppositely charged metal-containing polyelectrolytes creates regularly stacked, homogeneous electrostatic superlattices with essentially complete surface coverage[32].

Direct Visualization of PSS-PFS and PFS-PFS Multilayers

The electrostatically assembled multilayers can be directly visualized by using a gold coating/transmission electron microscopy (TEM) technique[33]. Multilayers of PSS-PFS were formed through LbL assembly on a gold substrate. Once the deposition was complete, a thin layer of gold was sputtered on top of the film. As there is no intrinsic electronic contrast between the multilayer and surrounding medium, the gold substrate and top sputtered layer provide the contrast needed for TEM visualization. The cross-section images of the multilayer films consisting of 5, 10, 20, and 30 bilayers of alternating PFS and PSS are shown in Figure 8[33]. The uneven coverage of the first several multilayers can be clearly seen in the 5 and 10 bilayer superlattices. The regular increase in multilayer thickness can be visually followed in the preparation of films composed of 20 and 30 bilayers.

The effect of charge density of the polyelectrolytes on superlattice thickness was also investigated by using the direct visualization technique[33]. A multilayer film composed of 20 bilayers of PSS-PFS was compared to that of 20 bilayers of PFS-PFS (Figure 9)[33]. As PSS contains the single negative charge per repeat unit in a smaller volume than the anionic PFS, less PSS was required to compensate the positive charges in the underlying layer, which ultimately resulted in a thinner PSS-PFS film in comparison to the PFS-PFS film.

LbL Assembly of PFS Polyelectrolytes in Silica Colloidal Crystals

An interesting application of the LbL assembly of PFS polyelectrolytes is in tuning the properties of photonic crystals at nanometer scale precision[39]. Photonic crystals are a class of materials which possess periodic variations in dielectric constants[40]. Because of the photonic band gaps existed in these materials, photons having energy within the band gap are not transmitted through. Photonic crystals are a powerful tool for the manipulation of light and are building blocks for devices in optical communications[41].

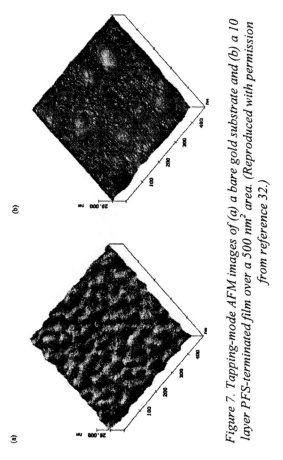

Figure 7. Tapping-mode AFM images of (a) a bare gold substrate and (b) a 10 layer PFS-terminated film over a 500 nm² area. (Reproduced with permission from reference 32.)

Figure 8. TEM images for (a) 5, (b) 10, (c) 20, (d) 30 bilayer films of PFS-PSS. (Reproduced with permission from reference 33.)

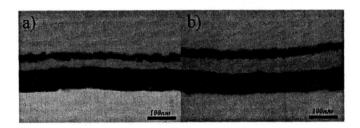

Figure 9. TEM images for 20 bilayers of (a) PSS-PFS and (b) PFS-PFS showing the effect of charge density on layer thickness. (Reproduced with permission from reference 33.)

The photonic crystals used in our study are 3-D ordered arrays of monodisperse silica colloidal spheres. The cationic and anionic PFS polyelectrolytes were alternately infiltrated in the opal structure, gradually filling the air voids with each step of deposition. The variation of the colloidal photonic crystal properties with the degree of polymer infiltration was monitored by optical transmission measurements, carried out by using a UV-vis spectrometer. As the number of deposition cycles increases, there is a gradual shift of the stop band peak to the longer wavelengths (Figure 10a, b)[39]. This is in accordance with the spectra obtained from theoretical fittings using a scalar wave approximation (SWA) approach.

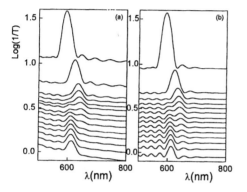 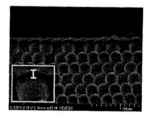

Figure 10. Experimental (a) and theoretical (b) optical spectra for a silica colloidal crystal film infiltrated of PFS-PFS. In (a), the corresponding number of deposition cycles increases from top to bottom by an increment of one bilayer between each curve. (c) Cross-sectional SEM of a silica-electrostatic multilayer composite colloidal photonic crystal. The inset displays a detail of the most external sphere layer, which is covered by a polymer layer. (Reproduced from reference 39 with permission.)

After a certain number of deposition cycles, a new peak starts to appear, blue-shifted relative to the original stop band peak frequencies. Theoretical simulations indicate that this is probably the result of an external layer deposited on top of the photonic crystal (Figure 10c)[39].

The results of this work have demonstrated that electrostatic assembly of polyelectrolytes occurs on the internal and external surfaces of a colloidal crystal film and provides a powerful tool in tuning the optical properties of the photonic lattice with nanometer scale precision.

Conclusions

Facile syntheses of a range of water-soluble PFS polyelectrolytes have been achieved. Direct nucleophilic substitution reactions with the chlorinated PFS **1** by dimethylaminoalkyl alcohol followed by quaternization of the dimethylamino groups afforded the cationic PFS **2**. Alternatively, ROP of [1]ferrocenophane **4** followed by quaternization of the amino groups gave the cationic PFS **5**. In addition, nucleophilic substitution of the chlorine atoms on PFS **1** by lithium acetylide $LiC \equiv CCH_2N(SiMe_2CH_2)_2$ **(6)** provided an intermediate PFS **7**, which can serve as a precursor to both the cationic PFSs **11** and **12** as well as the anionic PFSs **13** and **14**.

Both the organic-organometallic PSS-PFS and the novel all organometallic PFS-PFS ultra thin films have been formed through the electrostatic assembly of the oppositely charged polyelectrolytes. The data from QCM, ellipsometry, UV-vis, XPS, contact angle and AFM measurements of the PSS-PFS and PFS-PFS multilayers demonstrate that LbL assembly of oppositely charged metal-containing polyelectrolytes creates regularly stacked, homogeneous electrostatic superlattices with essentially complete surface coverage. And the electrostatically assembled ultra thin multilayers can be directly visualized by using a gold coating/transmission electron microscopy (TEM) technique.

We have also demonstrated that the electrostatic assembly of polyelectrolytes provides a powerful tool in tuning the optical properties of the photonic lattice with nanometer scale precision.

Acknowledgments

We acknowledge coworkers Madlen Ginzburg, Josie Galloro, Andre Arsenault, and John Halfyard and we thank the Natural Science and Engineering Research Council (NSERC) of Canada for funding. IM and GAO thank the Canadian Government for Canada Research Chairs.

References

1. See, for example: (a) *Water-Soluble Polymers: Synthesis, Solution Properties and Applications*; eds. Shalaby, S. W., Butler, G. B., McCormick, C. L.; ACS Symposium Series 467; American Chemical Society: Washington, DC, **1991**. (b) *Controlled Drug Delivery: Designing Technologies for the Future*, eds. Park, K., Mrsny, R. J.; ACS Symposium Series 752; American Chemical Society: Washington, DC, **2000**.
2. (a) Decher, G. *Science* **1997**, *277*, 1232. (b) Bertrand, P.; Jonas, A.; Laschewsky, A.; Legras, R. *Macromol. Rapid Commun.* **2000**, *21*, 319.

3. Decher, G. *Multilayer Thin Films: Sequential Assembly of Nanocomposite Materials*; Wiley-VCH: Weinheim, 2003.

4. See ref. 3 and the references cited therein.

5. Manners, I. *Synthetic Metal-Containing Polymers*, Wiley-VCH Verlag: Weinheim, **2004**.

6. For metal-containing polyelectrolytes, see, for example: (a) Knapp, R.; Schott, A.; Rehahn, M. *Macromolecules* **1996**, *29*, 478. (b) Neuse, E. W.; Khan, F. B. D. *Macromolecules* **1986**, *19*, 269. (c) Kurth, D. G.; Osterhout, R. *Langmuir* **1999**, *15*, 4842.

7. (a) Manners, I. *Science* **2001**, *294*, 1664. (b) Kulbaba, K.; Manners, I. *Macromol. Rapid Commun.* **2001**, *22*, 711.

8. Tanabe, M.; Manners, I. *J. Am. Chem. Soc.* **2004**, *126*, ASAP article.

9. Resendes, R.; Berenbaum, A.; Stojevic, G.; Jäkle, F.; Bartole, A.; Zamanian, F.; Dubois, G.; Hersom, C.; Balmain, K.; Manners, I. *Adv. Mater.* **2000**, *12*, 327.

10. MacLachlan, M. J.; Ginzburg, M.; Coombs, N.; Coyle, T. W.; Raju, N. P.; Greedan, J. E.; Ozin, G. A.; Manners, I. *Science* **2000**, *287*, 1460.

11. Ginzburg, M.; MacLachlan, M. J.; Yang, S. M.; Coombs, N.; Coyle, T. W.; Raju, N. P.; Greedan, J. E.; Herber, R. H.; Ozin, G. A.; Manners, I. *J. Am. Chem. Soc.* **2002**, *124*, 2625.

12. Ginzburg-Margau, M.; Fournier-Bidoz, S.; Coombs, N.; Ozin, G. A.; Manners, I. *Chem. Commun.* **2002**, 3022.

13. Berenbaum, A.; Ginzburg, M.; Coombs, N.; Lough, A. J.; Safa-Sefat, A.; Greedan, J. E.; Ozin, G. A.; Manners, I. *Adv. Mater.* **2003**, *15*, 51.

14. Chan, W. Y.; Berenbaum, A.; Clendenning, S. B.; Lough, A. J.; Manners, I. *Organometallics* **2003**, *22*, 3796.

15. Clendenning, S. B.; Han, S.; Coombs, N.; Paquet, C.; Rayat, M. S.; Grozea, D.; Brodersen, P. M.; Sodhi, R. N. S.; Yip, C. M.; Lu, Z. H.; Manners, I. *Adv. Mater.* **2004**, *16*, 291.

16. Clendenning, S. B.; Fournier-Bidoz, S.; Pietrangelo, A.; Yang, G. C.; Han, S. J.; Brodersen, P. M.; Yip, C. M.; Lu, Z. H.; Ozin, G. A.; Manners, I. *J. Mater. Chem.* **2004**, *14*, 1686.

17. Cheng, A. Y.; Clendenning, S. B.; Yang, G. C.; Lu, Z. H.; Yip, C. M.; Manners, I. *Chem. Commun.* **2004**, 780-781.

18. See, for example: (a) Ni, Y.; Rulkens, R.; Manners, I. *J. Am. Chem. Soc.* **1996**, *118*, 4102. (b) Massey, J. A.; Power, K. N.; Winnik, M. A.; Manners, I. *Adv. Mater.* **1998**, *10*, 1559. (c) Massey, J. A.; Winnik, M. A.; Manners, I.; Chan, V. Z.-H.; Ostermann, J. M.; Enchelmaier, R.; Spatz, J. P.; Möller, M. *J. Am. Chem. Soc.* **2001**, *123*, 3147. (d) Massey, J. A.; Temple, K.; Cao, L.; Rharbi, Y.; Raez, J.; Winnik, M. A.; Manners, I. *J. Am. Chem. Soc.* **2000**, *122*, 11577. (e) Raez, J.; Zhang, Y.; Cao, L.; Petrov, S.; Erlacher, K.; Wiesner, U.; Manners, I.; Winnik, M. A. *J. Am. Chem. Soc.* **2003**, *125*, 6010. (f) Wang, X. S.; Winnik, M. A.; Manners, I. *Angew. Chem. Int. Ed.* **2004**, *43*, 3703.

19. (a) Cheng, J. Y.; Ross, C. A.; Thomas, E. L.; Smith, H. I.; Vancso, G. J. *Adv. Mater.* **2003**, *15*, 1599. (b) Eitouni, H. B.; Balsara, N. P. *J. Am. Chem. Soc.* **2004**, *126*, 7446. (c) Kloninger, C.; Rehahn, M. *Macromolecules* **2004**, *37*, 1720.

20. (a) Lastella, S.; Jung, Y. J.; Yang, H.; Vajtai, R.; Ajayan, P. M.; Ryu, C. Y.; Rider, D. A.; Manners, I. *J. Mater. Chem.* **2004**,1791. (b) Hinderling, C.; Keles, Y.; Stöckli, T.; Knapp, H. F.; de los Arcos, T.; Oelhafen, P.; Korczagin, I.; Hempenius, M. A.; Vancso, G. J.; Pugin, R.; Heinzelmann, H. *Adv. Mater.* **2004**, *16*, 876.

21. Kulbaba, K.; Cheng, A.; Bartole, A.; Greenberg, S.; Resendes, R.; Coombs, N.; Safa-Sefat, A.; Greedan, J. E.; Stover, H. D. H.; Ozin, G. A.; Manners, I. *J. Am. Chem. Soc.* **2002**, *124*, 12522.

22. Galloro, J.; Ginzburg, M.; Miguez, H.; Yang, S. M.; Coombs, N.; Safa-Sefat, A.; Greedan, J. E.; Manners, I.; Ozin, G. A. *Adv. Funct. Mater.* **2002**, *12*, 382.

23. Arsenault, A. C.; Miguez, H.; Kitaev, V.; Ozin, G. A.; Manners, I. *Adv. Mater.* **2003**, *15*, 503.

24. Espada, L. I.; Shadaram, M.; Robillard, J.; Pannell, K. H. *J. Inorg. Organomet. Polym.* **2000**, *10*, 169.

25. (a) Lammertink, R. G. H.; Hempenius, M. A.; Vancso, G. J.; Chan, V. Z.-H.; Thomas, E. L. *Chem. Mater.* **2001**, *13*, 429. (b) Cao, L.; Massey, J. A.; Winnik, M. A.; Manners, I.; Riethmuller, S.; Banhart, F.; Spatz, J. P.; Moller, M. *Adv. Funct. Mater.* **2003**, *13*, 271.

26. Zechel, D. L.; Hultzsch, K. C.; Rulkens, R.; Balaishis, D.; Ni, Y.; Pudelski, J. K.; Lough, A. J.; Manners, I. *Organometallics* **1996**, *15*, 1972.

27. Jäkle, F.; Vejzovic, E.; Power-Billard, K. N.; MacLachlan, M. J.; Lough, A.; Manners, I. *Organometallics* **2000**, *19*, 2826.

28. Wang, Z.; Lough, A.; Manners, I. *Macromolecules* **2002**, *35*, 7669.

29. Power-Billard, K. N.; Manners, I. *Macromolecules* **2000**, *33*, 26.

30. Jäkle, F.; Wang, Z.; Manners, I. *Macromol. Rapid Commun.* **2000**, *21*, 1291.

31. Massey, J. A.; Kulbaba, K.; Winnik, M. A.; Manners, I. *J. Polym. Sci., Polym. Phys.* **2000**, *38*, 3032.

32. Ginzburg, M.; Galloro, J.; Jäkle, F.; Power-Billard, K. N.; Yang, S.; Sokolov, I.; Lam, C. N. C.; Neumann, A. W.; Manners, I.; Ozin, G. A. *Langmuir* **2000**, *16*, 9609.

33. Halfyard, J.; Galloro, J.; Ginzburg, M.; Wang, Z.; Coombs, N.; Manners, I.; Ozin, G. A. *Chem. Commun.* **2002**, 1746.

34. For analogous cationic and anionic PFS materials developed by Vancso *et al.* and their use in LbL assembly, see: (a) Hempenius, M. A.; Robins, N. S.; Lammertink, R. G. H.; Vancso, G. J. *Macromol. Rapid Commun.* **2001**, *22*, 30. (b) Hempenius, M. A.; Péter, M.; Robins, N. S.; Kooij, E. S.; Vancso, G. J. *Langmuir* **2002**, *18*, 7629. (c) Hempenius, M. A.; Vancso, G. J. *Macromolecules* **2002**, *35*, 2445.

35. (a) Lowe, A. B.; Billingham, N. C.; Armes, S. P. *Chem. Commun.* **1996**, 1555. (b) Le Moigne, J.; Gramain, P. *Eur. Polym.er J.* **1972**, *8*, 703.
36. Djuric, S.; Venit, J.; Mangnus, P. *Tetrahedron Lett.* **1981**, *22*, 1787.
37. Yoo, D.; Shiratori, S. S.; Rubner, M. F. *Macromolecules* **1998**, *31*, 4309.
38. Chen, W.; McCarthy, T. J. *Macromolecules* **1997**, *30*, 78.
39. Arsenault, A. C.; Halfyard, J.; Wang, Z.; Kitaev, V.; Ozin, G. A.; Manners, I. *Langmuir*, **2005**, *21*, 499; and unpublished results.
40. (a) John, S. *Phys. Rev. Lett.* **1987**, *58*, 2486. (b) Yablonovitch, E. *Phys. Rev. Lett.* **1987**, *58*, 2059. (c) Joannopoulos, J. D.; Villeneuve, P. R.; Fan, S. *Science* **1997**, *386*, 143.
41. See, for example: (a) Blanco, A.; Chomski, E.; Grabtchak, S.; Ibisate, M.; John, S.; Leonard, S. W.; Lopez, C.; Meseguer, F.; Miquez, H.; Mondia, J. P.; Ozin, G. A.; Toader, O.; van Driel H. M. *Nature* **2000**, *405*, 437. (b) Ogawa, S.; Imada, M.; Yoshimoto, S.; Okano, M.; Noda, S. *Science* **2004**, *305*, 227. (c) Yang, S.; Megens, M.; Aizenberg, J.; Wiltzius, P.; Chaikin, P. M.; Russel, W. B. *Chem. Mater.* **2002**, *14*, 2831.

Chapter 25

Synthesis and Bulk Morphology of Styrene–Ferrocenylsilane–Methyl Methacrylate Tri- and Pentablock Copolymers

Christian Kloninger, Uttam Datta, and Matthias Rehahn[*]

Ernst-Berl-Institute for Chemical Engineering and Macromolecular Science, Darmstadt University of Technology, Petersenstrasse 22, D–64287 Darmstadt, Germany

Systematic series of tri- and pentablock copolymers have been synthesized from styrene (S), [1]dimethylsilaferrocenophane (FS) and methyl methacrylate (MMA) via sequential anionic polymerization. In the first step of copolymer synthesis, mono- and dianionic PS, respectively, is generated. Direct grafting of FS follows. Then, 1,1-dimethylsilacyclobutane-mediated end-capping using diphenylethylene enabled us to graft the PMMA block(s) properly onto the living PFS chain termini. The obtained tri- and pentablock copolymers of various compositions were characterized with respect to their microphase behavior. The morphologies obtained as a function of copolymer composition and annealing conditions allow the conclusion that PS and PFS segregate only weakly while PMMA is highly incompatible with respect to PS and PFS.

Background

Block copolymers represent a class of macromolecular substances subjected to broad current interest.[1] In most cases, this interest is directed toward the specific properties of these materials which in turn are direct consequences of their characteristic self-organization behavior. The microphase morphologies developed by block copolymers depend on a number of microscopic and macroscopic parameters such as overall copolymer chain length, lengths of the individual blocks, distribution of the block lengths, distribution of the overall molecular weights, chain architecture, compatibility of the different blocks, goodness of the solvent for the respective blocks, temperature, and the method used for sample preparation like film casting or annealing. By creating appropriate micromorphologies, for example, the mechanical properties of thermoplastics can be tailored. In a more visionary perspective, block copolymers could develop into useful building blocks for highly innovative technologies:[2,3] when microphase-separated in an appropriate way they may combine excellent mechanical properties with, for example, electrical conductivity, optical effects or magnetism. If so, they will gain tremendous importance as antistatic coatings, in (electro-) optical devices or in nanolithography. Hence, they could be used for fabrication of semiconductors, chips, integrated circuits, microprocessors and memory storage devices.[4]

Consequently, there is a strongly increasing interest in functional block copolymers having electrically conducting sub-structures. In that concern, transition-metal containing block copolymers are particularly attractive.[5,6] Promising candidates are systems where organometallic polyferrocenyldimethylsilane (PFS)[7] segments as the functional subunits are combined with segments of classic thermoplastics or rubbers.[8-28] In order to broaden this concept of functional block copolymers further we developed an efficient synthesis of diblock copolymers from [1]dimethylsilaferrocenophane (FS) and methyl methacrylate (MMA)[29,30] where PMMA plays the thermoplastic's part with potential further benefit as a photoresist in lithographic processes. The synthesis of these materials was accomplished via classic sequential anionic polymerization, however, taking advantage of a novel procedure in order to ensure a high block efficiency: diphenylethylene (DPE) as end-capping agent was supported by dimethylsilacyclobutane (DMSB) as end-capping accelerator. By this combination of reagents, an almost quantitative blocking efficiency was reached of MMA onto the living PFS chain ends. For high-molecular-weight materials, here, only the three classic micromorphologies were observed, *i.e.* spheres, cylinders and lamellae. For low-molecular-weight materials and high annealing temperatures, on the other hand, we were able to observe the bicontinuous gyroidic morphology as well.[30] This gyroidic phase was, in principle, what we were looking for since bicontinuous phases should combine the PMMA's mechanical properties with the benefits of the functional PFS block in an ideal manner. A potential drawback of this bicontinuous phase, however, might be its quite high content of PFS which is rather expensive and colored: the gyroidic phase appears at a FPS volume fraction of $\phi_{PFS} \approx 0.40$. Consequently, it is the next important goal to lower the volume fraction of the functional PFS blocks drastically, without losing its well-defined micromorphology simultaneously.

Objective

A potential strategy that might allow reduction of the PFS content in bicontinuous functional block copolymers consists of the preparation of ABC triblock- or CBABC pentablock copolymers, having long thermoplastic A and C blocks together with two rather short functional B blocks formed by the PFS. At this particular composition, a variety of fascinating morphologies can be expected.[31] In some of them the functional PFS subunits might form a continuous phase despite of their considerably higher "dilution". For this purpose, we had to select an appropriate combination with A and C blocks. Based on published interaction parameters and the pK_a values of living chain ends we decided to use styrene as the third component. Here we report on the synthesis of PS-*b*-PFS-*b*-PMMA triblock- and PMMA-*b*-PFS-*b*-PS-*b*-PFS-*b*-PMMA pentablock copolymers having PFS volume fractions of around $\phi_{PFS} \approx 0.15$. Also, we describe some first results obtained by transmission electron microscopy (TEM) concerning the micromorphologies formed by these materials.

Experimental

Monomer syntheses and techniques were as described recently.[29,30] For block copolymer syntheses, in the following, two representative entries are given.

Synthesis of triblock copolymers

Styrene (1 g, 9.6 mmol) was polymerized in an ampoule for 3 h at 40 °C using *sec*-BuLi (30 μL, 0.06 mmol) as the initiatior and cyclohexane (10 mL) as the solvent. After allowing the reaction mixture to cool down to room temperature a solution of FS (0.7 g, 2.89 mmol) in THF (20 mL) was added. After complete monomer conversion (approx. 40 min) DPE (42 μL, 0.24 mmol) was added, immediately followed by DMSB (16 μl, 0.12 mmol; [*sec*-

BuLi]/[DMSB]/[DPE] = 1/2/4). In a second ampoule dry LiCl (26 mg, 0.6 mmol) was dissolved in THF (75 mL). Both ampoules were attached to the polymerization reactor. After combining the solutions, the reactor was immersed into a liquid nitrogen/isopropanol bath at −78 °C. For polymerization of the third block, MMA (2.8 g, 28 mmol) was rapidly introduced from a syringe into the reactor through a Teflon-coated silicon septum under vigorous stirring. The polymerization was terminated after stirring the solution for 90 min at −78 °C by adding degassed methanol. After precipitating in methanol and drying under vacuum, 4.4 g (98%) of crude PS-b-PFS-b-PMMA could be isolated. The pure triblock was obtained via selective precipitation: 0.8 g of the crude product were dissolved in THF (10 mL). This solution was dropped into hexane (100 mL) under stirring. After slowly adding another batch of hexane (about 40 mL) the formation of a gel was observed. The triblock copolymer could be isolated by ultracentrifuging the mixture for 7 min at 4000 rpm.

Synthesis of pentablock copolymers

Styrene was rapidly injected into a green solution of lithium naphthalide in THF at −50 °C. The styrene was polymerized for 35 min at this temperature. The reaction mixture was allowed to warm up to room temperature. FS, dissolved in THF, was added with stirring. After 45 min the living polymer was trapped by means of DMSB-mediated end-capping with DPE. A molar ratio of 1:4:2 for initiator:DPE:DMSB was employed for the trapping reaction. The red solution was cooled to −80 °C and MMA was added. After 75 min, the copolymerization was terminated by addition of a few drops of methanol. The block copolymer was isolated by precipitation into methanol. It was purified by three cycles of repetitive precipitation in cyclohexane: about 0.7 g of the crude pentablock were dissolved in 10 mL of THF. To this solution, cyclohexane was added drop wise. The precipitate was separated off using a centrifuge. By this method, around 0.66 g of the pentablock (containing approx. 15% of tetrablock) copolymer is obtained.

Results & Discussion

Due to synthetic constraints determined by the nucleophilicity of the living anionic PS-, PFS- and PMMA chain ends, the planned sequential anionic polymerization leading to the PS-*b*-PFS-*b*-PMMA triblock and pentablock copolymers is only possible in the order S → FS → MMA.

Syntheses

The reaction sequence selected here for the triblock copolymer synthesis, including the DMSB-mediated endcapping step, is shown in Scheme 1.

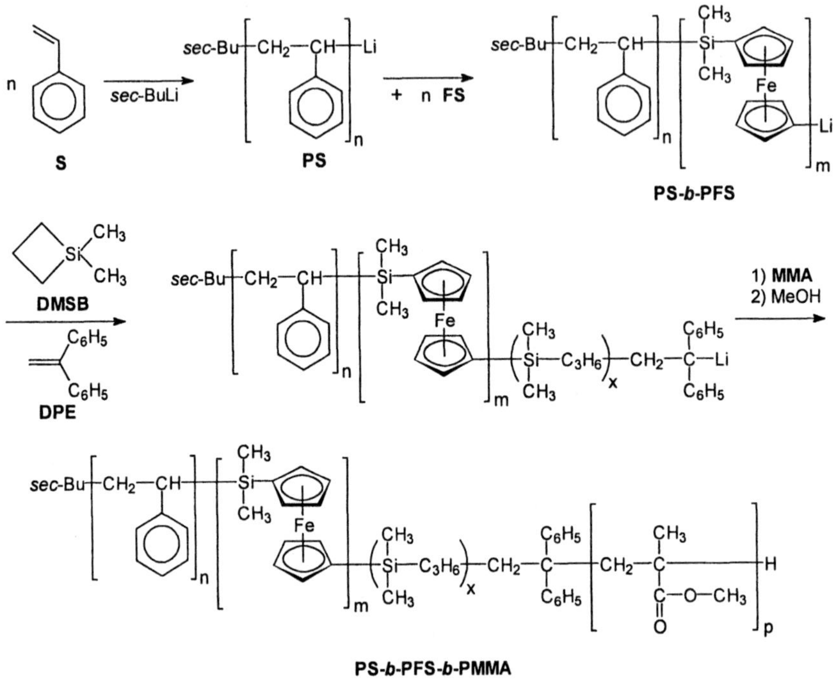

Scheme 1. Synthesis of PS-b-PFS-b-PMMA triblock copolymers.

In order to have a good control over the polymerization of styrene, the first reaction step was carried out in cyclohexane at slightly elevated temperature. Since sec-BuLi has a higher reactivity in apolar solvents as compared to n-BuLi, it was used as the initiator in order to ensure fast initiation and hence a narrow molecular-weight distribution for the PS block. Unfortunately, the subsequent FS polymerization cannot be carried out in an apolar solvent. Therefore, after complete conversion of styrene, a solution of FS in THF was added to the living polystyryl lithium precursor. Removal of the cyclohexane from the reaction mixture was found to be unneeded since FS polymerization occurs smoothly and quantitatively also in mixtures of an apolar and a polar solvent. After full conversion of FS the DMSB-mediated DPE endcapping follows. Finally, the polymerization of MMA was carried out at −78 °C.

Figure 1a displays representative SEC traces of the PS precursor, of the PS-*b*-PFS diblock intermediate, and of the finally obtained PS-*b*-PFS-*b*-PMMA triblock copolymer. All samples were taken after complete conversion of the respective monomer from the reaction flask and terminated using degassed methanol.

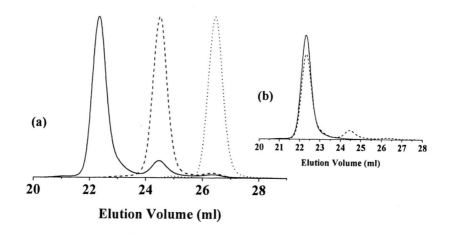

Figure 1. SEC traces of (a) the block copolymerization of styrene, silaferrocenophane and methyl methacrylate; PS: dotted line, PS-b-PFS: dashed line, PS-b-PFS-b-PMMA triblock copolymer: solid line; (b) the triblock copolymer prior to (dashed line) and after (solid line) selective precipitation.

While the PS precursor and the PS-*b*-PFS intermediate show a very narrow and almost monomodal molecular-weight distribution, some terminated PS-*b*-PFS diblock copolymer is evident in the final triblock system. This impurity originates from termination reactions of the living diblock intermediate which occur in particular after compled FS polymerization. The amount of diblock co-polymer in the final product is typically 20%. Fortunately, we were able to sepa-rate off the terminated precursors via selective precipitation nearly quantitatively (Figure 2b). By this procedure, several PS-*b*-PFS-*b*-PMMA triblock copolymer samples of approximately identical overall degrees of polymerization were pre-pared. Throughout the PMMA represents the main component, and the ratio be-tween PS and PFS volume fractions is varied systematically. The lengths of the individual blocks were determined using ^{1}H NMR spectroscopy. The molecular weights and the molecular-weight distributions were determined via SEC in THF as the eluent, using PMMA as a standard. Table 1 summarizes the molecular characteristics of three representative triblock copolymer samples.

Table 1: Molecular characteristics of PS-b-PFS-b-PMMA triblock copolymers.

Sample	$x_S^{a,b}$	w_S^b	ϕ_S^b	$x_F^{a,b}$	w_F^b	ϕ_F^b	$x_M^{a,b}$	w_M^b	ϕ_M^b	M_n^c (g/mol)	PDI^c
$S_{13}F_{13}M_{74}^{101}$	0.12	0.12	0.13	0.06	0.14	0.13	0.81	0.74	0.74	101,000	1.02
$S_{13}F_{19}M_{68}^{110}$	0.13	0.12	0.13	0.10	0.20	0.19	0.78	0.68	0.68	110,000	1.03
$S_{21}F_{13}M_{67}^{102}$	0.20	0.19	0.21	0.06	0.14	0.13	0.74	0.67	0.67	102,000	1.04

[a] determined by ^{1}H NMR with CD_2Cl_2.

[b] x, w and ϕ are the molar fraction, the weight fraction and the volume fraction, respectively, of the corresponding segments in the pure diblock copolymers. Volume fractions were calculated utilizing ρ_{PS} = 1.05 g·mL^{-1}, ρ_{PFS} = 1.26 g·mL^{-1}, and ρ_{PMMA} = 1.15 g·mL^{-1}.

[c] determined by SEC, using THF as the eluent and PMMA standards.

In the next step, the pentablock copolymers were preprared analogously according to Scheme 2:

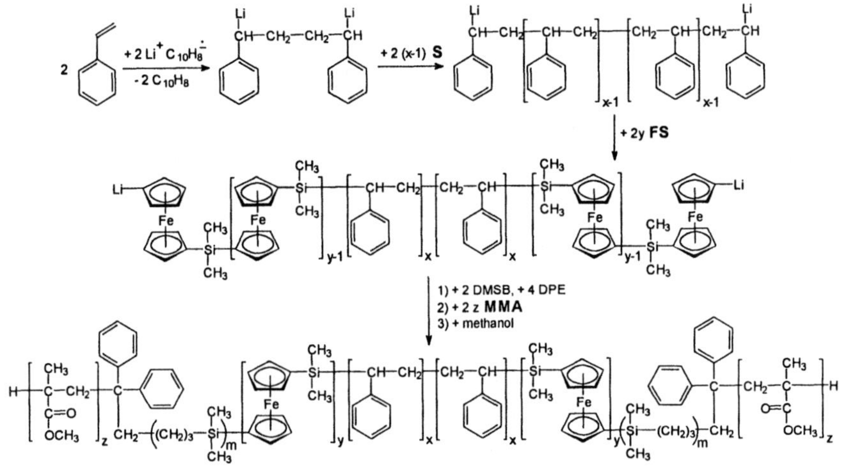

Scheme 2. Synthesis of PMMA-b-PFS-b-PS-b-PFS-b-PMMA pentablock copolymers.

At first, an appropriate quantity of styrene was added to the solution of lithium naphthalide in THF at -50 °C. When the polymerization was complete, the reaction mixture was allowed to warm up to room temperature. Then a solution of FS in THF was added to the reaction mixture. After complete conversion the living chain termini of the PFS-*b*-PS-*b*-PFS triblock copolymer were end-capped by adding a mixture of DPE and DMSB. Subsequently, MMA was added at -78 °C to grow onto the formed bifunctional macroinitiator in a controlled way. Finally, the whole reaction mixture was deactivated by degassed methanol. The final product as well as the samples taken during the synthesis were analyzed using SEC. It could be shown that there is almost no termination when FS is grafted onto the difunctional polystyrene chains. Only during DPE/DMSB end-capping and subsequent addition of MMA to the bifunctional PFS-b-PS-b-PFS macroinitiator, approx. one quarter of the living chain ends terminates. Unfortunately, the pentablock copolymer architecture makes efficient purification of the products difficult: the removal of tri- and tetrablock copolymer contaminations from pentablock systems is a well-known problem associated with multiblock copolymer syntheses. Nevertheless, we were able to develop an efficient procedure which allows, via repeated precipitation, the removal of all impurities except approx. 15% of the tetrablock species. Since we can expect that the material's micromorphologies are not influenced in an irreproducible way by that minor amount of tetrablock copolymers, we decided to continue our investigations using the thus available materials. In the following, these block copolymers are named $S_kF_lM_m^x$ where S, F and M are the abbreviations of the respective blocks, PS, PFS and PMMA. The subscripts k, l and m represent the volume fractions (in %) while the superscript x represents the overall number-average molecular weight (in kg·mol^{-1}).

Bulk Morphology of the Triblock Coplymers

In order to study the microphase behavior of the triblock copolymers, thin films were prepared by casting samples from methylene chloride followed by annealing at 180 °C. Figure 2 displays the TEM pictures of films made of (a) $S_{13}F_{13}M_{74}^{101}$ and (b) $S_{13}F_{19}M_{68}^{110}$. It is obvious that in both cases spherically or oval shaped objects of PS (grey) are formed in a continuous matrix formed by the PMMA (white). The PS sheres are moreover covered by "drops" of PFS (either black or dark grey). The structures appear nicely phase-separated but it is also obvious that the PS spheres are arranged rather at random in the PMMA matrix. In full agreement with the spherical morphology found for PFS-b-PMMA diblock copolymers reported recently[9] we assume that this limited long-range structural perfection is due to the rather high viscosity of the polymer melt even under annealing conditions and thus is a direct consequence of the high molecular weights of the samples under investigation. The morphologies found for $S_{13}F_{13}M_{74}^{101}$ and $S_{13}F_{19}M_{68}^{110}$ are fully consistent with the morphology of the corresponding SBM and SEBM triblock copolymers published by Breiner et al.[32] These authors report the formation of "spheres on spheres" morphologies (sos) for materials containing high volume fractions of the matrix polymer. This morphology is schematically drawn in Figure 2 as well.

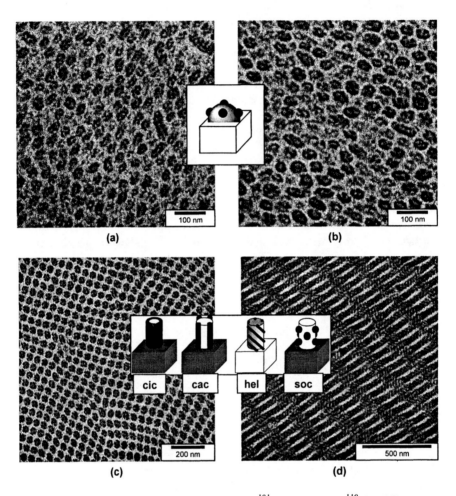

Figure 2. TEM pictures of (a) $S_{13}F_{13}M_{74}{}^{101}$, (b) $S_{13}F_{19}M_{68}{}^{110}$ [middle: schematical illustration of the "spheres on spheres" morphology (sos)], (c) $S_{21}F_{13}M_{67}{}^{102}$ [cut perpendicular the the cylinder axis] and (d) $S_{21}F_{13}M_{67}{}^{102}$ [angularly cutted cylinders; middle: schematically drawn cylinder structures in triblock copolymers as published by Breiner et al.[32]]

The third triblock copolymer studied in this work, i.e. $S_{21}F_{13}M_{67}^{102}$, is different in the sense that the volume fraction of PS is the largest. Because of this high content of PS, here, we do not observe a sphere-like morphology of the PS domains but instead a morphology with cylindrical arrangement. Figure 2c displays a typical picture of this structure when the sample is cut perpendicular to the cylinder axis. The cylinders are arranged in a surprisingly high order, and each cylinder again is surrounded by four ellipsoid phases formed by the PFS (black). This might be assigned to PFS drops or additional small PFS cylinders. In order to decide which explanation might be true, further projections of this phase have to be analyzed. But even by means of the projection shown in Figure 2d where the cylinders are cut angularly it is not very clear to see what the shape and order might be of the PFS phase that is arranged exactly between the PMMA matrix and the PS cylinders.

At this point, a comparison might be helpful of the observed structure with that observed for PS-*b*-PB-*b*-PMMA triblock copolymers of similar composition as was published by Breiner et al.:[32] in the range of composition of $\phi_A + \phi_B = 0.26 - 0.41$, these authors observed various cylinder morphologies which differ in the arrangement of the central block (see insert in Figure 2). For $\phi_B/\phi_A > 1.1$ one observes core-shell cylinders (cylinder in cylinder morphology, cic) where the center block covers, like a shell, the cylindric core formed by the end blocks of lower volume fraction, i.e. polymer A. Upon further reduction of the length of the central block B and simultaneous increase of the length of the A block (or upon reduction of the overall portion of A and B), first, a phase is formed in which four B cylinders are arranged along the cylindric A phase (cylinder at cylinder morphology, cac). This morphology is followed by a structure in which the central blocks B surround the A block cylinders in a helical shape (helix morphology, hel). This latter morphology, however, was only found for SBM triblock copolymers where PMMA forms the matrix. For very low values of ϕ_B/ϕ_A (< 0.5), finally, a morphology is reached in which the cylindric A block is surrounded by regularly arranged drops of the minor component B (spheres on cylinder morphology, soc).

Projections as shown in Figure 2c might be assigned to cylinder-at-cylinder-, helical- and spheres-on-cylinder morphologies. A more reliable assignment of the structure is, therefore, only possible by means of the projection parallel to the cylinder axis and enlargement of the pictures. In the corresponding TEM pictures (Figure 3) there is attributable a certain helical arrangement of the PFS phase, but at some other places a somewhat disconnected structure is formed where a rather drop-like soc morphology seems to be given of the central PFS block B. Hence it is reliable to assume that for $S_{21}F_{13}M_{67}^{102}$ we hit a triblock copolymer composition which is just at the boundary between helical und soc morphology.

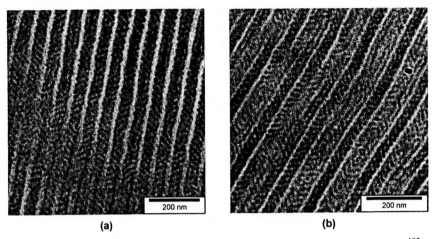

Figure 3. *Enlarged TEM pictures of the cylindric morphology of $S_{19}F_{14}M_{67}{}^{102}$ with parallel arrangement of the PS cylinders in the PMMA matrix. The PFS phase (black) has partially a helical and partially a disconnected structure.*

Bulk Morphology of the Pentablock Copolymers

The microphase morphology of the PMMA-*b*-PFS-*b*-PS-*b*-PFS-*b*-PMMA pentablock copolymers was studied analogously using TEM. The formation of the thermodynamically stable micromorphologies was forced by storing the films freshly cast from methylene chloride solution in the saturated atmosphere for 3 – 4 weeks. Then the films were stored for 2 days in vacuo at room temperature and subsequently annealed for 24h at elevated temperatures (160, 180, 200, 220 and 230 °C). After quenching in ice water, the films were cut into slices of approx. 50 nm thicknesses.

When the annealing was done at temperatures above 200 °C, apparently biphasic morphologies were observed. They were either lamellar or cylindrical, depending on the PMMA's volume fraction. Under these conditions, obviously, the rather high compatibility – and hence the low segregation tendency – of PFS and PS[33] results in a homogeneous mixture of these segments (grew phase in the TEM pictures). Only the PMMA which is incompatible with both, PFS and PS, creates its own phase (white phase in the TEM pictures). Consequently, a biphasic morphology is formed which assumes a lamellar structure when both, ϕ_{PMMA} and ϕ_{PS+PFS} are between 0.4 and 0.6, in full agreement with expectations

(Figure 4a). On the other hand, when ϕ_{PMMA} < 0.4, the PMMA segments form cylinders in a continuous, homogeneous blend matrix of PS and PFS (Figure 4b).

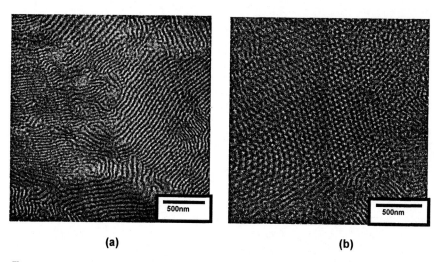

(a)   (b)

Figure 4. Transmission electron micrographs of (a) a lamellar morphology and (b) a cylinder morphology of pentablock copolymers annealed at 200 °C. Black regions represent the blend domains of PS and PFS while the PMMA phases appear in white. For details, see text.

A completely different picture was found when annealing was carried out at slightly lower temperatures. This lowering of temperature is equivalent to a higher value of the Flory Huggins interaction parameter, χ. Hence, phase separation of PFS and PS occurs as well under these conditions.[33] Consequently, complex morphologies arise. Best TEM pictures were obtained for samples annealed at 180 °C. In the samples thus prepared three distinct morphologies could be identified by TEM. On the one hand, a spheres-in-spheres ("ball in the box"; *sis*) morphology was observed (see Figure 5):[34,35] larger spheres or ellipsoids are formed by the PS (grew) wherein – next to the surface – small PFS spheres are integrated. The latter appear as black dots in the TEM picture. The PMMA majority component, on the other hand, represents the continuous (white) matrix. The *sis* morphology was found whenever ϕ_{PMMA} was approx. 0.60 while ϕ_{PS} was around 0.25 and $\phi_{PFS} \approx 0.15$. It is the favored morphology here because there is a lower interfacial tension, γ_{AB}, between the rather compatible PS and PFS phases as compared to the interfacial tensions γ_{AC} and γ_{BC} of these two phases with that of PMMA.

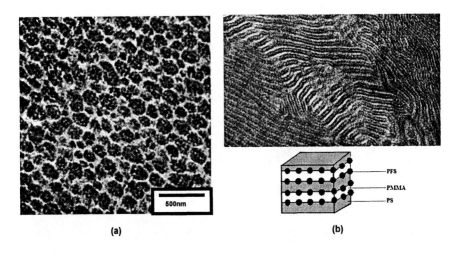

(a)   (b)

*Figure 5. Transmission electron micrograph of (a) a spheres-in-spheres
morphology and (b) a lamellar morphology of PS and PMMA with PFS droplets
at the interphase; the pentablock copolymer films were cast from methylene
chloride solution and annealed for 1 d at 180 °C.*

When the PS volume fraction is increased at the expense of PMMA, PMMA
and PS form a lamellar morphology. At their interphase little PFS spheres are
observable which seem to be integrated mainly into the PS lamellae – again for
interfacial tension reasons (Figure 5b).[36-39] This morphology is found for samples
where the volume fractions of the majority components, *i.e.* PMMA and PS, are
quite similar (ϕ_{PMMA}, ϕ_{PS} ≈ 0.4), leading to that lamellar arrangement. The minor-
ity component, PFS, forms its own microphase at the boundaries of the lamellar
phase and appears as droplets due to its low volume fraction (ϕ_{PFS} < 0.2). In
some TEM pictures, moreover, the PFS spheres seem to merge into cylinders.
This observation might be attributed to an incomplete phase separation between
PS and PFS in these samples, leading to a somewhat higher volume fraction of
the "PS-diluted" PFS domains and thus to the cylinder shape.

For pentablock copolymer samples where PS is the majority constituent (ϕ_{PS}
≈ 0.6), and where the volume fraction of PMMA is only ϕ_{PMMA} = 0.2 – 0.3 (ϕ_{PFS}
≈ 0.15), the TEM pictures allow to assume the occurrence of a spheres-on-
spheres morphology.[34] Here, the PS represents the continuous phase and PMMA
is the material that forms the bigger spheres while PFS gives the small droplets
sitting at the surface of the PMMA spheres.

Conclusions

In the present contribution we demonstrate that it is possible to combine the classical thermoplastic materials PS and PMMA with the organometallic polymer PFS in linear triblock and pentablock copolymers. The synthetic approach is based on the sequential anionic polymerization technique and allows the formation of well-defined samples with high molecular weights and narrow polydisperities. The key step to the copolymers is the DMSB-mediated endcapping step of the living PS-*b*-PFS precursor. Moreover it could be shown that these three different segments in the block copolymers possess a tendency towards microphase separation, resulting in an interesting morphological behavior. Already on the basis of the rather limited repertory of materials available so far one can estimate the enormous variety of microphase morphologies that should be accessible by these functional organometallic block copolymers when molecular parameters such as block length or compostion are varied further. Presently, we try to generate other morphologies containing PFS as a minor component by varying the film preparation conditions and the composition of the samples. Also, we employ SAXS for a more detailed microstructure analysis, and dielectric spectroscopy for studying the conductivity, magnetic susceptibility and permittivity of the materials before and after doping.

Acknowledgments

We thank the Fonds der Chemischen Industrie e.V. (FCI), the Otto Röhm Stiftung, and the Vereinigung der Freunde der TU Darmstadt e.V. for financial support of this work.

References

1. Hsieh, H. L.; Quirk, R. P. *Anionic Polymerisation: Principles and Practical Applications,* Marcel Dekker Inc., **1996**, New York.
2. Hamley, I. W. *Angew. Chem.* **2003**, *115*, 1730.
3. Park, C.; Yoon, J.; Thomas, E. L. *Polymer* **2003**, *44*, 6725.
4. Ni, Y.; Rulkens, R.; Manners, I. *J. Am. Chem. Soc.* **1996**, *118*, 4102.
5. Brandt, P. F.; Rauchfuss, T. B. *J. Am. Chem. Soc.* **1992**, *114*, 1926.
6. Manners, I. *Adv. Mater.* **1994**, *6*, 68.
7. Foucher, D. A.; Tang, B.-Z.; Manners, I. *J. Am. Chem. Soc.* **1992**, *114*, 6246.
8. Morrison, F. A.; Winter, H. H. *Macromolecules* **1989**, *22*, 3533.
9. Lammertink, R. G. H.; Hempenius, M. A.; Vancso, G. J.; Shin, K.; Rafailovich, M. H.; Sokolov, J. *Macromolecules* **2001**, *34*, 942.

10. Li, W.; Sheller, N.; Foster, M. D.; Balaishis, D.; Manners, I.; Annís, B.; Lin, J.-S. *Polymer* **2000**, *41*, 719.
11. Lammertink, R. G. H.; Hempenius, M. A.; Thomas, E. L.; Vancso, G. J. *J. Polym. Sci.: Part B: Polym. Phys.* **1999**, *37*, 1009.
12. Lammertink, R. G. H.; Versteeg, D. J.; Hempenius, M. A.; Vancso, G. J. *J. Polym. Sci.: Part A: Polym. Chem.* **1998**, *36*, 2147.
13. Lammertink, R. G. H.; Hempenius, M. A.; Vancso, G. J. *Langmuir* **2000**, *16*, 6245.
14. Raez, J.; Manners, I.; Winnik, M. A. *J. Am. Chem. Soc.* **2002**, *124*, 10381.
15. Massey, J. A.; Temple, K.; Cao, L.; Rharbi, Y.; Raez, J.; Winnik, M. A.; Manners, I. *J. Am. Chem. Soc.* **2000**, *122*, 11577.
16. Resendes, R.; Massey, J. A.; Dorn, H.; Power, K. N.; Winnik, M. A.; Manners, I. *Angew. Chem.* **1999**, *111*, 2738.
17. Gômez-Elipe, P.; Resendes, R.; Macdonald, P. M.; Manners, I. *J. Am. Chem. Soc.* **1998**, *120*, 8348.
18. Massey, J.; Power, K. N.; Manners, I.; Winnik, M. A. *J. Am. Chem. Soc.* **1998**, *120*, 9533.
19. Ni, Y.; Rulkens, R.; Manners, I. *J. Am. Chem. Soc.* **1996**, *118*, 4102.
20. Resendes, R.; Massey, J.; Dorn, H.; Winnik, M. A.; Manners, I. *Macromolecules* **2000**, *33*, 8.
21. Manners, I. *J. Polym. Sci.: Part A: Polym. Chem.* **2002**, *40*, 179.
22. Manners, I. *Macromol. Rapid Commun.* **2001**, *22*, 711.
23. Manners, I. *Chem. Commun.* **1999**, 857.
24. Temple, K.; Kulbaba, K.; Manners, I.; Leach, A.; Xu, T.; Russell, T. P.; Hawker, C. J. *Adv. Mater.* **2003**, *15*, 297.
25. Cheng, J. Y.; Ross, C. A.; Thomas, E. L.; Smith, H. I.; Vancso, G. J. *Appl. Phys. Lett.* **2002**, *81*, 3657.
26. Massey, J. A.; Winnik, M. A; Manners, I. *J. Am. Chem. Soc.* **2001**, *123*, 3147.
27. Cheng, J. Y.; Ross, C. A.; Chan, V. Z.-H.; Thomas, E. L.; Lammertink, R. G. H.; Vancso, G. J. *Adv. Mater.* **2001**, *13*, 1174.
28. Lammertink, R. G. H.; Hempenius, M. A.; van den Enk, J. E.; Chan, V. Z.-H.; Thomas, E. L.; Vancso, G. J. *Adv. Mater.* **2000**, *12*, 98.
29. Kloninger, C.; Rehahn, M. *Macromolecules* **2004**, *37*, 1720.
30. Kloninger, C.; Rehahn, M. previous chapter in this book.
31. Abetz, V. in *Comprehensive Polymer Science*: Allen, G.; Bevington, J. C. (Eds.), Pergamon: Oxford, **2003**, *Vol. 3*, p. 482.
32. Breiner, U.; Krappe, U.; Abetz, V.; Stadler, R. *Macromol. Chem. Phys.* **1997**, *198*, 1051.
33. Lammertink, R. G. H.; Hempenius, M. A.; Thomas, E. L.; Vancso, G. J. *J. Polym. Sci.: PartB: Polym. Phys.* **1999**, *37*, 1009.
34. Breiner, U.; Krappe, U.; Jakob, T.; Abetz, V.; Stadler, R. *Polymer Bulletin,* **1998**, *40*, 219.

35. Riess, G.; Schlienger, M.; Marti, S. J. *J. Macromol. Sci. Phys.* **1989**, *B17*, 355.
36. Breiner, U.; Krappe, U.; Thomas, E. L.; Stadler, R. *Macromolecules* **1998**, *31*, 135.
37. Auschra, C.; Stadler, R. *Macromolecules* **1993**, *26*, 2171.
38. Stadler, R.; Auschra, C.; Beckmann, J.; Krappe, U.; Voigt-Martin, I.; Leibler, L. *Macromolecules* **1995**, *28*, 3080.
39. Stocker, W.; Beckmann, J.; Stadler, R.; Rabe, J. P. *Macromolecules* **1996**, *29*, 7502.

Organometallic Polymers, Materials, and Nanoparticles

Non-Ferrocene-Containing Systems

Chapter 26

Evolution of Lowest Singlet and Triplet Excited States with Electronic Structure of Fluorene Group in Metal Polyyne Polymers

Wai-Yeung Wong[1,*], Ka-Ho Choi[1], Guo-Liang Lu[1], Li Liu[1], Jian-Xin Shi[1], and Kok-Wai Cheah[2]

[1]Department of Chemistry and Centre for Advanced Luminescence Materials and [2]Department of Physics, Hong Kong Baptist University, Waterloo Road, Kowloon Tong, Hong Kong

The design of metal acetylide complexes and polymers with unusual optoelectronic properties has aroused growing research interests. An identified problem in organic light-emitting diodes is the ratio of 3:1 for the generation of nonemissive triplet to emissive singlet excitons on the basis of spin statistics. In view of this, conjugated polymers containing transition metal atoms have been widely studied as model systems to explain aspects of the photophysics of excited states in such polymers and obtain a clear picture of the spatial extent of the singlet and triplet manifolds. The strong spin-orbit coupling associated with these heavy metals renders the spin-forbidden triplet emission (phosphorescence) partially allowed. Very recently, a comprehensive program was launched in our laboratory on the study of some novel organometallic polyyne polymers with fluorene-based auxiliaries. One of the merits here is that the 9-fluorenyl positions can be functionalized easily so that the solubility, the emission and electronic properties as well as the bandgaps of these materials can be chemically tuned. In this chapter, we report our investigations on the synthesis, characterization, structural, redox and photoluminescent properties of a series of transition metal polyyne polymers containing 9-functionalized fluorene units.

Background

In the past few decades, great efforts have been devoted to the design and synthesis of light-emitting organic polymers in the scientific community because of their potential applications in optoelectronic devices[1] such as light-emitting diodes (LEDs),[2] lasers[3] and photocells.[4,5] One of the major issues in organic LEDs is the ratio of 3:1 for the generation of nonemissive triplet to emissive singlet excitons based on the spin statistics.[6] Higher efficiencies can be realized by harvesting the energy of the nonemissive triplet excitons.[7–10] This requires detailed knowledge of the energy levels of the triplet state. To study the triplet excitons directly, conjugated polymers of the type $trans$-$[-PtL_2C\equiv CRC\equiv C-]_n$ have attracted the attention of a number of research groups. They can be used as good model systems to obtain a clear picture of the spatial extent of the singlet and triplet manifolds and provide important information on the photophysical processes that occur in organic conjugated polymers.[11–23] The prototype for much of this work is $trans$-$[-Pt(PBu_3)_2C\equiv C(p\text{-}C_6H_4)C\equiv C-]_n$, and a wide variety of derivatives have been prepared where alkynyl units are linked to a variety of aromatic ring systems. The electronic properties of these organometallic polymers are primarily governed by the nature of the metal, the auxiliary ligands L or the spacer R.[11–23] Importantly, the strong spin-orbit coupling associated with these heavy metals enables significant mixing of the singlet and triplet states, which renders the spin-forbidden emission from the triplet excited state partially allowed. In contrast to hydrocarbon conjugated polymers, the triplet excited state is experimentally accessible by various optical methods.

We are interested in the chemistry and photophysics of organometallic polyynes with fluorene-based auxiliaries.[20–23] The synthesis of polymeric derivatives of fluorene is an interesting approach to the development of efficient blue-light-emitting polymers.[24] The facile functionalization at C-9 imparts an appropriate control of both polymer solubility and potential interchain interactions in films. Here we have investigated the dependence of the first excited singlet and triplet electronic state on the electronic structure of the fluorene spacers of some platinum(II) polyynes. By varying the substituents at the C-9 position of the fluorene ring, the optical absorption and emission properties of these metal-containing polymers can be tuned. This line of research work has also been extended to the mercury(II) congener.

Results & Discussion

Preparations of Ligand Precursors

Schemes 1 and 2 show the syntheses of L_1–L_8. The TMS-substituted derivatives of L_1–L_6 were prepared via a Pd[II]/CuI-catalyzed cross-coupling reaction of the respective dibromo fluorene derivatives with trimethylsilylacetylene.[20,22] The ferrocenylfluorene complex was obtained by condensing the parent fluorene with ferrocenecarbaldehyde using LiN[i]Pr$_2$ in dry THF.[23] 9-Dicyanomethylene-2,7-bis(trimethylsilylethynyl)fluorene was prepared by condensation of 2,7-bis(trimethylsilylethynyl)fluoren-9-one with malononitrile in DMSO at 110 °C.[21] Except for the synthesis of L_8, which uses Bu$_4$NF in CH$_2$Cl$_2$ as the deprotecting agent, desilylation was achieved using K$_2$CO$_3$ in MeOH in other cases to afford L_1–L_7.

Platinum(II) Diynes and Polyynes

(a) Synthesis and Chemical Characterization

The ligands L_1–L_6 can be used to form platinum(II) monomeric (**M1–M6**) and polymeric complexes (**P1–P6**) by the dehydrohalogenation method using the CuI/[i]Pr$_2$NH catalytic system (Scheme 3).[20–23] The feed mole ratio of the platinum chloride precursors and the diethynyl ligands were 2:1 and 1:1 for the monomer and polymer syntheses, respectively These platinum complexes are air-stable and readily dissolve in THF, CH$_2$Cl$_2$ or CHCl$_3$. The polymers are easily processable from organic solvents and exhibit good film-forming properties. Table 1 lists the molecular weight values of these metal polyynes as determined by GPC.

The spectroscopic data agree with the chemical structures of our samples. The IR spectra of these new platinum(II) complexes display a single sharp $\nu(C\equiv C)$ absorption, which together with the singlet signal observed in each ^{31}P-$\{^1H\}$ NMR spectrum reveal a *trans* geometry of the Pt(PR$_3$)$_2$ unit. The $\nu(C\equiv C)$ frequencies for the polymers are lower than those for the ligand precursors, in line with a higher degree of conjugation in the former. NMR analyses clearly indicate a well-defined structure in each compound. Two distinct ^{13}C NMR signals for the individual *sp* carbons in these monomers and polymers were observed and they are shifted downfield with respect to the free ligands. The aromatic region of their ^{13}C NMR spectra also gives more precise information about the regiochemical structure of the main chain skeleton and reveals a high

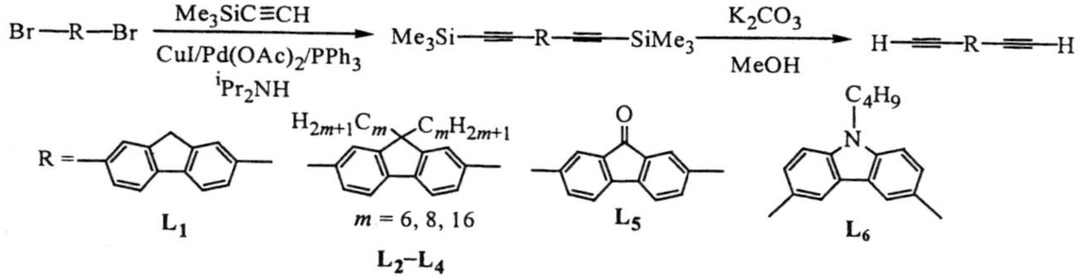

Scheme 1.

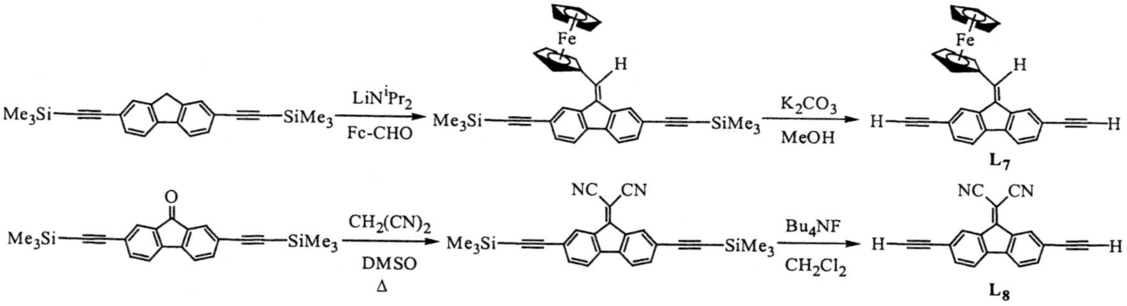

Scheme 2.

degree of structural regularity in the polymers. The structures of M1 and M4–M6 have been confirmed by single-crystal X-ray analyses (Figure 1).

Table 1. Structural and Thermal Properties of the Polymers.

Polymer	M_w	M_n	M_w/M_n	T_{decomp}(onset) (°C)
P1	38,960	21,410	1.82	295 ± 5
P2	28,900	16,900	1.71	349 ± 5
P3	76,460	38,820	1.97	294 ± 5
P4	34,540	17,970	1.92	250 ± 5
P5	116,910	101,600	1.15	352 ± 5
P6	19,100	8,920	2.14	315 ± 5

(b) Electrochemistry

Table 2 shows the results of cyclic voltammetry with reference to the Fc/Fc$^+$ couple. Two irreversible oxidation waves were observed for **M2** and **M6**, attributable to the stepwise two-electron oxidation at the platinum center [Pt^{2+} → Pt^{3+} + e$^-$ → Pt^{4+} + 2e$^-$]. The polymers **P2** and **P6** only display a single irreversible oxidation wave at about +0.40 and +0.27 V, respectively, corresponding to the one-electron oxidation of metal. The first oxidation potential was shown to be more anodic for the monomers than for the polymers. CV of **M4** and **P4** shows a reversible ferrocenyl oxidation and an irreversible oxidation wave at greater potentials due to the one-electron oxidation of platinum. The half-wave potential of the ferrocene moiety is slightly more anodic in **M4** and **P4** than in **L$_7$** because of the loss of electron density from the ferrocenyl donor to the platinum center through the acetylide bridge. **M5** and **P5** both show a reversible reduction wave, due to reduction of the 9-dicyanomethylenefluorene unit, and such half-wave potential is more cathodic in platinum compounds than in **L$_8$**, indicative of its electronic communication with the Pt(PBu$_3$)$_2$ group. **M5** and **P5** also display an irreversible one-electron oxidation wave of the platinum residue at +0.68 V.

Table 2. Electrochemical Data for Platinum Monomers and Polymers.

Monomer	E_{ox} (V)	E_{red} (V)	Polymer	E_{ox} (V)	E_{red} (V)
M2	+0.46, +0.70		P2	+0.40	
M4	+0.67, +1.16		P4	+0.71, +1.34	
M5	+0.67	−1.27	P5	+0.68	−1.26
M6	+0.30, +0.78		P6	+0.27	

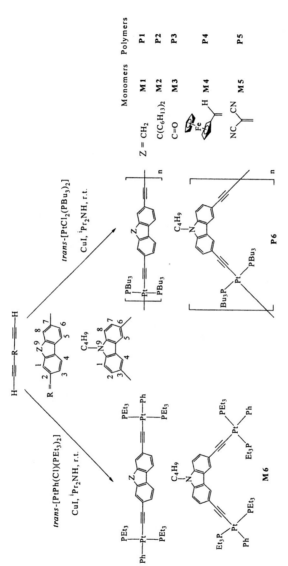

Scheme 3. Synthesis of platinum (II) monomers and polymers.

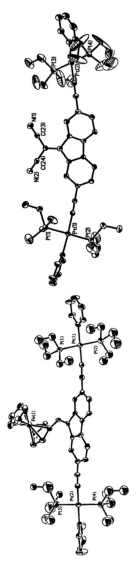

*Figure 1. X-ray crystal structures of **M4** and **M5**.*

(c) Thermal Analysis

The thermal properties of the polymers were examined by thermal gravimetry (TGA) and differential scanning calorimetry (DSC) under nitrogen (Table 1). The onset of decomposition is very similar for **P2** and **P5**, but **P1**, **P3**, **P4** and **P6** have lower decomposition temperatures. The decomposition step can be assigned to the removal of two PBu$_3$ groups from the polymers. As an example, we note a 42% weight loss between 350–545 °C for **P2** whereas 46% of the weight is lost between 315–515 °C for **P6**. Another endotherm observed at 213 and 200 °C may be attributed to the melting of **P2** and **P6**, respectively. The polymer **P2** also shows an additional exothermic phase transition at 137 °C.

(d) Optical Absorption Spectroscopy

Absorption data of **M2**, **P2**, **M6** and **P6** are collected in Table 3. Each of them shows absorptions in the near UV region due to π–π* transitions of the bridging ligand. We associate the lowest energy absorption peak with the S$_0$→S$_1$ transition from the HOMO to the LUMO, which are delocalized π- and π*-orbitals. The position of the lowest energy absorption band is red-shifted when the platinum group is introduced. This reveals that π-conjugation of the ligands extends through the metal center.[11–16] Likewise, the transition energies of the polymers are lowered relative to those of the corresponding monomers, suggesting that the first optical absorption transition in the polymers arises from an electronic excitation that is delocalized over more than one monomer unit. These results agree with other platinum polyynes with pyridine, phenylene or thiophene spacer.[11–16,20–23] but is in contrast to **P5** where the lowest singlet excited state is only confined to a single repeat unit.[21] The HOMO–LUMO energy gap (E_g) as measured from the onset wavelength in solid film state is 2.92 eV for **P2** and 3.10 eV for **P6**, which are larger than those in other fluorene-based materials in our study (Table 4).[20–23] According to the identity of the Z unit, the optical gap of these polyynes follows the experimental order: $N(C_4H_9) > C(C_6H_{13})_2 \geq CH_2 > C=O \approx C=C(H)Fc > C=C(CN)_2$. Due to the presence of strong electron-withdrawing cyano groups in **P5**, the lowest energy absorption peak is red-shifted by ca. 154–266 nm as compared to other fluorene derivatives. There is a dramatic decrease in the optical gap upon coordination of the cyano unit to the fluorene unit, featuring the importance of donor-acceptor interaction in this class of materials. The bandgap of **P5** is estimated to be 1.58 eV and, to our knowledge, this bandgap value is the lowest known for metal polyyne polymers. Replacing the fluorene spacer by a 9-dicyanomethylenefluorene unit lowers the optical gap substantially by 1.3 eV. Hence, the energy of the S$_1$ singlet state depends significantly on the nature of the Z moiety of the ring and is the highest for **P2** and **P6**.

(e) Photoluminescence Spectroscopy

At 290 K, **P2** and **P6** emit in the blue region ranging from 411 to 428 nm in CH$_2$Cl$_2$ solutions (Table 3). Analogous to the absorption spectra, these emission features are shifted to longer wavelengths from monomers to polymers. Upon cooling, two main emission bands appear in each case. For **P2**, we attribute the feature peaking at 433 nm to emission from the singlet excited state (fluorescence S$_1$→S$_0$) at 11 K, due to the small Stokes shift (ca. 0.20 eV) between the bands in the absorption and the emission spectra. The fluorescence quantum yields measured in CH$_2$Cl$_2$ at 290 K for the monomers lie in the range of 0.30–0.48 and the lower values associated with their polymeric counterparts are of the same order of magnitude as other reported Pt polyynes.[25] At 11 K, the intense feature at the low-energy regime of the emission spectrum of **P2** with the maximum at around 554 nm is due to emission from the triplet excited state (phosphorescence T$_1$→S$_0$) and such a triplet emission shows a vibronic structure at low temperature.[11-14] The large energy shift observed between absorption and emission features (about 1.0 eV) ascertains this emitting state as the T$_1$ excited state. This assignment can be further interpreted in terms of the observed temperature dependence of the emission data (Figure 2). The triplet emission is found to be strongly temperature dependent in contrast to the singlet emission. From 150 to 11 K, the singlet emission peak intensity increases only by a factor of 1.3 for **P2**. However, the intensity of the lower-lying emission increases by a factor of 63.5 and such an increase in emission intensity indicates a long-lived excited state that is quenched by thermally activated diffusion to dissociation sites at room temperature. The emission pattern for **P6** also consists of a S$_1$→S$_0$ transition at 426 nm and a T$_1$→S$_0$ peak at the lower energy side (462 nm) with a large Stokes shift of ca. 0.9 eV for the latter. These triplet emissive peaks were observed at 554 and 459 nm for **M2** and **M6**, respectively. In these systems, the lowest triplet excited state remains strongly localized, as can be inferred from the small energy difference between triplet emissions in the monomers and in the polymers. To evaluate the relative efficiency of intersystem crossing (ISC), S$_1$→T$_1$, triggered by the heavy metals, the peak height ratio from triplet emission to singlet emission at 11 K, ΔE(T$_1$→S$_0$, S$_1$→S$_0$), was taken as a good indicator (Table 3). Clearly, the ISC rate is higher for the polymers than that for the monomers in each system. These data attest to the importance of tuning the electronic properties of the central fluorene ring in governing the ISC efficiency in these polyynes. While triplet emission is also detected at 549 nm for **P1** at 11 K, there is no evidence of detecting any T$_1$→S$_0$ transitions over the measured range for **P3–P5** even at 11 K. One salient feature here is that a reduction in the HOMO–LUMO bandgap leads to a lowering of the ISC efficiency and hence the principal emission arises from the S$_1$→S$_0$ transition. This implies that reduced conjugation in these 9-anchored biphenyl structures effectively controls phosphorescence decays via the heavy-atom effect of platinum.

Table 3. Absorption and Emission Data for M2, P2, M6 and P6.

| | λ_{max} (nm)^a | λ_{em} (nm) | | |
	CH$_2$Cl$_2$	CH$_2$Cl$_2$ (290 K)^b	Film (11 K)	$\Delta E(T_1 \rightarrow S_o,$ $S_1 \rightarrow S_o)$
M2	309 (0.2), 358 (1.0), 376 (1.4)	411 (0.30)	396*, 418*, 554, 604*	6.6
P2	307, 399	421, 555* (0.01)	416*, 433*, 554, 589*, 606*	18.0
M6	260 (4.1), 304 (2.1), 333 (3.7)	410*, 426 (0.48)	407*, 425*, 459, 474*	11.0
P6	258, 301, 344	412*, 428 (0.01)	403*, 426*, 462, 497*, 511*	136.0

^{a} ε (10^4 M^{-1} cm^{-1}) values are shown in parentheses. ^{b} Φ_{FL} shown in parentheses are measured in CH$_2$Cl$_2$ relative to anthracene. * As shoulders or weak bands.

Table 4. Optical Data of Various Platinum(II) Fluorene-based Materials.

Compound Z =	Color	Largest λ_{max} in CH$_2$Cl$_2$ (nm)	E_g (eV)
CH$_2$	off-white	394	2.90
C(C$_6$H$_{13}$)$_2$	off-white	399	2.92
C=O	red	506	2.10
C=C(H)Fc	red	437	2.10
C=C(CN)$_2$	blue	660	1.58
N(C$_4$H$_9$)	off-white	344	3.10

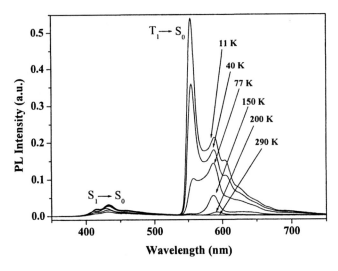

*Figure 2. Temperature dependence of the photoluminescence of **P2**.*

(f) Photocurrent Measurements

Fabrication of the single-layer ITO/polymer/Al photocells (ITO = indium-tin-oxide) was carried out for **P2**, **P5** and **P6** to study their photoconductive properties. The photocurrent of the devices is field-dependent and it increases with increasing bias voltage. Both **P2** and **P6** show a photocurrent quantum yield of approximately 0.01% at 400 and 370 nm, respectively, which is a common value for single layer devices.[12] There is no significant difference in the quantum efficiency with variation of the central fluorene structure and a similar quantum yield was also measured for **P3**.

Mercury(II) Diynes and Polyynes

In the past two decades, while much is known for the photoluminescent behavior of oligomeric and polymeric acetylides of platinum(II) and gold(I) complexes where spin-forbidden triplet emissions can be identified clearly,[11–16,26–28] we have relatively little understanding on the d^{10} mercury(II) system. To our knowledge, there is no literature report on the photophysics and structural properties of soluble mercury-based polyynes. Although linear polymeric copper and mercury acetylides of the form $[-M-C\equiv C(p\text{-}C_6H_4)C\equiv C-]_n$ were reported as early as 1960, these materials were often found to be insoluble and intractable which hampered their purification and characterization.[18] Group 12 heavy mercury atom, with its propensity to enhance spin-orbit coupling, could be an ideal candidate for harvesting the energy of triplet excitons. We have reported the first examples of soluble well-defined high molecular weight mercury(II) polyyne polymers with 9,9-dialkylfluorenes from which we have been able to directly trigger $^3(\pi\pi^*)$ emissions localized on the organic system through efficient ISC by ligation to the mercury(II) moiety.[29]

The synthesis of Hg(II) polyynes **P7–P9** is shown in Scheme 4. Mercuration of **L₂–L₄** with HgCl₂ using methanolic NaOH produced a pale yellow suspension in each case and the solids were collected by filtration to afford off-white air-stable products **P7–P9** in high yields and purity. Their excellent film-forming properties suggest macromolecular nature of the materials and their good solubilities in CH_2Cl_2 and $CHCl_3$ render them amenable to spectroscopic studies and readily solution-processable for optical characterizations. Estimates of the molecular weights using GPC in THF indicate a high degree of polymerization (DP = 24–47) in these polyynes (Table 5). Similar synthetic route using **L₁** only afforded an intractable solid and thus the use of long alkyl chains is crucial in achieving high solubility of these Hg polymers. **M8**, as a model complex of **P8**, was also synthesized by treatment of **L₃** with two equivalents of MeHgCl. Two distinct ^{13}C NMR signals for the sp carbons of the acetylenic units were observed for **P7–P9** and they are shifted downfield with respect to **L₂–L₄**, consistent with the formation of Hg–C(sp) bond. Only six peaks appeared in the aromatic region of their ^{13}C NMR spectra, related to the 12 aromatic carbons in the structures and these data indicate high structural regularity in the polymers. Monomer **M8** displayed a single ^{199}Hg NMR peak at $\delta = -453$ (c.f. $\delta = -847$ for MeHgCl), reflecting polarization of the Hg–C≡ bonds.[30] The X-ray structure of **M8** (Figure 3) also helped to establish polymer structures in the solid state and to correlate the photophysical properties with the structural data. The structure of **M8** consists of dinuclear molecules in which the mercury centers adopt a two-coordinate linear geometry in a rigid-rod manner. The lattice structure is highlighted by the presence of weak intermolecular noncovalent Hg···Hg interactions (3.738 and 4.183 Å), which result in a loose polymeric structure of **M8** in a 3-D network. These Hg···Hg contacts are toward the upper limit of those accepted as representing metallophilic interactions.[31,32]

Polymers **P7–P9** displayed moderate thermal stability with $T_{decomp} > 200$ °C (Table 5) and exhibited an exotherm coincident with mass loss due to decomposition. Polymer **P9** showed the lowest onset decomposition temperature, while **P7** and **P8** with shorter alkyl chains exhibited increasing decomposition temperatures with decreasing m value. Each showed two stepwise losses in the TGA curve, due to the removal of two alkyl groups.

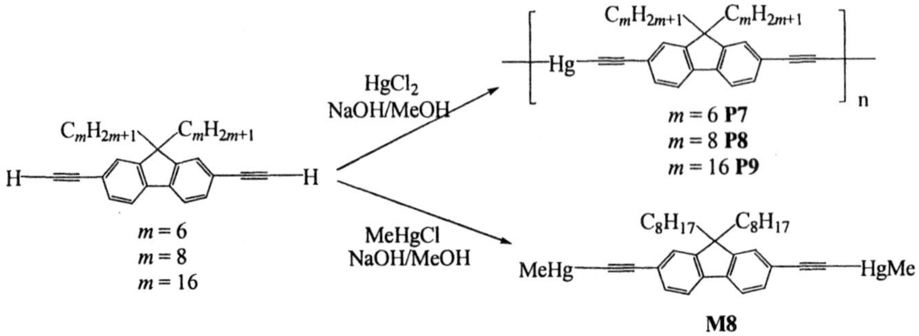

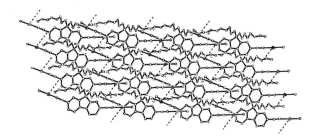

Scheme 4. Synthesis of mercury(II) diynes and polyynes.

*Figure 3. Crystal packing diagram for the model complex **M8**.*

Table 5. Structural and Thermal Properties of P7–P9.

Polymer	M_w	M_n	M_w/M_n	DP	T_{decomp} (onset) (°C)
P7	28720	27100	1.06	47	282 ± 5
P8	18320	15090	1.22	24	220 ± 5
P9	38650	36250	1.07	42	200 ± 5

The photophysical data of **P7–P9** are shown in Table 6. They all display similar structured absorption bands in the near UV region at room temperature. These bands arise mainly from the organic $^1(\pi\pi*)$ transitions, possibly with some admixture of metal orbitals, and the 0–0 absorption peak is assigned as the $S_0 \rightarrow S_1$ transition. We note a red-shift of absorption and emission bands in **P7–P9** after the inclusion of heavy metal atom which indicates an increase in π-conjugation. The C-9 substituents on the fluorene ring exert minimal effect on the absorption energies and comparable optical gaps (E_{gap}) are noted for **P7–P9** (Figure 4a). The transition energies of **P8** are lowered with respect to those of the monomer **M8**, suggesting a well-extended singlet excited state in the polymers. As compared to the solution absorption data, bathochromic shifts of ~10 nm are observed for the lowest energy absorption bands of **P7–P9** in the solid state. Examination of the absorption behavior of the polymers in $CHCl_3$, $CHCl_3$/MeOH solutions, and in the solid state, corroborates the presence of solid-state aggregates in thin films.[33] For instance, addition of a nonsolvent (MeOH) to a $CHCl_3$ solution of **P8** leads to the development of a new band centered at 365 nm, which corresponds to the strongest absorption observed in the solid state. Further support for this red-shift comes from the mercuriophilic interactions in the crystal structure of **M8**. The long side chains in **P7–P9** should suppress aggregation or at least delay its onset, and indeed, changes in absorptions are observed for **P8** only after addition of about 80% methanol.

Table 6. Photophysical Data for P7–P9 and M8.

	λ_{max} (nm)	λ_{em} (nm)	
	Film	CH_2Cl_2 (290 K)^a	Film (11 K)
P7	316, 345, 365	381, 406* (0.003)^b	427, 570, 610*
P8	316, 345, 364	382, 406* (0.004)^b	426, 583, 630*
P9	316, 345, 364	385*, 409 (0.007)^b	429, 590, 634*
M8	316, 339, 358	374 (0.02)^c	416, 438, 544*, 591*

^{a} Φ_{FL} in parentheses relative to quinine sulfate in 0.1 N H_2SO_4 ($\Phi = 0.54$). ^{b} λ_{ex} = 328 nm. ^{c} λ_{ex} = 340 nm. * As shoulders or weak bands.

The PL spectra display a red-shift on going from binuclear to polynuclear structures, implying that the excited state is stabilized by the greater degree of delocalization in the polymers. In dilute fluid solutions, we observe an intense $^1(\pi\pi*)$ emission peak near 400 nm for **P7–P9** due to fluorescence. The thin film singlet emissions in **P7–P9** appear broad at 290 K and are red-shifted which are consistent with their absorption features and are likely to be a result of interchain interactions due to aggregate formation. At 11 K, another lower-lying emissions emerge around 570–590 nm for **P7–P9** and the large Stokes shifts of these peaks from the dipole-allowed absorptions (ca. 1.2–1.3 eV), plus the long emission lifetimes at 11 K (28 ± 0.2 (**P7**), 35 ± 0.2 (**P8**) and 42 ± 0.2 µs (**P9**)) are indicative of their triplet parentage, and they are thus assigned to the $^3(\pi\pi^*)$ excited states of the diethynylfluorenes. The intensity of the phosphorescence relative to fluorescence increases as the value of m increases from **P7** to **P9**. We note that the order of S_1–T_1 crossover efficiency is **P9** > **P8** > **P7** (Figure 4b). The ISC rate is also higher in the polymers than in the monomers (e.g. **M8** vs. **P8**). To examine the spatial extent of the singlet and triplet excitons in our systems, values of $\Delta E(S_0$–$T_1)$ (energy gap between S_0 and T_1) were found to be 2.10–2.23 eV for **P7**–**P9**. The $\Delta E(S_1$–$T_1)$ values for **P7**–**P9** lie within the range of 0.73–0.79 eV and they match well the S_1–T_1 energy gap of 0.7 ± 0.1 eV for similar π-conjugated Pt and Au polyynes.[14,20,26,34]

Concluding Remarks

To study the nature of photoexcited states in organometallic conjugated polymers as a function of π-conjugation in the bridging ligand, a group of soluble, well-defined platinum-containing diyne and polyyne materials consisting of 9-functionalized fluorene and 9-butylcarbazole linking units was prepared and fully characterized. The polymers exhibit good thermal stability and photoconducting properties. A systematic correlation was made between the effective conjugation length (or conversely, bandgap) and the ISC rate in these polyynes. The larger the bandgap, the higher the efficiency of triplet emission. These polymers represent suitable model systems to investigate the relationship between chemical structure and the evolution of singlet and triplet excited states. Similar to the Pt(II) polyynes, the use of mercury in **P7**–**P9** together with their associated high optical gaps can offer a good way to enhance the ISC rate and we have shown for the first time that the organic triplet emissions of fluorenyleneethynylenes have been "illuminated" by the heavy-atom effect of mercury. The present investigation is desirable for applications that harvest the T_1 state for light emission and will provide impetus for future study.

Acknowledgments

This work was supported by CERG Grants from the Hong Kong Research Grants Council (Project Nos. HKBU 2048/01P and 2022/03P) and a Faculty Research Grant from Hong Kong Baptist University (FRG/01-02/II-48).

References

(1) *Conjugated Polymeric Materials: Opportunities in Electronics, Optoelectronics and Molecular Electronics*; Brédas, J.L., Chance, R.R., Eds.; Kluwer Academic Publishers: Dordrecht, 1990.

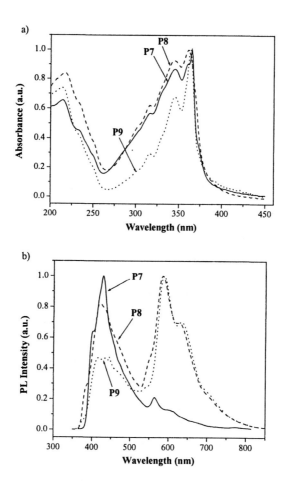

*Figure 4a. Absorption spectra of **P7–P9**. Figure 4b. Emission spectra of*
***P7–P9**.*

(2) Burroughes, J.H.; Bradley, D.D.C.; Brown, A.R.; Marks, R.N.; Mackay, K.; Friend, R.H.; Burn, P.L.; Holmes, A.B. *Nature*, **1990**, *347*, 539.

(3) Tessler, N.; Denton, G.J.; Friend, R.H. *Nature*, **1996**, *382*, 695.

(4) Halls, J.J.M.; Walsh, C.A.; Greenham, N.C.; Marseglia, E.A.; Friend, R.H.; Moratti, S.C.; Holmes, A.B. *Nature*, **1995**, *376*, 498.

(5) Yu, G.; Gao, J.; Hummelen, J.C.; Wudl, F.; Heeger, A.J. *Science*, **1995**, *270*, 1789.

(6) Brown, A.R.; Bradley, D.D.C.; Burroughes, J.H.; Friend, R.H.; Greenham, N.C.; Burn, P.L.; Holmes, A.B.; Kraft, A. *Appl. Phys. Lett.* **1992**, *61*, 2793.

(7) Baldo, M.A.; Thompson, M.E.; Forrest, S.R. *Nature* **2000**, *403*, 750.

(8) Baldo, M.A.; O'Brien, D.F.; You, Y.; Shoustikov, A.; Sibley, S.; Thompson, M.E.; Forrest, S.R. *Nature* **1998**, *395*, 151.

(9) Friend, R.H.; Gymer, R.W.; Holmes, A.B.; Burroughes, J.H.; Marks, R.N.; Taliani, C.; Bradley, D.D.C.; dos Santos, D.A.; Brédas, J.-L.; Lögdlund, M.; Salaneck, W.R. *Nature* **1999**, *397*, 121.

(10) Cleave, V.; Yahioglu, G.; Le Barny, P.; Friend, R.H.; Tessler, N. *Adv. Mater.* **1999**, *11*, 285.

(11) Wilson, J.S.; Köhler, A.; Friend, R.H.; Al-Suti, M.K.; Al-Mandhary, M.R.A.; Khan, M.S.; Raithby, P.R. *J. Chem. Phys.* **2000**, *113*, 7627.

(12) Chawdhury, N.; Köhler, A.; Friend, R.H.; Wong, W.-Y.; Lewis, J.; Younus, M.; Raithby, P.R., Corcoran, T.C.; Al-Mandhary, M.R.A.; Khan, M.S. *J. Chem. Phys.* **1999**, *110*, 4963.

(13) Chawdhury, N.; Köhler, A.; Friend, R.H.; Younus, M.; Long, N.J.; Raithby, P.R.; Lewis, J. *Macromolecules*, **1998**, *31*, 722.

(14) Wilson, J.S.; Chawdhury, N.; Al-Mandhary, M.R.A.; Younus, M.; Khan, M.S.; Raithby, P.R.; Köhler, A.; Friend, R.H. *J. Am. Chem. Soc.* **2001**, *123*, 9412.

(15) Wong, W.-Y.; Wong, C.-K.; Lu, G.-L.; Lee, A.W.-M.; Cheah, K.-W.; Shi, J.-X. *Macromolecules* **2003**, *36*, 983.

(16) Younus, M.; Köhler, A.; Cron, S.; Chawdhury, N.; Al-Mandhary, M.R.A.; Khan, M.S.; Lewis, J.; Long, N.J.; Friend, R.H.; Raithby, P.R. *Angew. Chem. Int. Ed.* **1998**, *37*, 3036.

(17) Long, N.J.; Williams, C.K. *Angew. Chem. Int. Ed.* **2003**, *42*, 2586.

(18) Nguyen, P.; Gómez-Elipe, P.; Manners, I. *Chem. Rev.* **1999**, *99*, 1515.

(19) Manners, I. *Synthetic Metal-Containing Polymers*, Chapter 5, Wiley, Weinheim, 2004.

(20) Wong, W.-Y.; Lu, G.-L.; Choi, K.-H.; Shi, J.-X. *Macromolecules* **2002**, *35*, 3506.

(21) Wong, W.-Y.; Choi, K.-H.; Lu, G.-H. *Macromol. Rapid Commun.* **2001**, *22*, 461.

(22) Lewis, J.; Raithby, P.R.; Wong, W.-Y. *J. Organomet. Chem.* **1998**, *556*, 219.

(23) Wong, W.-Y.; Wong, W.-K.; Raithby, P.R. *J. Chem. Soc., Dalton Trans.* **1998**, 2761.

(24) Neher, D. *Macromol. Rapid Commun.* **2001**, *22*, 1365.

(25) Khan, M.S.; Kakkar, A.K.; Long, N.J.; Lewis, J.; Raithby, P.; Nguyen, P.; Marder, T.B.; Wittmann, F.; Friend, R.H. *J. Mater. Chem.* **1994**, *4*, 1227.

(26) Chao, H.-Y.; Lu, W.; Li, Y.; Chan, M.C.W.; Che, C.-M.; Cheung, K.-K.; Zhu, N. *J. Am. Chem. Soc.* **2002**, *124*, 14696.

(27) Lu, W.; Xiang, H.-F.; Zhu, N.; Che, C.-M. *Organometallics*, **2002**, *21*, 2343.

(28) Che, C.-M.; Chao, H.-Y.; Miskowski, V.M.; Li, Y.; Cheung, K.-K. *J. Am. Chem. Soc.* **2001**, *123*, 4985.

(29) Wong, W.-Y.; Liu, L.; Shi, J.-X. *Angew. Chem. Int. Ed.* **2003**, *42*, 4064.

(30) Sebald, A.; Wrackmeyer, B. *Spectrochim. Acta, Part A* **1982**, *38*, 163.

(31) Faville, S.J.; Henderson, W.; Mathieson, T.J.; Nicholson, B.K. *J. Organomet. Chem.* **1999**, *580*, 363.

(32) Pyykkö, P. *Chem. Rev.* **1997**, *97*, 597.

(33) Bunz, U.H.F. *Chem. Rev.* **2000**, *100*, 1605 and references cited therein.

(34) Khan, M.S.; Al-Mandhary, M.R.A.; Al-Suti, M.K.; Ahrens, B.; Mahon, M.F.; M ale, L .; R aithby, P.R.; Boothby, C.E.; Köhler, A. *Dalton Trans.* **2003**, 74.

Chapter 27

Peptide Films on Surfaces: Preparation and Electron Transfer

G. A. Orlowski and H. B. Kraatz

Department of Chemistry, University of Saskatchewan, 110 Science Place, Saskatoon, Saskatchewan S7N 5C9, Canada

A new method for the electrodeposition of peptide disulfides onto gold surfaces is described together with a comparison of these surfaces to those prepared by conventional methods. For this purpose we are using acylic peptides of the general formula [Fc-Xxx-CSA]$_2$ (Fc = ferrocenoyl, Xxx = amino acid, CSA = cysteamine) and cyclic compounds Fc[Xxx-CSA]$_2$. These two classes of peptides show significant differences in their electron transfer behavior, which is most likely due to the ability of the cyclic systems to engage in two Au-S linkages.

Introduction

Understanding electron-transfer (ET) processes in proteins is of fundamental importance.[1-10] In a series of photophysical studies of well-behaved peptide model systems, it has become evident that the ET through the peptide spacer is greatly influenced by the separation between the acceptor (A) and the donor (D), the nature of the peptide backbone, the amino acid sequence, and the resulting flexibility.[11-13] In particular, it was suggested in the literature that the presence of H-bonding will increase the rate of ET, and there is experimental evidence (mostly in proteins) to suggest that H-bonding indeed increases the rate of ET.[14-17] We have been involved in electrochemical measurements of the ET process in redox-labeled peptide thiol films. The formation of the peptide film is achieved via a Au-thiolate linkage. ET has to proceed from the redox probe through the peptide spacer to the Au surface. Electrochemical measurements allow us to evaluate the forward and backward ET processes in these systems.

We have demonstrated before that short Fc-peptides form poorly ordered films and leave about 20% of the surface accessible to solvent and supporting electrolyte. Using hexanethiol it was possible to fill these holes and to evaluate the ET kinetics in great detail.[18] The question arose if the poor packing is a result of the self-assembly procedure, which usually entails soaking of a clean gold substrate for several days. Thus we embarked on a study aimed at developing a method that allows us to form densely packed and well-ordered peptide films from disulfides.

Here, we investigate the role of the attachment of the peptide on a gold surface and describe the results of an electrochemical study of ferrocenoyl (Fc)-labeled peptides. We made use of two classes of Fc-peptides: acylic ferrocenoyl (Fc)-peptide disulfides [Fc-CSA]$_2$ (**1a**), [Fc-Gly–CSA]$_2$ (**2a**), [Fc-Ala–CSA]$_2$ (**3a**), [Fc-Val–CSA]$_2$ (**4a**) and [Fc-Leu–CSA]$_2$ (**5a**) and cyclo-1,1'-Fc-peptide disulfides Fc[CSA]$_2$ (**1c**), Fc[Gly–CSA]$_2$ (**2c**), Fc[Ala–CSA]$_2$ (**3c**), Fc[Val–CSA]$_2$ (**4c**) and Fc[Leu–CSA]$_2$ (**5c**) (see Scheme 1). These compounds were prepared as described before.[18-20] In addition, we describe an electro-deposition method for the preparation of cyclic and acyclic Fc-peptide disulfides leading to tightly-packed films.

Experimental

Ellipsometry. Au on Si(100) (Platypus Technologies, Inc) wafers were incubated in a 1 mM Fc-peptide ethanolic solution for 5 days and finally rinsed with EtOH and H$_2$O. A Stokes ellipsometer LSE (Gaertner Scientific Corporation, Skokie, IL, fixed angle (70°), fixed wavelength (632.8 nm)) was used, and the data were collected and analyzed using LGEMP (Gaertner Ellipsometer Measurement Software) on a PC. Ellipsometry constants were as

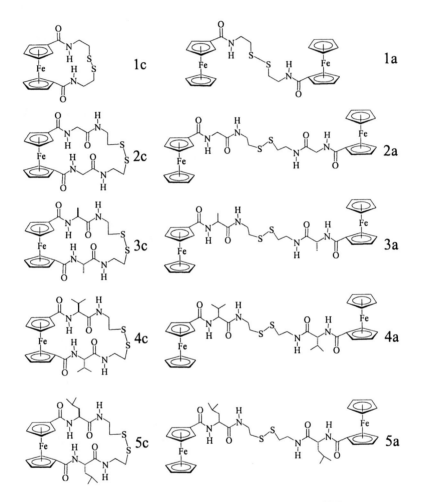

Scheme 1. Chemical drawings of all ferrocene-peptide disulfide conjugates.

follows: $n_S = 0.00$ and $K_S = 1.44$ was used as the refractive index of the monolayer.

Electrochemistry was performed on two types of potentiostats: home made and CHI660B, CHI instruments Inc. All potentials are given vs Ag/AgCl which was used as a reference electrode. Supporting electrolyte in all cases was 2 M NaClO$_4$. Calculation of electron transfer rate is described elsevere.[18] Experiments were performed on gold ultramicroelectrode with a diameter of 25 µm.

Electrodeposition. 10 mM solutions of Fc-peptide disulfide in pure EtOH were prepared and a potential of - 1.3 V (vs A g/AgCl) w as a pplied f or 3 0 m inutes. Lengthening of the electrodeposition time does not significantly change the film.

Results and Discussion

Film of cyclic Fc-peptides **1c-5c** and of the acylic systems **1a-5a** were prepared as schematically shown in Figure 1. The electrochemical properties are evaluated by cyclic voltammetry (CV) and chronoamperometry (CA). All films exhibit reversible one electron redox waves (Figure 2).

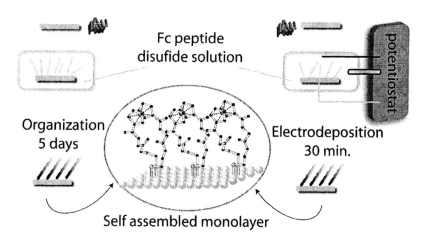

Figure 1. Schematic representation of the two methods used to prepare Fc-peptide films.

Integration of the Faradaic current provides the Fc surface concentration, from which a specific area per molecule can be calculated.[18] Surface concentration of the monolayer ranges from 2.3 – 4.6 * 10^{-10} mol/cm^{-2}. The theoretical area (calculated from crystal structure data of the acylic Fc-peptides and 1,1'-cyclo-Fc-peptides are ~30 and ~40 Å^2·molecule^{-1}, respectively. The thickness of the films prepared by electrodeposition and by soaking was determined by ellipsometry and gave values of 9(3) Å, which is in good agreement with data obtained from x-ray crystallography[20] and by molecular modeling (Spartan). The signal of the cylcopeptides **1c-5c** is shifted to higher potential as expected for disubstituted ferrocene derivatives. The monolayers prepared by the electrodeposition exhibited remarkable stability and showed about a 5% of monolayer loss after more than 100 cycles (100 mV to 1000 mV vs. Ag/AgCl) (Figure 3).

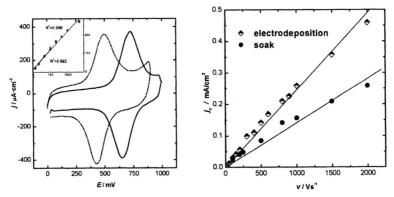

Figure 2. (left) Electrochemical response of immobilized films for acyclic [Fc-Gly-CSA]₂ (2a) and Fc[Gly-CSA]₂ (2c). Inset shows linear dependency between scan rate and peak current indicating successful immobilization of cyclic and acyclic compounds on gold surface; (right): Comparison of two methods: electrodeposition and incubation (soak). Graph presenting successful immobilization (linear dependency of scan rate and peak current) of cyclic Fc[Ala-CSA]₂ (3c) by application of both methods.

All CVs showed reversible Fc to Fc$^+$ redox reactions as measured by peak current ratios. All of the cyclic compounds showed a higher redox potential compared to the non-cyclic analog, which is typical when comparing mono- to 1,1'-di-substituted Fc. The electron withdrawing capability of the amides makes the disubstituted Fc more difficult to oxidize. All electrochemical parameters are included in Table 1 and Table 2.

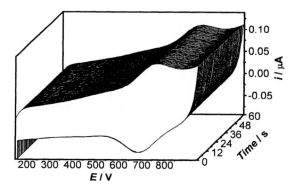

Figure 3. Multiple CVs of Fc[AlaCSA]₂ (3c) taken every 0.05 seconds for 60 seconds with a 12.5 μm radius Au-modified electrode, 2 M NaClO₄ supporting electrolyte and a Ag/AgCl/(3.5 M KCl) reference electrode.

Table 1. Summary of electrochemical parameters analyzed by CV and CA. Values in parentheses are the standard deviations from 5 electrode measurements

Comp.	Electrodeposition			Incubation		
	$E^{0'}/mV$	$k_{ET} \times 10^3 / s^{-1}$*	$Area / Å^2 \cdot molecule^{-1}$	$E^{0'}/mV$	$k_{ET} \times 10^3 / s^{-1}$*	$Area / Å^2 \cdot molecule^{-1}$
1-c	670(7)	9.5	45(7)	660(9)	8.0	120(9)
1-a	465(9)	8.0	40(7)	473(6)	7.0	50(3)
2-c	688(6)	14.0	47(8)	682(8)	13.0	150(20)
2-a	464(6)	13.5	50(8)	445(9)	12.0	78(8)
3-c	635(6)	12.0	68(9)	624(8)	11.0	141(20)
3-a	490(7)	6.0	36(5)	468(8)	6.9	53(9)
4-c	670(7)	12.0	60(9)	665(9)	11.0	130(25)
4-a	488(7)	9.5	65(8)	484(7)	10.0	70(10)
5-c	686(8)	17.0	60(8)	680(7)	14.0	220(10)
5-a	484(7)	11.0	72(8)	476(9)	11.0	101(9)

* Error for k_{ET} calculations was 1.5×10^3 s^{-1}

Electron transfer kinetics was assessed by CV and CA and both electrochemical methods yielded similar values (Table 1). The ET rate, k_{ET}, for all compounds are between 17(1.5) and 6(1.5) $\times$ 10^3 s^{-1}. In general, the cyclic Fc-peptide systems exhibit higher $k_{ET}s$ compared to the corresponding acyclic systems. The enhanced ET for the cyclic peptides is attributed to the double junction – double peptide wire of the 1,1'-Fc peptide system with both sulfur atom linked to the Au surface, compared to the single connection of acyclic systems. Intramolecular H-bonding in the cyclopeptide, as confirmed by X-ray structure analysis,[20] may contribute to the increased electron transfer rate for cyclic compounds.

Table 2. Electrochemical parameters calculated from CV experiments using the electrochemical deposition and incubation methods.

	Incubation			Electrodeposition		
	ΔE_p / mV	E_{fwhm} / mV	I_a/I_c	ΔE_p / mV	E_{fwhm} / mV	I_a/I_c
1c	90(8)	200(15)	1.00(9)	90(9)	190(10)	0.92(5)
1a	120(6)	240(15)	1.00(9)	120 (10)	220(10)	1.00(5)
2c	60(8)	180(10)	0.90(9)	60(7)	170(8)	1.00(5)
2a	65(5)	195(10)	1.00(9)	140(10)	175(8)	0.92(5)
3c	85(7)	190(15)	0.90(9)	55(7)	160(8)	0.94(5)
3a	90(5)	200(15)	1.00(9)	111(10)	190(10)	0.90(5)
4c	80(7)	210(8)	0.90(5)	55(7)	170(8)	0.98(5)
4a	85(5)	230(15)	0.90(5)	120(10)	190(10)	1.00(5)
5c	70(5)	200(10)	0.90(5)	62(7)	157(10)	0.90(3)
5a	95(7)	210(10)	0.90(5)	110(10)	190(10)	0.90(3)

In summary, we have presented an electrochemical method to form Fc-peptide films from Fc-peptide disulfides, giving raise to well-packed films on gold. This method should find wide-spread applications for the formation of films from disulfides. Our studies allowed a direct comparison of the ET kinetics of cyclic and acyclic Fc-peptide disulfide systems. Our results show faster ET kinetics for the cyclic systems compared to the acyclic systems, which may be result of the enhanced rigidity of the molecules or simple increase number of "conductive peptide wires" (one versus two) to the surface. We are now investigating this phenomenon in more detail.

Acknowledgment

The authors thank NSERC for funding. H.B.K. is the Canada Research Chair in Biomaterials.

References

1. Bobrowski, K.; Poznanski, J.; Holcman, J.; Wierzchowski, K. L. *J. Phys. Chem. B* **1999**, *103*, 10316-10324.
2. DeFelippis, M. R.; Faraggi, M.; Klapper, M. H. *J. Am. Chem. Soc.* **1990**, *112*, 5640-5642.
3. Farver, O.; Pecht, I. *J. Biol. Inorg. Chem.* **1997**, *2*, 387-392.
4. Gray, H. B.; Winkler, J. R. *Annu. ReV. Biochem.* **1996**, *65*, 537-561.
5. Gray, H. B.; Winkler, J. R. *J. Electroanal. Chem.* **1997**, *438*, 43-47.
6. Isied, S. S.; Ogawa, M. Y.; Wishart, J. F. *Chem. Rev.*. **1992**, *92*,381-394.
7. Koch, T.; Armitage, B.; Hansen, H. F.; Orum, H.; Schuster, G. B.*Nucleosides Nucleotides* **1999**, *18*, 1313-1315.
8. Lang, K.; Kuki, A. *Photochem. Photobiol.* **1999**, *70*, 579-584.
9. Langen, R.; Chang, I. J.; Germanas, J. P.; Richards, J. H.; Winkler,J. R.; Gray, H. B. *Science* **1995**, *268*, 1733-1735.
10. Langen, R.; Colon, J. L.; Casimiro, D. R.; Karpishin, T. B.; Winkler,J. R.; Gray, H. B. *J. Biol. Inorg. Chem.* **1996**, *1*, 221-225.
11. Zheng, Y. J.; Case, M. A.; Wishart, J. F.; McLendon, G. L. *J. Phys.Chem. B* **2003**, *107*, 7288-7292.
12. Sisido, M.; Hoshino, S.; Kusano, H.; Kuragaki, M.; Makino, M.;Sasaki, H.; Smith, T. A.; Ghiggino, K. P. *J. Phys. Chem. B* **2001**, *105*, 10407-10415.
13. Galka, M. M.; Kraatz, H. B. *ChemPhysChem* **2002**, *3*, 356-359.
14. Isied, S. S.; Ogawa, M. Y.; Wishart, J. F. *Chem. Rev.* **1992**, *92*, 381-394.
15. Williams, R. J. P. *J. Biol. Inorg. Chem.* **1997**, *2*, 373-377.
16. Beratan, D. N.; Skourtis, S. S. *Curr. Opin. Chem. Biol.* **1998**, *2*, 235-243.
17. Farver, O.; Kroneck, P. M. H.; Zumft, W. G.; Pecht, I. *J. Inorg. Biochem.* **2001**, *86*, 84-84.

18. Bediako-Amoa, I.; Sutherland, T.C.; Li, C.-Z.; Silerova, R.; Kraatz, H.B. *J. Phys. Chem. B.* **2004**, *108*, 704-714

19. Bediako-Amoa, I.; Silerova, R.; Kraatz, H.B. *Chem. Commun.* **2002**, 2430-2431.

20. Chowdhury, S.; Schatte, G.; Kraatz, H.B. *J. Chem. Soc., Dalton Trans.* **2004**, 1726-1730.

Chapter 28

Neutral and Cationic Cyclopentadienyliron Macromolecules Containing Azo Dyes

Alaa S. Abd-El-Aziz[*] and Patrick O. Shipman

Department of Chemistry, The University of Winnipeg, Winnipeg, Manitoba R3B 2E9, Canada
[*]Corresponding author: a.abdelaziz@uwinnipeg.ca

Abstract

A number of classes of organoiron polymers containing azo dyes pendent to, or in their backbone have been prepared. The first class includes cationic organometallic polyethers, polythioethers and polynorbornenes incorporating aryl and hetaryl azo dyes. These organometallic polymers were synthesized via nucleophilic aromatic substitution reactions or ring opening metathesis polymerization. The resultant polymers were all brightly colored and displayed excellent solubility in polar organic solvents. Thermal analysis indicated that the polymers were thermally stable with decomplexation of the metal moiety at approximately 235 °C. Photolytic demetallation of the polymers resulted in the decoordination of the cationic cyclopentadienyliron moieties and the formation of organic polymers. The organic polymers displayed lower glass transition temperatures than their cationic organoiron analogues. The second class of polymers is ferrocene based polymers with pendent cationic cyclopentadienyliron moieties. The reaction of arene complexes containing azo dye chromophores and terminal hydroxyl groups with 1,1'-ferrocenedicarbonyl chloride gave rise to triiron complexes with a substituted ferrocene center and two terminal arene complexes containing chloro groups. Nucleophilic aromatic substitution polymerization of these triiron complexes with O- and S-containing nucleophiles followed by photolytic demetallation led to the isolation of neutral ferrocene based polymers containing azo dyes in the backbone. Electrochemical studies of these complexes showed the reduction of the cationic iron moieties at −1.42 V, while the neutral iron species were oxidized at 0.89 V. UV-visible studies showed absorption at 419 nm and a bathochromic shift to 530 nm with the addition of HCl.

401

Background

Due to the large number of potential applications for organometallic macromolecules, there has been tremendous growth in this field of research over the past two decades.[1-4] Organometallic macromolecules can be used as electrocatalysts, chemical sensors, and photoactive molecular devices. Based on the nature of the organic ligand as well as the metallic moiety, organometallic macromolecules can be divided into a number of different classes. Metals can be either σ- or π-bonded to the carbon skeleton of the polymer backbone or the side chains. The most widely studied type of organometallic polymer is that which incorporates metallocenes into their structures.[5-9]

Cationic cyclopentadienyliron chloroarene complexes have been utilized in the synthesis of novel classes of organoiron monomers and polymers.[10-15] Sequential nucleophilic aromatic substitution reactions of dichlorobenzene complexes with hydroquinone has given rise to organoiron oligomers and polymers with well-defined molecular weights and molecular weight distributions.[16] A large variety of poly(aromatic ethers) have been synthesized using this methodology, including those containing up to 35 cyclopentadienyliron moieties pendent to the alternating arene rings in the backbone.[17] Nucleophilic aromatic substitution polymerizations of dichloroarene complexes with O- and S-containing nucleophiles allowed for one-step syntheses of cationic cyclopentadienyliron polyethers and polythioethers.[18,19] Ring opening metathesis polymerization (ROMP) of cationic organoiron norbornenes using Grubbs' catalyst has allowed for the production of polynorbornenes with pendent cationic cyclopentadienyliron moieties.[20]

Azo dye containing polymers, commonly referred to as photoresponsive polymers are interesting for numerous reasons. One reason that stands out is the *cis-trans* isomerism of the azo dye.[21-24] It is thought, and in some cases proven that this isomerism in conjunction with the special environment of the polymer matrix, can control the chemical and physical properties of the polymer, such as conductivity, glass transition temperature, and metal ion capture ability.[24] Incorporation of azo dyes into polymers can occur by two main methodologies; in the first method, the azo dye is incorporated through a diffusion or adsorption pathway into a pre-made polymer.[24] The second method uses a reactive azo dye, which can be reacted to form either the side chains or the backbone of the polymer.[24] The incorporation of azo dyes into organometallic polymers is relatively new. These polymers are particularly interesting due to the combination of the physical properties of the azo dyes with those of the organometallic complex.[10, 21-24]

This chapter will provide an overview of recent developments in the field of organoiron polymers containing azo dyes. Synthetic routes to obtain these polymers, their physical and chemical properties, and spectral studies will also be discussed.

Linear Organometallic Polymers Containing Azo Dyes as Side Chains

Aryl Azo Dye Containing Organoiron Polymers:

Recently, the synthesis of the first examples of cationic organoiron macromolecules with azo dyes incorporated in their side chains has been reported.[25-27] The synthesis of the valeric acid based diiron complex (**1**), followed by the condensation reaction of the bimetallic complex with a number of azo dyes containing terminal hydroxyl groups (**2a-c**) allowed for the formation of the cationic diiron complexes with pendent azo dyes (**3a-c**) (Scheme 1).

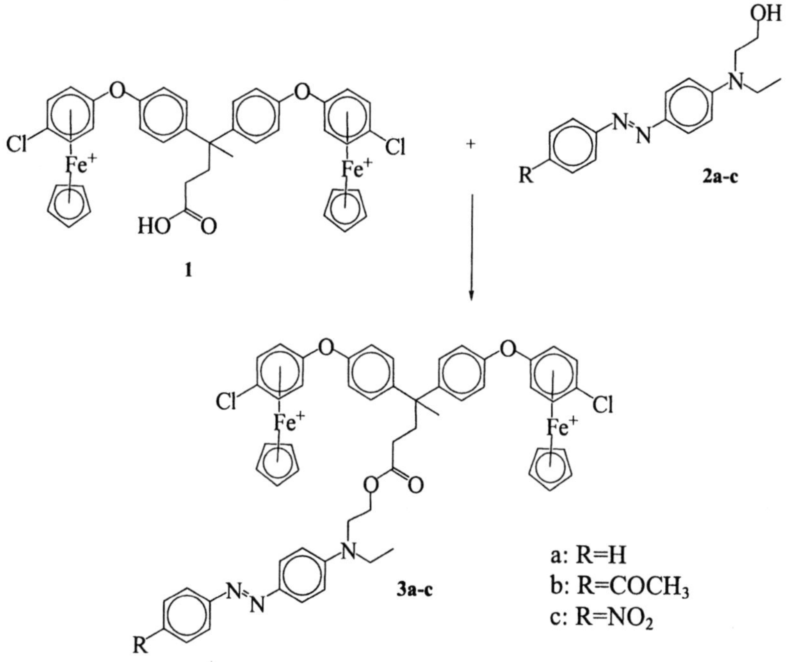

a: R=H
b: R=COCH$_3$
c: R=NO$_2$

Scheme 1

Nucleophilic aromatic substitution polymerization reactions of these azo dye functionalized complexes with a number of O- and S-containing nucleophiles has led to the formation of polymeric materials with pendent azo dyes in very good yields (Scheme 2). Polymers 4-6 exhibited excellent solubility in polar organic solvents a nd w ere f ound to be thermally stable with the decoordination of the cyclopentadienyliron moiety between 220-240 °C and the degradation of the backbone starting at 450 °C. Differential scanning calorimetry (DSC) showed that the polymers 4a-c exhibited glass transition temperatures (T_g) that ranged from 110 to 125 °C.[25]

Photolysis of these polymers in an acetonitrile/dimethylformamide mixture allowed for decomplexation of the cationic cyclopentadienyliron moieties, which yielded the organic polyethers or polythioethers with the azo dyes intact (7a-c - 9a-c). Due to the interaction of the cationic organoiron moieties with the gel

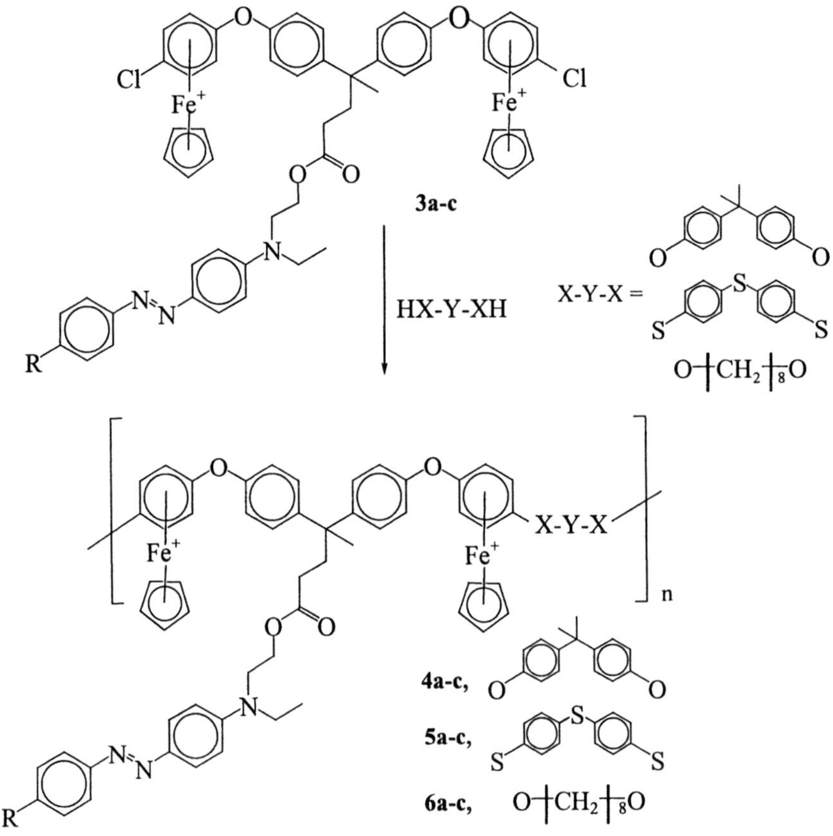

Scheme 2

permeation chromatography (GPC) column, the molecular weights of the metallated polymers were estimated from the molecular weights of their corresponding demetallated analogues. Mw's for the metallated polymers **4-6** ranged from 13 400 to 31 600 with polydispersities from 1.2 to 2.6. Polymers **7-9** exhibited much lower glass transition temperatures than those of their organometallic analogues. The glass transition temperature for polymers **9a-c** was between 34 and 64 °C and was between 110-122 °C for polymers **7a-c** and **8a-c**.

UV-visible studies indicated that the polymers exhibited similar UV spectra to their respective monomers. The peaks obtained in various organic solutions of varying pH are indicative of the n$\rightarrow$ π* and π$\rightarrow$ π* transitions of the azo dye.[26] These polymeric materials displayed bathochromic shifts when the R group was altered to a more electron withdrawing group. The spectra of **4a-c** are shown in Figure 1 as an example. The λ_{max} for the polymers with a weakly electron withdrawing group ranged from 418 to 420 nm, whereas polymers containing a highly electron withdrawing group displayed λ_{max} between 451 to 454 nm. Bathochromic shifts were also seen with changing pH, thus increasing the pH of the solution showed a shift from 418 to 520 nm for polymer **4a** due to the formation of the azonium ion.

The second class of linear organometallic polymers is organoiron polynorbornenes functionalized with aryl azo dyes.[26] Preparation of monomers **14a, b** was accomplished via the reaction of azo dyes **10a,b** with the organoiron complex (**11**) followed by a condensation reaction with 5-norbornene-2-carboxylic acid (**13**) as describe in Scheme 3. ROMP of these monomers using Grubbs' catalyst in dichloromethane led to the isolation of substituted polynorbornenes (**15a, b**) in very good yields.

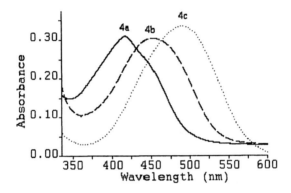

*Figure 1: UV-visible spectra for organoiron polymers **4a-c**.*

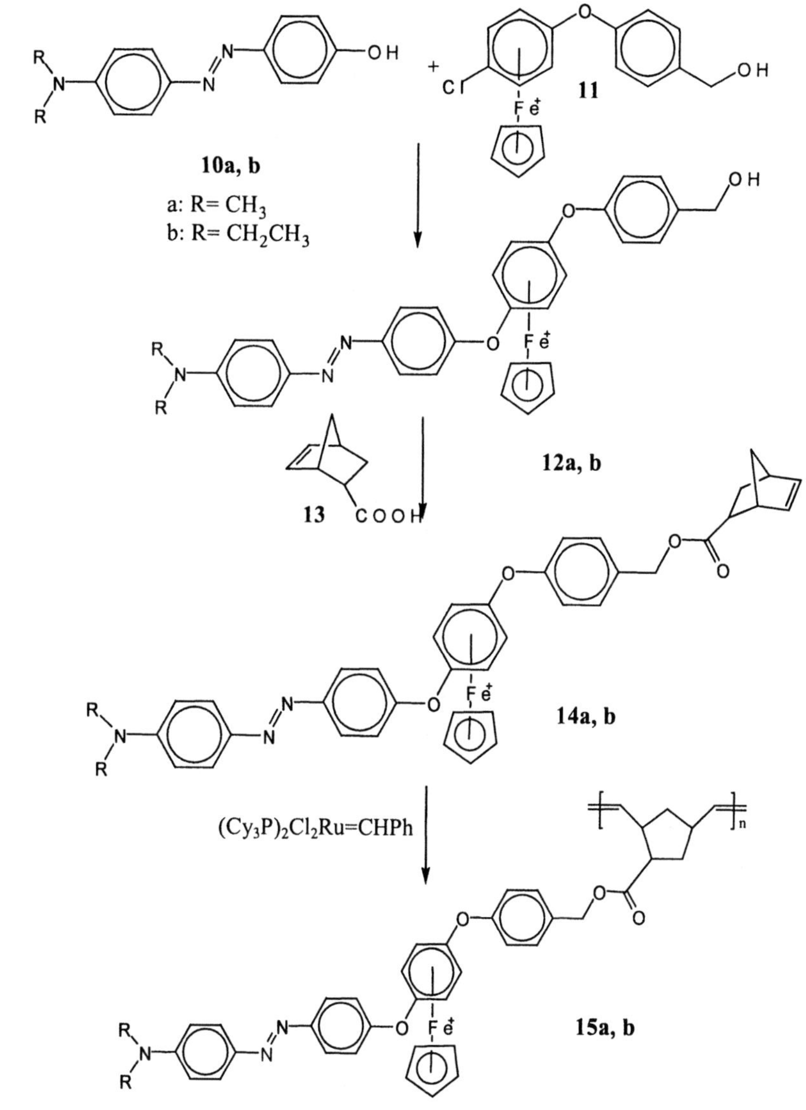

Scheme 3

These brightly colored polymers (**15a, b**) showed λ_{max} values between 420 and 430 nm in DMF solutions with a bathochromic shift upon addition of HCl solutions to 518 nm. Polymers **15a, b** displayed excellent solubility in polar organic solvents and had molecular weights between 16 800 and 32 000 with a polydispersity index between 1.10 and 1.16. Electrochemical studies of polymers **15a, b** showed that the cationic cyclopentadienylironarene complexes underwent a reversible reduction between −1.2 to −1.4 V, with the production of the nineteen-electron iron complex from the eighteen-electron complex. The cathodic and anodic peak potentials occurred at −1.51 and −1.19 V respectively (Figure 2).

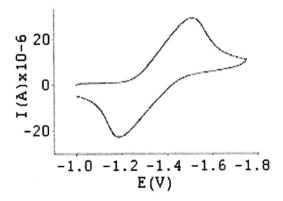

Figure 2: Cyclic Voltammogram of Polymer 15a.

Thermogravimetric analysis (TGA) showed that the cationic organoiron polynorbornenes **15a, b** were thermally stable, with the cleavage of the cyclopentadienyliron moiety and partial decomposition of the side chains starting between 225-231 °C. Degradation of the backbone occured between 400-450 °C. Organic azo dye functionalized polynorbornenes (**16a, b**) were isolated by the photolytic cleavage of the cyclopentadienyliron moiety of the organoiron polymers. Complete thermal data for these polymers are listed in Table 1.

a = CH$_3$
b = CH$_2$CH$_3$

16a, b

Table 1. Thermal analysis of azo dye containing organoiron
polynorbornenes and their organic analogues.

Polymer	First Weight Loss		Second Weight Loss		T_g °C
	T °C	%	T °C	%	
15a	226	12	450	25	178
15b	225	16	437	36	172
16a	230	12	400	31	147
16b	231	23	424	23	145

Polymers **16a, b** exhibited lower glass transition temperatures than those for their metallated analogues. For example polymer **16a** showed a glass transition temperature of 147 °C, whereas polymer **15a** exhibited a glass transition temperature of approximately 178 °C.

Hetaryl Azo Dye Containing Organoiron Polymers:

New research has delved into azo dyes containing heterocyclic rings due to the color enhancement and increased spectral range of these dyes.[23, 24] Azo dyes based on thiazol rings are of particular interest since thiazole ring systems have shown antibacterial and anti-inflammatory activity.[23] The first examples of organoiron polynorbornenes containing hetaryl azo dyes based on benzothiazole (**17, 18**) have recently been reported.[27]

The synthesis of organoiron polynorbornenes containing hetaryl azo dyes followed the same methodology as previous organoiron polynorbornenes containing azo dyes. UV-visible studies showed that polymer **18** exhibited a λ_{max} at 423 nm and polymer **17** displayed a λ_{max} at 520 nm. Upon the addition of an HCl solution, the polymers underwent a bathochromic shift to 522 and 608 nm respectively. The UV-visible spectrum for polymer **17** is shown in Figure 3 as an example.

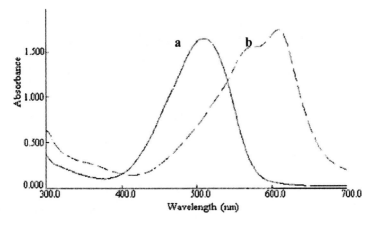

Figure 3: UV-visible spectrum of polymer 17, a) before addition of HCl solution, b) after addition of HCl.

Electrochemical studies at low temperatures showed that these polymers exhibited the reversible reduction process of the cationic iron centers at –1.08 V. The weight average molecular weights of polymers **17** and **18** were 24 500 to 40 900, respectively. Thermal analysis showed that polymers **17** and **18** possessed glass transition temperatures of 146 and 161 °C respectively.[27]

Ferrocene Based Polymers Containing Azo Dyes in the Backbone

Nucleophilic aromatic substitution reactions of p-dichloro-arenecyclopentadienyliron complex (**19**) with azo dyes possessing terminal phenolic and alcoholic groups (**20a, b**) gave rise to organoiron azo dye complexes (**21a, b**) with terminal alcoholic groups. These complexes **21a, b** were further reacted with 1,1'-ferrocenedicarbonyl chloride (**22**) to yield monomers **23a, b** which contain a bridging ferrocene unit, two backbone azo dye units, and two pendent chloro functionalized cyclopentadienyliron units (Scheme 4).[28]

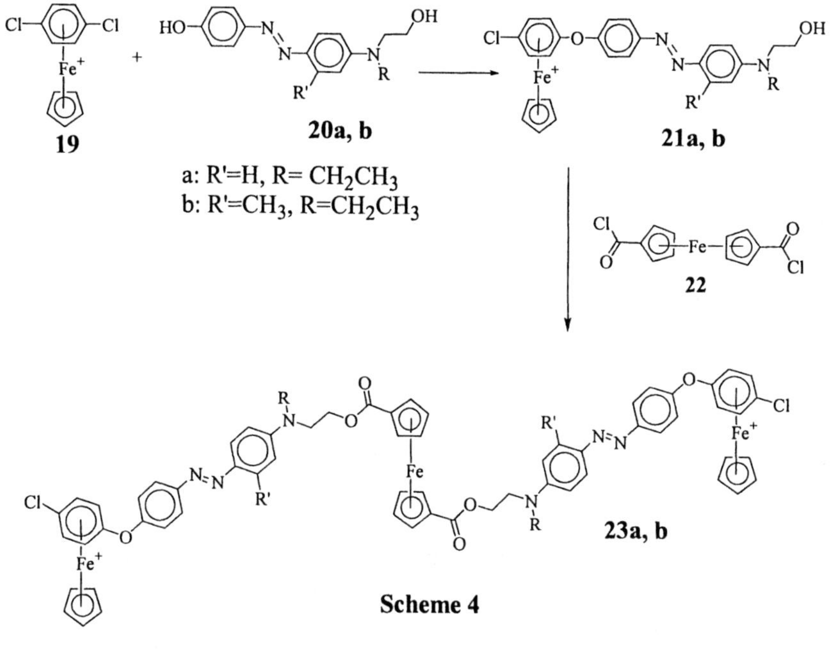

Scheme 4

Reaction of the monomers **23a, b** with various S- or O-containing dinucleophiles led to the synthesis of organometallic polymers (**24-26**) as described in Scheme 5. These brightly colored organoiron polymers (**24-26**) showed good solubility in polar organic solvents and exhibited weight averaged molecular weights between 11 000 to 16 000. Thermogravimetric analysis showed that polymers **24-26** were thermally stable with cleavage of the cationic cyclopentadienyliron moiety between 210-300 °C.

Neutral ferrocene based polymers containing azo dyes in the backbone (**27-29**) were formed through the photolysis of their corresponding cationic organoiron polymers (Scheme 5). Polymers **27-29** exhibited lower glass

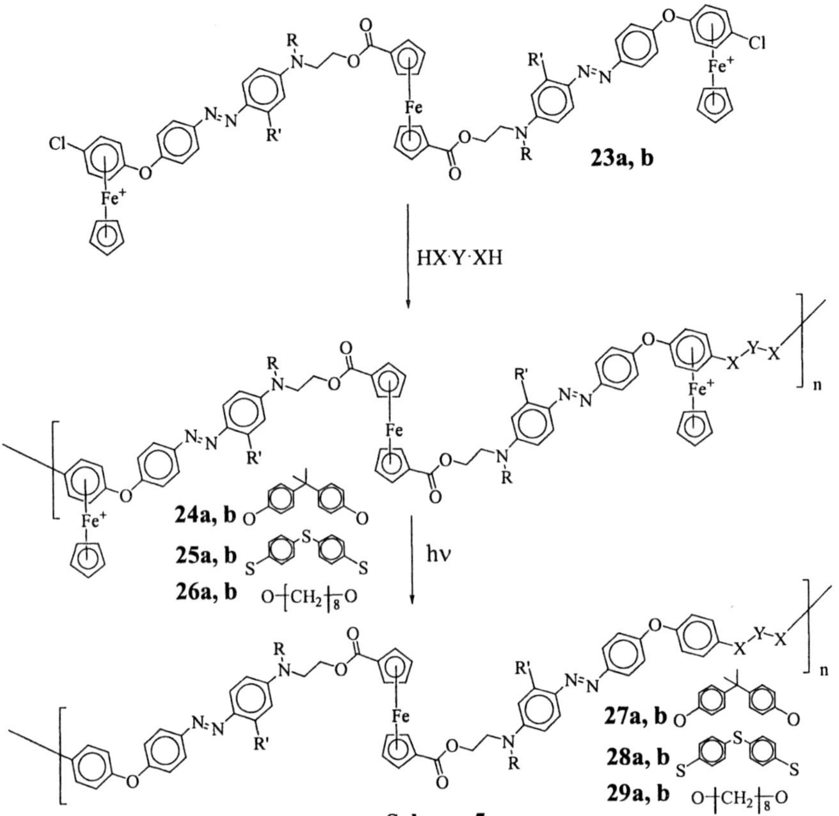

Scheme 5

transition temperatures (85 to 92 °C) than those of their cationic analogues polymers **24-26** (126 to 164 °C). Electrochemical studies of the cationic polymers **24-26** displayed two redox processes. The first process showed the oxidation of the neutral iron centers at 0.89 V, while the second corresponded to the reduction of the cationic iron center at -1.42V. UV-visible studies of these polymers (**27-29**) showed that the polymers displayed a λ_{max} similar to those of the corresponding starting azo dye material. Polymer **27a**, as an example, showed a λ_{max} at 419 nm in a DMF solution. Upon addition of HCl, a bathochromic shift occurred at a λ_{max} of 530 nm due to the protonation of the azo group (Figure 4).

Organoiron polymers containing azo dyes in the backbone have also been prepared via condensation reactions of 1,1'-ferrocenedicarbonyl chloride (**22**) with disubstituted cyclopentadienyliironarene complexes (**30a, b**) as shown in Scheme 6.

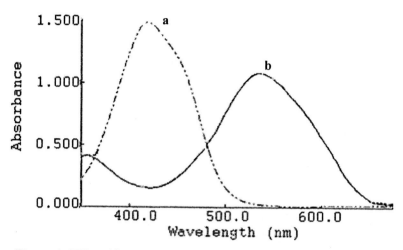

Figure 4: UV-visible spectrum of polymer27a: a) before addition of HCl solution, b) after addition of HCl.

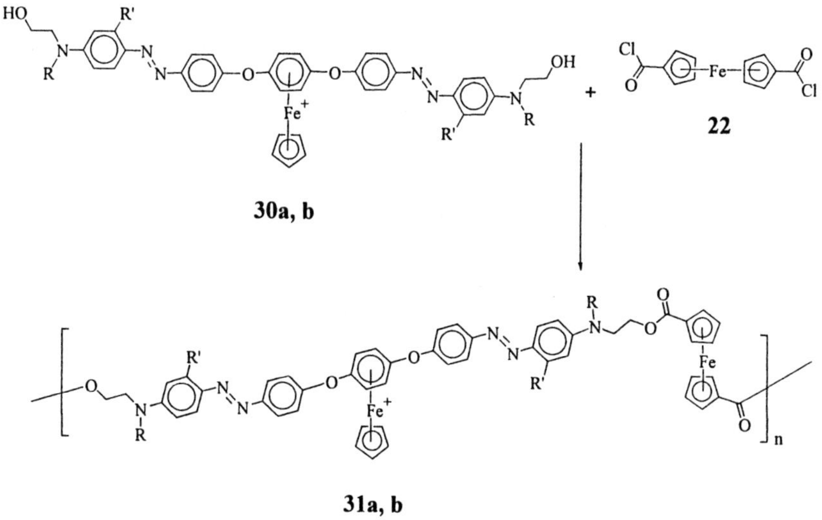

30a, b

22

31a, b

Scheme 6

Conclusion

An efficient route to the synthesis of aryl and hetaryl azo dye containing organoiron polymers has been developed. These polymers incorporate the azo dye chromophores into the backbone or the side chains. These polymers possess many different properties, which could be useful for numerous applications. While it is clear that the area of organometallic polymers containing azo dyes is still being explored, there have been significant steps taken in the study of these unique and interesting materials.

References

1. Abd-El-Aziz, A.S.; Carraher, C.E. Jr.; Pittman, C.U. Jr.; Sheats J.E.; Zeldin, M., *Macromolecules Containing Metal and Metal-like Elements* Volume 1, Wiley & Sons Inc., New Jersy, **2003**.
2. Abd-El-Aziz A.S., "Metal-Containing Polymers" in: *Encyclopedia of Polymer Science and Technology*, 3rd Edition J. I. Kroschwitz, Ed., John Wiley & Sons, New York, **2002**.
3. Archer R.D., in *"Inorganic and Organometallic Polymers"* Wiley-VCH, New York, **2001**.
4. Astruc, D., *Electron Transfer and Radical Processes in Tansition Metal Chemistry* VCH Publishers Inc., New York, **1995**.
5. Manners, I., *Angew. Chem.* **1996**, *35*, 1603.
6. Abd-El-Aziz, A.S., *Macromol. Rapid Commum.* **2002**, *23*, 995.
7. Abd-El-Aziz, A.S.; Todd, E.K., *Coord. Chem. Rev.* **2003**, *246*, 3.
8. Manners, I., *J. Polym. Sci Part A: Polym. Chem.* **2002**, *40*, 179.
9. Nguyen, P.; Gomez-Elipe, P.; Manners, I., *Chem. Rev.* **1999**, *99*, 1515.
10. Mallakpour, S. E.; Nasr-Isfahani, H., *J. App. Poly. Sci.* **2001**, *82*, 3177.
11. Abd-El-Aziz, A.S.; de Denus, C.R.; Todd, E.K.; Bernardin, S.A., *Macromolecules* **2000**, *33*, 5000.
12. Abd-El-Aziz, A.S.; Edel, A.L.; May, L.J.; Epp, K.M.; Hutton, H.M., *Can. J. Chem.* **1999**, *77*, 1797.
13. Pearson, A.J.; Gelormini, A.M., *J. Org. Chem.* **1994**, *59*, 4561.
14. Pearson, A.J.; Gelormini, A.M., *Macromolecules* **1994**, *27*, 3675.
15. Pearson, A.J.; Sun, L., *J. Polym. Sci. Part A: Polym. Chem.* **1997**, *35*, 447.
16. Abd-El-Aziz, A.S.; de Denus, C.R.; Zaworotko, M.J.; MacGillivray, L.R., *J. Chem. Soc. Dalton Trans.* **1995**, 3375.
17. Abd-El-Aziz, A.S.; Armstrong, D.A.; Bernardin, S.; Hutton, H.M., *Can. J. Chem.* **1996**, *74*, 2073.
18. Abd-El-Aziz, A.S.; Todd, E.K.; Epp, K.M., *J. Inorg. Organomet. Polym.* **1998**, *8*, 127.

19. Abd-El-Aziz, A.S.; Todd, E.K.; Ma, G.Z., *J. Polym. Sci. Part A : Polym. Chem.* **2001**, *39*, 1216.
20. Abd-El-Aziz, A.S.; May, L.J.; Hurd, J.A.; Okasha, R.M., *J. Polym. Sci. Part A: Polym. Chem.* **2001**, *39,* 2716.
21. Xie, S.; Natanshon, A.; Rochon, P., *Chem. Mater.* **1993**, *5*, 403.
22. Natansohn, A.; Rochon, P., *Chem. Rev.* **2002**, *102,* 4139.
23. Metwally, M.A.; Abdel-latif, E.; Khalil, A.M.; Amer, F.A.; Kaupp, G., *Dyes and Pigments* **2004**, *62*, 181.
24. Yen, M.S.; Wang, I.J., *Dyes and Pigments* **2004**, *62*, 173.
25. Abd-El-Aziz, A.S.; Afifi, T.H.; Budakowski, W.R.; Friesen, K.J.; Todd, E. K., *Macromolecules* **2002**, *35*, 8929.
26. Abd-El-Aziz, A.S.; Okasha, R.M.; Afifi, T.H.; Todd, E.K., *Macromol. Chem. Phys.* **2003**, *204*, 555.
27. Abd-El-Aziz, A.S.; Okasha, R.M.; Afifi, T.H., *J. Inorg. Organomet. Polym.* (accepted for publication).
28. Abd-El-Aziz, A.S.; Okasha, R.M.; Shipman, P.O.; Afifi, T.H., *Macromol. Rapid Commun.* **2004**, *25*, 1497.

Chapter 29

Ring-Collapsed Alternating Copolymerization of Organoarsenic Homocycles and Acetylenic Compounds

Kensuke Naka, Tomokazu Umeyama, Akiko Nakahashi, and Yoshiki Chujo

Department of Polymer Chemistry, Graduate School of Engineering, Kyoto University, Katsura, Nishikyo-ku, Kyoto 615–8510, Japan

The polymerization between pentamethylpentacycloarsine (*cyclo*-(MeAs)$_5$) or hexaphenylhexacycloarsine (*cyclo*-(PhAs)$_6$) with phenylacetylene (**1a**) in the presence of a catalytic amount of AIBN gave the corresponding poly(vinylene-arsine)s. Without any radical initiator at 25 °C, *cyclo*-(MeAs)$_5$ caused cleavage of the arsenic-arsenic bond spontaneously and copolymerized with **1a** in chloroform to give poly(vinylene-arsine)s. The measurement of the conversion of acetylenic compounds with various substituents during the copolymerization gave an evidence to support the assumption that the formation of vinyl radicals by addition of arsenic radicals to acetylenic compounds was the rate-determining step. The copolymer obtained showed fluorescence properties which were influenced by the substituents of the acetylenic compounds.

Background

Polymers containing inorganic elements in the main chain structure are of current interest as a result of their unique properties. For example, phosphorus containing polymers have found a variety of important uses including flame retardants[1], ionic conducting materials[2], and easily separable supports for metal catalysts[3]. Recently, conjugated organophosphorus polymers were synthesized and showed interesting features such as significant extension of conjugation along the polymer backbone via the lone pair on P, fluorescent properties, and electron-donating character[4-8].

On the other hand, the organoarsenic chemistry has a long history that dates back to the synthesis and discovery in 1760 of the first organometallic compound, $Me_2AsAsMe_2$, by L. C. Cadet de Gassicourt[9]. The discovery of the medicinal action of organoarsenicals on syphilis in 1910 led to a rapid expansion of the work on arsenic derivatives. Since an alternative cure was developed in 1940s, less attention has been paid to chemotherapy roles and more to structures, stereochemistry and donor properties of organoarsenic compounds. In the abundant accumulation of the organoarsenic chemistry, however, the incorporation of arsenic into polymer backbones has been limited[10-12]. Although the history of the organoarsenic homocycles[13-14] dates back at least to the synthesis of cyclo-$(PhAs)_6$ by Michaelis and Schulte in 1881[15] and the chemotherapic effects of "Salvarsan" and their derivatives were discovered[16] and studied intensively in the early part of the 20th century, only a few reactions of the organoarsenic homocycles with organic compounds[17-19] has been developed compared to the extensive studies on coordination chemistry of them as ligands in transition metal complexes[20]. No radical reactions of the cyclooligoarsines had been reported before our previous study. Revaluation of the organoarsenic homocycles would open a unique chemistry.

The development of a new polymerization reaction makes it possible to create polymers having a unique structure which is difficult to construct by conventional procedures. To develop new polymerization methods for heteroatom-containing unsaturated polymers is of considerable interest because such polymers show unusual properties. Among various types of such polymers, the simplest one is heteroatom-containing polyvinylenes. The properties of these polymers might be attractive. However, no example of these polymers has been reported except poly(vinylene-sulfide)[21-23], because of synthetic difficulties.

Objective

The objectives of this study were to develop new polymerization methods for heteroatom-including polyvinylenes. Inorganic polymers or oligomers are known whose chains or ring skeletons are made up of only one inorganic element such as silicon, germanium, phosphorus, arsenic, sulfur and selenium. These compounds often give homolysis of element-element bonds by a stimulation such as light and heat. We speculated that poly(vinylene-arsine)s would be obtained by a radical copolymerization if arsenic radicals obtained by the homolytic cleavage of cyclooligoarsines add to acetylenic compounds.

Experimental

Cyclooligoarsines. *cyclo*-(MeAs)$_5$ and *cyclo*-(PhAs)$_6$ were prepared according to ref. 24 and 25, respectively. *cyclo*-(MeAs)$_5$. ^{1}H-NMR (δ, ppm) 1.63, 1.64, 1.67 (lit.[13] 1.62, 1.63, 1.66). *cyclo*-(PhAs)$_6$. mp 208 – 211 °C (lit.[20] mp 204 – 208 °C)

Copolymerization in the Absence of AIBN. A typical experimental procedure was conducted as follows. To a chloroform solution (3.0 mL) of *cyclo*-(MeAs)$_5$ (0.556g, 1.24 mmol), **1a** (0.631 g, 6.18 mmol) was added all at once at room temperature. The reaction mixture was stirred for 12 h and then poured into *n*-hexane to precipitate a polymeric material. After twice reprecipitation from toluene into *n*-hexane, the resulting precipitate was freeze-dried for 10 h. A bright yellow powder was obtained (0.765 g) in 64% yield. ^{1}H-NMR (δ, ppm): 0.80 (As-CH$_3$), 6.21 ((*E*)-C=CH), 6.43 ((*Z*)-C=CH), 6.81 (ArH$_o$), 7.24 (ArH$_m$, ArH$_p$). ^{13}C-NMR (δ, ppm): 10.3 (As-CH$_3$), 126-129 (C_{Ar}H$_o$, C_{Ar}H$_m$, C_{Ar}H$_p$), 137-140 (C=*C*H), 140.1 (C_{Ar}-C), 155-159 (*C*=CH).

Copolymerization in the Presence of AIBN. A typical experimental procedure was conducted as follows. To a chloroform solution (3.0 mL) of *cyclo*-(MeAs)$_5$ (0.549g, 1.22 mmol), AIBN (0.030 g, 0.018 mmol) and PA (0.623 g, 6.10 mmol) were added at room temperature. The mixture was stirred at 60 °C for 2 h and then poured into *n*-hexane to precipitate the polymeric material. After twice reprecipitation from toluene into *n*-hexane, the resulting precipitate was freeze-dried for 10 h. A bright yellow powder was obtained (0.588 g) in 50% yield. The ^{1}H and ^{13}C NMR spectra of the resulting polymer were identical with those of the copolymer obtained without AIBN.

Results & Discussion

Synthesis and Reactions of Cyclooligoarsines. First, we discuss here the formation and reactions of the cyclooligoarsines. The methyl-substituted cyclooligoarsine, *cyclo*-(MeAs)$_5$, was synthesized by the reduction of sodium methylarsonate with hypophosphorus acid[24]. The five-membered arsenic ring was obtained exclusively, suggesting that this ring structure is stable compared to other forms containing arsenic-arsenic bonds. The obtained *cyclo*-(MeAs)$_5$ is a yellow liquid with high viscosity. After *cyclo*-(MeAs)$_5$ was left under a nitrogen atmosphere at 25 °C for several days, a purple black precipitate appeared, which was easily removed by a filtration under a nitrogen atmosphere. The resulting precipitate was insoluble in any solvent and fumed in the air, according to the literature[26-27], should be a linear poly(methylarsine) with a ladder structure. The arsenic-arsenic bond of the ring compound was cleaved spontaneously, and the open-chain oligoarsine stacked with each other to form the ladder structure (Scheme 1). Solutions of *cyclo*-(MeAs)$_5$ in organic solvents such as benzene, toluene, and chloroform also produced the purple-black precipitates after standing at 25 °C for several days. The formation of the precipitate was accelerated by heating in the presence of AIBN or by the irradiation with an incandescent lamp.

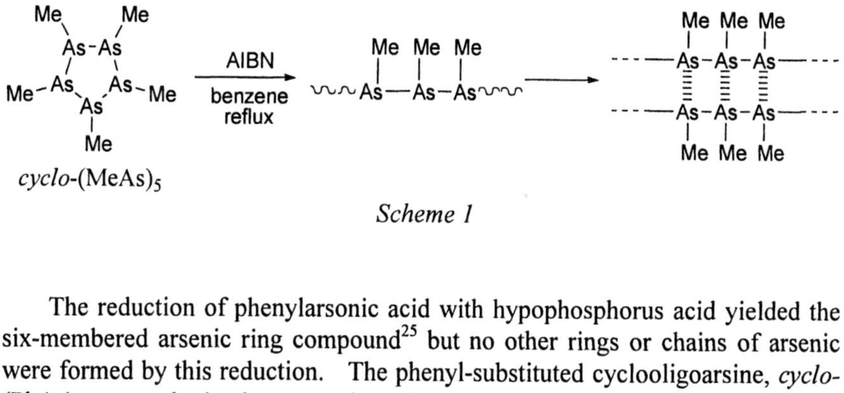

cyclo-(MeAs)$_5$

Scheme 1

The reduction of phenylarsonic acid with hypophosphorus acid yielded the six-membered arsenic ring compound[25] but no other rings or chains of arsenic were formed by this reduction. The phenyl-substituted cyclooligoarsine, *cyclo*-(PhAs)$_6$, was obtained as a pale yellow crystal after recrystallization from chlorobenzene. The crystal showed poor solubility in benzene. When *cyclo*-(PhAs)$_6$ was treated with 3 mol% of AIBN in refluxing benzene, the heterogeneous reaction mixture became clear within 30 minutes. The stable ring structure was collapsed and open-chain arsenic oligomers or arsenic atomic biradicals might be formed in the solution (Scheme 2). By cooling the clear solution to room temperature, *cyclo*-(PhAs)$_6$ was regenerated as a light yellow powder. The catalytic amount of AIBN was enough to homogenize the mixture, in other words, to collapse all of the stable six-membered ring structure, suggesting that the produced arsenic radial also contributed to the destruction of cyclo-(PhAs)$_6$. In contrast to cyclo-(MeAs)$_5$, cyclo-(PhAs)$_6$ was stable at room temperature in the air atmosphere in the solid state, and showed no reaction even in refluxing benzene for several hours.

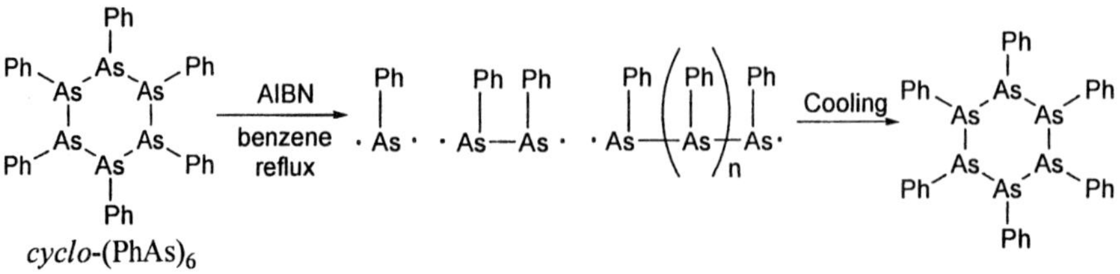

Scheme 2

Alternating Copolymerization of cyclo-(MeAs)$_5$ with 1a. A typical polymerization procedure was conducted as follows (Scheme 3). Under a nitrogen atmosphere, a benzene solution of a catalytic amount of 2,2'-azobisisobutyronitrile (AIBN) was added to a refluxing solution of cyclo-(MeAs)$_5$ and phenylacetylene (**1a**) in benzene. After stirring for 12 h, the reaction mixture was poured into n-hexane to precipitate the product, which was purified three times by reprecipitation from benzene to n-hexane. After freeze-drying for 10 h, polymer **2a** was obtained as a bright-yellow powder. From gel permeation chromatographic analysis (CHCl$_3$, PSt standards), the number-average molecular weight of **2a** was estimated to be 11 500. In the case of using cyclo-(PhAs)$_6$ instead of cyclo-(MeAs)$_5$, corresponding poly(vinylene-arsine) (**3a**) was obtained as a white powder. Both polymers were readily soluble in common organic solvents such as THF, chloroform and benzene.

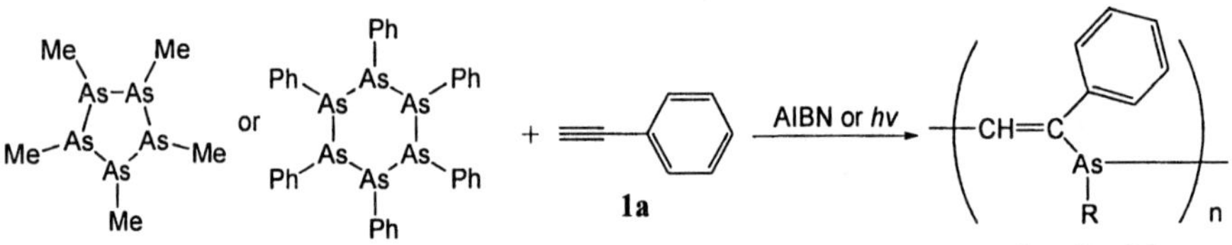

2a: R = Me
3a: R = Ph

Scheme 3

Structural characterization of the polymers was provided by ^{1}H and ^{13}C NMR spectroscopy. In the ^{1}H NMR spectrum of **2a**, the integral ratio of two peaks in a vinyl region (6.1 to 6.4 ppm) confirmed that the trans isomer was predominantly obtained. From the peak area ratio of an aromatic (6.6 to 7.5 ppm) and a methyl region in the ^{1}H NMR and the elemental analysis, a copolymer composition of **2a** was nearly 1 : 1 (phenylacetylene : methylarsine). Analysis of **2a** by a ^{13}C NMR spectroscopy showed only one sharp resonance for the methyl carbon at 11.6 ppm, suggesting that the arsenic in **2a** existed in a trivalent state and no arsenic-arsenic bond or no oxidized arsenic was present. These results revealed that the copolymerization proceeded alternatingly.

Spontaneous Alternating Copolymerization of *cyclo*-(MeAs)₅ with 1a. The arsenic ring compound with methyl-substitution, *cyclo*-(MeAs)₅, reacted with **1a** in chloroform at 25 °C in the absence of any radical initiators or catalysts to produce the corresponding poly(vinylene-arsine) in moderate yield[29]. The number-average molecular weight of the copolymer was estimated to be 11 100 by the GPC analysis (vs polystyrene standards). The structure of the resulting copolymer was determined by ^{1}H NMR and ^{13}C NMR spectra.

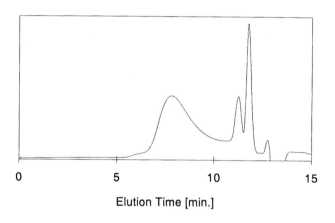

0   5   10   15

Elution Time [min.]

Figure 1. GPC trace (chloroform) of the reaction mixture after the copolymerization of cyclo-(MeAs)₅ with **1a** in chloroform at 25 °C in the absence of AIBN before the reprecipitation.

After the copolymerization of *cyclo*-(MeAs)₅ with **1a** in chloroform was carried out for 12 h, the reaction mixture was analyzed by GPC before the reprecipitation (Figure 1). Besides a peak assigned to the poly(vinylene-arsine) with high molecular weight, the GPC chart shows two peaks in the region of the molecular weight of several hundreds. By the recycling HPLC analysis using chloroform as an eluent, it was revealed that both of the two peaks in the lower molecular weight region consisted of several peaks. The low molecular weight compounds were stirred in toluene at 25 °C for 12 h, and then no polymer was obtained. This result suggests that the polymer with high molecular weight was *not* formed via the compounds responsible for these peaks. Similar peaks were also observed in the GPC trace when we employed AIBN as a radical initiator of the copolymerization. The formation of these compounds decreases the isolated yield of the polymer with high molecular weight. We assume from the 1H NMR analysis that these peaks are attributable to the cyclic compounds with the ring structure of vinylene-arsine.

Table 1.   Copolymerization of *cyclo*-(MeAs)₅ with 1a in the Absence of AIBN : Effect of the Feed Ratios of Monomers

^{a} Copolymerizations were carried out in chloroform at 25 °C for 12 h. ^{b} MeAs represents methylarsine (CH_3As). ^{c} Estimated by GPC analysis in chloroform on the basis of polystyrene standards. ^{d} Isolated yield after reprecipitation from toluene to methanol. ^{e} (weight of polymer)/(total weight of monomers). ^{f} Based on the molar amount of **1a**. ^{g} Based on the molar amount of *cyclo*-(MeAs)₅.

| | feed ratio | resulting polymer | | |
run^a	MeAs^b : 1a	$M_n{}^c$	$M_w/M_n{}^c$	Yield (%)^d
1	1 : 0.33	5 200	2.6	17^e (32)^f
2	1 : 0.67	5 200	2.7	42^e (52)^f
3	1 : 1	11 100	2.1	64^e
4	1 : 1.5	13 100	2.3	52^e (65)^g
5	1 : 1.5	11 200	2.1	30^e (61)^g

Spontaneous Copolymerization in Various Monomer Feed Ratios. The copolymerization of *cyclo*-(MeAs)₅ with PA was conducted in chloroform at 25 °C in various monomer feed ratios (Table 1).   In all cases the copolymerization resulted in the corresponding alternating copolymer, poly(vinylene-arsine). Thus, the 1 : 1 alternating structure of MeAs and **1a** is controlled very strictly under a wide variety of copolymerization conditions. Since the vinyl radical reacts with arsenic-arsenic bond or arsenic radical immediately after the production due to the instability, almost all of the propagating radical in the copolymerization system are arsenic radical and the concentration of the vinyl radical is extremely low.   Thus, the vinyl radical never reacts with another vinyl radical.   It rarely attacks PA either because of the low homopolymerizability of **1a**.   These are also the case for runs 4 and 5 in which the excess amount of **1a** was employed.   Therefore, the resulting copolymers have no consecutive **1a** units in the main chain in all the cases.   The other propagating radical, the arsenic radical, can cause the recombination reaction as a termination.   The resulting diarsenic linkage, however, is unstable and can bring about its homolytic cleavage to reproduce the arsenic radical as a propagating end-group. Consequently, even when the excess amount of *cyclo*-(MeAs)₅ was employed, the copolymerization yielded the copolymer with no consecutive MeAs units in the backbone.

While the excess amount of **1a** (Table 1, runs 4 and 5) was found to remain unchanged after the completion of the reaction, the excess amount of MeAs (Table 2, runs 1 and 2) decreased the molecular weights and yields of the obtained polymers in comparison with those in the equivalent case (Table 1, run 3). Since almost all the propagating radical are the arsenic radical even in the case where the excess **1a** was employed, the copolymerization in runs 4 and 5 proceeded comparably with the equivalent case. When the lesser amount of **1a** than that of MeAs was employed all **1a** was consumed before the sufficient growth of the copolymer chain. Thus, the copolymerization in runs 1 and 2 yielded the alternating copolymer with low molecular weight, and hence in low yield after the removal of the oligomer by reprecipitation.

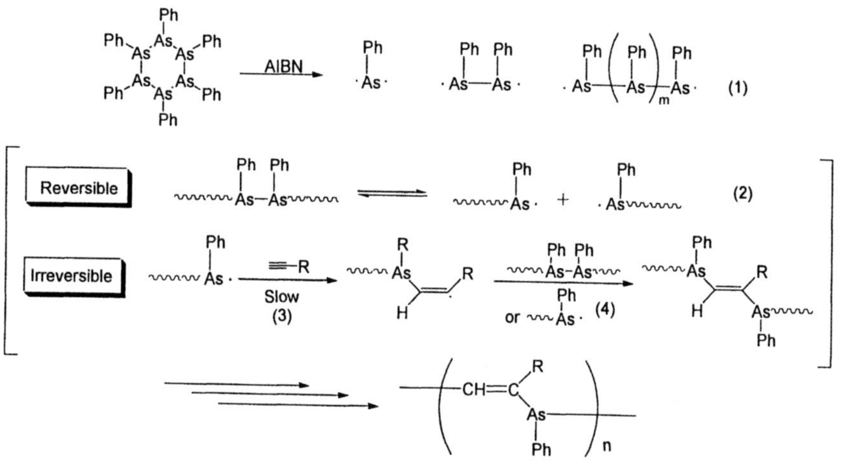

Scheme 4

Discussion of the Copolymerization Mechanism. The following reaction mechanism was proposed for the radical alternating copolymerization[28]. First, AIBN cleaved arsenic-arsenic bonds of the cyclooligoarsine to produce arsenic radicals. Second, the homolysis of the other arsenic-arsenic bonds proceeded spontaneously due to their instability by the destruction of the quite stable six-membered ring structure. In competition with this reaction, an arsenic radical added to an acetylenic compound to give a vinyl radical. Next, a vinyl radical reacted immediately with an arsenic-arsenic bond or with an arsenic radical to form a new carbon-arsenic bond. Although the carbon-arsenic bond formation seems to result in the lose of a growing radical and no chain growth proceeds anymore, the labile As-As bonds in the product caused homolysis easily to produce a new arsine radical, and the chain growth could restart. In this manner, repeating production of the arsine radical by the As-As bond cleavage and its addition to the acetylenic compound leads to a polymer with a simple main-chain structure, poly(vinylene-arsine). Since the vinyl radicals are more unstable and reactive than the arsenic radicals, formation of the vinyl radicals (eq. 3) should be relatively slower than creation of the arsine radical (eq. 2) and of a carbon-arsenic bond (eq. 4). One of the propagating radicals in this copolymerization system is the vinyl radical, which might not cause recombination due to the low concentration. The vinyl radical might not react with the acetylenic compound to produce a new vinyl radical because of its instability, but reacts with an arsenic radical or an arsenic-arsenic bond. The other propagating radical is the arsenic radical which cannot cause disproportionation as a termination reaction. A recombination of the arsenic radical generates an As-As bond which is easily cleaved under the benzene refluxing condition to reproduce the propagating arsenic radicals.

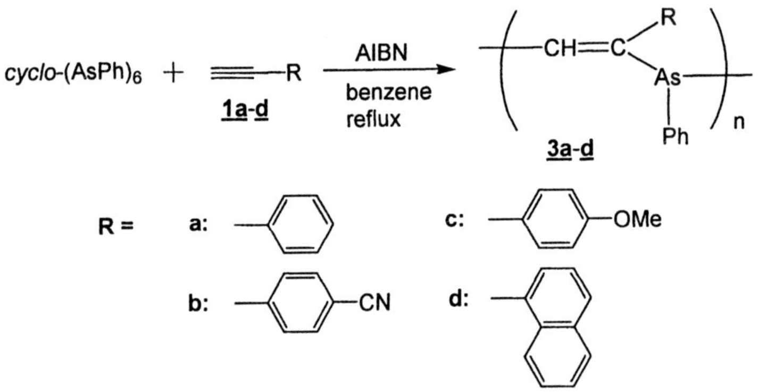

Scheme 5

The aromatic stabilization of the vinyl radical is essential for the copolymerization. We speculate that the more effectively aromatic stabilization acts on the vinyl radical, the more smoothly the copolymerization progresses. Here are shown the examination of the rates of conversion of monomers **1a-d** during the copolymerization with $cyclo$-(PhAs)$_6$[30] (Scheme 5). The results of the polymerization are summarized in Table 2. The radical copolymerization of **1a-d** with $cyclo$-(PhAs)$_6$ was carried out in the presence of n-alkane as a reference material to monitor the amount of consumption of monomers **1a-d** by using gas chromatography (GC). In each experiment, the same concentration of AIBN (0.006 mol/L), $cyclo$-(PhAs)$_6$ (0.060 mol/L), acetylenic compound (0.36 mol/L), and n-alkane (0.04 g/L) was used in benzene at 78 °C. The moment of feeding AIBN to the mixture of $cyclo$-(PhAs)$_6$ and acetylenic compounds **1a-d** in refluxing benzene was defined as initiating point (0 min) of the copolymerization. In every copolymerization, the consumption of acetylenic monomers **1a-d** was induced smoothly by addition of AIBN, and almost ceased in 4 h.   In the case of using cyano-substituted phenylacetylene **1b**, the polymerization proceeded to reach nearly quantitative conversion.   The rate increased in the order of **1c** < **1a** < **1d** < **1b**.   The copolymerization of *cyclo*-(PhAs)$_6$ with **1b** or **1d** was faster than that with **1a** because the aromatic stabilization of the vinyl radical by 4-cyanophenyl or naphthyl group is stronger than that by phenyl group.   The copolymerization of *cyclo*-(PhAs)$_6$ with **1c** was slightly slower than that with **1a**, suggesting that electrostatic effect of the substituent might affect the reaction between arsenic radical and carbon-carbon triple bond.   The arsenic radical may prefer to react with electron-accepting monomers rather than electron-donating monomers due to the lone pair on the arsenic atom.   Both conjugative and electrostatic effect of 4-cyanophenyl group made the fastest consumption of **1b** during the copolymerization with *cyclo*-(PhAs)$_6$ among the acetylenic monomers employed here.   The GC analyses suggest that the lower activation energy of the reaction eq. 3 (formation of the vinyl radical) results in more rapid progress of the total copolymerization. These results are consistent with our speculation and therefore support the mechanism of the copolymerization as proposed above.

Stability.   To investigate the thermal stability of the obtained polymers, a TGA analysis was carried out under nitrogen and air.   The methyl-substituted poly(vinylene-arsine) (**2a**) showed a 10% weight loss at 265 °C (under N$_2$) and 205 °C (under air), and the phenyl-substituted one (**2a**) showed them at 284 °C (under N$_2$) and 250 °C (under air).   The poly(vinylene-arsine) with phenyl-substitution was thermally more stable than that with the methyl-substitution. The glass transition temperatures (T_g) of the methyl- and phenyl-substituted polymers were 58.2 and 92.9 °C respectively determined by DSC analysis.   Both of the resulting copolymers are stable in the solid state at room temperature.   No decrease of the molecular weight or no change of the structure was observed even after exposing them to air for several months.   In order to examine the stability toward air moisture and oxidation more precisely, the chloroform solution of the methyl-substituted poly(vinylene-arsine) was stirred in air atmosphere at 50 °C, and the molecular weight change was monitored by GPC. The GPC analysis showed no decrease of the molecular weight after stirring for 24 h.   The ^{1}H NMR spectrum of the recovered copolymer indicated no change of the polymer structure.

Optical properties.   The electronic structures of the polymers **3a-d** were studied by UV-vis spectroscopy.   Figure 2 shows an absorption spectrum of **3b** recorded in chloroform at room temperature.   Not only strong absorption in the UV region derived from π−π* transition of the benzene ring, but also small absorption in the visible region was observed.   The absorption edge located at around 480 nm.   We assume that n-π* transition in the main chain brought about the lower energy absorption as seen in poly(phenylene-phosphine)s[7].

The polymers **3a-d** showed fluorescent properties. The fluorescence and excitation spectra of a dilute chloroform solution of **3b** measured at room temperature are shown in Figure 2 as a typical representative example. The emission was observed in the visible blue-purple region with a peak at 443 and 466 nm. The emission peak maximum was independent of the concentration of polymer **3b**. In the excitation spectrum of **3b** monitored at 470 nm, the absorption was not observed in the shorter wavelength region but in the longer wavelength region with a peak at 394 nm. This means that the absorption of **3b** in the higher energy region and the absorption in the lower energy region are originated from the different absorbing species; π–π* and n-π* transition. The emission of **3b** results from only the absorption of the latter transition. Table 2 also summarizes emission and excitation spectral data. Each emission peak maximum of **3a-c** was red-shifted in the order of **3c** < **3a** < **3b**, which coincides with the order of strength of electron-withdrawing properties of the substituents. This indicates that the donor-acceptor (arsenic atom with a lone pair and vinylene unit with electron-accepting group) repeating units made the band gap narrower and resulted in the lower energy of the emission. The optical properties of poly(vinylene-arsine)s were tuned by changing the substituents of the acetylenic compounds. In the case of **3d**, the naphthyl group itself has a fluorescent property which is stronger than that derived from the n-π* transition in the main chain of the polymer, and therefore the emission peak due to the main chain was hidden. The absorption peaks in the excitation spectra of **3a-d** showed the same order as the emission peaks because of the same reason discussed above.

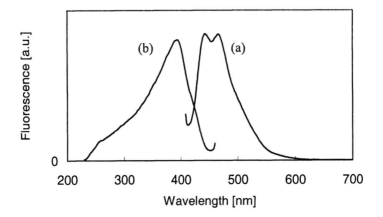

Figure 2. *(a) Fluorescence spectrum of **3b** excited at 394 nm. (b) Excitation spectrum of **3b** emitted at 466 nm. Both measurements were carried out in CHCl₃ at room temperature.*

Table 2. Syntheses and Optical Properties of Polymers 3a-d

Run	1	$M_w{}^a$	$M_n{}^a$	$M_w/M_n{}^a$	Yield (%)[b]	DP^c	Emission λ_{max} (nm)[d]	Excitation λ_{max} (nm)[d]
1	1a	5 600	3 900	1.4	35	15	437	375
2	1b	15 100	8 400	1.8	29	30	443, 466	394
3	1c	5 100	3 700	1.4	22	13	402	341
4	1d	9 900	7 600	1.3	33	25	350	315

[a] GPC ($CHCl_3$). Polystyrene standards. [b] Isolated yields after reprecipitation into *n*-hexane. [c] Degree of polymerization. [d] Recorded in dilute $CHCl_3$ solutions at room temperature.

Conclusions

We have described here the synthesis of the novel organoarsenic polymers, poly(vinylene-arsine)s, by ring-collapsed radical alternating copolymerization (*RCRAC*) of arsenic homocycles with acetylenic compounds. Besides the cyclooligoarsines, a number of cyclic compounds or polymers are known of which ring or chain skeletons are made up of only one inorganic element such as silicon, germanium, phosphorus, antimony, sulfur. These compounds often cause homolytic cleavage by stimulation such as light or heat. The present methodology, which involves the homolytic cleavage of element-element bond of inorganic homocycles or homochains and the addition of the produced inorganic radicals to the organic monomers, would be applicable to the preparation of various inorganic polymers containing inorganic elements in the main chain. Studies on the expansion of the scope of this methodology are now in progress.

References

1. Shobha, H. K.; Johnson, H.; Sankarapandian, M.; Rangarajanm, P; Baird, D. J..; McGrath, J. E. *J. Polym. Sci. Polym. Chem.* **2001**, *39*, 2904-2910.
2. Allcock, H. R.; Olmeijer, D. L.; O'Connor, S. J. M. *Macromolecules* **1998**, *31*, 75-759.
3. Leadbeater, N. E.; Marco, M. *Chem. Rev.* **2002**, *102*, 3217-3273.
4. Mao, S. S. H.; Tilley, T. D. *Macromolecules* **1997**, *30*, 5566-5569.

428

5. Hay, C.; Fischmeister, C.; Hissler, M.; Toupet, L.; Reau, R. *Angew. Chem. Int. Ed.* **2000**, *39*, 1812-1813.
6. Morisaki, Y.; Aiki, Y.; Chujo, Y. *Macromolecules* **2003**, *36*, 2594-2597.
7. Lucht, B. L.;St Onge, N. O. *Chem. Commun.* **2000**, 2097-2098.
8. Smith, R. C.; Chen, X., Protasiewicz, J. D. *Inorg. Chem.* **2003**, *42*, 5468-5470.
9. Krannich, L. K.; Watkins, C. L. *Encyclopedia of Inorganic Chemistry King*, R. B., Ed.; John Wiley & Sons: Chichester, England, 1994; Vol. 1, p 200, and references cited therein.
10. Kraft, M. Y. *Dokl. Akad. Nauk SSSR* **1960**, *131*, 1342-1344.
11. Kallenbach, L. R.; Irgolic, K. J., Zingaro, R. A. *Eur. Polym. J.* **1970**, *6*, 479-485.
12. Carraher, C. E.; Moon, W. G. *Eur. Polym. J.* **1976**, *12*, 329-331.
13. Smith, L. R.; Mills, J. L. *J. Organomet. Chem.* **1975**, *84*, 1-15.
14. Krannich, L. K.; Watkins, C. L. In *Encyclopedia of Inorganic Chemistry*; King R. B., Ed.; Wiley: Chichester, 1994; Vol. 1, p 19.
15. Michaelis, A.; Schulte, C. *Chem. Ber.* **1881**, *14*, 1912.
16. Ehrlich, P. *Lancet* **1907**, *173*, 351-353.
17. Sennyey, G.; Mathey, F.; Fischer, J. Mischler, A. *Organometallics* **1983**, *2*, 298-304.
18. Thiollet, G.; Mathey, F. *Tetrahedron Lett.* **1979**, 3157-3158.
19. Schmidt, U.; Boie, I.; Osterroht, C.; Schoer, R.; Grutzmacher, H.-F. *Chem. Ber.* **1968**, *101*, 1381-1397.
20. DiMaio, A.-J.; Rheingold, A. L. *Chem. Rev.* **1990**, *90*, 169-190.
21. Ikeda, Y.; Ozaki, M.; Arakawa, T. *J. Chem. Soc., Chem. Commun.* **1983**, *24*, 1518-1519.
22. Ikeda, Y.; Ozaki, M.; Arakawa, T.; Takahashi, A.; Kambara, S. *Polym. Commun.* **1984**, *25*, 79-80.
23. Ikeda, Y.; Ozaki, M.; Arakawa, T. *Mol. Cryst. Liq. Cryst.* **1985**, *118*, 431-434.
24. Elmes, P. S.; Middleton, S.; West, B. O. *Aust. J. Chem.* **1970**, *23*, 1559-1570.
25. Reesor, J. W. B.; Wright, G. F. *J. Org. Chem.* **1957**, *22*, 382-385.
26. Kraft, M. Y.; Katyshkina, V. V. *Dokl. Akad. Nauk. SSSR* **1949**, *66*, 207-209.
27. Rheingold, A. L.; Lewis, J. E.; Bellama, J. M. *Inorg. Chem.* **1973**, *12*, 2845-2850.
28. Naka, K.; Umeyama, T.; Chujo, Y. *J. Am. Chem. Soc.* **2002**, *124*, 6600-6603.
29. Umeyama, T.; Naka, K.; Chujo, Y. *Macromolecules* **2004**, *37*, 5952-5958.
30. Umeyama, T.; Naka, K.; Nakahashi, A.; Chujo, Y. *Macromolecules* **2004**, *37*, 1271-1275.

Chapter 30

Photodegradable Polymers Containing Metal–Metal Bonds along Their Backbones: Mechanistic Study of Stress-Induced Rate Accelerations in the Photochemical Degradation of Polymers

Rui Chen and David R. Tyler[*]

Department of Chemistry, University of Oregon, Eugene, OR 97403

A novel poly(vinyl chloride)-based polymer (1) containing Mo-Mo bonds along the backbone was prepared to facilitate a mechanistic study investigating the origin of the rate enhancements in the photochemical degradation of polymers subjected to tensile stress. When irradiated with visible light, the metal-containing polymer photodegrades, even in the absence of oxygen. Infrared spectroscopic analysis demonstrated that the chlorine atoms along the polymer backbone act as built-in traps for Mo-centered radicals formed by photolysis of the Mo-Mo bonds. The presence of the internal radical trap permitted the polymer samples to be irradiated in the absence of oxygen, thus eliminating the kinetically complicating effects of rate-limiting oxygen diffusion found in the photodegradation of "regular" polymers The effect of stress on the degradation quantum yield of 1 and its plasticized analogue, made by addition of 20 wt% dioctyl phthalate (DOP) to 1, was studied. Results from both polymers showed that stress initially increased the quantum yields for degradation but the quantum yields reached a maximum value and then decreased with higher stress. These results support the "decreased radical recombination efficiency" (DRRE) hypothesis, one of several hypotheses that have been proposed in the literature to explain the effect of stress on polymer photodegradation rates and efficiencies. The DRRE hypothesis proposes that the function of stress is to increase the initial separation of the photochemically generated radical pair, which has the effect of decreasing their recombination efficiency and thus increasing the degradation efficiency. The hypothesis predicts an eventual downturn in degradation efficiency because of polymer chain ordering; the increased order hinders diffusion apart of the radicals and thus increases their probability of recombination.

Background

Virtually every commercial polymer is subjected to either natural or artificial weathering tests. The purpose of these tests is to determine the useful lifetime of the polymer and to identify the various experimental parameters that can affect a polymer's lifetime. Among the parameters that have been identified as affecting polymer lifetime are temperature, exposure to ultraviolet radiation, oxygen diffusion rates in the polymer, tensile stress, compressive stress, chromophore concentration, and polymer morphology.[1-4] There is an interesting synergism between some of these parameters. For example, it has been reported that tensile stress and shear stress will accelerate the effects of exposure to UV, with a resulting increase in the rate of photodegradation.[5-8] An intriguing fundamental question, and one that has enormous practical implications as well,[9] is *why does stress cause changes in the photodegradation rate?* Although several hypotheses have been advanced to explain the effect of stress on photodegradation rates, there is little experimental support for any of the hypotheses. The reason for the dearth of experimental support is that several challenging experimental difficulties complicate the rigorous mechanistic investigation of the hypotheses. As part of our investigation on the mechanistic aspects of how stress affects polymer photodegradation rates, we have devised ways to overcome these problems, and, as described below, the outcome is that we now have a basic mechanistic understanding of how stress affects the photochemical degradation rates in certain, selected polymers.

Mechanistic Hypotheses to Explain Stress-Accelerated Photodegradation

Mechanistic hypotheses to explain stress-accelerated photodegradation fall into three main categories,[10] illustrated by reference to Scheme 1. In one category, it is proposed that stress leads to an increase in the quantum yields for bond photolysis, i.e., it is proposed that $\phi_{homolysis}$ increases with stress. The second category attributes the increased degradation rates to a decrease in the efficiency of radical recombination following homolysis, i.e., $k_{recombination}$ is proposed to decrease as stress increases. The third category attributes the effects of stress to changes in the rates of reactions that occur subsequent to formation of the radicals, i.e., to changes in the rate of the radical trapping reactions (k_{trap}). The key points of these theories are outlined below.

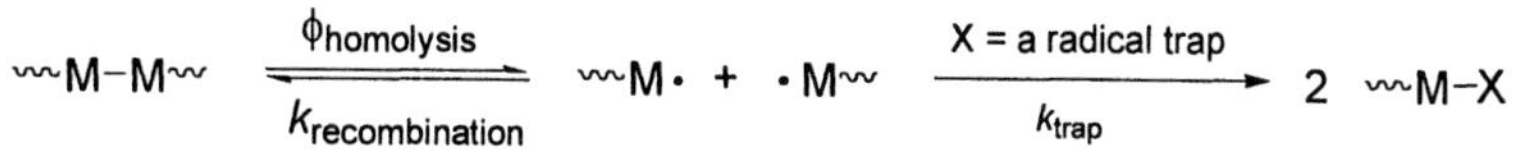

$$\text{M-M} \underset{k_{recombination}}{\overset{\phi_{homolysis}}{\rightleftharpoons}} \text{M}\cdot + \cdot\text{M} \quad \overset{X = \text{a radical trap}}{\underset{k_{trap}}{\longrightarrow}} \quad 2 \text{ M-X}$$

Scheme 1. A generalized reaction scheme showing photolysis of a bond along the backbone in a polymer (M represents a generic atom, carbon or otherwise)

Stress-induced changes in $\phi_{homolysis}$; the Plotnikov hypothesis

For a direct photochemical bond cleavage, the photochemical step in Scheme 1 labeled "$\phi_{homolysis}$" can be broken down into the set of elementary steps shown in eq 1. (The asterisk in eq 1 is used to indicate an excited state of the molecule.)

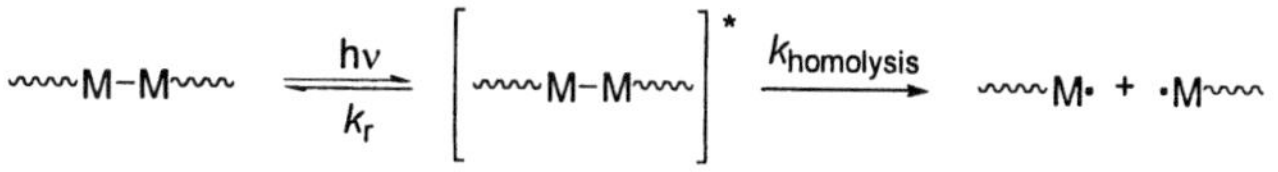

$$\text{M-M} \underset{k_r}{\overset{h\nu}{\rightleftharpoons}} \left[\text{M-M} \right]^* \overset{k_{homolysis}}{\longrightarrow} \text{M}\cdot + \cdot\text{M} \qquad (1)$$

Equation 2 shows the value of $\phi_{homolysis}$ in terms of the rate constants in eq 1.

$$\phi_{homolysis} = \frac{k_{homolysis}}{k_{homolysis} + k_r} \qquad (2)$$

Clearly, if stress affects either k_r or $k_{homolysis}$ then $\phi_{homolysis}$ will vary with stress. No studies have investigated the effect of stress on k_r, but Plotnikov derived an equation for the stress dependence of $k_{homolysis}$.[11] His theory attributes the increase in degradation rates with applied stress to a decrease in the activation barrier for bond dissociation in the excited state (Figure 1). His predicted relationship of quantum yield to stress is shown in Figure 2a.

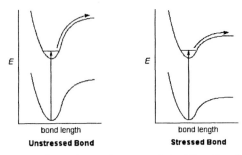

Figure 1. The photophysical origin of the Plotnikov hypothesis

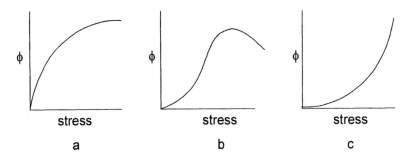

Figure 2. A plot of quantum yield for degradation vs. stress according to a) the Plotnikov equation; b) the DRRE hypothesis; c) the Zhurkov equation.

Stress-induced changes in $k_{recombination}$: the "decreased radical recombination efficiency" (DRRE) hypothesis

Busfield,[12] Rogers,[13,14] Baimuratov,[15] and Shlyapnikov[16] proposed explanations for stress effects that are based on decreases in radical recombination efficiencies in stressed systems (Figure 3). In their models (which are reasonably similar, so they are combined here for ease of discussion), the effect of stress is divided into four stages. Stage one represents the low stress domain. In this stage, there is only slight deformation of the original polymer structure and the rate of photodegradation is not greatly affected. In stage two, higher stress causes significant morphological changes, including the straightening of the polymer chains in the amorphous regions. When bonds in the taut tie molecules are cleaved by light, the probability of radical-radical recombination is decreased relative to non-stressed samples because entropic relaxation drives the radicals apart and prevents their efficient recombination. At slightly higher stresses (stage three), the chains are not only straightened but "stretched," and recoil aids in their separation. According to this model, the diminished ability of the radicals to recombine is the primary reason that tensile stress will increase the rate of photodegradation. An increased separation leads to slower radical-radical recombination, which increases the probability of radical trapping and thus of degradation. Finally, in stage four, a strong stress is present which gives the polymer a fibrillar structure with a higher degree of orientation and crystallinity. Diffusion in a crystalline structure is retarded relative to the amorphous material, and the rate of degradation is expected to decrease because of decreased movement apart of the radicals. In summary, the DRRE hypothesis predicts that tensile stress will initially increase the quantum

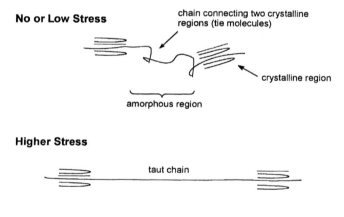

Higher Stress

Homolysis of a bond in the taut chain leads to radical recoil:

Figure 3. The proposed effect of stress in the several stages of the "decreased radical recombination efficiency" (DRRE) hypothesis.

yield of degradation and then further increases in stress will decrease the quantum yield (Figure 2b).

Stress-induced changes in the rates of radical reactions subsequent to radical formation

The effect of stress on the rates and efficiencies of reactions that occur subsequent to radical formation is complicated. Examples have been noted where the rates decrease because of decreased oxygen diffusion in the stressed (more ordered) sample.[17-19] In other systems, it was proposed that the rates increased because of an increase in oxygen diffusion. In yet other systems, it has been suggested that stress changes the conformations of the C-C bonds in the polymer chains; depending on the system this leads to either a rate increase or a rate decrease.[20-22] The quantitative relationship of quantum yields to stress in all of these cases is generally considered to be given by the so-called Zhurkov equation. According to this equation, the effect of stress on the thermal degradation rates of polymers can be fit to an empirical Arrhenius-like equation:[23] rate = A exp[-(ΔG-Bσ) /RT] where ΔG is an "apparent" activation energy, σ is the stress, and A and B are constants. It has been suggested[9] that an equation similar to the Zhurkov equation might apply in photodegradation reactions: ϕ_{obs} = A I_a exp[-(ΔG-B$I_a\sigma$)/RT], where σ is the stress, I_a is the absorbed light intensity, and A and B are again constants characteristic of a particular polymer. The Zhurkov equation is empirical and does not fall strictly into any of the three mechanistic categories discussed above because it deals with an "effective" activation energy, which (in the case of a photochemical reaction) is a composite of the activation barriers for the $k_{homolysis}$, $k_{recombination}$, and k_{trap} steps. The relationship of quantum yield to stress for a system that follows the Zhurkov equation is shown in Figure 2c.

Objective

The experimental objective of this study was to determine the effect of tensile stress on the quantum yields for the photochemical degradation of selected polymers. Mechanistic insights into how stress affects the photodegradation rates of polymers will come from a comparison of the results to the predictions of three different mechanistic hypotheses proposed in the literature.

Experimental

Synthesis of polymer 1, [-(OCH₂CH₂C₅H₄(CO)₃Mo-Mo(CO)₃C₅H₄CH₂CH₂O)(C(O)CHClCH₂)(CHClCH₂)ₘ -]ₙ.

Synthesis of polymer 1, [-(OCH$_2$CH$_2$C$_5$H$_4$(CO)$_3$Mo-Mo(CO)$_3$C$_5$H$_4$CH$_2$CH$_2$O)(C(O)CHClCH$_2$)(CHClCH$_2$)$_m$ -]$_n$.

PVC-COOH (2 g, equivalent to 0.8 mmole of carboxyl groups) was added to a 500 mL Schlenk flask, which was then capped with a septum and filled with nitrogen. THF (80 mL) was added by syringe and the solution stirred until all of the polymer was dissolved (4 h). Thionyl chloride (1.2 mL, 16.4 mmole), which was either freshly distilled or previously distilled and kept under nitrogen at -4 °C, was then added by syringe. The reaction mixture was refluxed in an oil bath for 9 h under N$_2$. At this point, the color of the solution was yellow to light brown. The solution was concentrated (to ca. 20 mL) under reduced pressure, and 150 mL of hexanes was slowly added by cannula or syringe. On addition of hexanes, an off-white to light yellow material precipitated out. After the addition of hexanes was complete, the reaction mixture was allowed to sit for 30 min, which facilitated the easy removal of the supernatant solution by cannula. The residual solid was rinsed with hexanes (50 mL) and then dried under vacuum overnight. The acyl chloride-activated PVC (0.4 g, 0.157 mmol –COCl) was dissolved in dry, degassed THF (50 mL) in a 100mL flask equipped with a magnetic stirring bar. A solution of (HOCH$_2$CH$_2$C$_5$H$_4$)$_2$Mo$_2$(CO)$_6$ (0.045 g, 0.077 mmol) in a small amount of THF was added dropwise into the solution containing the acyl chloride-activated PVC, followed by the addition of triethylamine (1 mL). The reaction was stirred for two days before separation. At that time, the solid salt of triethylamine hydrochloride (observed on the wall of the flask) was removed by filtration. The filtrate was concentrated by evaporating half of the solvent. The solution was then poured onto a Teflon-coated pan to cast a film. The film was dried for 2 days in the glovebox and then placed under vacuum to dry it completely. IR (of thin film): ν(C=O) 1770s and 1730m cm^{-1}; ν(C≡O) of [-Mo$_2$(CO)$_6$], 2005m, 1950s, and 1907s cm^{-1}. UV-vis (in THF): λ$_{max}$ = 393nm and 511nm. T$_g$ = 65 °C; M$_w$ > 528,000 as measured by GPC (gel permeation chromatography) with a light scattering detector. (Some photochemical decomposition occurred during the molecular weight measurement so this M$_w$ value is a lower limit.)

1

Preparation of plasticized polymer 1.

The procedure for polymer **1** was followed except that, after the filtration to remove the triethylamine hydrochloride salt, dioctyl phthalate (DOP; 0.08 g, 20 wt% of acyl chloride-activated PVC) was added to the filtrate with stirring. Half of the solvent was then evaporated and the concentrated solution was poured onto a Teflon-coated pan to form a liquid film. The film was dried in the glovebox for 2 days, and the final solvent residue removed in vacuo. IR (of thin film): $v(C=O)$, 1770m and 1730s cm^{-1} (the intensities were skewed by DOP); $v(C\equiv O)$ of [-Mo$_2$(CO)$_6$], 2005m, 1950s, and 1907s cm^{-1}. UV-vis (in THF): λ_{max} = 393 nm and 511 nm. T_g = 22 °C.

Irradiation of polymers 1 and 2.

Thin films of polymers 1 and 2 were photochemically reactive (λ = 546 nm) in the absence of oxygen. Thin films were irradiated (λ = 546 nm) under nitrogen using the output from a high pressure Hg arc lamp. Infrared spectroscopic monitoring of the photochemical reaction showed the disappearance of the $v(C\equiv O)$ bands of [-Mo$_2$(CO)$_6$] at 2005, 1950, and 1907 cm^{-1} and the appearance of bands attributed to the [-Mo(CO)$_3$Cl] unit at 2048 and 1967 cm^{-1}. Changes in absorbance as a function of time were recorded and the quantum yields calculated accordingly. The films were stressed by hanging masses from the films.

Results & Discussion

Approach to the Problem

As indicated in the Background section, several challenging experimental problems hinder the rigorous experimental exploration of stress and light intensity effects on polymer photodegradation rates. One of the difficulties is that polymer degradations are mechanistically complicated. This is not to say that the mechanisms are not understood; in fact, they are understood in detail. Rather, the mechanisms are intricate, often involving multiple steps, cross-linking, and side-reactions; this makes pinpointing the effects of stress difficult. Another complication is that oxygen diffusion is the rate-limiting step in many photooxidative degradations, the primary degradation mechanism in most polymers. This can add to the intricacy of the kinetics analysis because cracks and fissures develop in the polymer as degradation proceeds; these fractures provide pathways for direct contact of the polymers with oxygen, which will then no longer degrade at a rate controlled by oxygen diffusion. To circumvent these experimental and mechanistic complexities and therefore make it less difficult to interpret data and obtain fundamental insights, we used three key experimental strategies in our investigation. First, we studied the problem using special photodegradable polymers of our own design that contain metal-metal bonds along the backbone.[24] These polymers degrade with visible light by a straightforward mechanism involving metal-metal bond homolysis followed by capture of the metal radicals with an appropriate radical trap (typically an organic halide or molecular oxygen; Scheme 2). By studying the effect of stress on these model systems, we were able to extract information without the mechanistic complications inherent in the degradation mechanisms of organic radicals. (For example, metal radicals do not lead to crosslinking, so we can avoid this complicating feature found with organic radicals.) The second key experimental strategy was to use polymers that have built-in radical traps, namely C-Cl bonds. By eliminating the need for external oxygen to act as a trap, we excluded the complicating kinetic features of rate limiting oxygen diffusion. The third experimental strategy was to use the distinctive M-M bond chromophore to spectroscopically monitor the photodegradation reactions of the polymers. This allowed us to compare the efficiencies of the photodegradations by measuring the quantum yields of the reactions. (The quantum yield, Φ, is defined as the rate of a photoreaction divided by the absorbed light intensity; i.e., Φ = rate/absorbed intensity.) The use of quantum yields to quantify and compare the various degradation rates is a crucial advance because polymer degradation reactions have typically been monitored by stress testing, molecular weight measurements, or attenuated total reflection (ATR) spectroscopy, all of which can be laborious and time consuming. Relative to these techniques, quantum yield measurements are straightforward. To further expedite our quantum yield measurements, we built a computerized apparatus that automatically measures the quantum yields on thin film polymer samples.[25]

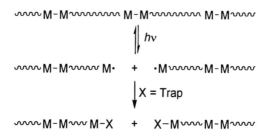

Scheme 2. Photochemical reaction of a polymer with metal-metal bonds along its backbone.

Quantum Yields as a Function of Stress

Thin films of polymer **1** were photochemically reactive ($\lambda = 546$ nm) in the absence of oxygen. Infrared spectroscopic monitoring of the photochemical reaction showed the disappearance of the $Cp_2Mo_2(CO)_6$ units and the appearance of $CpMo(CO)_3Cl$ moieties, indicative of polymer chain cleavage (Figure 4). Quantum yields for the degradation of polymer **1** as a function of stress were measured in an anaerobic environment and are shown in Figure 5 (solid curve). Note that stress initially caused an increase in the quantum of degradation but that a further increase in stress eventually caused a decrease in the quantum yield. These results are consistent with the predictions of the DRRE hypothesis (Figure 2b).

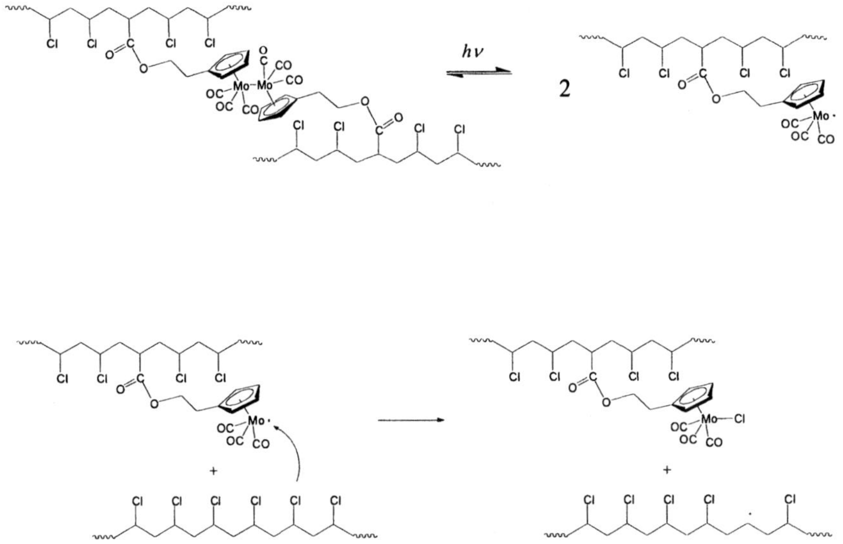

Figure 4. A scheme showing the photochemical reaction of polymer 1 under anaerobic conditions. The photogenerated Mo-centered radicals abstract Cl atoms from a PVC section of the polymer. The scheme shows Cl abstaction from an adjacent polymer chain but the abstraction could also occur intramolecularly from the chain to which the Mo-centered radical is attached. The fate of the carbon radicals produced in these abstraction reactions was not studied, but further decomposition of the PVC polymer chain may result.

Effect of Added Plasticizer

When DOP is added (25%) to polymer **1**, the quantum yield at any given stress is increased. Furthermore, the stress at which the maximum in the curve occurs is shifted to the left, i.e., to lower stress (Figure 5). Both of these features are consistent with the DRRE hypothesis. Thus, the overall higher quantum yields are consistent with facilitated chain movement in the plasticized polymer. Likewise, the shift in the maximum is consistent with increased ordering of the chains at lower stresses.

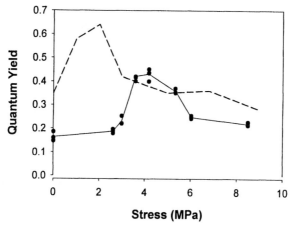

Figure 5. Quantum yields for degradation of polymer 1 (solid curve) and plasticized 1 (dashed curve) versus applied tensile stress.

Evidence for an Increase in Order with Increasing Stress; Infrared Spectroscopy and X-ray Diffraction

A key tenet of the DRRE hypothesis is that stress causes ordering of the amorphous polymer chains and this ordering causes a decrease in the quantum yields of the polymer decomposition. Infrared spectroscopy provided a convenient way to monitor changes in crystallinity in polymers **1** and **2** as a function of stress. Bands in the 1400-1450, 1200-1285, and 580-720 cm^{-1} regions have been assigned to the CH_2 bending, CH bending, and C-Cl stretching modes of PVC, respectively.[26,27] As discussed by a number of authors, the frequencies of these modes are sensitive to chain conformation and crystallinity.[28] For example, it is generally agreed that the band at 610-615 cm^{-1} arises from a C-Cl stretch in amorphous regions and the band at 635-638 cm^{-1}

arises from C-Cl stretching of a long syndiotactic sequence in crystalline regions. The absorbance ratio A_{637}/A_{615} can therefore be used to measure the relative amounts of crystallinity in stressed PVC.[29] In fact, a plot of A_{637}/A_{615} as a function of stress showed that ordering and crystallinity increased with increasing stress (Figure 6). Other infrared crystallinity indices are A_{1425}/A_{1436}, A_{1253}/A_{1240}, and A_{604}/A_{615}. In each case, these absorbance ratios also increased with increasing stress on polymers 1 and 2, a result consistent with increased stress causing increased crystallinity. X-ray diffraction also showed an increase in polymer crystallinity under stress. In particular, the data showed that with increasing stress on the film there was enhanced intensity in the 16 - 18° 2θ region, indicative of increased two-dimensional order perpendicular to the chain direction.[30] More precisely, the intensity changes of the (200) and (110) reflections showed an increase in the lateral crystallite dimensions in stressed PVC in both polymers 1 and 2.

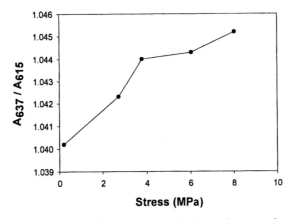

Figure 6. The ratio of A_{637}/A_{615} versus applied tensile stress for polymer 1.

Conclusions

This study demonstrates that metal-containing polymers can be useful in solving mechanistic problems related to polymer chemistry. Specifically, polymers containing metal-metal bonds were successfully used to investigate the origin of tensile stress-induced rate increases in polymer photodegradation reactions. Previous attempts to investigate this problem with "regular" polymers were unsuccessful because the photochemistry of these polymers is too complex for detailed mechanistic interpretation. In contrast, the photochemistry of the metal-containing polymers is relatively straightforward and uncomplicated, and this allowed a thorough interpretation of the results. The results are consistent with the "decreased radical recombination efficiency" hypothesis, from which it can be concluded that the role of stress is to increase the separation of the photochemically generated radical pair, which decreases their probability of recombination. Evidence for an increase in chain order in the stressed PVC polymer used in these experiments comes from infrared spectroscopic data and x-ray scattering. The absorbance ratio of the infrared peak at 635-638 cm^{-1} to that at 610-615 cm^{-1} has been used to assess crystallinity in PVC,[29] and our data shows an increase in the A_{637}/A_{615} ratio for the samples in these experiments. As reported elsewhere, wide angle x-ray diffraction patterns in the 16-18° 2θ region[30] are also consistent with an increase in crystallinity with increasing stress. Further mechanistic evidence consistent with the DRRE hypothesis comes from the effect of added plasticizer on the curve in Figure 5.

Acknowledgments

Acknowledgment is made to the National Science Foundation (DMR-0096606) for the support of this research.

References

1. Grassie, N.; Scott, G. *Polymer Degradation and Stabilization*; Cambridge University Press: New York, 1985.
2. Guillet, J. *Polymer Photophysics and Photochemistry: An Introduction to the Study of Photoprocesses in Macromolecules*; Cambridge University Press: New York, 1985.
3. Rabek, J. F. *Mechanisms of Photophysical Processes and Photochemical Reactions in Polymers*; Wiley: New York, 1987.
4. Hamid, S. H.; Editor *Handbook of Polymer Degradation: Second Edition, Revised and Expanded. [In: Environ. Sci. Pollut. Control Ser., 2000; 21]*, 2000.
5. Tong, L.; White, J. R. *Polym. Degrad. Stab.* **1996**, *53*, 381-396.
6. Popov, A.; Rapoport, N.; Zaikov, G. *Oxidation of Stressed Polymers*; Gordon and Breach Science Publishers: New York, 1991.
7. DeVries, K. L.; Hornberger, L. E. *Polym. Degrad. Stab.* **1989**, *24*, 213-240.
8. Popov, A. A.; Blinov, N. N.; Krisyuk, B. E.; Zaikov, G. E. *Polym. Degrad. Stab.* **1984**, *7*, 33-39.
9. White, J. R.; Rapoport, N. Y. *Trends Polym. Sci. (Cambridge, UK)* **1994**, *2*, 197-202.
10. Tyler, D. R. *Journal of Macromolecular Science. Polymer Reviews* **2004**, C44, 351-388.

442

11. Plotnikov, V. G. *Dok. Akad. Nauk SSSR* **1988**, *301*, 376-379.
12. Busfield, W. K.; Monteiro, M. J. *Mat. Forum* **1990**, *14*, 218-223.
13. Benachour, D.; Rogers, C. E. *ACS Symposium Series* **1981**, *151*, 263-274.
14. Nguyen, T. L.; Rogers, C. E. *Polym. Mater. Sci. Eng.* **1985**, *53*, 292-296.
15. Baimuratov, E.; Saidov, D. S.; Kalontarov, I. Y. *Polym. Degrad. Stab.* **1993**, *39*, 35-39.
16. Shlyapnikov, Y. A.; Kiryushkin, S. G.; Marin, A. P. *Antioxidative Stabilization of Polymers*; Taylor and Francis: Bristol, PA, 1996.
17. Baumhardt-Neto, R.; De Paoli, M. A. *Polym. Degrad. Stab.* **1993**, *40*, 59-64.
18. Baumhardt-Neto, R.; De Paoli, M. A. *Polym. Degrad. Stab.* **1993**, *40*, 53-58.
19. Rabello, M. S.; White, J. R. *Polym. Degrad. Stab.* **1997**, *56*, 55-73.
20. Bellenger, V.; Verdu, J.; Martinez, G.; Millan, J. *Polym. Degrad. Stab.* **1990**, *28*, 53-65.
21. Rapoport, N. Y.; Zaikov, G. E. *Eur. Polym. J.* **1984**, *20*, 409-414.
22. Rapoport, N. Y.; Shibryaeva, L. S.; Zaikov, V. E.; Iring, M.; Fodor, Z.; Tudos, F. *Polym. Degrad. Stab.* **1985**, *12*, 191-202.
23. Zhurkov, S. N.; Zakrevskii, V. A.; Korsukov, V. E.; Kuksenko, V. S. *J. Polym. Sci., Polym. Phys. Ed.* **1972**, *10*, 1509-1520.
24. Tyler, D. R. *Coord. Chem. Rev.* **2003**, *246*, 291-303.
25. Male, J. L.; Lindfors, B. E.; Covert, K. J.; Tyler, D. R. *J. Am. Chem. Soc.* **1998**, *120*, 13176-13186.
26. Krimm, S.; Folt, V. L.; Shipman, J. J.; Berens, A. R. *J. Polym. Sci., Pt. A* **1963**, *1*, 2621-2650.
27. Tasumi, M.; Shimauouchi, T. *Spectrochim. Acta* **1961**, *17*, 731-754.
28. Zerbi, G.; Ciampelli, F.; Zamboni, V. *Conference on the Vibrational Spectra of High Polymers*; G. Natta and G. Zerbi, Eds.; Interscience: New York, 1964; p 3.
29. Chartoff, R. P.; Lo, T. S. K.; Harrell, E. R.; Roe, R. J. *J. Macromol. Sci.-Phys.* **1981**, *B20*, 287-303.
30. Chen, R.; Yoon, M.; Smalley, A.; Johnson, D. C.; Tyler, D. R. *J. Am. Chem. Soc.* **2004**, *126*, 3054-3055.

Chapter 31

Spectroscopic and Liquid Crystal Properties of Phthalocyanine Macromolecules with Biomedical Applications

Ernie H. G. Langner, Wade L. Davis, Rebotsamang F. Shago, and Jannie C. Swarts[*]

Department of Chemistry, University of the Free State, Nelson Mandela Drive, Bloemfontein 9300, South Africa

The influence of alkyl substituent chain length, peripheral versus non-peripheral substitution positions, and coordinated central metal (2H, Cu^{2+}, Ni^{2+}, Zn^{2+} and $Cl-Al^{3+}$) on mesophase behaviour and spectroscopic properties of phthalocyanine complexes are described

Background

Phthalocyanines are used industrially as blue and green pigments.[1] Other areas of phthalocyanine study include their application as non-linear optical materials,[2] as catalysts in a variety of chemical reactions including oxidations of mercaptans[3] and hydrogenation of multiple bonds,[4] as converters of solar energy to other forms of energy[5] and as photodynamic cancer drugs.[6]

As photodynamic cancer drugs, phthalocyanines have many intrinsic advantageous over porphyrins such as photophrin, the first porphyrin approved for photodynamic cancer therapy in the United States.[7] Phthalocyanines are attractive as photodynamic cancer drugs because of the wavelength at which they absorb red or infrared light, and the efficiency by which they do so. Light of wavelength 630 nm or longer (i.e. red or infrared light) penetrates deepest through body tissue. The Q-band peak maximum for most metallated phthalocyanines is at 660 nm or longer, a wavelength range that is especially suitable for photodynamic cancer therapy. The most intense light absorbtion for photophrin is observed in the 400 nm Soret band range, while the Q-band absorbtion between 500 and 650 nm are much less intense (ε_{630nm} = 3500 dm^3 mol^{-1} cm^{-1}).[7,8] In contrast, extinction coefficients associated with the Q-band of phthalocyanines are large; sometimes exceeding[9] 200 000 dm^3 mol^{-1} cm^{-1}. The coordinated metal determines the success of phthalocyanines in photodynamic cancer therapy. The photophysics of zinc and aluminium phthalocyanines are among those best suited for photodynamic cancer therapy.[8] Different coordinated metals, however, have almost no influence on the wavelength at which complexes absorb light.

Limitations of phthalocyanines as photodynamic cancer drugs are nested in their notorious insolubility and the loss of optimum quantum yields upon radiation due to aggregation. To overcome both these issues, researchers substitute bulky substituents on the phthalocyanine macrocycle. Linear long chain alkyl substituents enhance organic solubility and lower aggregation, and Cook[10] has demonstrated that C$_5$ and C$_{10}$ non-peripheral octa-alkylated zinc phthalocyanines have photodynamic cancer activity. In an alternative approach, Van Lier[6] demonstrated that peripheral sulphonated aluminium phthalocyanines posess aqueous solubility as the sodium salts. Another possibility in enhancing aqueous solubility of phthalocyanine macrocycles, is to make use of water-soluble polymeric drug carriers, as demonstrated for some ferrocene derivatives.[11]

The introduction of long chain linear alkyl substituents and mixtures of peripheral and non-peripheral substituents impose interesting physical properties on phthalocyanine macrocycles. These include less aggregation than that observed for short chain substituents,[12] liquid crystalline properties[13] and differences in the wavelength where Q-band peak absorbtion is observed.[14] Phthalocyanine liquid crystal phases are normally discotic columnar.[13, 15] Interactions between the aromatic macrocyclic cores maintain columnar packing, while melted long side chains provide columnar mobility. Up to three[15] different mesophases have thus far been reported for phthalocyanine liquid crystals. Higher temperature mesophases normally has a disordered hexagonal symmetry, while relative lower temperature mesophases have been found to be either of hexagonal or rectangular symmertry.

Objective

The objective of this study was to determine the effect chain length and substitution position have on the liquid crystalline properties of alkylated and ferrocene-containing alkoxylated phthalocyanines with $2H^+$, Zn^{2+}, Cu^{2+}, Ni^{2+} and $Cl-Al^{3+}$ coordinated in the macrocyclic cavity by variable temperature optical microscopy and differential scanning calorimetric techniques. A UV-vis spectroscopic study also demonstrated the relationship between Q-band peak maxima and degree of non-peripheral substitution (as compared to peripheral substitution).

Experimental

Materials. All reagents were used without further purification. Dichloromethane was refluxed over calcium hydride and freshly distilled prior to use. Phthalocyanines without metallocene side chains were synthesized as described[9, 16, 17] before.

Synthesis of phthalocyanines with ferrocenylalkoxy side chains. The synthesis of 1,4,8,11,15,18-hexa(tetradecyl)-23-(4-ferrocenylbutoxy) phthalocyanine may serve as an example:[18] 3,6-Di(tetradecyl)phthalonitrile (0.940 g, 1.804 mmol) and 4-(4-ferrocenylbutoxy)phthalonitrile (0.090 g, 0.256 mmol) were dissolved in warm (80 °C) pentanol (12 cm^3). An excess of clean lithium metal (0.2 g, 0.0288 mol) was added in small portions and the mixture was heated thereafter for 16 hours at 110 °C. The cooled, deep green suspension was stirred with acetone (20 cm^3), the solution was filtered and the solids washed with acetone (50 cm^3) before the combined acetone solutions were concentrated to *ca.* 20 cm^3. Acetic acid (20 cm^3) was added and the heterogeneous mixture stirred for 30 minutes and the precipitated collected to afford the crude product (0.320 g) after recrystallisation from THF/methanol. The crude product was chromatographed on silica gel at different ratios of hexane/toluene as eluents. Three fractions could be clearly separated and identified. They were isolated and recrystalized from THF/methanol. The first fraction was recovered from the column with hexane/toluene (20:1), and was identified as 1,4,8,11,15,18,22,25-octa(tetradecyl)phthalocyanine (58 mg, 10.70 %, R_f = 0.99). The second fraction isolated with hexane/toluene as mobile phase in a 20:3 ratio, and was identified as 1,4,8,11,15,18-hexa(tetradecyl)-23-(4-ferrocenylbutoxy)phthalocyanine (38.3 mg, 6.16 %, R_f = 0.48); δ_H (CDCl$_3$); 0.87 (18 H, t, 6 x CH$_3$), 1.27 (132 H, m, 6 x -(CH$_2$)$_{11}$-), 1.93 (2H, m, 1 x (-CH$_2$-CH$_2$-CH$_2$-CH$_2$-Fc)), 2.18 (14 H, m, 1 x (-CH$_2$-CH$_2$-CH$_2$-CH$_2$-Fc), 6 x (Ar-CH$_2$-CH$_2$-)), 2.60 (2 H, m, 1 x O-CH$_2$-CH$_2$-CH$_2$-CH$_2$-Fc), 4.18 (14 H, m,1 x (-CH$_2$-CH$_2$-CH$_2$-CH$_2$-Fc), 6 x (Ar-CH$_2$-CH$_2$-), 4.44 (9 H, m, C$_{10}$H$_9$), 7.50 (1 H, d, Ar-H), 7.82 (6 H, m, 3 x Ar-H$_2$), 8.38 (1 H, s, Ar-H), 8.82 (1 H, d, Ar-H). The third fraction isolated with hexane/toluene (20:5) and was identified as tetra(tetradecyl)-di(ferrocenylbutoxy)phthalocyanine (7.10 mg, 1.51%, R_f = 0.19); δ_H (CDCl$_3$); 0.80 (12 H, t, 4 x CH$_3$), 1.24 (88 H, m, 4 x -(CH$_2$)$_{11}$-), 1.79 (4 H, m, 2 x (-CH$_2$-CH$_2$-CH$_2$-CH$_2$-Fc)), 1.98 (4H, m, 2 x (-CH$_2$-CH$_2$-CH$_2$-CH$_2$-Fc), 2.17 (8 H, m, 4 x (Ar-CH$_2$-CH$_2$-)), 2.64 (4 H, m, 2 x O-CH$_2$-CH$_2$-CH$_2$-CH$_2$-Fc), 4.00 – 4.60 (30 H, m, 2 x (-CH$_2$-CH$_2$-CH$_2$-CH$_2$-Fc), 4 x (Ar-CH$_2$-CH$_2$-), 2 x C$_{10}$H$_9$), 7.4 – 8.9 (10 H, m, 2 x Ar-H$_3$, 2 x Ar-H$_2$).

Metallation experiments. Phthalocyanine metallations were performed following the general procedures described in references 9 and 19.

Instrumentation and Techniques. Proton NMR-spectra at 298 K were recorded at 300 MHz on a Bruker Advance DPX 300 NMR spectrometer, chemical shifts are referenced to SiMe$_4$ at 0.00 ppm. Solution electronic spectra were recorded in THF or cyclohexane on a Varian Cary 50 UV-vis dual beam spectrophotometer at 298 K.

Phase transitions between crystal, liquid crystal and isotropic-liquid states were monitored by DSC (for enthalpy changes, ca. 7 mg samples at heating and cooling rate of 10 °C min^{-1} between –70 °C and a convenient maximum temperature at least 30 °C higher than the melting point of the compounds were used) on a TA Instruments DSC 10 thermal analyser fitted with a Du Pont Instruments mechanical cooling accessory and a TA Instruments Thermal Analyst 2000 data processing unit. To obtain accurate transition temperatures, visual measurements were performed on an Olympus BH-2 polarising microscope in conjunction with a Linkam TMS 92 thermal analyser with a Linkam THM 600 cell. The heating and cooling rates used were either 5 °C min^{-1} or 2 °C min^{-1}.

Results & Discussion

Synthesis and UV-vis spectroscopy

Metal-free and metallated phthalocyanines were synthesised as shown in Scheme 1. The reaction involves cyclisation of four phthalonitrile molecules to obtain the octa-aza phthalocyanine macrocycle. The Cl-Al phthalocyanines were obtained directly from the condensation of 3,6-didecyldiimino-isoindoline in the presence of AlCl$_3$.[20]

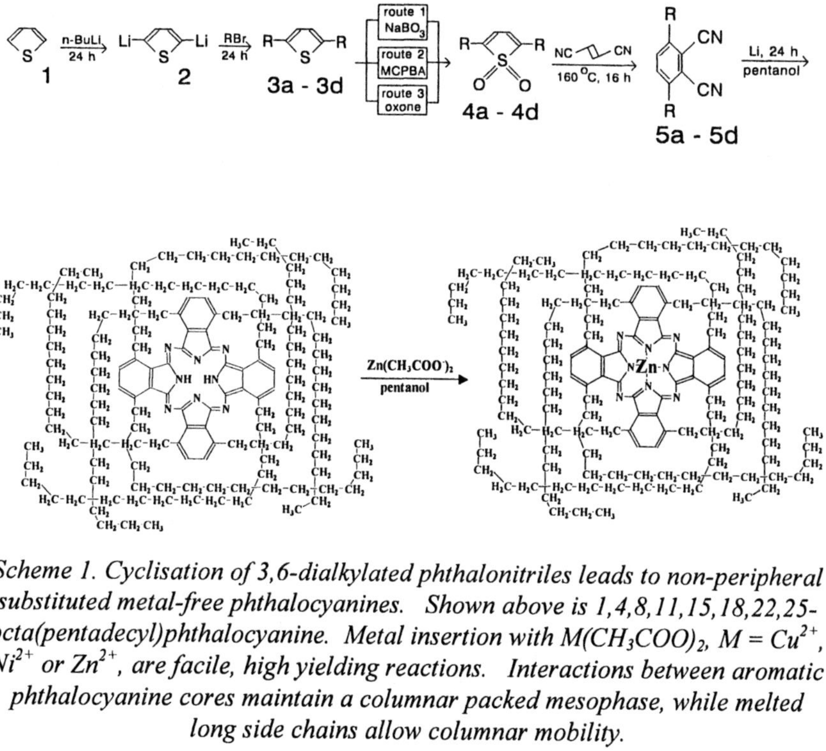

Scheme 1. Cyclisation of 3,6-dialkylated phthalonitriles leads to non-peripheral substituted metal-free phthalocyanines. Shown above is 1,4,8,11,15,18,22,25-octa(pentadecyl)phthalocyanine. Metal insertion with M(CH$_3$COO)$_2$, M = Cu^{2+}, Ni^{2+} or Zn^{2+}, are facile, high yielding reactions. Interactions between aromatic phthalocyanine cores maintain a columnar packed mesophase, while melted long side chains allow columnar mobility.

Normally, one would expect alkyl groups to align in a linear fashion in the plane of the phthalocyanine macrocyclic core. However, the flat structure of the phthalocyanine macrocycle causes non-peripheral substituents to collide. Hence, Cook[21] showed in a crystal structure of the C$_6$ compound that six of the eight side chains of non-peripheral octa-hexylated phthalocyanines are orientated in the plane of the macrocycle, pointing away from it. Four of these have their

C-C bonds staggered in the macrocyclic plane, while two are staggered out of the plane and make a sharp turn at the first C-atom adjacent to the macrocyclic core to minimise steric hindrance. The remaining two hexyl groups make an immediate sharp turn at the first C-atom to align themselves approximately perpendicular to the plane of the macrocyclic core. In the solid state, they act as spacers between two adjacent macrocycles.

Figure 1 shows the Soret and Q-bands in the UV-vis spectra of the non-peripheral substituted compounds 1,4,8,11,15,18,22,25-octa(dodecyl)-phthalocyanine (the Q-band shows two peaks with peak maxima at 730 and 700 nm), 1,4,8,11,15,18,22,25-octa(dodecyl)phthalocyaninatozinc(II) (the Q-band is a single peak with peak maxima at 700 nm), and the peripheral substituted tetra(^{t}butyl)-phthalocyaninatocobalt(II) which shows a single Q band peak maximum at 660 nm. Moving from peripheral to non-peripheral substituted metal-containing phthalocyanines leads to a red shift of 40 nm in Q-band peak maximum. The switch from a cobalt to zinc complex was not the reason for this shift as different coordinated metals have almost no influence on the wavelength at which complexes absorb light.

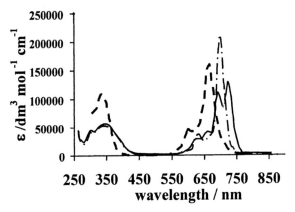

Figure 1. UV-vis spectra of 1,4,8,11,15,18,22,25-octa(dodecyl)phthalocyanine (solid line), its zinc complex(— — — — —), and the peripheral substituted complex tetra(^{t}butyl)phthalocyaninatocobalt(II) (- - -).

Metal-free and metallated phthalocyanines differ in having D_{2h} and D_{4h} symmetry and this is manifested in differences especially in the Q-band region. The Q-band twin peak of the metal-free system converges into a single peak when a metal is inserted in the phthalocyanine because metal insertion leads to a degeneracy of the phthalocyanine's lowest energy singlet state.

By increasing the alkyl chain length of the non-peripheral substituents on the phthalocyanines, less aggregation is observed. Aggregation manifests in deviations from the Beer-Lambert law. Without any aggregation, a linear dependence exists between phthalocyanine concentration and absorbance. Deviations from this linear relationship indicate aggregation of the flat macrocyclic molecules. Aggregation involves the stacking of phthalocyanines on top of each other much like a deck of cards. Thus, aggregated phthalocyanine molecules do not exist independently of each other in solution, and the different degrees of association in these non-ideal solutions lead to deviations from the Beer-Lambert law.

Figure 2 demonstrates the validity of the Beer-Lambert law for the Q-band of non-aggregated solutions of 1,4,8,11,15,18,22,25-octa(dodecyl)phthalo-cyanine derivatives as well as the concentration at which aggregation sets in for 1,4,8,11,15,18,22,25-octa(alkylated)phthalocyaninatozinc(II) complexes with different alkyl chain lengths.

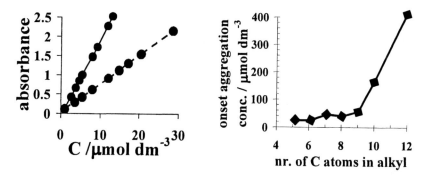

Figure 2. Left: The Beer-Lambert law, A = εcl, for 1,4,8,11,15,18,22,25-octa(decyl)phthalocyanine at 730 nm (- - -) and for 1,4,8,11,15,18,22,25-octa(decyl)phthalocyaninatozinc(II) at 700 nm (——) in THF at 25 °C. Right: The concentration where the Beer-Lambert law breaks down for 1,4,8,11,15,18,22,25-octa(alkylated)phthalocyaninatozinc(II) complexes dissolved in cyclohexane at 18 °C.

Of special importance is the observation that from an alkyl substituent chain length of C_{10}, the onset concentration of aggregation suddenly begins to increase dramatically. This is consistent with a view that, as in the solid state, separation of the phthalocyanine macrocycles by their own side chains, or even encapsulation of the macrocyclic core in its own side chains, is possible in solution, provided the substituents are long enough.

Phthalocyanines with ferrocenylalkoxy side chains were synthesised as shown in Scheme 2. The statistical condensation using a mixture of two different phthalonitriles invariably gives an entire range of different phthalocyanines designated as AAAA, AAAB, AABB, ABAB, ABBB and BBBB. In these acronyms, A represents a fragment of the macrocycle that originated from a 3,6-dialkylated phthalonitrile and B represents a fragment that originated from a 4-alkoxylated phthalonitrile bearing a terminal ferrocenyl group on the alkoxy chain. The ratio of 9:1 for reacting phthalonitriles A:B was chosen because this allowed a substantial yield[16] of the AAAB product while minimising the products AABB, ABAB, ABBB and BBBB. The B-fragment can condense in one of two configurations during phthalocyanine formation. This leads to a number of regio-structural isomers. The different phthalocyanine products having different numbers of 3-ferrocenylpropyloxy side chains were all separated from each other by intensive column chromatography, but the regio-structural isomers of each product could not be separated. Neither could the regio-isomeric products having general structure AABB and ABAB be separated from each other. The structures of all the products that can be obtained during this type of statistical condensation are shown in Figure 3.

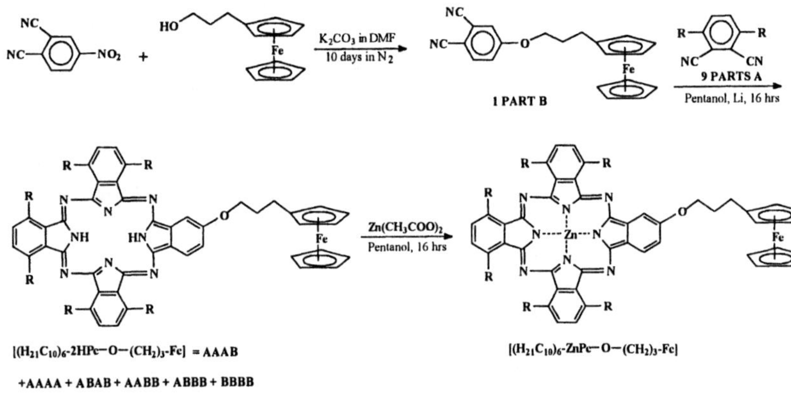

Scheme 2. Syntheses of phthalocyanines having non-peripheral alkyl substituents and a peripheral 3-ferrocenylpropyloxy side chain. The choice of mixing the two reacting phthalonitriles in a ratio of 9:1 allowed isolation of AAAA and AAAB as the main reaction products.

The UV-vis spectra of each ferrocene-containing product showed the sharp single peak associated with metallated phthalocyanines, or the twin peaks associated with metal-free phthalocyanines. The peak maxima of each product, however, did become more blue-shifted (i.e. peak maxima shifted to shorter wave lengths) as more peripheral side chains were introduced into each product. This blue shift is demonstrated in Figure 3 below.

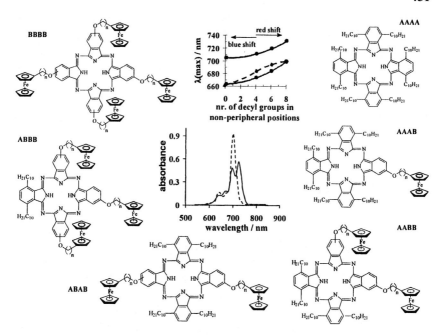

Figure 3. The different metal-free products arising from the reaction between 3,6-di(decyl)phthalonitrile and 4-alkoxyphthalonitriles having a ferrocenyl group on the terminal position. Most products are also regio-isomers. The Q-band in the UV-vis spectra for the all non-peripheral C_{10} alkyl-substituted metal-free (——) and zinc-containing phthalocyanine (- - -) is shown in the middle. The manner in which the various peak maxima shift to shorter wave lengths with increasing amount of non-peripheral substituents is shown on top in the centre. The broken line is associated with the single peak of the zinc-containing derivative, while the two solid lines are associated with the two maximums of the two Q-band peaks of the metal-free phthalocyanines.

Phase Studies

Phase studies on the present compounds of study were performed utilizing differential scanning calorimetry. Shown below (Figure 4) is a DSC trace for 1,4,8,11,15,18,22,25-octa(tetradecyl)phthalocyaninatozinc(II).[18] Four meso-phases were detected between the clearing point (conversion of mesophase to isotropic liquid) at 192 °C on the heating cycle and the melting point (transition from a mesophase to a crystalline solid) at 64 °C on the cooling cycle.   The longer alkyl substituents showed more mesophase transitions than compounds having shorter chain substituents, with four the largest amount of observed mesophases.   In this work each discotic mesophase is labelled D_1, D_2, D_3 or D_4 with D_1 assigned to the mesophase arising from the transition between isotropic liquid to the first observed (i.e. highest temperature) mesophase.

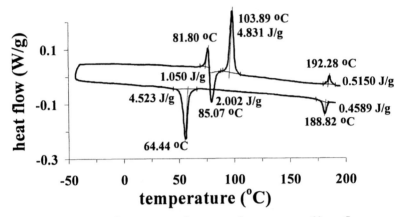

Figure 4.   Differential scanning calorimetry thermogram of heat flow versus temperature of 1,4,8,11,15,18,22,25-octa(tetradecyl)phthalocyaninatozinc(II). A heating and cooling rate of 10 °C per minute were used. The temperature at which the phase change occurs and the phase change enthalpy associated with it is indicated at each peak.

The temperature range in which the present complexes exhibited mesophase behavior was found to be dependent on three variables:
  a)   The type of metal coordinated in the macrocyclic cavity
  b)   The type (alkyl or ferrocene-containing alkoxy) and position of substitution (peripheral or non-peripheral) for each substituent.
  c)   The length of each side chain, here C_5-C_{18}.
Figure 5 shows the temperature range in which non-peripheral octadecyl substituted phthalocyanines exhibit mesophase behaviour   as a function of the coordinated metals Cu^{2+}, 2H (metal free), Ni^{2+}, $Cl-Al^{3+}$ and Zn^{2+}. It is apparent that zinc stabilizes the mesophases of the compounds of this study over the widest temperature range, while nickel is the least effective metal for liquid crystal stabilization.

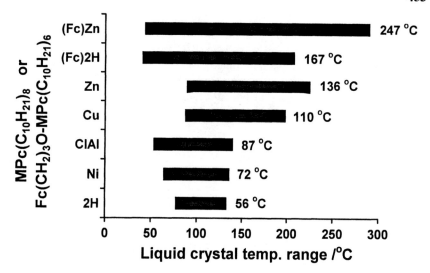

Figure 5. Bottom 5 bars: Temperature range in which 1,4,8,11,15,18,22,25-octa(decyl)phthalocyaninato complexes exhibit liquid crystal behaviour as a function of coordinated metal. Top two bars: Liquid crystal temperature range for two complexes in which the two non-peripheral decyl substituents in positions 22 and 25 has been replaced with a single peripheral 3-ferrocenylpropyloxy substituent in position 23. The numbers next to each bar represent the temperature difference between the melting and clearing point.

Replacement of two of the non-peripheral decyl substituents on an annulated benzene ring with a peripheral 3-ferrocenylpropyloxy substituent increased the temperature range in which these phthalocyanines exhibited liquid crystal behaviour dramatically. In the case of the metal-free complex, mesophase behaviour was observed over a 167 °C temperature range. This temperature range is 167 – 56 = 111 °C wider than for the non-ferrocene-containing compound. The increase observed for the zinc complex was (accidently) also 111 °C.

The effect of alkyl side-chain length on the temperature range in which the phthalocyanines exhibited mesophase behaviour is shown in Figure 6 for the metal-free and zinc complexes. Two key observations can be made. The first is that C_5 is, in the case of the zinc complexes, the shortest alkyl chain length that can support mesophase behaviour. The second is that the C_{12} substituent provides the largest temperature range for mesophase behaviour. The mesophase temperature range increases steadily with increasing side chain length until it reaches for the ZnC_{12} complex a maximum: from 60 to 205.6 °C, that is 145.6 °C. It then decreases again steadily to values of 126 °C for the ZnC_{14} species, 110.9 °C for the ZnC_{15} and 94.6 °C for the ZnC_{18} species. This decrease in mesophase temperature range is attributed to the wax-like properties of very long alkyl groups that begins to dominate over the columnar discotic liquid crystalline properties of the phthalocyanine complexes.[9]

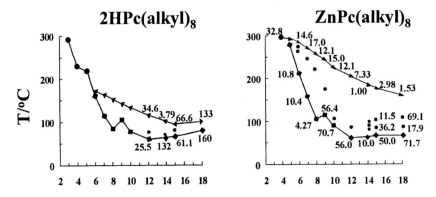

number of carbon atoms in the alkyl chain

Figure 6. Comparison of the effect alkyl substituent chain length has on temperature range in which metal-free and zinc phthalocyanines exhibit mesophase behaviour. The bottom line represents the temperature at which the crystalline solid to mesophase transition occurs, while the top line shows the temperature at which the mesophase converts to an isotropic liquid. The dots show some of the transitions of one mesophase to another. The numbers represent molar phase transition enthalpies in units of kJ/mol.

The introduction of 3-ferrocenylpropyloxy side chains to the phthalocyanine macrocyclic core have enabled us to produce phthalocyanine materials that have liquid crystalline properties at high temperatures (290 °C for the C_{10} alkyl compound series). The lower temperature limit that supports mesophase behaviour is, however, still above room temperature. The search is in progress to find phthalocyanine materials that exhibit mesophase behaviour well below room temperature. Although the present compounds are well soluble in organic media, they are totally insoluble in aqueous media. This makes them less ideal in a photodynamic cancer therapy application. To overcome this, a research effort is also in progress to covalently bind these phthalocyanine complexes to water-soluble, polymeric drug carriers.

Conclusions

From this ongoing work, it has been found that non-peripheral alkyl substituted phthalocyanines has liquid crystal properties that are dependent on the length of the alkyl side chain. The type of metal coordinated in the macrocyclic cavity has a large influence on the temperature range in which each phthalocyanine complex exhibits mesophase behaviour. Zinc was the cation that supported mesophase behaviour over the largest temperature range. By introducing a 3-ferrocenylpropyloxy side-chain in a peripheral position, mesophase behaviour was observed over much larger temperature ranges. Introduction of peripheral side chains caused the peak maxima of the Q-band in the UV-vis spectra of the phthalocyanines to become shifted to shorter wave lengths. The extent of blue shifting increased with increasing amount of peripheral substituents.

Acknowledgments

This research was supported by the National Research Foundation of South Africa, the Medical Research Council of South Africa as well as by the University of the Free State. WLD and RFS acknowledge the Mellon Foundation for financial support. The authors also acknowledge Prof. M. J. Cook from the University of East Anglia, UK, for access to his differential scanning calorimetry equipment.

References

1. Kasuga, K.; Tsutsui, M. *Coord. Chem. Rev.* **1980**, *32*, 67.
2. De la Torre, G.; Vazquez, P.; Agullo-Lopez, F.; Torres, T. *J. Mater. Chem.* **1998**, *8*, 1671.
3. Buck, T.; Wöhrle, D.; Schulz-Ekloff, G.; Andreef, A. *J. Mol. Catal.* **1991**, *70*, 259.
4. Eckert, H.; Kiesel, Y. *Angew. Chem., Int. Ed. Engl.* **1981**, *20*, 473.
5. Wöhrle, D.; Schlettwein, D.; Kischenmann, M.; Kaneko, M.; Yamada, A. *J. Macromol. Sci. Chem.* **1990**, *A27*, 1239; Eichhorn, H. *J. Porphyrins Phthalocyanines* **2000**, *4*, 88; Wöhrle, D.; Meissner D. *Adv. Mater.* **1991**, *3*, 129.

456

6. Ali, H; Van Lier J. E. *Chem. Rev.* **1999**, *99*, 2379.
7. MacDonald, I. J.; Dougherty, T. J. *J. Porphyrins Phthalocyanines.* **2001**, *5*, 105.
8. Allen C. M.; Sharman, W. M.; Van Lier, J. E. *J. Porphyrins Phthalocyanines.* **2001**, *5*, 161.
9. Swarts J. C.; Langner, E. H. G.; Krokeide-Hove, N.; Cook, M. J. *J. Mater. Chem.* **2001**, *11*, 434.
10. Fabris, C.; Ometto, C.; Milanesi, C.; Jori, G.; Cook, M. J.; Russel, D. A. *J. Photohem. Photobiol. B: Biol.* **1997**, *39*, 279; Ometto, C.; Fabris, C.; Milanesi, C.; Jori, G.; Cook, M. J.; Russel, D. A. *Br. J. Cancer* **1996**, *74*, 1891.
11. Swarts, J. C.; Swarts, D. M.; Maree, D. M.; Neuse, E. W; La Madeleine, C.; Van Lier J. E. *Anticancer. Res.* **2001**, *21*, 2033.
12. Cook, M. J.; Chambrier, I.; Cracknell, S. J.; Mayes, D. A.; Russel, D. A. *Photochem Photobiol.* **1995**, *62*, 542.
13. Cook, M. J. *J. Mater. Sc.: Mat. Electr.* **1994**, *5*, 117.
14. Kobayashi, N; Nakajima, S.; Osa, T. *Inorg. Chim. Acta* **1993**, *210*, 131.
15. Cherodian, A. N.; Davies, A. N.; Richardson , R. M.; Cook, M. J.; McKeown, N. B.; Thomson, A. J.; Feijoo, J.; Ungar, G.; Harrison, K. J. *Mol. Cryst. Liq. Cryst.* **1991**, *196*, 103.
16. McKeown, N. B.; Chambrier, I.; Cook, M. J. *J. Chem. Soc., Perkin Trans.* **1990**, 1169.
17. Chambrier I.; Cook M. J.; Cracknell, S. J.; McMurdo, J. *J. Mater. Chem.* **1993**, *3*, 841.
18. Swarts, J. C.; Langner, E. H. G.; Shago, R. F.; Davis, W. L. *Polym. Prepr.* **2004**, *45(1)*, 452.
19. Cook, M. J.; Cracknell, S. J.; Harrison, K. J. *J. Mater. Chem.* **1991**, *1*, 703.
20. Linsky, J. P.; Paul, T. R.; Nohr, R. S.; Kenney, M. E. *Inorg. Chem.* **1980**, *19*, 3131.
21. Chambrier, I.; Cook, M. J.; Helliwell, M.; Powell, A. K. *J. Chem. Soc., Chem. Commun.* **1992**, 444.

Chapter 32

Synthesis, Reactivity, and Ring-Opening Polymerization of Sila-Metallacyclobutanes

Hemant K. Sharma and Keith H. Pannell[*]

Department of Chemistry, University of Texas at El Paso,
El Paso, TX 79968-0513

Sila-metallacyclobutanes are a novel class of transition metal complexes which undergo insertions of CO, CO_2, RCHO, NO, CN-R into the M-C bond and ring-opening polymerization to high molecular weight polymers.

1-Sila-2(3)-metallacyclobutanes of the transition metals[1-10] have been reported and constitute a potentially important class of electrophilic organometallic reagents. Insertions into the M-C bond(s) of the metallacyclobutane ring; C-C reductive coupling, formation of new classes of metallacycles and recently reported ring-opened polymers are some examples of their chemistry. In this chapter, we will briefly overview the chemistry of these interesting complexes.

Synthesis of sila-metallacyclobutanes.

1. 1-sila-3-metallacyclobutanes.
A Salt-elimination reactions.

Early transition sila- metallacyclobutanes[1-4] were synthesized by the salt-elimination reaction of the magnesium reagent, $(XMgCH_2SiMe_2CH_2MgX)$, with a variety of metallocene dihalides Cp_2MX_2, where $Cp = \eta^5\text{-}C_5H_5$, M = Ti (1), Zr (2), Hf (3), Nb (4); X = Cl; M = Mo (5) and X = I, equation 1.

$$Cp_2MX_2 \quad + (XMg\text{-}CH_2\text{-}SiMe_2\text{-}CH_2\text{-}MgX) \xrightarrow[\text{RT}]{\text{THF}} Cp_2M \diagup^{\diagdown SiMe_2} + 2\ MgX_2 \quad (1)$$

1-5

The presence of the β-Si atom in the four-membered MC_2Si ring greatly enhances the thermal stability of the 1-sila -3- metalla- cyclobutanes in comparison with the corresponding metallcyclobutanes which decompose in the solution, presumably due to β-hydrogen eliminations or degradation of metal carbene-olefin complexes.

B γ-elimination.

1-Sila-3-metallacyclobutanes of Ru, Os and Rh were synthesized by Wilkinson et al. with an unusual loss of a γ-hydrogen from acyclic $M\text{-}CH_2SiMe_3$ complexes.[5,6] Similarly, 1-sila-3-irridacyclobutane (6) was obtained by the thermolysis of the acyclic iridium complex, $Cp^*(PPh_3)Ir(CH_2SiMe_3)_2$ for a week at 50 °C in cyclohexane, equation 2.[7]

$$\cdots \xrightarrow{\Delta} \cdots + SiMe_4 \quad (2)$$

6

No metallacycle was obtained from the thermolysis of analogous acyclic rhodium complex Cp*(PPh$_3$)Rh(CH$_2$SiMe$_3$)$_2$ which underwent decomposition.

C Base-mediated ring closure reactions.

Recently, we were successful in synthesizing 1-sila-3-ferracyclobutanes[8] **9** by treating either (η^5-C$_5$H$_5$)Fe(CO)$_2$-SiR$_2$CH$_2$Cl, **7** (R$_2$ = Me$_2$ (**a**), ^{n}BuMe (**b**), ^{n}Bu$_2$ (**c**), PhMe (**d**)) containing a direct Fe-Si bond or their rearranged isomers, (η^5-C$_5$H$_5$)Fe(CO)$_2$-CH$_2$SiR$_2$Cl, **8** (**a-d**), containing Fe-C bond, with lithium diisopropylamide, LDA, in THF, equation 3.

$$\text{LDA / -25}^0\text{C} \tag{3}$$

The metallacycles **9** (**a-d**) from **7** results from the initial metallation of the cyclopentadienyl ring followed by silyl group migration to the ring, generating an iron-based anionic species that performs the final intramolecular ring closing salt-elimination. The formation of the metallacycles **9** from **8** presumably results from metallation of the cyclopentadienyl ring followed by a direct ring-closing, Scheme 1.

The metallacycles **9a**, **9b**, **9d** are thermally labile and upon evaporation of the solvent, they undergo ring-opening polymerization (*vide infra*). On the other hand, when R$_2$ = ^{n}Bu$_2$, the 1-sila-3-metallacycle **9c** is thermally stable and can be photochemically transformed into its phosphine substituted-metallacycle **10**, equation 4, which can be characterized by single crystal X-ray diffraction.

$$(4)$$

D Intramolecular alkylation.

Treatment of the platinum complex $(Ph_2MeP)_2Pt(Cl)CH(SiMe_3)_2$ with t-BuLi in toluene yielded 1-sila-3-metallaplatinacyclobutane, **11**, through intramolecular carbanion attack on the metal center[9], equation 5.

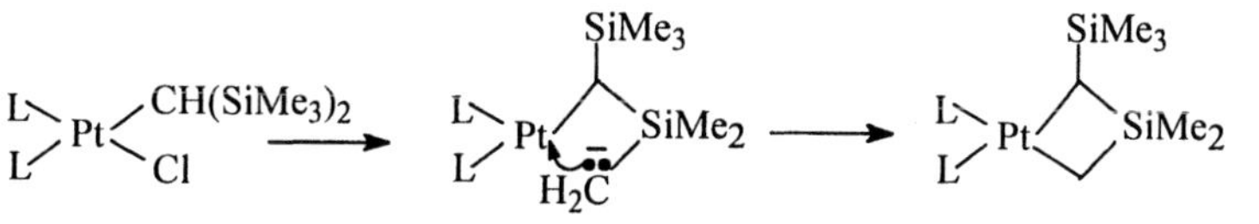

$$(5)$$

11

2. 1-sila-2-metallcyclobutanes.

A Addition of allenes.

The oxidative addition of arylallene by a bis-silyl-platinum complex led to a thermally stable 1-sila-2-platinacyclobutane[10] **12**, eq 6.

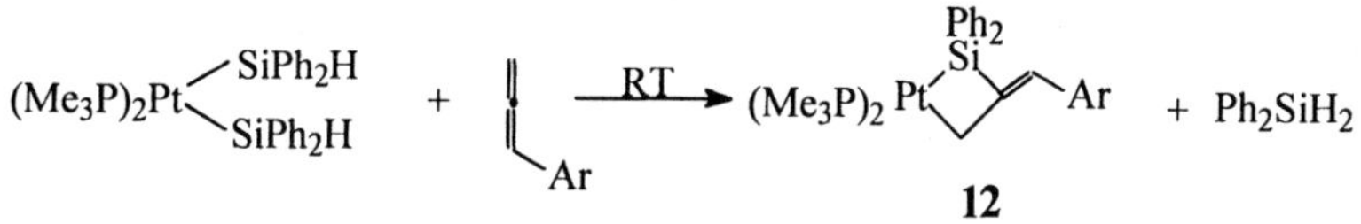

$$(6)$$

12

B Activation of arene C-H bonds.

In contrast to the well-known examples of aryl C-H bond activation by aryl phosphines, the C-H activation of arylsilyl ligands leading to cyclometallations are rare. For example, the 1-sila-2-platinacyclobutane[11] **13** was shown to be susceptible to traces of moisture and undergo hydrolysis of the Si-C bond to give ring-opened silanol, equation 7.

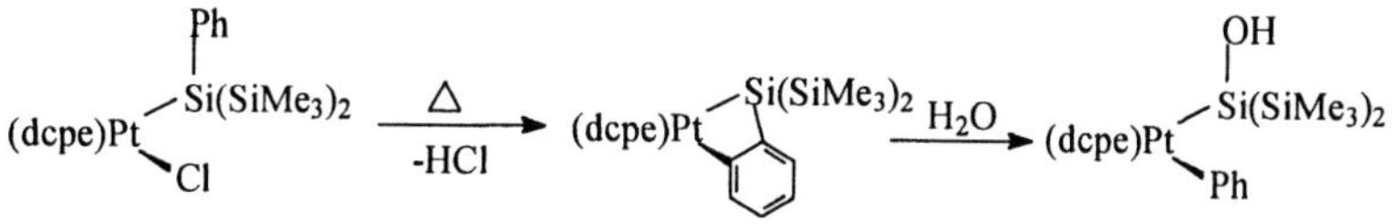

$$(7)$$

13
dcpe =bis(cyclohexylphosphino)ethane

The cationic iridium (III) complexes $[Cp^*(PMe_3)Ir(Me)]^+$ are reported to exhibit C-H activation reactions with alkanes under mild conditions. The reaction with Si-H bonds, e.g. Ph_3SiH, resulted in an oxidative addition/reductive elimination cycle, and a subsequent 1,2 phenyl shift from silicon to iridium followed by *ortho*-metallation produced an iridium (V) 1-sila-iridacycle[12] **14**, equation 8.

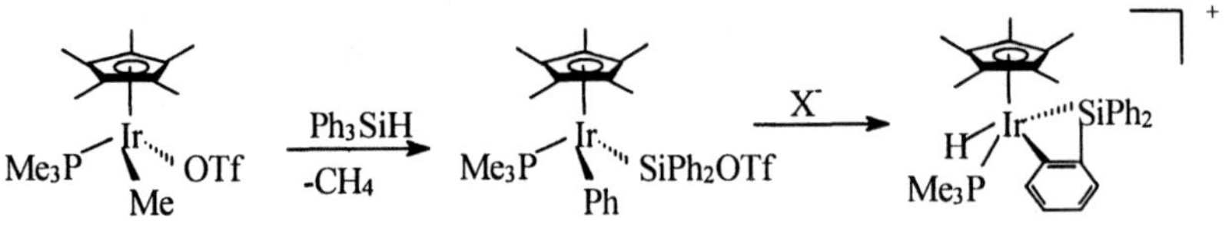

$$(8)$$

14

This 4-membered metallacycle **14** undergoes ring-opening in the presence of coordinating solvent CH_3CN to give an acyclic Ir (III) complex $[Cp^*(PMe_3)Ir(SiPh_3)CH_3CN]^+$ by reductive elimination of aryl C-H bond. This chemistry represents a very rare example of Ir (III) — Ir(V) — Ir (III) cycle.

Another example of a 1-sila-2-iridacyclobutane, $(Me_3P)_3Ir(H)(o\text{-}C_6H_4SiPh_2)$ **15**, was isolated from the oxidative addition of Ph_3SiH to the neutral Ir(III) silyl complex $(Me_3P)_3IrCH_3(H)(SiPh_3)$. The initially isolated Ir-silyl complex underwent loss of CH_4 to form the very reactive Ir (I) silyl complex, $(Me_3P)_3Ir(SiPh_3)$, which thermally cyclometallates.[13]

Synthesis of 1,2-disila-3-metallacyclobutanes.

There are two well-characterized examples of 1,2-disila-3 metallacyclobutanes of Fe, **16**[14], and Re, **17**[15]. Both these metallacycles involve base-mediated metal-to-cyclopentadienyl ring silatropic shifts which subsequently cyclize to disilametallacycles, equation 9.

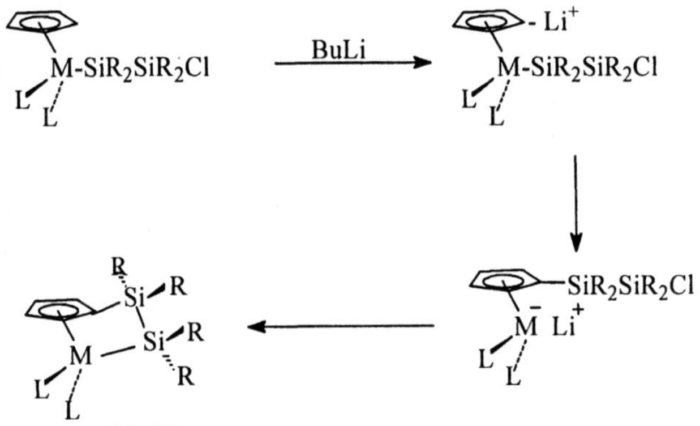

$$(9)$$

16, 17

($M = Fe$, $L = CO$, $R = {}^nBu$, **16**; $M = Re$, $L = NO$, PPh_3, $R = Me$, **17**)

Both disila-metallacycles are thermally stable and their chemistry is yet unexplored.

Structural features.

An examination of structural data in Table 1 reveals that the sila-metallacyclobutanes exhibit close to planar ring configurations and distorted ring bond angles. In the series of metallacycles **2**, **4**, **5**, where the metal is Zr, Nb, Mo, it was noted that the metal's electronic configuration exhibit a significant influence upon the structural features of the metallacyclic ring.[1,2] As one goes across in a period from Zr to Mo, a continuous increase in the M-C bond length from 2.240(5)Å to 2.301(5)Å has been observed. The increase in M-C bond length is accompanied by gradual reduction in C-M-C bond angle from 81.0(2)⁰ for M = Zr to 76.1(2)⁰ for Nb to 72.4(2)⁰ for Mo respectively. The reduction of C-M-C bond angles influences the C-Si-C bond angles that also undergo decrease accompanied by increase in M-C-Si bond angles. The ring strain introduced by the decrease of C-M-C bond angles is relieved by the puckering of the ring and subsequently the folding angles of ring are increased from 4.7⁰ for Zr to 14.3⁰ for Mo in these systems.

Spectroscopic features.

A special spectroscopic feature of the 1-sila-3-metallacycles is the unusual high field chemical shift of methylene carbon resonances, Table 2. Methylene carbon resonances for sila- metallacyclobutanes **1** and **2** are shifted ca. 12.9 and 20.5 ppm upfield from the corresponding metallacyclobutanes. Similarly, the methylene carbon resonances for metallacycles **5** at –34.5 and **9**, **10** at ca. –53 ppm are 14.5 ppm and 26 ppm upfield shifted respectively from the corresponding metallacyclobutanes and acyclic complexes of the type **8**.

Reactivity.

A Insertion of CO.

The insertion of CO into the metal-carbon bond represents one of the fundamental reaction in organometallic chemistry. This reaction is widely studied for metal-alkyl and metal-silyl complexes but there are relatively few examples in metallacyclic systems. The insertion of CO into Zr-C bond(s) of 1-sila-3-zirconacylclobutane **2** at room temperature proceeds with stepwise insertion of 2 equivalents of CO. The first insertion of CO is followed by 1,2-silyl shift then the insertion of second equivalent of CO again followed by 1,2-silyl shift to form six-membered zirconacyl dienolate with two exocyclic methylene groups **18**. However, if the reaction of **2** is carried with excess of CO at low temperature, reductive coupling of carbonyls occurs with the formation of entirely different di-insertion bicyclic compound **19**,[16] equation 10.

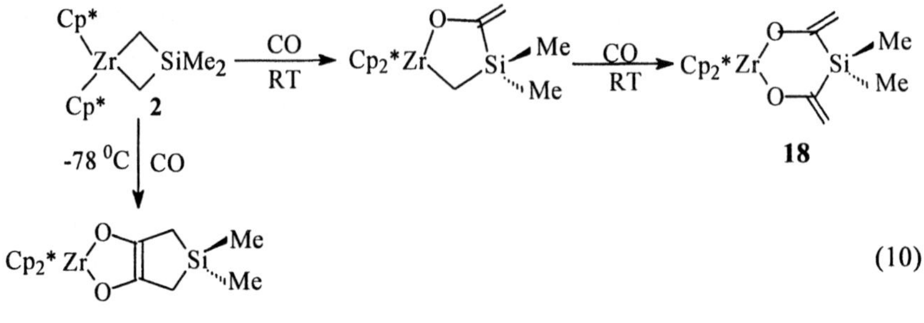

(10)

19

B Insertion of HCHO.

The insertion of formaldehyde into the Zr-C bond of 1-sila-3-zirconacylclobutane **2** at 70 ^{0}C leads to ring expansion with formation of a 6-membered ziconacyclic ring **20**,[17] equation 11.

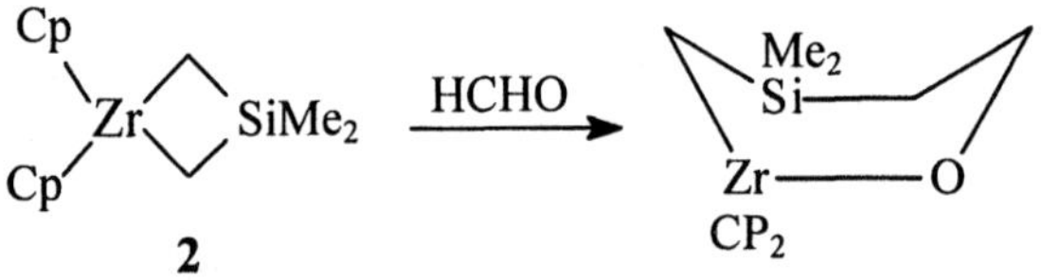

(11)

20

1-Oxa-4-sila-6-zirconacyclohexane, **20** adopts a distorted boat conformation with short Zr-O bond of 1.941(2) Å and obtuse Zr-O-C angle of 143.1(2)^0. The Zr-O bond is about 0.21 Å shorter than a Zr-O single bond of 2.15-2.20 Å. These structural features indicate some degree of multiple bond character between Zr and oxygen which probably arises from the π-donation from formaldehyde into the empty d orbital of Zr.

C Insertion of isocyanide.

The steric effects of the alkyl groups significantly influence the insertion chemistry of the alkyl isocyanides. The reaction of alkyl isocyanides, RNC, R = CH$_3$, to 1-sila-3-metallacyclobutanes **2** and **3** follows essentially identical pathway as previously observed for CO. The addition of 2 equivalents of alkyl isocyanide to metallasilacyclobutanes undergo stepwise addition of isocyanide to

M-C bond(s) followed by 1,2-silyl shift to produce six-membered cycle dienamide with two exocyclic CH_2 groups as the final product 21, equation 12.

$$(12)$$

The reaction at low temperature with excess of alkyl isocyanide with 2 produces the bicyclic enediamide 22 by the reductive coupling of isocyanide followed by addition across the metal-carbon bonds. However, when the reaction of 2 is carried with bulky alkyl isocyanides, for example where R = *tert*-butyl or 2,6-xylyl, a different chemistry is observed. Nucleophilic attack by isocyanide at the electrophilic metal center produces partial insertion of isocyanide in the M-C bond to afford five-membered η^2-iminoacyl complex 23 without 1,2-silyl shifts in which nitrogen is donating an electron pair to electrophilic metal center forming an additional three-membered ring comprising MCN. The three-membered ring MCN, where M =Zr, has an acute N-Zr-C bond angle of 33.2 $(1)^0$ and a Zr-N bond of 2.227 (3) Å which is longer than the Zr-C bond of 2.164 (4) Å. The addition of one more equivalents of isocyanide to η^2-iminoacyl complex 23 results in the ring expansion due to the reductive coupling of isocyanides to form unusual six-membered η^2-iminoacyl complex 24 in which the three-membered ring composed of MCN remains intact. The angle N-Zr-C in the three-membered ring of Zr, C and N remains essentially unperturbed (33.1 $(1)^0$) with slightly elongation of the Zr-N bond to 2.231 (3) Å. The latter complex 24 on heating at 120 ^{0}C rearranges into a bicyclic metallacyle 25,[4,18-21] essentially identical to the bicyclic hetrocycle, 22, equation 13.

$$(13)$$

D Insertion of silanes.

The 1-sila-2-platinacyclobutane **12** undergoes ring expansion by oxidative – addition of Ph_2SiH_2 into Pt–C bond forming the disilaplatinacyclopentane **26**,[10] equation 14.

$$(14)$$

E Ring -opening by weak acids: Generation of cations.

The protonation of a 1-sila-3-metallacyclic rings **1** and **2** with a weak acid like trialkylammonium cation, R_3NH^+ containing a poor donating conjugate base provides a convenient method for generating metallocene–trimethylsilylmethyl cations **27**,[22] equation 15.

$$(15)$$

The cations have been characterized by spectroscopy and X-ray diffraction studies. The X-ray structure of the cation $[Cp_2^*Zr-CH_2SiMe_3.THF]^+$ **27** ($Cp^* = \eta^5-C_5Me_5$), reveals a weakly coordinated THF molecule to the coordinatively unsaturated metal center with a relatively long Zr-O bond of 2.243 (3) Å in comparison with a typical Zr-O bond length of 2.15-2.20 Å. Attempts to isolate solvent-free cations resulted in decomposition. The cations behave as ethylene polymerization catalysts under mild conditions.

Ring-opening polymerization of sila-metallacyclobutanes.

The parent silacyclobutanes have the capacity to undergo ring-opening polymerization due to the ring strain while, presumably due to the larger size of the transition metal, the early transition metal silametallacyclobutanes do not exhibit such chemistry. However, the recently obtained 1-sila-3-ferrametallacycles, **9**, are thermally labile and indeed even at room temperature after removal of the solvent, they produce ring-opened polymers **28** in almost quantitative yield[8], equation 16.

R = Me$_2$ (a); Me^{n}Bu(b); MePh(d)

The resulting yellow-brown polymers can be readily dissolved in THF and precipitated into hexanes; however, **28d** polymer became insoluble after precipitation into hexane.

The most significant spectral feature for the polymers is the low field shift of the methylene carbon resonance from ~ –53 ppm for metallacycles **9** to more conventional value of ~ –27 ppm for polymers **28**, a value comparable with (η^5-C$_5$H$_5$)Fe(CO)$_2$-CH$_2$R complexes. The silicon resonance for the polymer **28a** is also shifted to lower field, from –5.6 ppm in the metallacycle **9a** to 4.9 ppm in the polymer **28a**. Wide angle laser light scattering analysis of the **28a** polymer revealed a bimodal molecular weight distribution with a predominant high molecular weight fraction; (M$_w$ = 1. 75 × 10^5, M$_n$ = 7.2 × 10^4). The molecular weight for **28b** polymer was found to be relatively low, M$_w$ = 9500, M$_n$ = 1500 with polydispersity of 6.3 suggesting that the introduction of the butyl group significantly retards the propagation of a polymer chain. The polymers are amorphous as determined by WAXS.

The polymers are stable to the atmosphere and films can be cast from THF solutions. Over a period of time their solubility is decreased. The glass transition temperature for **28a** was detected to be 3 ^{0}C by DSC.

Thermogravimetric analysis (TGA) of both **28a,b** indicated that they undergo ~45 % weight loss between 100-200 ^{0}C and that a 25 % residue remains at 700 ^{0}C.

Ring-opening polymerization of disila- metallacyclobutanes.

Disilacyclobutanes are attractive candidates for ring-opened polymers with intact Si-Si bonds. Polymers containing unsaturated or aromatic groups bridged by disilyl units are photoactive and show conducting properties when appropriately doped. Based upon the ring-opening of the 1-sila-3-ferracyclobutanes **9** discussed above, related 1,2-disila-3-ferracyclobutanes **16** are promising candidates for ring-opening and the formation of molecular wire materials. Indeed, recently we synthesized the first such disila-metallapolymer **29** by the base treatment of $(\eta^5\text{-}C_5H_5)Fe(CO)_2SiMe_2SiMe_2Cl$, with LDA in THF at -5 ^{0}C. The reaction resulted in direct and quantitative formation of the polymer **29**,[14] eq 17.

Presumably, the polymer formation occurs through the initial formation of four-membered disila-metallacycle which is thermally labile and undergoes ring–opening polymerization; however, we observed no evidence for such a species. Light brown polymer **29** was precipitated directly from THF solution but is soluble in hot THF. Cross-polarization solid state ^{29}Si NMR of the polymer **29** exhibits two narrow resonances at 22.0 ppm assigned to Fe-Si and at –8.2 ppm due to (Cp-Si) indicating the regular alternating arrangement of the building blocks in the polymer backbone and the structure is essentially retained in the solution as indicated by the solution NMR spectroscopy. Wide angle laser light scattering analysis of the polymer revealed molecular weight M_w = 6400, M_n = 3200 with polydispersity of 2.0. Thin films can be cast from the THF solution of the polymer. The morphology of the polymer was investigated by WAXS; the scattering pattern indicated distinct 2θ peaks at 13.3^0 and 14.2^0 which confirms that the polymer was partially crystalline. Thermogravimetric analysis (TGA) of the polymer under N_2 atmosphere showed that the polymer underwent two successive weight losses, 6 % between 50 –75 ^{0}C and 16 % between 200-300 ^{0}C and about 50 % residue remained at 500 ^{0}C.

Table 1. Structural data for sila-metallacyclobutanes

Comp	< M (^{0})^a	< C (^{0})	< Si (^{0})	M-C bond length (Å)	M-Si bond length (Å)	Folding angle (^{0})	Ref
1	A 84.1 (2)	87.1 (1)	101.0 (2)	2.146 (3)		7.7	1
	B 83.7 (2)	87.0 (2)	101.6 (2)	2.169 (4)		0	
		87.7 (2)	102.2 (3)	2.152 (5)			
2	81.0 (2)	88.3 (2)	102.2 (3)	2.240 (5)		4.7	2
4	76.1 (2)	92.1 (1)	98.7 (2)	2.275 (3)		10.4	2
5	72.4 (2)	94.9 (2)	95.8 (3)	2.301		14.3	2
6	77.3 (4)	92.4 (4)	94.5 (5)	2.158 (10)		18.9 (6)	7
		92.5 (4)		2.160 (11)			
10	80.04 (12)	92.45 (11)		2.116 (3)		5.9	8
		93.20 (13)					
12	68.1 (3)	103.2 (7)	85.3 (3)	2.14 (1)	2.367 (3)		10
14	64.2 (2)	113.4 (5)	85.3 (3)	2.086 (8)	2.479 (2)		12
15	66.9 (3)	108.1 (7)	85.1 (3)	2.16 (1)	2.404 (3)		13
30^b	A 77.0 (4)	91.9 (3), 92.2 (3)	98.9 (4)	2.241 (9) 2.253 (9)		1.6 (3)	5
	B 77.0 (4)	92.2 (3) 92.3 (3)	98.5 (4)	2.236 (9) 2.243 (9)		0.5 (3)	

^{a}Endocyclic angles of silacyclobutane ring. A and B refers to two molecules in asymmetric unit. ^{b}1-sila-3-osmacyclobutane

Table 2. Methylene carbon resonances for sila-metallacycles

Compd	^{13}C CH$_2$ resonance (ppm)	Ref
1	70.5	1
2	46.0	2
3	48.3 ($^1J_{CH}$ = 123 Hz)	20
5	-34.6	2
9, 10	~ -53	8

The characterizable disila-ferracyclobutanes, **16** obtained by treatment of $FpSiMe_2Si^nBu_2Cl$ or $FpSiBu_2SiBu_2Cl$, with LDA has thus far resisted attempts at ring-opening polymerization[14].

Acknowledgements

This research has been supported by the NIH-MARC and SCORE programs. We also thank Dr. K. Rahimian from Sandia National Laboratories, Albuquerque, NM for molecular weight determinations and DSC and TGA measurements.

References

1. Tikkanen, W. R.; Liu, J. Z.; Egan, J. W.; Petersen, J. L., *Organometallics* **1984**, *3*, 825.
2. Tikkanen, W. R.; Liu, J. Z.; Egan, J. W.; Petersen, J. L., *Organometallics* **198**, *3*, 1646.
3. Kabi-Satpathy, A.; Bajgur, C.S.; Reddy, K.P.; Petersen, J.L., *J. Organomet. Chem.* **198**, *364*, 105.
4. Kloppenburg, L.; Petersen, J.L., *Polyhedron* **199**, *14*, 69.
5. Behling, T.; Girolami, G. S.; Wilkinson, G.; Somerville, R. G.; Hursthouse, M. B., *J. Chem. Soc., Dalton Trans.* **198**, 877.
6. Andersen, R.A.; Jones, R.A.; Wilkinson, G., *J. Chem. Soc., Dalton Trans.* **197**, 446.
7. Andreucci, L.; Diversi, P.; Ingrosso, G.; Lucherini, A.; Marchetti, F.; Adovasio, V.; Nardelli, M., *J. Chem. Soc., Dalton Trans.* **198**, 477.
8. Sharma, H.K.; Lee-Cervantes, F.; Pannell, K. H., *J. Amer. Chem. Soc.* **2004**, *126*, 1326.
9. Suzuki, H.; Tsukui, T.; Moro-oka, Y., *J. Organomet. Chem.* **1986**, *299*, C35.
10. (a) Tanabe, M.; Yamazawa, H.; Osakada, K., *Organometallics* **2001**, *20*, 4453. (b) Silaplatinacyclobutene was recently isolated and characterized by X-ray crystallography. See: Tanbe, M.; Osakada, K., *J. Amer. Chem. Soc.* **2002**, *124*, 4550.
11. Change, L. S.; Johnson, M. P.; Fink, M., *Organometallics* **1991**, *10*, 1219.
12. (a) Klei, S.R.; Tilley, T.D.; Bergman, R.G., *J. Amer. Chem. Soc.* **2000**, *122*, 1816. (b) Klei, S.R.; Tilley, T.D.; Bergman, R.G., *Organometallics* **2002**, *21*, 3376.
13. (a) Aizenberg, M.; Milstein, D., *Angew. Chem. Int. Ed.* **1994**, *33*, 317. (b) Aizenberg, M.; Milstein, D., *J. Amer. Chem. Soc.* **1995**, *117*, 6456.
14. Sharma, H.K.; Pannell, K.H., *Chem. Commun* **2004**, 2556.

15. Crocco, G. L.; Young, C. S.; Lee, K. E.; Gladysz, J. A., *Organometallics* **1988**, *7*, 2158.
16. Peterson, J.L.; Egan, J.W., *Organometallics* **1987**, *6*, 2007.
17. Tikkanen, W.R.; Petersen, J. L., *Organometallics* **1984**, *3*, 1651.
18. Berg, F. J.; Petersen, J. L., *Organometallics* **1989**, *8*, 2461.
19. Berg, F. J.; Petersen, J. L., *Organometallics* **1991**, *10*, 1599.
20. Berg, F. J.; Petersen, J. L., *Organometallics* **1993**, *12*, 3890.
21. Valero, C.; Grehl, D.; Wingbermuhle, D.; Kloppenburg, L.; Carpenetti, D.; Erker, G.; Petersen, J. L., *Organometallics* **1994**, *13*, 415.
22. Amorose, D. M.; Lee, R. A.; Petersen, J. L., *Organometallics* **1991**, *10*, 2191.

Chapter 33

Luminescence Properties of Phosphine–Isocyanide Cu(I)- and Ag(I)-Containing Oligomers in the Solid State

Pierre D. Harvey and Éric Fournier

Département de Chimie, Université de Sherbrooke,
Sherbrooke, Québec J1K 2R1, Canada

The luminescence properties of the binuclear complexes $M_2(dmpm)_3^{2+}$ (M = Cu, Ag; dmpm = *bis*(dimethylphosphino)-methane), and $Cu_2(dmpm)_3(CN\text{-}t\text{-}Bu)_2^{2+}$ (as BF_4^- salts), as well as the oligomers described as $\{Cu_2(dmpm)_3(dmb)_{1.33}^{2+}\}_3$ and $\{Ag_2(dmpm)_2(dmb)_{1.33}^{2+}\}_3$ (dmb = 1,8-diisocyano-p-menthane), were investigated and compared to the well-known $\{M(dmb)_2^+\}_n$ polymers (M = Cu, Ag). These compounds exhibit emission maxima ranging from 445 to 485 nm with emission lifetimes found in the μs regime in the solid state at 298 K. The time-resolved emission spectra for the oligomers and polymers exhibit blue-shifted emission bands at the early stage of the photophysical event after the excitation pulse, which red-shift with delay times. The decay traces are non-exponential, and their analysis according to the Exponential Series Method (ESM) exhibit a distribution of lifetimes that is fairly broad, consistent with an exciton phenomenon. A qualitative correlation between the number of units and the distribution width is reported.

Background

The syntheses and the applications of metal-containing polymers where the metal is located in the backbone are becoming the subject of more and more recent research projects.[1] While these polymers are built with various assembling ligands, the nature of the M-L coordinations is largely dominated by M-N and M-C bonding. On the other hand, organometallic/coordination polymers built upon diphosphines and diisocyanides are more rare.[1] Among the sought after properties for these new and original materials, luminescence has become more predominant,[2-11] since these materials may find applications for digital displays, light-emitting diodes, and sensors.

This group recently reported a series of works on luminescent Cu(I) and Ag(I)-containing polymers built with assembling diisocyanides and diphosphines, as well as mixed-ligand species.[12-20] Examples include the doubly bridged homo- and mixed-ligand polymers $\{M(dmb)_2^+\}_n$[16-20] and $\{M(dmb)(dppm)^+\}_n$[13] (M = Cu, Ag; dmb = 1,8-diisocyano-p-menthane; dppm = bis(diphenylphosphino)methane), and the recently reported oligomers $\{Cu_2(dmpm)_3(dmb)_{1.33}^{2+}\}_3$ and $\{Ag_2(dmpm)_2(dmb)_{1.33}^{2+}\}_3$ (dmpm = bis(diphenylphosphino)methane; Schemes 1 and 2).[12]

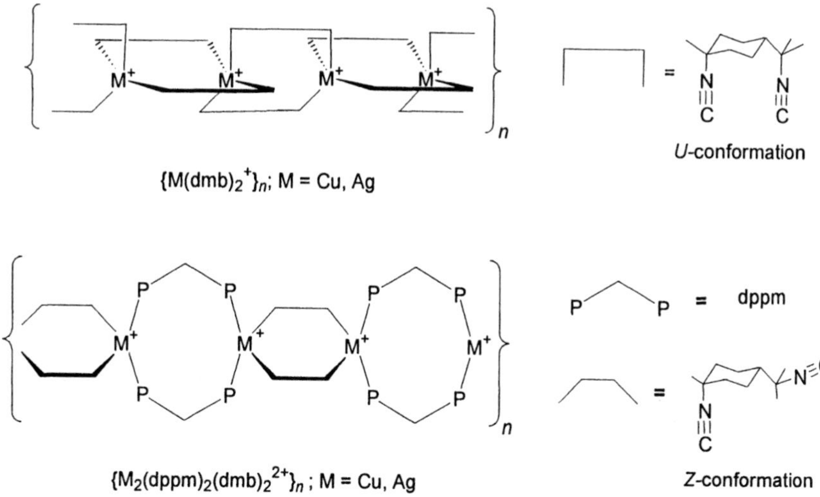

$\{M(dmb)_2^+\}_n$; M = Cu, Ag

U-conformation

$\{M_2(dppm)_2(dmb)_2^{2+}\}_n$; M = Cu, Ag

P⌒P = dppm

Z-conformation

Scheme 1

One important particularity is that the various $M(CNR)_4^+$, $M(CNR)_2P_2^+$, $Cu(CNR)_3P^+$ and $Ag(CNR)P_2^+$ chromophores (M = Cu, Ag; P = phosphine) are separated by saturated chains such as methylene groups, included within the assembling ligands. Consequently, no electron delocalisation is possible between the chromophores, and so interactions or "communications" between these, if any, must occurs through space. During the earlier investigations, the presence of an intra-chain excitonic process (Scheme 3) was reported for the $\{M(dmb)_2^+\}_n$ polymers (M = Cu, Ag).[19] While the exciton phenomena is well known for organic materials,[21-24] it is less so for organometallic and coordination polymers due, in part, to the recent development of this field.

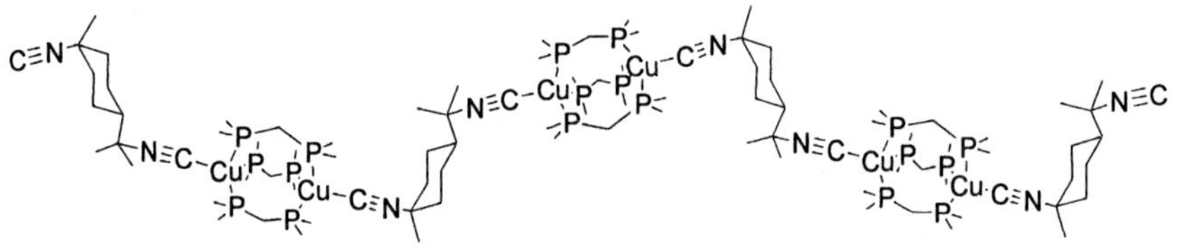

$\{Cu_2(dmpm)_3(dmb)_{1.33}^{2+}\}_3$

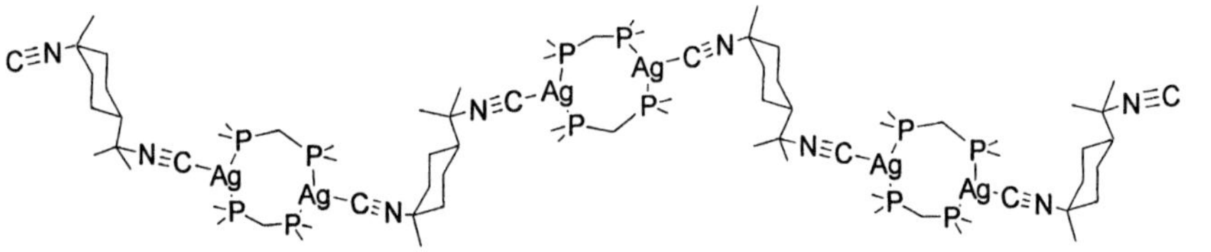

$\{Ag_2(dmpm)_2(dmb)_{1.33}^{2+}\}_3$

Scheme 2

Numerous striking features were apparent for the $\{M(dmb)_2^+\}_n$ polymers. For example, the luminescence decay traces were non-exponential for all investigated media (solid state, crystal, solution), the decay traces were also super-imposable for both solution and solid state (indicating an intra-chain phenomenon), and the emission light was depolarized. Finally and more importantly, the time-resolved emission spectroscopy exhibited a continuous red-shift of the luminescence band with delay time after the excitation pulse, all strongly contrasting with the typical linear logarithm decay traces, polarized emission light and uniqueness of the emission maximum with delay times for non-interacting chromophores. Between these two extreme series of features, there is no data, so one cannot tell how these properties change with the polymer length.

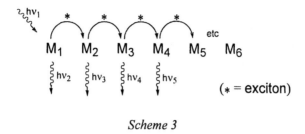

Scheme 3

Objective

The objective of this study was to provide some information on a possible correlation between the chain length and the photophysical properties. In this respect, the newly synthesized coordination oligomers $\{Cu_2(dmpm)_3(dmb)_{1.33}^{2+}\}_3$ and $\{Ag_2(dmpm)_2(dmb)_{1.33}^{2+}\}_3$ were investigated in some detail in the solid state. The solid state was selected to avoid dissociation phenomena in solution often encountered in d^{10} metallic species.

Experimental

Materials. $[Cu_2(dmpm)_3](BF_4)_2$ (**1**), $[Ag_2(dmpm)_3](BF_4)_2$ (**2**), $[Cu_2(dmpm)_3-(CN-t-Bu)_2](BF_4)_2$ (**3**), $\{[Cu_2(dmpm)_3(dmb)_{1.33}](BF_4)_2\}_3$ (**4**), $\{[Ag_2(dmpm)_2-(dmb)_{1.33}](BF_4)_2\}_3$ (**5**) and $\{[M(dmb)_2]BF_4\}$ (M = Cu (**6**, crystalline form), Ag (**7**)) were synthesized according to literature procedures.[12,19]

Apparatus: The continuous wave emission and excitation spectra were obtained using a SPEX Fluorolog II spectrometer. The emission lifetimes were measured with a nanosecond N_2 laser system from PTI model GL-3300. The time-resolved emission spectra were acquired on the same instrument used for the lifetime measurements. The excitation wavelength was 311 nm for all experiments.
Procedures. The average molecular weight in number (M_n) of the crystalline $\{[Cu(dmb)_2]BF_4\}_n$ polymer was obtained from the measurements of the intrinsic viscosity using polymethyl methacrylate standards from Aldrich (M_n = 12000,

15000, 120000, and 320000). The evaluated M_n is 24000 (i.e. about 45 units). The emission lifetimes were analyzed using the ESM (Exponential Series Method),[25,26] which consisted of calculating a decay curve composed of 200 exponentials. The results were checked to ensure that they did not depend on input parameters. Typically, the results were presented as a distribution of lifetimes. For single exponential decays, the traditional deconvolution (1-4 components) and ESM methods gave the same results, and the distribution of lifetimes was narrow. For data giving a large distribution with ESM, the deconvolution method failed to find satisfactory fits with 2, 3, or 4 decays. The quality of the fit between the experimental and calculated curves was addressed using the parameter χ (goodness of fit), which approached the diagnostic value of 1 for an acceptable fit for all reported data, and from the analysis of the residual.

Results & Discussion

The photophysical data of the two targeted oligomers **4** and **5** are compared with the binuclear complexes $[M_2(dmpm)_3](BF_4)_2$ (M = Cu (**1**), Ag (**2**)) and $[M_2(dmpm)_3(CN-t-Bu)_2](BF_4)_2$ (**3**; Scheme 4), and polymers $\{[M(dmb)_2]BF_4\}_n$ (M = Cu (**6**), Ag (**7**)). The nature of the excited states of the closely related lumophores $M(CNR)_4^+$,[19] $Cu_2(dppm)_2(O_2CCH_3)^+$,[27] MP_4^+ (P = phosphine),[28,29] and $M_2(dmpm)_3^{2+}$ (M = Cu, Ag)[30-35] was established to be triplet metal-to-ligand-charge-transfer (MLCT) for the first two, and metal-ligand $[t_2(d\sigma^*)]^5[t_2(\pi^*,p\sigma^*)]^1$ and metal-metal $d\sigma^*p\sigma$ for the last two, respectively, based upon experiments and DFT calculations.

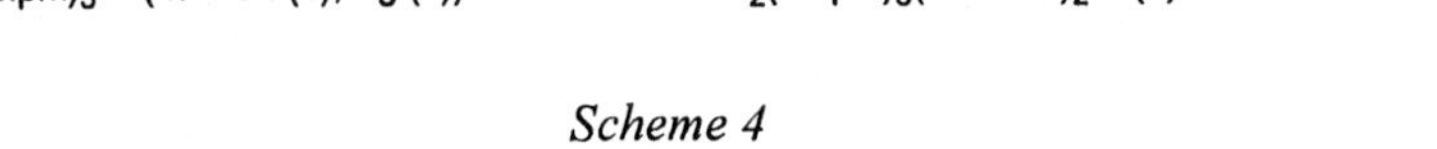

$M_2(dmpm)_3^{2+}$ (M = Cu (**1**), Ag (**2**)) $Cu_2(dmpm)_3(CN-t-Bu)_2^{2+}$ (**3**)

Scheme 4

As stated, the photophysical properties of the $\{M(dmb)_2^+\}_n$ polymers were previously reported by us, and were compared to the tetrahedral mononuclear model complexes $M(CN-t-Bu)_4^+$ (M = Cu, Ag).[19] The spectroscopic and photophysical properties were found to be drastically different between the mononuclear species and polymers. The key features are as follows: the λ_{max} for the $M(CN\text{-}t\text{-}Bu)_4^+$ species are blue-shifted with respect to the corresponding polymers (up to 40 nm), and the fwhm of the emission bands are smaller as well under continuous wave excitation. The decay traces are rigorously mono-exponential for the mononuclear complexes, while they are non-exponential in the polymers. Time-resolved emission spectra indicate that at the early event after the light pulse, both λ_{max} and slope of the emission decay traces (equivalent of a lifetime) compare favorably with those of the corresponding mononuclear chromophores $M(CN\text{-}t\text{-}Bu)_4^+$ (M = Cu, Ag). At longer delay times, the recorded emission bands are significantly red-shifted (up to 65 nm). Finally, another notable difference is that the emission arising from the polymers is depolarized. All these features are typical of energy transfer exciton phenomena (Scheme 3). One of the interesting features is that the decay traces in the polymers are found to be independent of the medium (solution vs solid), indicating that the process is primarily intramolecular.

This latter feature is related to the long intermolecular N···N distances (8.25 and 8.67 Å)[16-20] which indicate that the chains are relatively isolated from one another in the solid state, while the intramolecular N···N separations are in the order of 4.5 Å (dmb in its U-conformation). Despite the fact that no X-ray data is available for **4** and **5**, it is possible to anticipate what the approximate interchromophore distances in the solid state are, by examining the X-ray data for the building blocks **1** and **2**. For these species, the closest intermolecular P···P distances are 6.683 Å for **1**, and 6.328, 6.585 and 6.982 Å for **2**.[12] These distances are greater than the intramolecular N···N separation in the linking dmb ligand (Z-conformation) in the recently reported computed model compound $Cu_6(dmpm)_9(dmb)_2(CN\text{-}t\text{-}Bu)_4^{6+}$ (~5.8 Å; PC-Model; Figure 1)[12] and in the polymer $\{[Pd_4(dmb)_4(dmb)]^{2+}\}_n$ (5.549 Å; X-ray).[36] If one accepts that the rate for exciton hopping (or energy transfer; k_{ET}) varies as $k_{ET} \propto 1/r^6$ (r = interchromophore distance; for a i.e. Förster mechanism),[37] then the contribution of the intermolecular process is minor.

Figure 2 shows the time resolved emission spectra (20-2000 µs) for **4** and **5** in the solid state. **4** and **5** exhibit blue emission maxima at 482 and 447 nm, respectively, when submitted to continuous wave excitation. At the early event after the light pulse, the recorded emission band is blue-shifted with respect to the emission band measured in continuous wave mode. As the delay time increases, the observed emission band red-shifts constantly and the intensity decreases. All in all, the emission bands measured with continuous wave light are composed of a number of blue- and red-shifted components. The maximum band shifts are ~10 and ~34 nm for **4** and **5**, respectively, which are smaller than those observed for the longer $\{M(dmb)_2^+\}_n$ polymers (up to 50 nm).[19] The 298 K decay traces are found to be linear for **1-3** with emission lifetimes of 251 (472),

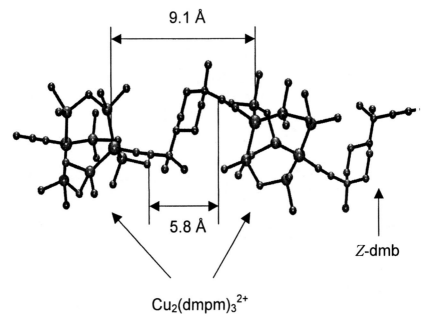

Figure 1. Representation of a fragment of the $\{[Cu_2(dmpm)_3(dmb)_{1.33}](BF_4)_2\}_3$ oligomer (4), where the computed intrachain interchromophore distances are indicated (as 5.8 and 9.1 Å are for the N···N and Cu···Cu separations, respectively).

41 (445) and 291 µs (476 nm), respectively. These results indicate for the first time that qualitatively the amplitude of red-shift increases as the polymer length increases. The shorter lifetimes normally encountered for the Ag species with respect to the Cu homologues is due to the larger spin-orbit coupling of the heavier element, commonly called "heavy atom effect".[37]

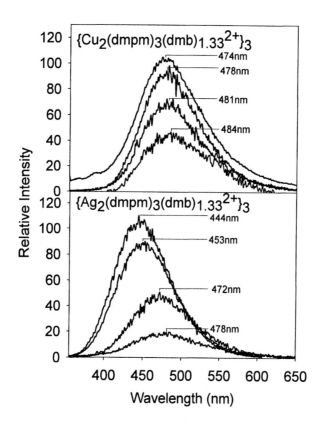

Figure 2. Time resolved emission spectra for 4 and 5 in the solid state at 298 K. The measurements were made in the following time frames: for 4: 474 nm, 20-70; 478, 500-600; 481, 1000-1300; 484; 2000-2500µs; for 5: 444 nm, 20-70; 453, 300-400; 472, 500-600; 478, 1000-1300 µs.

As stated, the decay traces are non-exponential for 4-7. A typical example is shown in Figure 3, where a straight line is observed for the model compound 3 and a curve is measured for the polymer 6 in the log plot of the emission intensity vs delay time after the excitation pulse.

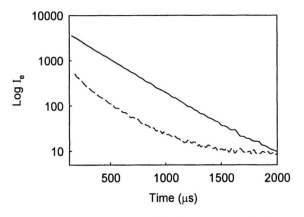

Figure 3. Solid state decay trace for the emission of 3 (——) versus
6 (----) at 298K.

In these cases, the data are analyzed using the ESM, and the results are plotted as population distribution (or relative amplitude) vs lifetimes in Figures 4 and 5. The maximum of probability represents an average lifetime or the most probable lifetime, and these are 257 (**4**) and 31 μs (**5**). The width of the distribution is related to the curvature of the decay traces (log plot); as the width increases, the curvature increases. The data are not dependent on the excitation intensity (using neutral density filters), indicating that local heating has little or no effect on the results.

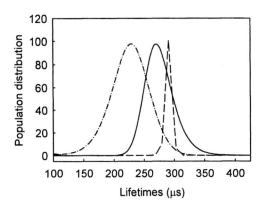

Figure 4. Comparison of the distribution of lifetimes as a function of lifetime
fitting the emission decay traces for 1 (---), 4 (—) and 6 (- - -) in the solid
state at 298 K.

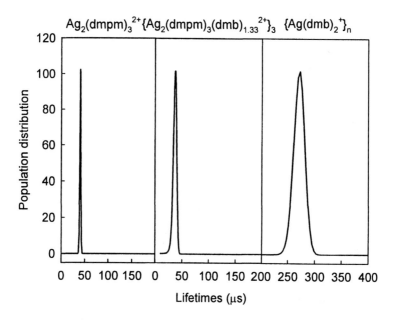

*Figure 5. Comparison of the distribution of lifetimes as a function of lifetimes fitting emission decay traces for **2**, **5** and **7** in the solid state at 298 K.*

However one interesting question arises. The number of units is 1, 3 and ~ 45 (here as evaluated by the measurement of the intrinsic viscosity), for **1**, **4** and **6**, and 1, 3, and "very large" for **2**, **5**, and **7**, respectively. The width of the population distribution plots in Figures 4 and 5 does not follow the trend proportionally, particularly for **4** and **6** as a large change in the number of units should be accompagnied by a large change in width. This observation cannot be explained straightforwardly. First, one has to consider that the exciton process is reversible (Scheme 5). So a small chain can exhibit the same spectroscopic feature as a long chain depending on the efficiency of exciton hopping.

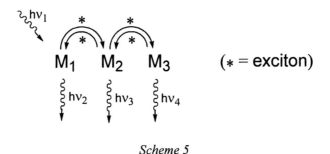

Scheme 5

Secondly, it is not necessarily true that the extent of exciton migration is as extensive for one chromophore vs another, simply because the atomic distribution of the HOMO and LUMO is not identical for all chromophores. Therefore, the probability of energy transfer cannot be the same. In addition, the distance between the chromophores measured as $MC \equiv N \cdots N \equiv CM$ or $M \cdots M$ or $P \cdots P$ are different as well.

One interesting and important remark concerning this point is the comparison of the photophysical properties of the Cu(I) and Ag(I) species described above with Pd-Pd and Pt-Pt bond-containing oligomers and polymers also built with dmb and diphosphine bridging ligands, such as $\{Pd_2(dmb)_2(diphos)^{2+}\}_n$, $\{Pd_4(dmb)_4(dmb)^{2+}\}_n$ and $\{Pt_4(dmb)_4(diphos)^{2+}\}_n$ (Scheme 5).[36,38,39] While the former polymers exhibit non-exponetial decay traces and depolarized broad emissions, the latter ones exhibit mono-exponential luminescence decays, indicating the presence of non-interacting chromopohores within the chain, and polarized emissions, strongly suggesting that the exciton process is either very weak or absent in these cases. A close examination of the atomic contributions of the HOMO and LUMO reveals that a large contribution of the C and N p orbitals (π and π^* orbitals) are computed for the Cu(I) and Ag(I) chromophores,[19] while these are minor or non-existent for the Pd and Pt species.[39] In other words, the electronic density is more spread out in the former two series, and more localized around the M-M bond in the latter compounds. In this respect, the distance for energy transfer is greater for the Pd-Pd- and Pt-Pt-containing chromophores.

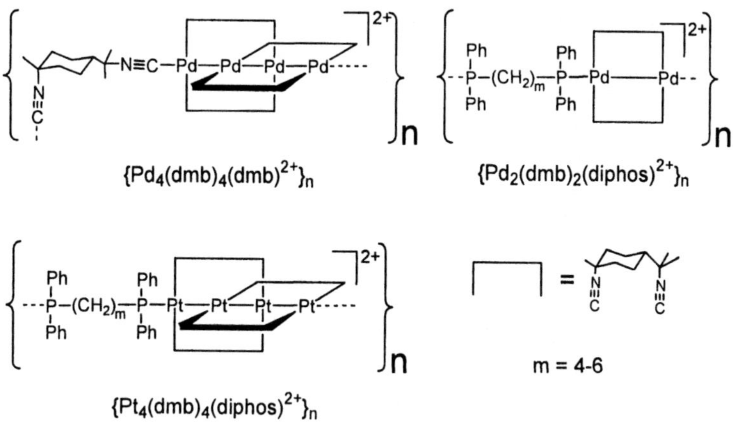

$\{Pd_4(dmb)_4(dmb)^{2+}\}_n$

$\{Pd_2(dmb)_2(diphos)^{2+}\}_n$

$\{Pt_4(dmb)_4(diphos)^{2+}\}_n$

$m = 4{-}6$

Scheme 6

Conclusion

It is arguable that the investigated systems are not ideal for rigourous analysis to fine probe the exciton migration across the chain, but the tight control of chain length for coordination polymers is not yet achieved. This work has, however, unquestionably demonstrated that a qualitative relationship between the number of chromophores in the oligomer/polymer chain and the extent of exciton migration measured as the width of the population distribution of emission lifetimes, or the curvature of the decay traces in the log scale, exists. Further research in this area are in progress, including the use of the $\{M(dmpm)(dmb)\}_n$ polymers (M = Cu, Ag) which represent other suitable materials for comparison purposes with the materials described in this work.

Acknowledgment. This research was supported by the Natural Sciences and Engineering Research Council of Canada (NSERC). PDH also thanks the students that contributed to the various aspects of the research on organometallic/coordination polymers over the years. Their names are listed within the references below.

References

1. *Macromolecule Containing Metal and Metal-like Elements*, Eds. Abd-El-Aziz, A.S., Carraher, C. E., Jr., Pittman, C. U., Jr., and Zeldin, M., eds. Wiley Intersceince, John Wiley and Sons Inc., New York, Vol. 5, 2004.
2. Wu, C.-D.; Ngo, H. L.; Lin, W., *Chem. Comm.* **2004**, 1588.
3. Dong, Y.-B.; Wang, P.; Huang, R.-Q.; Smith, M. D. *Inorg. Chem.* **2004**, *43*, 4727.
4. Fernandez, E. J.; Lopez-de-Luzuriaga, J. M.; Monge, M.; Montiel, M.; Olmos, M. E.; Perez, J.; Laguna, A.; Mendizabal, F.; Mohamed, A. A.; Fackler, J. P., Jr. *Inorg. Chem.* **2004**, *43*, 3573.
5. Miller, T. A.; Jeffery, J. C.; Ward, M. D.; Adams, H.; Pope, S. J. A.; Faulkner, S. *Dalton Trans.* **2004**, 1524.
6. Song, J.-L.; Zhao, H.-H.; Mao, J.-G.; Dunbar, K. R. *Chem. Materials*, **2004**, *16*, 1884.
7. Fleming, C. N.; Jang, P.; Meyer, T. J.; Papanikolas, J. M. *J. Phys. Chem. B* **2004**, *108*, 2205.
8. Dong, Y.-B.; Zhao, X.; Tang, B.; Wang, H.-Y.; Huang, R.-Q.; Smith, M. D.; zur Loye, H.-C. *Chem. Comm.* **2004**, 220.
9. Song, D.; Wang, S. *Eur. J. Inorg. Chem.* **2003**, 3774.
10. Dong, W.; Zhu, L.-N.; Sun, Y.-Q.; Liang, M.; Liu, Z.-Q.; Liao, D.-Z.; Jiang, Z.-H.; Yan, S.-P.; Cheng, P. *Chem. Comm.* **2003**, 2544.
11. Seward, C.; Chan, J.; Song, D.; Wang, S. *Inorg. Chem.* **2003**, *42*, 1112.
12. Fournier, É.; Decken, A.; Harvey, P. D. *Eur. J. Inorg. Chem.* **2005**, 1011.
13. Fournier, É; Lebrun, F.; Drouin, M.; Decken, A.; Harvey, P. D. *Inorg. Chem.* **2004**, *43*, 3127.
14. Fournier, É.; Sicard, S.; Decken, A.; Harvey, P. D. *Inorg. Chem.* **2004**; *43*, 1491.
15. Mongrain, P.; Harvey, P. D. *Can. J. Chem.* **2003**, *81*, 1246.
16. Turcotte, M.; Harvey, P. D. *Inorg. Chem.* **2002** *41*, 2971.
17. Fortin, D.; Drouin, M.; Harvey, P. D. *Inorg. Chem.* **2000**, *39*, 2758.
18. Fortin, D.; Drouin, M.; Harvey, P. D. *J. Am. Chem. Soc.* **1998**, *120*, 5351.
19. Fortin, D.; Drouin, M.; Turcotte, M.; Harvey, P. D. *J. Am. Chem. Soc.*, **1997**, *119*, 531.
20. Perreault, D.; Drouin, M.; Michel, A.; Harvey, P. D. *Inorg. Chem.* **1992**, *31*, 3688.
21. Martini, I. B.; Smith, A. D.; Schwartz, B. J. *Physical Rev.* **2004**, *69*, 035204-2.
22. List, E. J. W.; Leising, G. *Synth. Metals* **2004**, *141*, 211.
23. Gunaratne, T.; Kennedy, V. O.; Kenney, M. E.; Rogers, M. A. J. *J. Phys. Chem. A* **2004**, in press.
24. Mirzov, O.; Cichos, F.; von Borczyskoswky, C.; Scheblykin, I. G. *Chem. Phys. Lett.* **2004**, *386*, 286.

25. Siemiarczuk, A.; Wagner, B. D.; Ware, W. R. *J. Phys. Chem.* **1990**, *94*, 1661.

26. Siemiarczuk, A.; Ware, W. R. *Chem. Phys. Lett.* **1989**, *160*, 285.

27. Harvey, P. D.; Drouin, M.; Zhang, T. *Inorg. Chem.* **1997**, *36*, 4998.

28. Harvey, P. D.; Schaefer, W. P.; Gray, H. B. *Inorg. Chem.* **1988**, *27*, 1101.

29. Orio, A. A.; Chastain, B.B. ; Gray, H. B. *Inorg. Chim. Acta* **1969**, *3*, 8.

30. Leung K. H.; Phillips, D. L.; Mao, Z.; Che, C.-M.; Miskowski, V. M., Chan, C.-M. *Inorg. Chem.* **2002**, *41*, 2054.

31. Zhang, H.-X.; Che, C.-M. *Chem. Eur.* **2001**, *7*, 4887.

32. Fu, W.-F.; Chan, K.-C.; Cheung, K.-K.; Che, C.-M. *Chem. Eur. J.* **2001**, *7*, 4656.

33. Leung, K. H.; Phillips, D. L.; Tse, M.-C.; Che, C.-M.; Miskowski, V.M. *J. Am. Chem. Soc.* **1999**, *121*, 4799.

34. Fu, W.-Fu; Chan, K.-C.; Miskowski, V. M.; Che, C.-M. *Angew. Chem. Int. Ed.* **1999**, *38*, 2783.

35. Piché, D.; Harvey, P. D. *Can. J. Chem.* **1994**, *72*, 705.

36. Zhang, T.; Drouin, M.; Harvey, P. D. *Inorg. Chem.* **1999**, 38, 1305.

37. Turro, N. J., *Modern Molecular Photochemistry*, Benjamen / Cummings Pub. Co., Menlo Park, **1978**.

38. Zhang, T.; Drouin, M.; Harvey, P. D. *Inorg. Chem.* **1999**, 38, 957.

39. Sicard, S.; Berubé, J.-F.; Samar, D.; Messaoudi, A.; Fortin, D.; Lebrun, F.; Fortin, J.-F.; Decken, A.; Harvey, P. D. *Inorg. Chem.* **2004**, *43*, 5321.

Chapter 34

Facile Synthesis of High-Quality Large-Pore Periodic Mesoporous Organosilicas Templated by Triblock Copolymers

Wanping Guo, Jin-Woo Park, and Chang-Sik Ha[*]

Department of Polymer Science and Engineering, Pusan National University, Pusan 609–735, Korea

One exciting development in the surfactant-templated synthesis strategies is the discovery of a novel class of hybrid materials called periodic mesoporous organosilicas (PMOs) through surfactant-templated condensation of organosilanes with two organically bridged trialkoxysilyl groups. We report a facile synthesis of high-quality PMOs with large pores using commercially available poly(ethylene oxide)-*b*-poly(propylene oxide)-*b*-poly (ethylene oxide) (PEO-PPO-PEO) triblock copolymers as the structure directing agents under strongly acidic media in the presence of inorganic salts. These PMO materials exhibit not only highly ordered hexagonal (*p6mm*) or cubic (*Im3m*) pore structures, but also characteristic external morphologies.

Background

In recent years, surfactant-templated synthesis strategies have been successfully applied to the preparation of a variety of mesoporous materials.[1] One of the most exciting new developments was the discovery of a novel class of organic-inorganic hybrid materials called periodic mesoporous organosilicas (PMOs) through surfactant-templated condensation of organosilanes with two organically bridged trialkoxysilyl groups.[2-4] To date, PMO materials have been prepared using various bridged organic groups including methane,[5] ethane,[2,3] ethylene,[3,4a] benzene,[4b,6] thiophene,[4b] biphenylene,[7] etc. In addition, successful syntheses of PMOs have been achieved under a wide range of pHs from highly basic to strongly acidic conditions using cationic,[2-7] anionic,[8] neutral,[9] and nonionic oligomeric[10] surfactants. Furthermore, PMO materials were already found to have such potential applications as novel catalysts,[11] selective adsorbents,[12] and hosts for nanocluster synthesis.[13] At present, particular interest is focused on large-pore PMO materials for the immobilization and encapsulation of large molecules. There are, however, only a few reports[14-18] on the synthesis of PMOs with large pores. Two communications[14,15] first described the use of triblock copolymer P123 ($EO_{20}PO_{70}EO_{20}$) in strongly acidic media to prepare large-pore PMOs with poorly ordered mesostructures. Later, an attempt to synthesize large-pore PMOs using triblock copolymer B50-6600 ($EO_{39}BO_{47}EO_{39}$) under low-acidic conditions led to the formation of PMOs with large cagelike pores in limited long-range order.[16]

A very recent report demonstrated the preparation of well-ordered phenylene-bridged PMOs using P123 triblock copolymer.[17] The other report on the synthesis of highly ordered large-pore PMOs involved direct crystal templating of P123 triblock copolymer.[18] In spite of a little success mentioned above, a general and facile route to the synthesis of highly ordered large-pore PMO materials using commercial triblock copolymer surfactants and bridged organosilanes has not been realized yet.

Objective

In this paper, we report a facile method for the synthesis of high-quality PMOs with large pores using commercially available poly(ethylene oxide)-*b*-poly(propylene oxide)-*b*-poly(ethylene oxide) (PEO-PPO-PEO) triblock copolymers as the structure directing agents under strongly acidic media in the presence of inorganic salts, based on our preliminary works published previously.[19] Pluronic P123 ($EO_{20}PO_{70}EO_{20}$) and Pluronic F127 ($EO_{106}PO_{70}EO_{106}$) were employed to prepare hexagonal PMO material (designated PMO-SBA-15) and cubic PMO material (designated PMO-SBA-16), respectively, as did in the preparation of silica-based SBA-15 and SBA-16 mesoporous materials.[20] [1,2-bis(trimethoxysilyl)ethane] (BTME) was chosen as the organically bridged silica source only because of its commercial availability.

Experimental

Synthesis

The following procedure was the typical preparation of PMO-SBA-15[19a]: 1.2 g of P123 (Aldrich) and 3.5 g of NaCl (Aldrich) were dissolved in 10 g of water and 30 g of 2.0 M HCl solution (Wako, Japan) with stirring at 40 ^{0}C. To this homogeneous solution was added 1.6 g of BTME (Aldrich) and then the mixture was stirred for 24 h at the same temperature. Subsequently, the resulting mixture was transferred into a Teflon-lined autoclave and heated at 80 ^{0}C for an additional 24 h under static conditions. The final reactant molar composition was BTME / P123 / HCl / NaCl / H_2O = 0.5 : 0.017 : 5.07 : 5.07 : 178. The solid products were obtained by filtration, washed thoroughly with water, and air-dried at room temperature. The surfactant was removed by stirring 1.0 g of as-synthesized sample in 150 mL of ethanol with 3.8 g of 36% HCl aqueous solution at 50 ^{0}C for 6 h. The resulting solid was recovered by filtration, washed with ethanol, and dried in air. This extraction process was repeated in order to remove the surfactant thoroughly. The complete extraction was confirmed by the disappearance of the characteristic IR adsorptions of the surfactant and by the absence of surfactant carbon signals in the ^{13}C CP MAS NMR spectrum. According to the same procedure, PMO-SBA-16 was synthesized using F127 triblock copolymer as the template under strongly acidic media in the presence of K_2SO_4, following a molar ratio of 0.5 BTME / 0.004 F127 / 4.51 HCl / 2.61 K_2SO_4 / 116 H_2O[19b]. Two control samples (P123 blank and F127 blank) corresponding to PMO-SBA-15 and PMO-SBA-16, respectively, were prepared according to the protocols described above, without the addition of inorganic salts.

Characterization

SAXS measurements were carried out using 4C2 beam lines with Co Kα synchrotron radiation (λ = 0.1608 nm) at 2.5 GeV and 140 mA in the Pohang Accelerator Laboratory, POSTECH, Korea. Nitrogen adsorption-desorption isotherms were obtained using a Quantachrome Autosorb-1 apparatus at liquid nitrogen temperature and the pore size distributions were calculated by the BdB method. TEM images were acquired on a JEOL JEM-2010 microscope operating at 200 kV. SEM images were collected with a Hitachi S-4200 field emission scanning microscope. ^{29}Si CP MAS NMR and ^{13}C CP MAS NMR spectra were recorded on a Bruker DSX400 spectrometer at a ^{29}Si resonance frequency of 59.63 MHz and a ^{13}C resonance frequency of 75.47 MHz with tetramethylsilane as the reference.

Results & Discussion

Small angle X-ray scattering

The framework structures of obtained samples were investigated by small-angle X-ray scattering (SAXS) experiments. The SAXS pattern of the solvent-extracted PMO-SBA-15 (Figure 1c) shows three well-resolved peaks with interplanar d spacings of 10.5, 6.06, and 5.25 nm, which can be assigned to (100), (110), and (200) reflections of the two-dimensional hexagonal space group (*p6mm*), similar to that reported for the silica-based mesoporous counterpart SBA-15.[20a] The unit cell parameter of the hexagonal lattice is 12.1 nm. Analogously, three well-resolved peaks at very small scattering angles, with interplanar d spacings of 12.8, 9.03, and 7.38 nm, are observed in the SAXS pattern of the solvent-extracted PMO-SBA-16 (Figure 1d). These three peaks can be indexable as (110), (200), and (211) reflections corresponding to the body-centered cubic space group (*Im3m*)[20b] with the unit cell parameter as large as 18.1 nm.

The well-resolved SAXS data demonstrate that highly ordered large-pore hexagonal and cubic PMO materials have been obtained from strongly acidic media in the presence of inorganic salts. By contrast, the SAXS patterns of two solvent-extracted control samples P123 blank (Figure 1b) and F127 blank (Figure 1a), which were prepared in the absence of inorganic salts, exhibit the formation of either disordered mesoporous powder with a single broad peak or amorphous gel.

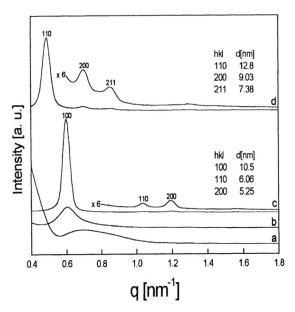

Figure 1. Small-angle X-ray scattering (SAXS) patterns of solvent-extracted samples: a) F127 blank, b) P123 blank, c) PMO-SBA-15, and d) PMO-SBA-16.

Nitrogen adsorption

Nitrogen adsorption technique was used to characterize the pore structures of PMO materials. Figure 2 shows nitrogen adsorption-desorption isotherms and the corresponding pore size distributions calculated by the BdB (Broekhoff and de Boer) method[21] for solvent-extracted PMO-SBA-15 and PMO-SBA-16 samples. The isotherm of PMO-SBA-15 (Figure 2a) is of type IV with a clear H_1-type hysteresis loop at high relative pressure characteristic of large-pore mesoporous materials with one-dimensional cylindrical channels.[20] The pore sizes determined from the adsorption and desorption branches of the isotherm (see Figure 2a inset) using the cylindrical model give the nearly same values (6.4 nm and 6.6 nm, respectively), which attests the pore structures of PMO-SBA-15 are certainly cylindrical.[21] The solvent-extracted PMO-SBA-15 has a BET surface area of 737 m^2/g and a pore volume of 0.88 cm^3/g. The pore wall thickness of PMO-SBA-15 evaluated from the unit cell parameter and the pore size data is around 5.5-5.7 nm. In comparison, the solvent-extracted PMO-SBA-

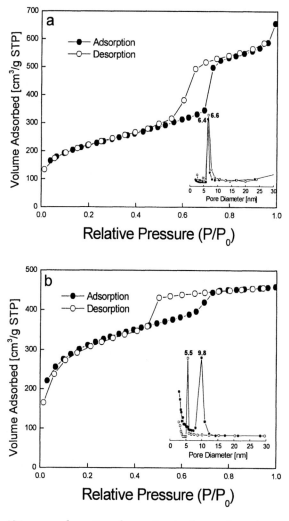

Figure 2. Nitrogen adsorption-desorption isotherms for solvent-extracted a) PMO-SBA-15, and b) PMO-SBA-16. The insets show the pore size distributions calculated by the BdB method.

16 yields a type IV isotherm with a large H_1 hysteresis loop (Figure 2b),[22] indicating bottle-shaped pore structures.[23] This PMO material has a BET surface area of 989 m^2/g and a pore volume of 0.65 cm^3/g, with a cell size of 9.8 nm determined from the adsorption branch of the isotherm and a window size of 5.5 nm determined from the desorption branch of the isotherm (see Figure 2b inset).[21]

The combination of SAXS and nitrogen adsorption data provides strong evidence of the high quality of PMO-SBA-15 and PMO-SBA-16 materials. The structural order of these two PMO materials is analogous to that observed in silica-based SBA-15 and SBA-16 counterparts.[20] Furthermore, the high-quality large-pore PMO materials exhibit high hydrothermal stability. For example, if PMO-SBA-15 with a pore size of 6.4 nm, a BET surface area of 737 m^2/g, and a pore volume of 0.88 cm^3/g was treated in boiling water for 6 days, it then showed a pore size of 6.7 nm, a BET surface area of 772 m^2/g, and a pore volume of 0.92 cm^3/g. The SAXS pattern of the PMO material was essentially unchanged after hydrothermal treatment. The high hydrothermal stability of PMO materials is believed to result from the thick pore walls and hydrophobicity imparted by the organic components in the PMO framework.

Electron microscopy

Transmission electron microscopy (TEM) images shown in Figure 3 provide direct visualization of the PMO pore structures. The TEM images of the solvent-extracted PMO-SBA-15 (Figure 3a,b) further corroborate well-ordered hexagonal *p6mm* arrays of one-dimensional mesoporous channels. A detailed analysis results in an estimated pore size of 6.5 nm and a pore wall thickness of about 5.5 nm. These values are in good agreement with those determined from the nitrogen adsorption measurement. The TEM images of the solvent-extracted PMO-SBA-16 recorded along the [100] and [110] directions (Figure 3c,d) clearly show well-ordered domains of three-dimensional cubic mesostructrues.[24] Thus, the assignment of *Im3m* space group to PMO-SBA-16 in SAXS experiments is strongly supported by the TEM images. It should be noted that the PMO materials synthesized under strongly acidic media in the presence of inorganic salts have characteristic external morphologies. As seen in Figure 4, The NaCl-assisted PMO-SBA-15 exhibits a rod-like morphology with the diameter of around 5 μm, whereas the PMO-SBA-16 synthesized with K_2SO_4 shows a cauliflower-type morphology.

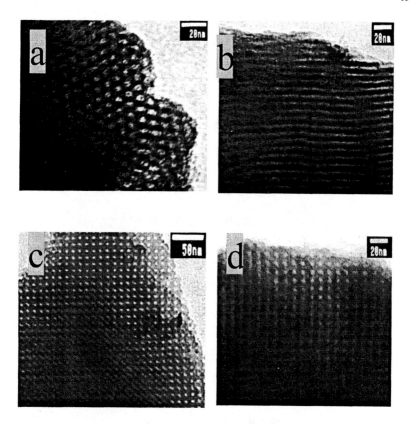

Figure 3. TEM images of solvent-extracted PMO-SBA-15 recorded along the a) [100], b) [110] directions, and solvent-extracted PMO-SBA-16 recorded along the c) [100], d) [110] directions.

NMR spectroscopy

The basic structural unit ethane-silica (Si-CH$_2$-CH$_2$-Si) in the PMO materials was confirmed by solid-state NMR spectroscopy. The ^{13}C cross-polarization (CP) MAS NMR spectrum (not shown) of the solvent-extracted PMO-SBA-15 shows a strong resonance at 6.8 ppm that is attributed to ethane

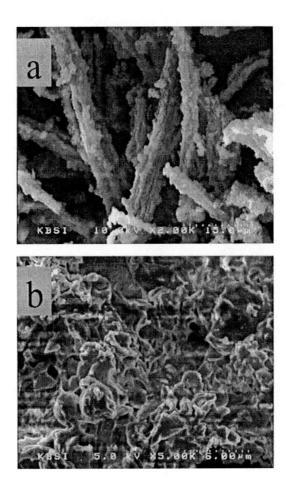

Figure 4. SEM images of as-synthesized a) PMO-SBA-15; b) PMO-SBA-16.

carbon atoms in the PMO framework. The ^{29}Si CP MAS NMR spectrum of the solvent-extracted PMO-SBA-15 (Figure 5b) exhibits two signals at –56.8 and – 64.2 ppm corresponding to T^2 [RSi(OSi)$_2$OH] and T^3 [RSi(OSi)$_3$] resonances, respectively. Alike, one signal at –57.6 ppm assigned to T^2 resonance and the other signal at –64.6 ppm corresponding to T^3 resonance are observed in the ^{29}Si CP MAS NMR spectrum of the solvent-extracted PMO-SBA-16 (Figure 5a). In addition, the control sample synthesized using P123 triblock copolymer as the template without NaCl also presents similar two signals T^2 at –57.8 ppm and T^3 at –64.9 ppm in the ^{29}Si CP MAS NMR spectrum (Figure 5c). The absence of signals due to Q^n [Si(OSi)$_n$(OH)$_{4-n}$] species between –90 and –120 ppm indicates that all silicon atoms are covalently connected to carbon atoms in the PMO materials, and that no carbon-silicon bond cleavage occurred under the salt-assisted strongly acidic synthesis conditions used herein.

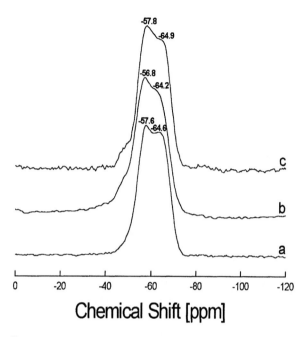

Chemical Shift [ppm]

Figure 5. ^{29}Si CP MAS NMR spectra of solvent-extracted a) PMO-SBA-16, b) PMO-SBA-15, and c) P123 blank.

Inorganic salts were used to improve the hydrothermal stability,[25] control the morphology,[26] extend the synthesis domain,[27] and to tailor the framework porosity[28] during the formation of mesoporous materials, which could be attributed to the specific effect of inorganic salts on the self-assembly interaction between surfactant headgroups and inorganic species.[27] Our study confirms a similar effect of inorganic salts on the formation of high-quality large-pore PMO materials. In addition, the well-ordered PMO-SBA-15 can also be prepared in the presence of other neutral inorganic salts such as KCl, Na_2SO_4 and K_2SO_4. Instead of K_2SO_4, Na_2SO_4 can also be used to synthesize the well-ordered PMO-SBA-16 materials. However, the use of NaCl or KCl resulted in the deterioration of the long-range order of PMO-SBA-16, suggesting that highly charged salts favor the formation of the well-ordered PMO-SBA-16 materials. Therefore, we believe that the addition of inorganic salts, especially highly charged salts, can increase the self-assembly interaction between the headgroups of triblock copolymer surfactant and the organosilane species. The following consideration gives further elucidation. PEO-PPO-PEO triblock copolymers form micelles in water with the core of the PPO block surrounded by the shell of hydrated PEO end blocks.[29] The addition of inorganic salts causes dehydration of ethylene oxide units from the hydrated PEO shell remaining adjacent to the PPO cores, leading to an increase of hydrophobicity in the PPO moieties and a reduction of hydrophilicity in the PEO moieties.[28a] By the counterion-mediated $(S^0H^+)(X^-I^+)$ pathway[20b] for the formation of the PMO materials under strongly acidic conditions, the low-hydrophilic PEO headgroups in the positively charged triblock surfactant are expected to have increased interaction with the positively charged organosilane species with low hydrophilicity due to the organic components. This enhanced self-assembly interaction can result in long-range ordered domain of organosilica-surfactant mesostructures.

On the other hand, the more condensed pore walls could make contribution to the long-range structural order of the PMO materials, which is demonstrated by the ^{29}Si CP MAS NMR spectra shown in Figure 5. After deconvolution, the T^3/T^2 ratio of the solvent-extracted PMO-SBA-15 sample synthesized with NaCl is calculated to be 0.69, whereas the T^3/T^2 ratio of the solvent-extracted corresponding control sample synthesized without NaCl (see Figure 5c) is found to be 0.48. The higher T^3/T^2 ratio indicates that there is higher degree of organosilane cross-linking during the formation of the PMO materials, which is comparable to that reported for the synthesis of silica-based SBA-15.[30] Moreover, the T^3/T^2 ratio of the solvent-extracted PMO-SBA-16 is as high as 0.82, indicating the presence of much more condensed pore walls. In short, the increased self-assembly interaction between the headgroups of triblock copolymer surfactant and the organosilane species, and the much high degree of organosilane cross-linking, which are two kinds of effects initiated by the addition of inorganic salts, result in the highly ordered large-pore PMO materials.

Conclusions

In conclusion, we have provided a facile route to the first synthesis of highly ordered large-pore hexagonal (*p6mm*) and cubic (*Im3m*) PMO materials under strongly acidic media in the presence of inorganic salts using P123 triblock copolymer and F127 triblock copolymer as the templates, respectively. These PMO materials with well-ordered large pores show not only characteristic external morphologies, but also much high hydrothermal stability. Although only two important two-dimensional hexagonal and three-dimensional cubic PMO materials were researched in this paper, we believe that the salt-assisted synthesis strategy could be extended to the preparation of other types of high-quality PMO materials with large pores using other kinds of triblock copolymers as the templates. Moreover, the resulting highly ordered large-pore PMO materials are expected to have promising applications involved in large molecules.

Acknowledgments

This work was supported by the National Research Laboratory Program, the Center for Integrated Molecular Systems, and the Brain Korea 21 Project. We thank Prof. R. Ryoo (KAIST, Korea) for his help in nitrogen adsorption measurements, and Prof. D. Zhao (Fudan University, China) for providing BdB program. The Pohang Accelerator Laboratory, POSTECH, Korea is also acknowledged for SAXS measurements. Dr. W. Guo is now working as a research fellow at Department of Chemical and Biomolecular Engineering, National University of Singapore.

References

1. a) Kresge, C.T.; Leonowicz, M.E.; Roth, W. J.; Vartuli, J.C.; Beck, J.S. *Nature* **1992**, *359*, 710-712. b) Ying, J.Y.; Mehnert, C.P.; Wong, M.S. *Angew. Chem. Int. Ed.* **1999**, *38*, 56-77. c) Schüth, F.; Schmidt, W. *Adv. Mater.* **2002**, *14*, 629-638.
2. Inagaki, S.; Guan, S. ;Fukushima, Y.; Ohsuna, T.; Terasaki, O. *J. Am. Chem. Soc.* **1999**, *121*, 9611-9614.
3. Melde, B.J.; Holland, B.T.; Blanford, C.F.; Stein, A. *Chem. Mater.* **1999**, *11*, 3302-3308.
4. a) Asefa, T.; MacLachlan, M.J.; Coombs, N.; Ozin, G.A. *Nature* **1999**, *402*, 867-871. b) Yoshina-Ishii, C.; Asefa, T.; Coombs, N.; MacLachlan, M.J.; Ozin, G.A. *Chem. Commun.* **1999**, 2539-2540.
5. Asefa, T.; MacLachlan, M.J.; Grondey, H.; Coombs, N.; Ozin, G.A. *Angew. Chem. Int. Ed.* **2000**, *39*, 1808-1811.
6. Inagaki, S.; Guan, S.; Ohsuna, T.; Terasaki,O. *Nature* **2002**, *416*, 304-307.
7. Kapoor, M.P.; Yang, Q.; Inagaki, S. *J. Am. Chem. Soc.* **2002**, *124*, 15176-15177.
8. Lu, Y.; Fan, H.; Doke, N.; Loy, D.A.; Assink, R.A.; LaVan, D.A.; Brinker, C.J. *J. Am. Chem. Soc.* **2000**, *122*, 5258-5261.

498

9. McInall, M.D.; Scott, J.; Mercier, L.; Kooyman, P.J. *Chem. Commun.* **2001**, 2282-2283.
10. a) Burleigh, M.C.; Markowitz, M.A.; Spector, M.S.; Gaber, B.P. *J. Phys. Chem. B* **2002**, *106*, 9712-9716. b) Hamoudi, S.; Kaliaguine, S. *Chem. Commun.* **2002**, 2118-2119. c) Sayari, A.; Yang, Y. *Chem. Commun.* **2002**, 2582-2583.
11. Yang, Q.; Kapoor, M.P.; Inagaki, S. *J. Am. Chem. Soc.* **2002**, *124*, 9694-9695.
12. Burleigh, M.C.; Dai, S.; Hagaman, E.W.; Lin, J.S. *Chem. Mater.* **2001**, *13*, 2537-2546.
13. Fukuoka, A.; Sakamoto, Y.; Guan, S.; Inagaki, S.; Sugimoto, N.; Fukushima, Y.; Hirahara, K.; Iijima, S.; Ichikawa, M. *J. Am. Chem. Soc.* **2001**, *123*, 3373-3374.
14. Muth, O.; Schellbach, C.; Fröba, M. *Chem. Commun.* **2001**, 2032-2033.
15. Burleigh, M.C.; Markowitz, M.A.; Wong, J. S.; Lin, E.M.; Gaber,B.P. *Chem. Mater.* **2001**, *13*, 4411-4412.
16. Matos, J.R.; Kruk, M.; Mercuri, L.P.; Jaroniec, M.; Asefa, T.; Coombs, N.; Ozin, G.A.; Kamiyama, T.; Terasaki, O. *Chem. Mater.* **2002**, *14*, 1903-1905.
17. Goto, Y.; Inagaki, S. *Chem. Commun.* **2002**, 2410-2411.
18. Zhu, H.; Jones, D.J.; Zajac, J.; Rozière, J.; Dutartre, R. *Chem. Commun.* **2001**, 2568-2569.
19. a) Guo, W.P.; Park, J.Y.; Oh, M.O.; Jeong, H.W.; Cho, W.J.; Kim, I.; Ha, C.S. *Chem. Mater.* **2003**,*15*, 2295-2298. b) Guo, W.P.; Kim, I.; Ha, C.S. *Chem. Commun.* **2003**, 2692-2693.
20. a) Zhao, D.; Feng, J.; Huo, Q.; Melosh, N.; Fredrickson, G.H.; Chmelka, B.F.; Stucky, G.D. *Science* **1998**, *279*, 548-552. b) Zhao, D.; Huo, Q.; Feng, J.; Chmelka, B.F.; Stucky, G.D. *J. Am. Chem. Soc.* **1998**, *120*, 6024-6036.
21. Lukens, Jr., W.W.; Schmidt-Winkel, P.; Zhao, D.; Feng, J.; Stucky, G.D. *Langmuir* **1999**, *15*, 5403-5409.
22. Zhao, D.; Yang, P.; Melosh, N.; Feng, J.; Chmelka, B.F.;Stucky, G.D. *Adv. Mater.* **1998**, *10*, 1380-1385.
23. a) Kim, J.M.; Stucky, G.D. *Chem. Commun.* **2000**, 1159-1160. b) Kim, J.M.; Sakamoto, Y.; Hwang, Y.K.; Kwon, Y.U.; Terasaki, O.; Park, S.E.; Stucky, G.D. *J. Phys. Chem. B* **2002**, *106*, 2552-2558.
24. Sakamoto, Y.; Kaneda, M.; Terasaki, O.; Zhao, D.Y.; Kim,J.M.; Stucky, G.D.; Shin, H.J.; Ryoo, R. *Nature* **2000**, *408*, 449-453.
25. a) Ryoo, R.; Jun, S. *J. Phys. Chem. B* **1997**, *101*, 317-320. b) Kim, J.M.; Kim, S.K.;Ryoo, R. *Chem. Commun.* **1998**, 259-260. c) Kim, J.M.; Jun, S.; Ryoo, R. *J. Phys. Chem. B* **1999**, *103*, 6200-6205.
26. a) Zhao, D.; Yang, P.; Chmelka, B.F.; Stucky, G.D. *Chem. Mater.* **1999**, *11*, 1174-1178. b) Zhao, D.; Sun, J.; Li, Q.; Stucky, G.D. *Chem. Mater.* **2000**, *12*, 275-279. c) Yu, C.; Tian, B.; Fan, J.; Stucky, G.D.; Zhao, D. *J. Am. Chem. Soc.* **2002**, *124*, 4556-4557.

27. Yu, C.; Tian, B.; Fan, J.; Stucky, G.D.; Zhao, D. *Chem. Commun.* **2001**, 2726-2727.
28. a) Newalkar, B.L.; Komarneni, S. *Chem. Mater.* **2001**, *13*, 4573-4579. b) Newalkar, B.L.; Komarneni, S. *Chem. Commun.* **2002**, 1774-1775.
29. Mortensen, K.; Pedersen, J.S. *Macromolecules* **1993**, *26*, 805-812.
30. Yu, C.; Fan, J.; Tian, B.; Zhao, D.; Stucky, G.D. *Adv. Mater.* **2002**, *14*, 1742-1745.

Chapter 35

Assembling Prussian Blue Nanoclusters along Single Polyelectrolyte Molecules

Anton Kiriy[1], Vera Bocharova[1], Ganna Gorodyska[1], Paul Simon[2], Ingolf Mönch[3], Dieter Elefant[3], and Manfred Stamm[1]

[1]Leibniz Institute of Polymer Research, Hohe Strasse 6, 01069 Dresden, Germany
[2]Max Planck Institut for Chemical Physics of Solids, Nöthnitzer Strasse 40, 01187 Dresden, Germany
[3]Leibniz Institute for Solid State and Materials Research Dresden, IFW Dresden, Helmholtzstrasse 20, D–01069 Dresden, Germany

A simple method for the preparation of charge-stabilized Prussian Blue nanocrystals (PBNs) of readily adjustable size is reported. PBNs have been purified by addition of non-solvents and redispersed in water without aggregation. PBNs may be electrostatically arranged along *isolated polycation (PC) chains* adsorbed onto flat surfaces. PC-PBNs nanohybrids constitute useful materials for the manufacture of electrooptical devices. PBNs can be also used as contrasting reagent to improve AFM visualization of positively charged polymer chains deposited on substrates of relatively high roughness.

Background

The design of new materials with desired properties from organic macromolecules and inorganic building blocks involving non-covalent interactions is a newly emerging and rapidly developed area of research. The properties of thus fabricated materials (e.g., optical, electronic, mechanical, etc.) are defined by the properties of the components used and can be easily tuned by altering the composition (varying the kind, number and ratio of the components). Along this line various macromolecules containing chelating ligands were already utilized to immobilize different metallic ions via complexation reactions.[1] On the other hand, combination of polyelectrolytes with oppositely charged metal ions or clusters constitutes an alternative and highly universal route to a number of metal-containing nanostructural materials.

Prussian Blue (PB),[2] an old pigment, is a coordination polymer formed by reaction of either hexacyanoferrate(II) (HCF-II) anions with ferric (Fe(III)) cations, or hexacyanoferrate(III) (HCF-III) anions with ferrous (Fe(II)) cations.[3] According to X-ray diffraction analysis, PB is a three-dimensional crystal of ferric and ferrous ions which alternate at the sites of a cubic lattice.[4] The ferric ion is coordinated to the nitrogen atoms, and the ferrous ion to the carbon atoms, of the bridging cyanide ligands. The remaining charge is balanced either by potassium ions in the so-called "soluble" PB, or by ferric ions in the "insoluble" PB. The term "soluble", however, does not refer to the true solubility but only to the tendency of PB to form colloidal solutions.[5]

A large family of cyano-bridged compounds with a cubic structure (PB analogues; PBs) are known for interesting physical properties.[3] Neff et al. reported that the oxidation state of the iron centers could be controlled electrochemically, making dramatic color changes possible[5] that could be used in electrochromic devices.[6] PB shows a long-range ferromagnetic ordering at 5.6 K, whereas few PBs undergo magnetization at room temperature and even higher.[7] Hashimoto et al. showed that the magnetic properties of PBs could be modulated not only by the chemical composition but also by an optical or electrical stimulation (photomagnetism and electromagnetism, respectively).[8-10] Finally, PBs exhibit remarkable ion-sieving properties as result of an open pore zeolite-like structure.[11]

For the unique properties of PBs to be exploited, PBs must be deposited properly onto a solid support. It is highly desirable to prepare mechanically robust PBs films with controlled thickness, chemical composition and crystallinity, having ion-sieving membranes and electrochromic devices in mind,[6] or to create regular patterns of PB-based single molecule magnets.[12]

Classical methods of PB immobilization by casting,[13] dip-coating[5] or electrochemical deposition do not allow film thickness, composition and/or mechanical properties to be controlled accurately. This problem has recently been overcome by the Langmuir-Blodget technique[14] and multiple sequential adsorption techniques[11,15] based on the stepwise adsorption of HCF anions and ferric (or ferrous) cations. On the other hand, utilization of larger size PB *nanoparticles* instead of small precursor *ions* would have a favorable effect on film assembly. However, only few methods to produce well-defined PB nanoparticles have been reported.[16,17] Uemura et. al succeeded in the preparation of PB nanoparticles stabilized by poly(vinylpyrrolidone).[16a] Mann et al. have developed the approach to crystalline nanoparticles of some PB analogues combining water-in-oil microemulsions of appropriate reactants.[16b] In that case growth of the nanoparticles occurred through the aggregation of primary clusters inside micelles. Thus, all of the methods mentioned above lead to PB particles stabilized by either polymeric or small-molecule surfactants.

Objective

Here we report on the preparation of surfactant-free, water-dispersible Prussian Blue nanoparticles (PBNs) by mixing solutions of ferric chloride and excess of potassium ferrocyanide. PBNs display a remarkable stability against aggregation because of an uncompensated negative charge. The average size of PBNs was readily controlled by the molar ratio of the two reagents. Thus formed PB nanoparticles are crystalline and display optical and magnetic properties similar to the properties of PB bulk. PBNs have been electrostatically arranged along the individual polycation molecules. We also developed a simple contrasting procedure to improve the AFM visualization of positively charged polymer chains deposited on substrates of relatively high roughness via counter ion exchange between *small* Cl⁻ anions and *bulky* HCF anions or negatively charged nanoclusters of Prussian Blue.

Experimental

Synthesis of poly(methacryloyloxyethyl dimethylbenzyl) ammonium chloride (PMB) (M_W = 6000 kg/mol, polydispersity index (PDI) = 1.6) was reported elsewhere.[18]

Substrates. Si-wafers (Wacker-Chemitronics), patterned Si-wafers and glasses were first cleaned with dichloromethane in a ultrasonic bath for 5 min (3 times), followed by cleaning with a solution of NH_4OH and H_2O_2 at 60 °C for one hour. This $NH_4OH:H_2O_2$ solution must be handled cautiously because of violent reaction with organic compounds. Samples were finally exposed to 50% sulfuric acid for 15 min and then rinsed repeatedly with water purified through a Millipore (18 MΩxcm) filter.

Layer-by-layer assembly on isolated polyelectrolyte molecules. PMB molecules were deposited onto freshly cleaned Si-wafer in a stretched conformation by spin-coating of a 0.005 g/L solution in acidified water (pH 2, HCl) at 10000 rpm. The substrate with the PMB molecules was then immersed in solutions of $K_4Fe(CN)_6$ (or $K_3Fe(CN)_6$) solutions (0.5-15 g/L) and KCl (0-50 g/L) at the same pH and for the same period of time as in the first part of the cycle. The substrate was analyzed by AFM after either half of cycle, a complete cycle or several cycles.

Deposition of PB clusters. Dispersion of PB clusters was prepared by mixing vigorously a solution of $K_4Fe(CN)_6 \cdot 3 H_2O$ (HCF-II, 1.18 mmol/L) in acidified water (HCl, pH 2.0) and an equal volume of a solution of $FeCl_3$ at the same pH. Concentration of the $FeCl_3$ solution was either 0.148 mmol/L for the preparation of smaller PB1 clusters (3.7 nm), or 0.296 mmol/L for larger PB2 clusters (4.8 nm). A small amount of tetrahydrofuran (THF, Aldrich) was added to the freshly prepared dispersions of PB in order to separate the PB clusters from the unreacted HCF-II and KCl. PB clusters were collected by filtration, washed with a water-THF solution (2:1), and redispersed in acidified water (HCl, pH 2). A Si-wafer onto which PMB molecules were deposited was dipped in the freshly prepared dispersion of PB clusters for 3 min at 25 °C, followed by washing with water and drying under an argon flow.

AFM measurements. A multimode AFM instrument and a NanoScope IV-D3100 (Digital Instruments, Santa Barbara) were operated in the tapping mode. Silicon tips with a radius of 10-20 nm, a spring constant of 30 N/m and a resonance frequency of 250-300 kHz were used.

TEM measurements. TEM images were recorded with a Philips CM 200 FEG at 200 kV. They were processed by the Digital Micrograph program (Gatan, USA).

Magnetisation measurements. The magnetisation measurements were performed in a SQUID magnetometer (MPSM-Quantum Design).

Results & Discussion

Layer-by-layer (LBL) deposition of the Prussian Blue precursors on isolated polyelectrolyte molecules. Prussian Blue was tentatively deposited onto isolated polycations (PC) by the well-known LBL (or multiple sequential adsorption) technique. The stepwise adsorption of HCF anions and ferric (or ferrous) cations was actually effective in fabricating PB films.[11] In this work, the first adsorption step of HCF anions along positively charged PMB chains was successful as confirmed by the increase of the chain thickness by approximately 0.7 nm (Figure 1a). However, whenever the HCF pre-adsorbed PMB chains on the Si-substrate were dipped into the ferric chloride solution, the HCF anions were completely removed leaving bare PMB molecules undetectable at the rough Si-wafer surface. The complete removal of the previously deposited HCF layer was systematically observed whatever the salt, e.g., $FeCl_2$, $NiCl_2$, $CoCl_2$, $CuCl_2$, $PdCl_2$, and $AuCl_3$, in a broad range of pH and ionic strength. When the deposition cycle of HCF and $FeCl_3$ onto pre-adsorbed PMB molecules was repeated several times, local growth of relatively big clusters was occasionally observed (Figure 1b).

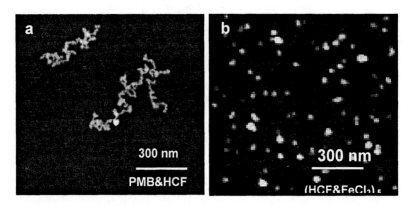

Figure 1. (a) AFM image of PMB molecules adsorbed onto Si-wafer after dipping into a 5 g/L solution of $K_4Fe(CN)_6$ in acidic water and washing with water (the height of the worm-like structure is about 0.6 nm). Further dipping into a $FeCl_3$ solution removes the HCF anions and the PMB molecules are unobserved (image not shown). (b) AFM image of the sample shown in (a) after the fifth cycle of the sequential dipping into $K_4Fe(CN)_6$ and $FeCl_3$ solutions ($FeCl_3$ being the outermost deposit layer): PB clusters (10 nm height) are randomly located and the PMB molecules are unobserved.

The failure of the LBL deposition of PB clusters can be explained as follows. The HCF anions interact expectedly with the positively charged units of the PMB molecules. In the next step, an excess of ferric cations interact with the pre-adsorbed HCF particles and overcharge them. The accordingly formed PB nanoparticles are positively charged by a shell of ferric cations, which facilitate their detachment from the similarly charged PMB chains. The observations in this work are consistent with Tieke et al. who reported on the non-linear increase of the PB film thickness with the number of dipping cycles.[11] This irregular growth was pronounced up to the 6-th cycle and whenever the dipping time was relatively long. Thus, although the LBL method is useful to prepare PB films onto *polyelectrolyte multilayers*, it is ineffective in the case of deposition onto *isolated polycationic molecules*.

Deposition of water-dispersible Prussian Blue nanoclusters onto polycations. In order to prepare clear dispersions of PB nanoclusters stable for several months, the method of mixing diluted acidic solutions (pH = 2) of $FeCl_3$ and $K_4Fe(CN)_6$ (used in excess) proved to be effective.[17] The characteristic blue color and broad signal in the UV-vis spectrum with λ_{max} at 695 nm are consistent with an intermetal charge-transfer band from Fe^{2+} to Fe^{3+} and reflect the formation of PB (Figure 2).

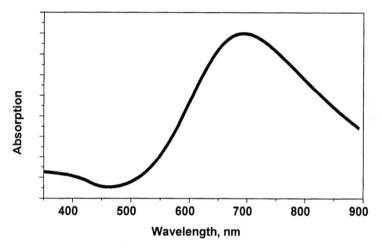

Figure 2. UV-vis spectrum of Prussian Blue dispersion (prepared at molar ratio $K_4Fe(CN)_6/FeCl_3$ - 4/1).

Upon addition of THF until a H_2O-THF volume ratio of 2, the PB clusters precipitated quantitatively and selectively. They were filtrated and washed with a water-THF solution (2:1). Remarkably enough, they are easily redispersed in water with a constant size.

A drop of a PB dispersion in water (0.5 mg/L) was deposited on spin-stretched PMB chains pre-adsorbed onto a Si-wafer. After washing and drying, the sample was analyzed by AFM in the tapping mode. Figures 3a,d show typical AFM images of the beads-on-string morphology at the surface of the PMB chains. PB clusters appear to be selectively and regularly attached along the polycation chains, which indicates that this method should be successful anytime a substrate is covered by a tiny amount of polycations. A range of PB dispersions were prepared by changing the mixing ratios and the stirring time.

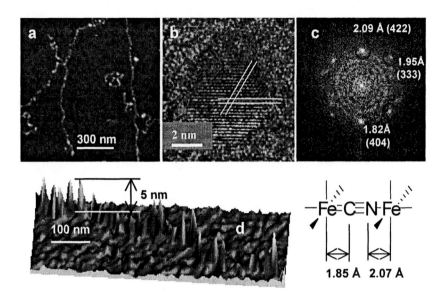

Figure 3. Two-dimensional (a) and three-dimensional (d) AFM images of PB clusters adsorbed along PMB chains deposited onto a Si-wafer. HR TEM image (b) and selected area diffraction pattern (c) reveal crystalline structure of the PB nanoparticle.

The PB clusters were analyzed by AFM and the images were statistically treated. As an example, Figure 4 shows histograms of the height distribution of PB1 and PB2 clusters dissolved in water (0.5 mg/L) and stirred before the deposition at different time (from 3 minutes to 100 hours). In all the cases, a narrow size distribution is observed (PDI = 1.1-1.2), which depends on the $K_4Fe(CN)_6/FeCl_3$ mixing ratio. The average diameter of the PB particles is approximately 3.7 nm for PB1 and 4.8 nm for PB2. As a rule, the size of the PB clusters remains quasi constant with time.

High resolution TEM images and selected area diffraction patterns reveal the crystalline structure of the PB2 clusters (mixing ratio of 4/1, Figures 3b-c). The lattice of the PB nanodots is resolved with spacings of 2.09 Å, 1.95 Å and 1.82 Å that correspond to (422), (333) and (404) reflections.[19] The first reflection can be assigned to the Fe-N interatomic distance, and the last one to the Fe-C distance.

Magnetic properties. The magnetic properties of the dry powder of the PB nanoparticles (PB2, ~ 4.8 nm), compacted by epoxy glue, were investigated. Figure 5 shows the field-cooled magnetization curve at external magnetic field

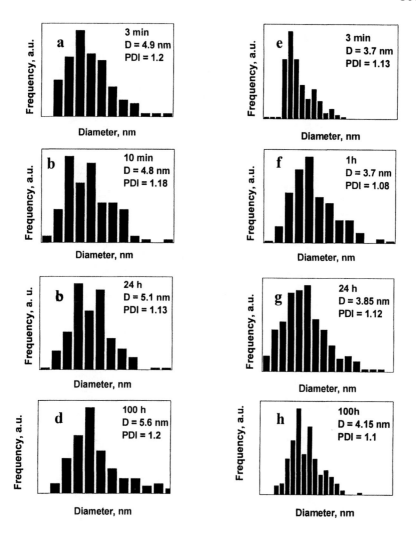

Figure 4. Histograms indicating the size distribution of the PB1 (a-d) and PB2 (e-h) clusters stirred before the deposition during the corresponding time shown in the inset of each histogram.

of 10 G. The critical temperature (T_c) where PB clusters turn to a ferromagnetic compound was found to be ~ 5 K that is slightly less then the value for bulk PB (5.5 K). This result is qualitatively consistent with earlier reported observations of Zhou et al.[14b] and Uemura et al.[14a] for PB nanoclusters and PB nanowires embedded into either a polymer or alumina matrix, respectively. It is believed that the decrease of Curie temperature in PB nanoparticles comparably to PB bulk comes from the diminution of the average number of nearest magnetic interaction neighbors.[14] In our case the decrease of the T_c is less pronounced, because the data represents the magnetic properties of *aggregated* PB2 nanoclusters. Investigation of the magnetic properties of separated PB nanoclusters (embedded into polyelectrolyte matrixes) is under way and will be reported elsewhere.

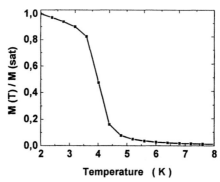

Figure 5. Field-cooled magnetization versus temperature curve for 5 nm PB clusters at an external magnetic field of 10 G.

Visualization of single adsorbed polycation molecules. Interaction of HCF-anions and PB clusters with oppositely charged polycation molecules can be used for analytical purposes to enhance AFM contrast by counter ion exchange between *small* Cl⁻ anions and *bulky* HCF anions or negatively charged nanoclusters of Prussian Blue (Figure 6a). For successful visualization the grain size of the contrasting agent should be larger than the mean size of surface features.

To demonstrate the applicability of the approach we used in this study three different PC molecules: poly(methacryloyloxyethyl dimethylbenzyl ammonium chloride) (PMB) M_w = 6130 kg/mol, poly(2-vinylpyridine) (P2VP) M_W= 385 kg/mol, and the star-shaped heteroarm block copolymer with seven polystyrene M_W= 20 kg/mol and seven P2VP M_W= 56.5 kg/mol arms (PS₇-P2VP₇).[20] Both PMB and P2VP molecules deposited onto Si-wafer are not resolved in the tapping mode (Fig. 6b). Although PS₇-P2VP₇ molecules adsorbed onto mica from acid water (pH 2) solution display a clear core-shell morphology (Figure 6e),[20] only the core of unimers with the height of 5 nm can be resolved on the Si-wafer (Fig. 6f). We were successful in visualizing all these polymers on the Si-wafer substrate with the following staining procedure. The polymer molecules were deposited on the freshly cleaved mica or Si-wafers from very diluted (0.0005 mg/mL) acid solution (pH 2.5-3).

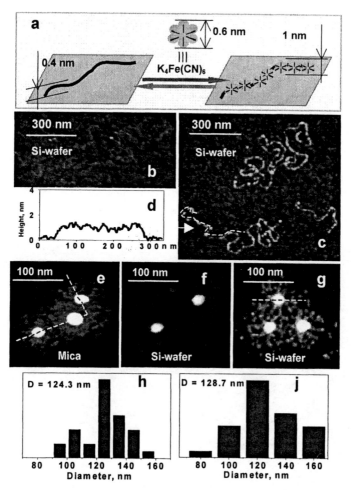

Figure 6. Scheme of the contrasting of adsorbed polycations (a), AFM topography images (b,c) and a cross-section (d) of PMB, and AFM images of PS₇-P2VP (e-g) molecules before (b,e,f), and after (c,g) contrasting with HCF (Z-range 5 nm). All images on Si-wafers, but (e) on mica. Histograms of molecular diameter distribution for PS₇-P2VP₇ adsorbed onto the mica, no contrasting (h), and onto the Si-wafer after contrasting with HCF (j).

510

The drop of the examining solution was set on the substrate for 60 s and afterwards it was removed with a centrifugal force. The molecules were examined in a dry state with AFM instrument. Details of the AFM experiments were reported elsewhere.[21] Then, the samples were stained upon exposure for 3 min to $K_4Fe(CN)_6$ acid solution bath. The samples were rinsed in water and dried for the AFM experiment. Figures 6c,g present AFM images of PMB and PS_7-$P2VP_7$ molecules contrasted with HCF. In all cases we observed a 0.6-0.7 nm increase of height of resulting structures, that roughly correspond to the size of a HCF-anion (Figure 6a). We found a strong effect of pH on the contrasting process. No attachment of HCF-anions was observed at pH higher than 4, i. e. above iso-electric point of Si-wafer (pH 3.8). We may speculate that at high pH the negatively charged Si-wafer suppresses the interaction of the polycations with HCF-anions. The statistical analysis of the molecular diameter from AFM images of the PS_7-$P2VP_7$ unimers before the contrasting on mica and after the contrasting on the Si-wafer provides evidence that the contrasting procedure introduces no changes in the conformation (Figure 6h,j).

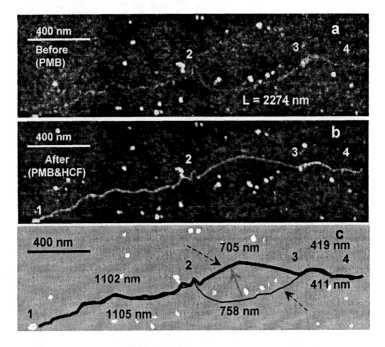

Figure 7. AFM topography images of the PMB molecule on Si-wafer (RMS=0.6±0.4 nm) before (a), and after (b) contrasting with HCF (Z-range 5 nm). Snapshot (c) demonstrates the transition of the 2-3 segment induced by the contrasting procedure (see explanations in the text).

Similarly, we detected no changes of the dimensions of PMB molecules upon staining with $K_4Fe(CN)_6$. We found that deposited HCF can be removed simply upon rinsing the sample with either acid (HCl, 5%) or basic (NH_3, 3%) water solution for several minutes.

We performed a special experiment to precisely monitor the effect of the staining on molecular details of the PC conformations deposited onto the Si-wafer by scanning always the same place and visualizing the same molecule before and after the contrasting. The detailed analysis of the AFM data proved that the location and fine conformational details of most of the PMB molecules remain unchanged upon treatment with $K_4Fe(CN)_6$. We also found that minor part of molecules appears to be partially distorted during the procedure. A "negative" example is shown in Figure 7. The 2.2 μm long PMB molecule deposited on the Si-wafer in the extended conformation is visualized before (Figure 7a) and after (Figure 7b) staining. After the contrasting with $K_4Fe(CN)_6$ it is thicker and nicely visible. Although the segments **1-2** and **3-4** of the molecule with the length of 1100 nm and 410 nm, respectively, remained unchanged, the central part **2-3** with the initial length of 758 nm has been moved up of about 250 nm, as it is shown by the green arrow on the snapshot (Figure 7c).

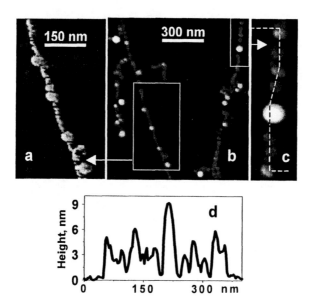

Figure 8. AFM phase (a) and topography (b-c) images of the PMB molecules deposited onto the Si-wafer by spin-stretching and then contrasted with PB nanoparticles (prepared at molar ratio ($K_4Fe(CN)_6/FeCl_3$ - 4/1 (PB2). The dash line in (c) indicates the locus of the cross-sections (d).

The transformation is accompanied by about 7% shortening of the **2-3** fragment.[22] Thus, we see the important role of the attractive interaction between the surface and PC molecule. If a PMB chain (or its fragment) bearing HCF-anions loses the contact with the surface it collapses due to the counter ion condensation effect.

Thus, we demonstrated that the deposition of HCF-anion (diameter of about 0.7 nm) is sufficient for visualization of various PC deposited onto surfaces with a RMS less than 0.5 nm. For visualization of PC chains onto surfaces with even higher roughness PB clusters can be used. The attachment of PB clusters occurs selectively and homogeneously along whole PC chain and significantly improves both the topography (Figure 8b-c) and the phase AFM images (Figure 8a). The distance between adjacent clusters depends on the diameter of PB clusters and usually equal to 10-20 nm.

Conclusions

The layer-by-layer deposition of hexacyanoferrate anions and ferric cations onto isolated polycation molecules failed to form Prussian Blue, because the first deposited layer was desorbed upon the deposition of the next one. In an alternative approach, water-dispersible, charge-stabilized, surfactant-free Prussian Blue nanoparticles were prepared by mixing solutions of ferric chloride and excess of potassium ferrocyanide. The average size was readily controlled by the molar ratio of the two reagents. The so formed PB nanoparticles are crystalline and display optical and magnetic properties similar to the properties of PB bulk. They can be electrostatically deposited onto polycation chains pre-adsorbed individually onto flat surfaces. PB nanocrystals have potential in the manufacture of electrooptical devices.

Acknowledgments

Financial support from the "Deutsche Forschungsgemeinschaft" (DFG) within the ESF EUROCORES/SONS program (02-PE-SONS-092-NEDSPE) and DFG/CNRS German-French bilateral program (STA 324/1) is gratefully acknowledged. We thank Constantinos Tsitsilianis and Werner Jaeger for polymer samples.

References

1. Schubert, U. S.; Eschbaumer, C. *Angew. Chem.* **2002**, *114*, 3016.
2. *Miscellanea Berolinensia ad incrementum scientiarum;* Berlin, 1710, p. 377.
3. Sharpe, A. G. *The chemistry of Cyano Complexes of the Transition Metals;* Academic Press: New York, 1976.

4. Keggin, J. F.; Miles, F. D. *Nature* **1936**, *137*, 577.
5. Neff, V. D. J. *Electrochem. Soc.* **1978**, *125*, 886.
6. DeLongchamp, D. M.; Hammond, P. T. *Adv. Funct. Mater.* **2004**, *3*, 224-232.
7. Garde, R.; Villain, F.; Verdaguer, M. *J. Am. Chem. Soc,* **2002**, *124*, 10531-10538.
8. Sato, O.; Iyoda, T.; Fujishima, A.; Hashimoto, K. *Science* **1996**, *271*, 49.
9. Moore, G. J.; Lochner, E. J.; Ramsey, C.; Dalal, N. S.; Stiegman, A. E. *Angew. Chem.* **2003**, *115*, 2847.
10. Ohkoshi, S. Fujishima, A.; Hashimoto, K. *J. Am. Chem. Soc,* **1998**, *120*, 5349-5350.
11. Pyrash, M.; Toutianoush, A.; Jin, W.; Schnepf, J.; Tieke, B. *Chem. Mater.* **2003**, *15*, 245-254.
12. Sessoli, R.; Gatteschi, Caneschi, A.; Novak, M. A. *Nature* **1993**, *365*, 141.
13. Tosshima, N; Liu, K.; Kaneko, M. *Chem. Lett.* **1990**, 485.
14. Ravaine, S.; Lafuente, C.; Mingotaud, C. *Langmuir,* **1998**, *14*, 6347.
15. Millward, R. C.; Madden, C. E.; Sutherland, I.; Mortimer, R. J.; Fletcher, S.; Marken, F. *Chem. Commun.* **2001**, 1994.
16. (a) Uemura T.; Kitagawa S. *J. Am. Chem. Soc.* **2003**, *125*, 7814. (b) Zhou, P.; Xue, H.; Luo, H.; Chen, X. *Nano Letters* **2002**, *2*, 845. (c) Vaucher, S.; Li, M.; Mann, S. *Angew. Chem.* **2000**, *112*, 1863. Catala, L.; Gacoin, J.-P. B.; Rivere, E.; Paulsen, C.; Lhotel, E.; Mallah, T. *Adv. Mater.* **2003**, *15*, 826.
17. Kiriy, A.; Gorodyska, A.; Minko, S.; Tsitsilianis, C.; Jaeger, W., Stamm, M. *J. Am. Chem. Soc.* **2003**, *125*, 11202.
18. Kiriy, A.; Gorodyska, A.; Minko, S.; Jaeger, W., Štěpánek, P.; Stamm, M. *J. Am. Chem. Soc.* **2002**, *124*, 13454.
19. Buser, H. J.; Schwarzenbach, D; Petter, W; Ludi, A. *Inorg. Chem.* **1977**, *16*, 2704-2710. Robin, M. B. *Inorg. Chem.* **1962**, *1*, 337.
20. Gorodyska, G.; Kiriy, A.; Minko, S; Tsitsilianis, C.; Stamm, M., *Nano Letters* **2003**, *3*, 365. Kiriy, A.; Gorodyska, G.; Minko, S.; Stamm, M. *Macromolecules* **2003**, *36*, 8704-8711.
21. Minko, S., Kiriy, A., Gorodyska, G., Stamm, M., *J. Am. Chem. Soc.* **2002**, *124*, 3218. Kiriy, A.; Minko, S.; Gorodyska, G.; Stamm, M.; Jaeger W. *Nano Letters* **2002**, *2*, 881. Minko, S.; Kiriy, A.; Gorodyska, A.; Stamm, M., *J. Am. Chem. Soc.* **2002**, *124*, 10192.
22. The contour length was measured by processing of the image with home-made software dragging a cursor along the contour of the molecule and automatically recording the point coordinates.

Chapter 36

Reactive Polymers Possessing Metallacycles in the Main Chain

Ikuyoshi Tomita

Department of Electronic Chemistry, Interdisciplinary Graduate School of Science and Engineering, Tokyo Institute of Technology, Nagatsuta-cho 4259–G1–9, Midori-ku, Yokohama 226–8502, Japan

Synthesis and reactions of organometallic polymers containing reactive metallacycles such as cobaltacyclopentadiene and titanacyclopentadiene in the main chain are described. Metallacycle-containing polymers were prepared by the reactions of diynes with low-valent organometallic complexes such as $CpCo(PPh_3)_2$ and $(^iPrO)_2Ti(CH_2=CHCH_3)$. Their polymer reactions which involve the chemical conversion of the main chain structure yielded organic polymers containing versatile functional groups in their main chain repeating units.

Background

Reactive polymers are important synthetic intermediates for diverse functional materials. Nevertheless, their reactive functional groups are generally located not in the main chain but in the pendant groups of the polymers. Thus, it is difficult to modify the main chain structure of the reactive polymers. Polymers possessing functional groups in the main chain have been prepared by the polymerization of the functional monomers. If we could create polymers possessing reactive groups in the main chain, an alternative synthetic approach for polymers containing versatile main chain functional groups will be realized. Reactive organometallic systems are attractive candidates to be incorporated in the main chain repeating unit to create new reactive polymers that can yield organic polymers with versatile main chain functionality. Also, unique macromolecular design will be attainable by the incorporation of organometallic cores with unique geometry and properties.

Objective

On the basis of these research backgrounds, we have been working on the synthesis of polymers with organometallic repeating units such as cobaltacyclopentadiene and titanacyclopentadiene moieties in the main chain. That is, new reactive polymers possessing reactive organometallic systems in the main chain have been prepared and their reactions have been performed so as to obtain organic polymers having versatile main chain functionality and other organometallic polymers.

Synthesis of Reactive Cobaltacyclopentadiene-Containing Polymers

Using the metallacyclization process of acetylene derivatives with a low-valent cobalt complex (1) described by Yamazaki et al.,[1] polymers possessing cobaltacyclopentadiene units (3) were obtained in high yields by the polymerization of CpCo(PPh$_3$)$_2$ (1) and diynes (2) in toluene at 50-60 °C.[2] Most of the polymers (3) are soluble in organic solvents and can be isolated by reprecipitation under air. The number-average molecular weight of the polymers reached over 2×10^5 if the cobalt complex (1) were purified carefully. Some of the polymers can provide brown-colored transparent film by casting from their solution. They are occasionally contaminated with cyclobutadienecobalt units (~10%). In terms of the regiochemistry of the main chain connections at each cobaltacyclopentadiene unit, the ratio of the possible three regioisomeric units (i.e., the connections through 2,5-, 2,4-, and 3,4-positions of the metallacycles) can be somewhat controlled by the substituents on the diyne monomers. For example, polymers from diynes with less sterically hindered lateral substituents such as 2b and 2c have a higher content of the 2,5-linkage (3b: ~70% and 3c: ~100%). However, the solubility of the polymers in organic solvents decreased as the 2,5-content increased.

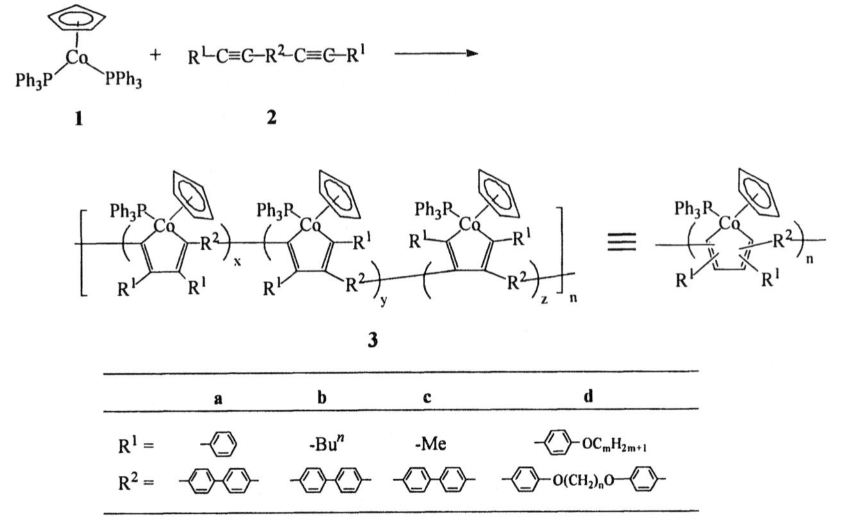

Scheme 1

In accordance with the decomposition temperature of the cobaltacyclopentadiene derivatives (e.g., 193-194 °C for a tetraphenyl substituted cobaltacyclopentadiene[1a]), the decomposition of the organocobalt polymers (3) are observable at approximately 200 °C in their thermogravimetric analyses (TGA) which can be attributed to the rearrangement of the cobaltacyclopentadiene into the cyclobutadienecobalt units by the elimination of triphenylphosphine (vide infra). No peak for either the glass transition (T_g) or the melting temperature (T_m) is observable for 3a in its differential scanning calorimetric (DSC) analysis. Using diynes with flexible spacers (2d), it is possible to design organocobalt polymers (3d) which exhibit phase transition.[3] The polymers (3d) have T_g at -20~130 °C, depending upon the length of the aliphatic spacers. Some of the polymers also exhibit T_m in the range of 60~120 °C.

Conversion into Organic Polymers with Various Main Chain Structures

As mentioned above, the organocobalt polymers are stable under air. However, they serve as novel type of reactive polymers whose main chain can be reconstructed by the polymer reactions under appropriate conditions. Organic polymers possessing various functional groups in the main chain could be produced through the chemical conversion of the metallacycle units in the organocobalt polymers (3) into organic functional groups. For example, polymers having pyridone units in the main chain (4) were produced from 3 by the reaction with isocyanates at 120 °C.[4] The content of the 2-pyridone reaches about 70% with respect to the starting cobaltacyclopentadiene units. In this case, the remaining 30% was found to be η^4-cyclobutadienecobalt units as a result of the rearrangement reaction.

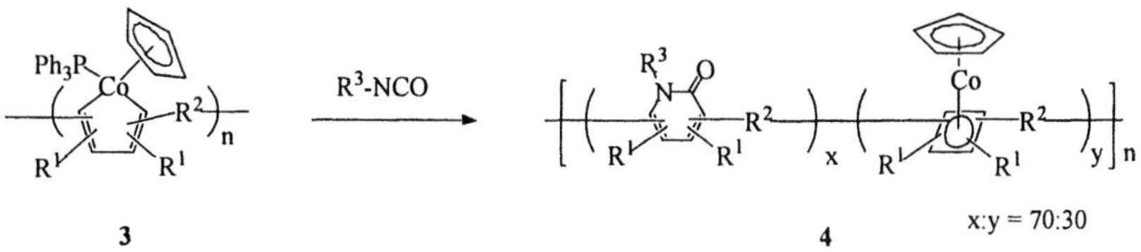

x:y = 70:30

Scheme 2

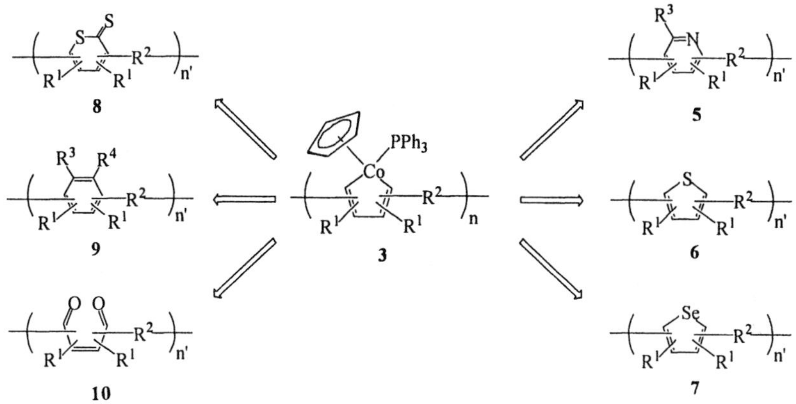

Scheme 3

As summarized in Scheme 3, the organocobalt polymers can be converted into various organic polymers with versatile functional groups in the main chain. That is, polymers containing pyridine (5), thiophene (6), selenophene (7), dithiolactone (8), phenylene (9), and diketone units (10) were obtained by the reaction with nitriles, sulfur, selenium, carbon disulfide, acetylenes, and oxygen, respectively.[5-10] The efficiency of these polymer reactions is affected by the reagents and the reaction conditions, which ranged from 70-100%. Even in the cases of polymer reactions that do not proceed in a quantitative fashion, no distinct decrease of the molecular weight of the produced polymers was observed because the lower efficiency of the polymer reactions does not mean the scission of the main chain of the polymers but the conversion into other structural units such as the cyclobutadienecobalt. In the case of the reaction that also proceeds in a catalytic fashion (e.g., the reaction with nitrile), the molecular weight of the resulting polymers increased probably by the catalytic ring forming process at the terminal acetylene groups in **3**. Starting from the organocobalt polymers with fully aromatic main chain systems, the resulting organic polymers (e.g., the thiophene-containing polymers from **3a-c**) exhibited the properties of π-conjugated oligomers judging from their UV-vis spectra and electrochemical properties, probably due to the regio-irregular main chain connection of the organocobalt polymers.

Conversion into Other Organocobalt Polymers

On heating the organocobalt polymers (**3**) in the presence of appropriate ligands such as trialkylphosphines, triphenylphosphine on the organocobalt polymers (**3**) can be replaced quantitatively by the added ligands, by which the properties such as the solubility of the polymers can be modified (Scheme 4).[11]

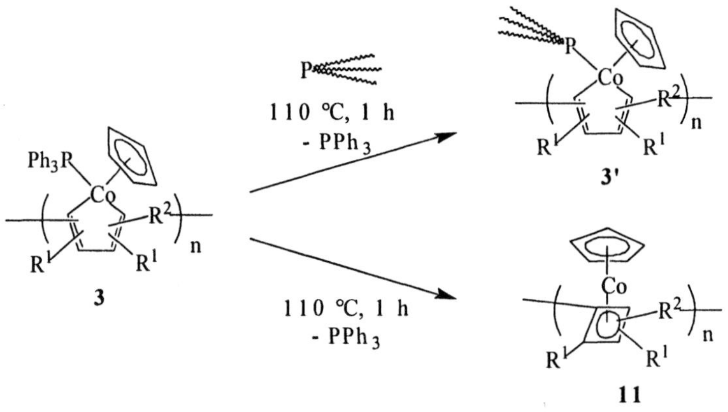

Scheme 4

In the absence of added ligands, the thermal rearrangement reaction of the polymers (**3**) took place under similar heating conditions to give yellow-colored polymers (**11**) containing η^4-cyclobutadienecobalt units.[12] This reaction seems to proceed by the dissociation of the ligand followed by the elimination of the cobalt. The cyclobutadienecobalt-containing polymers (**11**) are also stable under air and soluble in organic solvents. The polymer (**11a**) produced from **3a** exhibited good thermal stability and the 5% weight loss was observed at 480 °C in its TGA. The cobaltacyclopentadiene-containing polymers with flexible spaces (**3d**) can also be converted to cyclobutadienecobalt-containing polymers

(11d) by the thermal rearrangement. The resulting cyclobutadienecobalt-containing polymers (11d) also exhibited T_g and T_m.[3]

Similar to the cases of the conversion of the organocobalt polymers (3) into organic polymers, 3 can also be converted into other organometallic polymers by the reactions with appropriate reagents. Polymers having (η^5-cyclopentadienyl)(η^4-iminocyclopentadiene)cobalt moieties in the main chain (12) were obtained by the reaction with isocyanides (Scheme 5).[13] It is of note that the efficiency of the polymer reaction was quantitative and the polymers produced exhibit a unique solvatochromism. That is, a polymer solution exhibits a reversible color change from purple to red by varying the nature of the solvent (e.g., purple in benzene and red in methanol). This color change might be ascribable to the structural change between the neutral (η^4-iminocyclopentadiene)cobalt (12) and the zwitterionic cobalticenium unit (12'). The subsequent polymer reaction of 12 with alkyl halides gave polymers containing cobalticenium units (13) although the efficiency of the alkylation was not quantitative due to the precipitation of the polymer during the reaction. The electrochemical analysis of 13 suggested the presence of the electronic interaction between the plural organometallic centers.

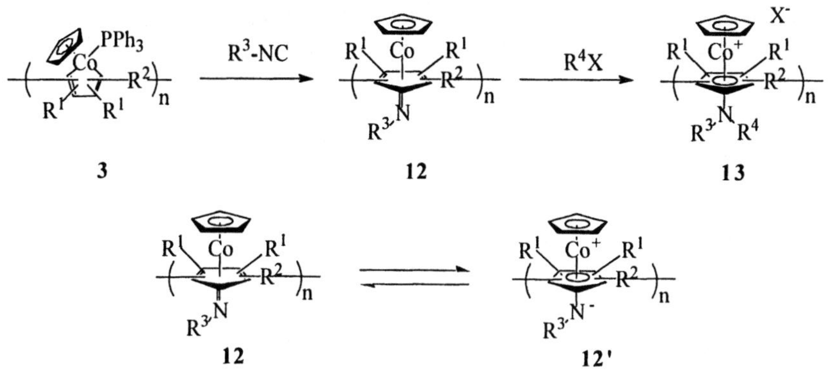

Scheme 5

It is reported that the reaction of cobaltacyclopentadiene derivatives with carbon disulfide gives unsaturated dithiolactones in moderate yields[1k] by which dithiolactone-containing polymers (8) were produced from 3 as described above. On the contrary, the addition of an equimolar amount of Co(I) complexes such as CpCo(cod) to the reaction system provided an entirely different result. In this case, the dithiolactones were not detected at all but (η^4-cyclopentadiene)cobalt complexes could be obtained in excellent yields.[14] On the basis of this reaction, analogous cobalticenium-containing polymers (14) were also obtained by the reaction of the polymers (3) with carbon disulfide in the presence of a Co(I) complex, followed by the S-alkylation (Scheme 6).[15]

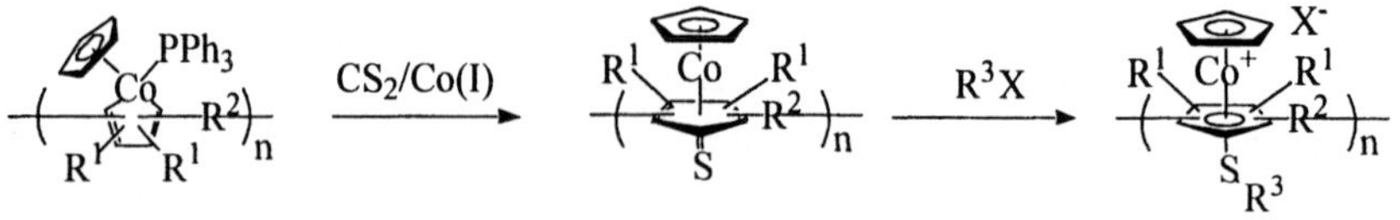

Scheme 6

Synthesis and Reactions of Titanacyclopentadiene-Containing Polymers

As reported by Tilley et al., zirconacyclopentadiene-containing polymers were prepared by means of the analogous metallacyclization process of a low-valent *bis*(cyclopentadienyl)zirconium and diynes The zirconium-containing polymers also show interesting reactivity to produce organic polymers with diene and heterocycle systems in the main chain.[16] Polymers containing other metallacycle units are attractive for creation of novel reactive polymers that can be converted to more versatile organic functional polymers. Besides, the organometallic polymers with high regioregularity must be of importance to design regioregular organic polymers through the polymer reactions. For example, organic polymers with π-conjugated backbone would be prepared from the main chain regioregular organometallic polymers.

Judging from the diverse chemistry of titanacycles which is still now in progress,[17] it is quite attractive for us to prepare polymers possessing titanacycles such as titanacyclopentadiene units in the main chain. Thus, titanacyclopentadiene-containing polymers were prepared by the titanacycle formation process described recently by Takahashi et al.[17a] That is, the polymerization of diynes (2) and a low-valent titanocene derivative (15) generated from *bis*(cyclopentadienyl)titanium dichloride gave polymers (16) which are stable at ambient temperature under argon atmosphere (Scheme 7).[18] The hydrolytic work-up of the polymers (16) gave diene-containing polymers (17, M_n = 2000~4000) in moderate yields (50-75%). The regioisomeric linkage of the main chain of 16 at the titanacycle systems must be dependent upon the diynes used. The polymers (16a and 16e) may have statistical distribution of 2,5-, 2,4-, and 3,4-linkages although these units are difficult to be distinguished by the spectroscopic methods. In the case of the polymer (16f), the regiochemistry could be determined by the model experiment. That is, the titanacycles obtained from 1-heptynylbenzene, two phenyl substituents are located at the 2,4- and the 2,5-positions (2,4-:2,5- = 90:10) judging from their hydrolysis products. Accordingly, the main chain of the polymer (16f) was supposed to be connected through the 2,4- and the 2,5- positions of the metallacycle moieties. Because of the rather low content of the 2,5-connection, an effective linkage for the π-conjugation, the polymer (**17f**) produced exhibited a relatively small red shift in the UV-vis spectrum in comparison with the model dienes. The result can also be taken to mean that the polymer (**16f**) produced from the low-valent titanocene derivative and diyne (**2f**) has the major 2,4- and the minor 2,5- connections through the titanacycle units.

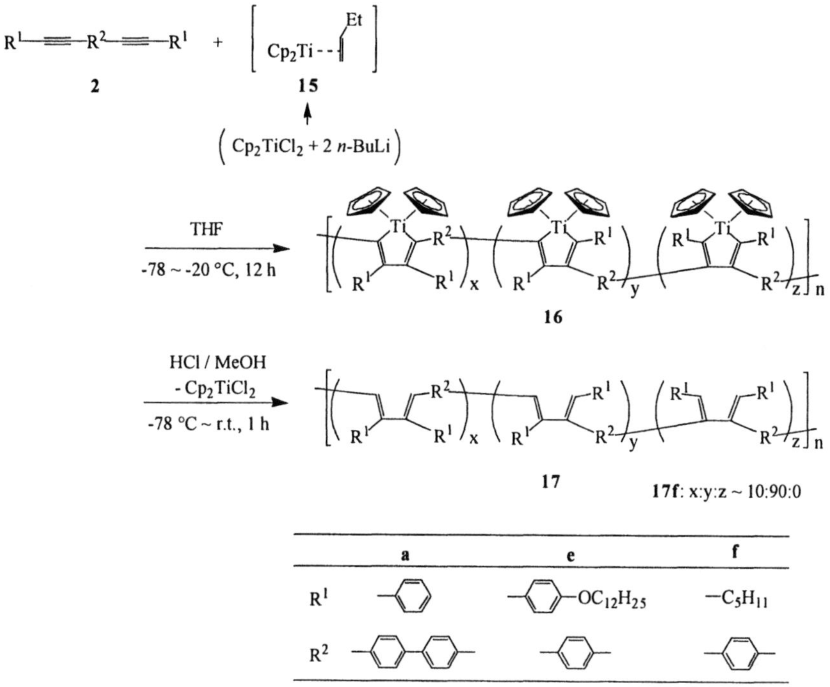

Scheme 7

Titanacyclopentadiene-containing polymers with a regiospecific main chain connection could be obtained by the polymerization of terminal diynes (**2g**) and a low-valent titanium generated from titanium(IV) isopropoxide (Scheme 8).[19] The polymerization took place at $-78 \sim -50$ °C and the polymers produced should be converted into organic polymers without isolation because they are not stable at ambient temperature. According to the report of Yamaguchi et al.,[17b] 1,4-disubstituted dienes were produced in the reaction of terminal acetylenes such as phenylacetylene followed by the hydrolysis. In accordance with their report, the polymerization of terminal diynes such as 1,4-diethynyl-2,5-dioctyloxybenzene (**2g**) gave polymers with regioregular backbone (**18**). The hydrolysis or the iodination of the titanacycle units yielded polymers with diene units (**19** and **20**, respectively). For example, the treatment of the polymer (**18**) with iodine provided an iodinated diene-containing polymer (**20**) in 81% yield (M_n = 7,700). Owing to the alkoxy substituents, both the polymers (**19** and **20**) are soluble in organic solvents.

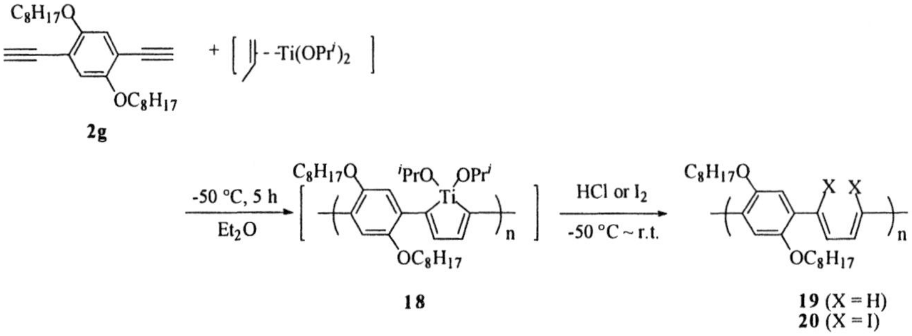

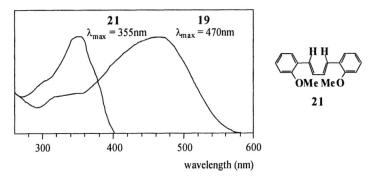

Scheme 8

*Figure 1. UV-vis spectra of the diene-containing polymer (**19**) and its model compound (**21**) (measured in $CHCl_3$).*

The regioregular main chain connection could be convinced from the properties of the diene-containing polymers. The polymer (**19**) produced by the hydrolysis of the titanacyclopentadiene-containing polymer exhibits a clear bathochromic shift of the UV-vis absorption compared to that of a model compound (**21**) (Figure 1). That is, the absorption maximum (λ_{max}) of the polymer (λ_{max} = 470 nm) appeared at longer wavelength by 115 nm than that of the model compound (λ_{max} = 355 nm). The polymers also exhibited the luminescence upon irradiation of the UV-vis light.

A poly(p-phenylene) derivative (**22**) was produced by the reaction of the titanacyclopentadiene-containing polymer (**18**) with propargyl bromide (Scheme 9).[20] Similar to the case of the chemical conversion into diene-containing polymers, the polymer reaction proceeded smoothly under very mild conditions (-50 °C to ambient temperature) and a yellow powdery polymer produced is soluble in organic solvents (~80% yield, M_n ~5,000). Because of the π-conjugated backbone, the poly(p-phenylene) derivative (**22**) obtained in this study has the UV-vis absorption in longer wavelength range (λ_{max} = 329 nm) compared to that of a model compound, a p-terphenyl derivative produced via the titanacyclopentadiene (2,5-*bis*(2-methoxyphenyl)toluene) (λ_{max} = 276 nm). The polymer also exhibited the photoluminescence whose emission maximum (E_{max}) also shifted to the longer wavelength region (E_{max} = 445 nm) with respect to that of the model compound (E_{max} = 407 nm).

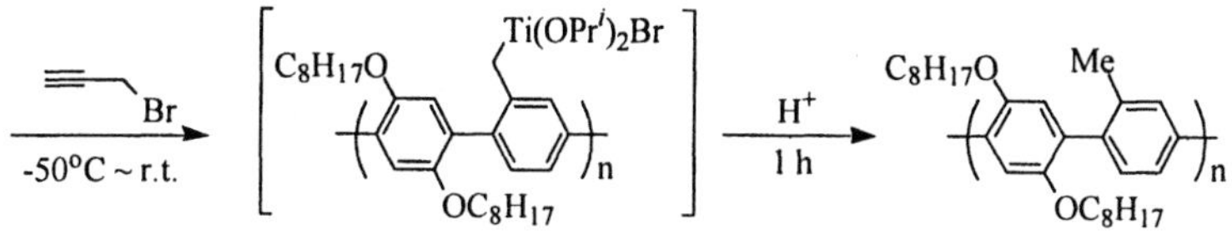

Scheme 9

A thiophene-containing polymer (**23**) is likewise produced by the reaction of the titanacyclopentadiene-containing polymer (**18**) with sulfur monochloride under mild conditions (Scheme 10).[21] The yield of the soluble fraction was a little lower (~ 50%), because the produced polymer is partially insoluble in organic solvents. The brown powdery polymer thus obtained exhibited a substantial bathochromic shift in the UV-vis absorption spectrum (λ_{max} = 386 nm) in comparison with that of a model compound, a 2,5-*bis*(2-methoxyphenyl)thiophene (**24**, λ_{max} = 287 nm). Both the polymer (**23**) and the model compound (**24**) exhibited the photoluminescence (E_{max} = 345 nm and 482 nm, respectively).

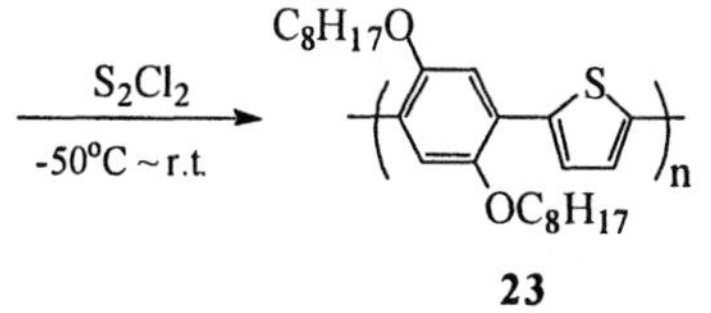

Scheme 10

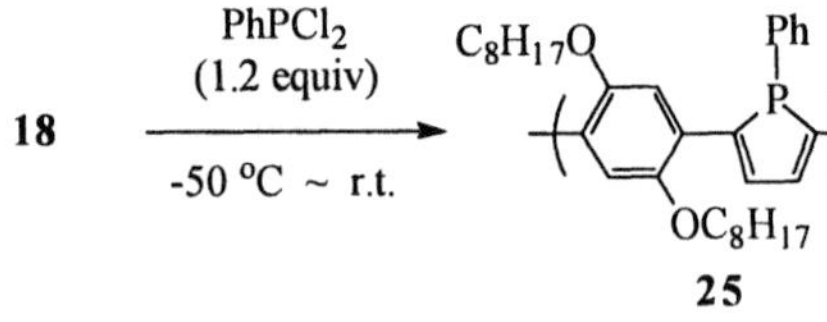

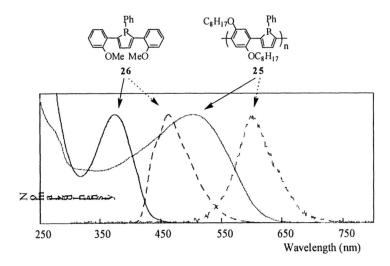

Scheme 11

Figure 2. UV-vis (——) and photoluminescence (·····) spectra of the phosphole-containing polymer (25) and its model compound (26) (measured in CHCl₃).

The titanacyclopentadiene-containing polymer (18) can be converted into a π-conjugated phosphole-containing polymer (25) by the reaction with dichlorophenylphosphine (Scheme 11).[22] Similar to the case of the above-mentioned conversion reaction into thiophene-containing polymer, the produced polymer (25) is partially insoluble in organic solvents. The polymers thus obtained revealed λ_{max} at 504 nm in its UV-vis spectrum which is bathochromically shifted by 130 nm in comparison with that of a model compound (26), supporting the π-conjugated character of the resulting phosphole-containing polymer (Figure 2). In the photoluminescence spectra, the polymer emitted a yellow light (E_{max} = 601 nm) with a quantum yield (Φ) of 0.1 while the model compound showed a blue emission (E_{max} = 463 nm, Φ = 0.55).

Conclusions

Organometallic polymers possessing reactive sites in the main chain have been successfully prepared by the reactions of diynes with low-valent transition metal complexes such as cobalt(I) and titanium(II). The cobaltacyclopentadiene-containing polymers are stable under air and can be handled as if they are conventional organic polymers. However, they exhibit versatile reactivity under appropriate reaction conditions. That is, the organocobalt polymers serve as reactive precursors for a variety of polymers containing functional groups such as heterocycles, benzene rings, and unsaturated ketones in their repeating units. The organocobalt polymers can also be converted to other organocobalt polymers with unique building blocks such as cyclobutadienecobalt and cobalticenium units. The titanacyclopentadiene-containing polymers are likewise obtained by the metallacyclization process of diynes with low-valent titanium complexes. Although the two series of the titanacyclopentadiene-containing polymers are not stable under air, they also serve as reactive precursors for organic polymers with versatile main chain functionality. Especially, the titanacycle-containing polymers from titanium(IV) isopropoxide with regioregular main chain connection at each titanacyclopentadiene unit are found to serve as precursors of π-conjugated organic polymers.

References

1. (a) Yamazaki, H.;, Wakatsuki, Y. *J. Organomet. Chem.* **1977**, *139*, 157. (b) Wakatsuki, Y.; Nomura, O.; Kitaura, K.; Morokuma, K.; Yamazaki, H. *J. Am. Chem. Soc.* **1983**, *105*, 1907. (c) Wakatsuki, Y.; Yamazaki, H. *J. Chem. Soc., Dalton Trans.* **1982**, 1923. (d) Hong, P.; Yamazaki, H. *Synthesis* **1977**, 50. (e) Yasufuku, K.; Hamada, A.; Aoki, K.; Yamazaki, H. *J. Am. Chem. Soc.* **1980**, *102*, 436. (f) Yamazaki, H.; Hagihara, N. *J. Organomet. Chem.* **1967**, *7*, P22. (g) Wakatsuki, Y.; Yamazaki, H. *Tetrahedron Lett.* **1973**, 3383. (h) Wakatsuki, Y.; Yamazaki, H. *Synthesis* **1976**, 26. (i) Wakatsuki, Y.; Yamazaki, H. *J. Chem. Soc., Dalton Trans.* **1978**, 1278. (j) Yamazaki, H.; Wakatsuki, Y. *Bull. Chem. Soc. Jpn.* **1979**, *52*, 1239. (k) Wakatsuki, Y.; Yamazaki, H. *J. Chem. Soc., Chem. Commun.* **1973**, 280.
2. (a) Tomita, I.; Nishio, A.; Igarashi, T.; Endo, T. *Polym. Bull.* **1993**, *30*, 179. (b) Lee, J.-C.; Nishio, A.; Tomita, I.; Endo, T. *Macromolecules* **1997**, *30*, 5205.
3. Rozhanskii, I. L.; Tomita, I.; Endo, T. *Macromolecules* **1996**, *29*, 1934.
4. Tomita, I.; Nishio, A.; Endo, T. *Macromolecules* **1995**, *28*, 3042.
5. Lee, J.-C.; Tomita, I.; Endo, T. *Polym. Bull.* **1997**, *39*, 415.
6. Lee, J.-C.; Tomita, I.; Endo, T. *Macromolecules* **1998**, *31*, 5916.

7. (a) Lee, J.-C.; Tomita, I.; Endo, T. *unpublished results.* (b) Lee, J.-C.; Tomita, I.; Endo, T. *Polym. Prepr. Jpn.* **1996**, *45*, 1463.
8. (a) Tomita, I.; Lee, J.-C.; Nishio, A.; Endo, T. *unpublished results.* (b) Tomita, I.; Nishio, A.; Endo, T. *Pacific Polym. Prepr.* **1993**, *3*, 661. (c) Lee, J.-C.; Tomita, I.; Endo, T. *Polym. Prepr. Jpn.* **1997**, *46*, 1617.
9. (a) Lee, J.-C.; Tomita, I.; Endo, T. *unpublished results.* (b) Lee, J.-C.; Tomita, I.; Endo, T. *Polym. Prepr. Jpn.* **1998**, *47*, 1760.
10. (a) Lee, J.-C.; Tomita, I.;, Endo, T. *unpublished results.* (b) Lee, J.-C.; Tomita, I.; Endo, T. *Polym. Prepr. Jpn.* **1997**, *46*, 1619.
11. Tomita, I.; Nishio, A.; Endo, T. *Appl. Organometal. Chem.* **1998**, *12*, 735.
12. Tomita, I.; Nishio, A.; Endo, T. *Macromolecules* **1994**, *27*, 7009.
13. Tomita, I.; Lee, J.-C.; Endo, T. *J. Organomet. Chem.* **2000**, *611*, 570.
14. Lee, J.-C.; Tomita, I.; Endo, T. *Chem. Lett.* **1998**, 121.
15. (a) Lee, J.-C.; Tomita, I.; Endo, T. *unpublished results.* (b) Lee, J.-C.; Tomita, I.; Endo, T. *Polym. Prepr. Jpn.* **1997**, *46*, 1617.
16. (a) Mao, S. S. H.; Tilley, T. D. *J. Am. Chem. Soc.* **1995**, *117*, 5365. (b) Mao, S. S. H.; Tilley, T. D. *Macromolecules* **1997**, *30*, 5566.
17. (a) Sato, K.; Nishihara, Y.; Huo, S.; Xi, Z.; Takahashi, T. *J. Organomet. Chem.* **2000**, *633*, 18. (b) Yamaguchi, S.; Jin, R. Z.; Tamao, K.; Sato, F. *J. Org. Chem.* **1998**, *63*, 10060. (c) Block, E.; Birringer, M.; He, C. *Angew. Chem. Int. Ed. Engl.* **1999**, *38*, 1604. (d) Suzuki, D.; Tanaka, R.; Urabe, H.; Sato, F. *J. Am. Chem. Soc.* **2002**, *124*, 3518. (e) Tanaka, R.; Hirano, S.; Urabe, H.; Sato, F. *Org. Lett.* **2003**, *5*, 67.
18. (a) Ueda, M.; Tomita, I. *unpublished results.* (b) Ueda, M.; Tomita, I. *Polym. Prepr. Jpn.* **2002**, *51*, 1259.
19. (a) Tomita, I. Atami, K.; Endo, T. Ueda, M.; Utsumi, T. *unpublished results.* (b) Tomita, I.; Atami, K.; Endo, T. *Polym. Prepr. Jpn.* **1999**, *48*, 341. (c) Tomita, I.; Atami, K.; Endo, T. *Polym. Prepr. Jpn.* **2000**, *49*, 1641.
20. (a) Utsumi, T.; Tomita, I. *unpublished results.* (b) Utsumi, T.; Tomita, I. *Polym. Prepr. Jpn.* **2002**, *51*, 267.
21. (a) Utsumi, T.; Tomita, I. *unpublished results.* (b) Utsumi, T.; Tomita, I. *Polym. Prepr. Jpn.* **2002**, *51*, 1323.
22. (a) Ueda, M.; Tomita, I. *unpublished results.* (b) Ueda, M.; Tomita, I. *Polym. Prepr. Jpn.* **2003**, *52*, 1255.

Chapter 37

Coordination Compounds for Functional Nonlinear Optics: Enhancing and Switching the Second-Order Nonlinear Optical Responses

Inge Asselberghs[1], Michael J. Therien[2], Benjamin J. Coe[3], Jon A. McCleverty[4], and Koen Clays[1,*]

[1]Department of Chemistry, Katholieke Universiteit Leuven, Celestijnenlaan 200D, B3001 Leuven, Belgium
[2]Department of Chemistry, University of Pennsylvania, Philadelphia, PA 19104–6323
[3]Department of Chemistry, University of Manchester, Oxford Road, Manchester M13 9PI, United Kingdom
[4]School of Chemistry, University of Bristol, Cantock's Close, Bristol BS8 1TS, United Kingdom

In this work, we describe the second-order nonlinear optical (NLO) properties of a number of chromophores that feature transition metal ions in classic coordination environments. We focused our attention on the advantages of these species over standard hyperpolarizable chromophores based on conventional all-organic frameworks. For example, studies of Ruthenium(II)-based electron donor-acceptor (D-A) polyenes illustrate that transition metal-based compounds can show atypical conjugation-length dependences of the observed hyperpolarizability relative to closely related organic NLO chromophores. Likewise, the second-order NLO responses observed for highly conjugated (polypyridyl)metal-(phorphinato)zinc(II) chromophores can be extraordinarily large, illustrating how coupled oscillator photophysics can be exploited to design materials with record hyperpolarizabilities at telecommunication-relevant wavelengths. Finally, our work demonstrates that the presence of a transition metal ion in NLO chromophores makes possible new strategies to switch and gate NLO responses voltammetrically.

527

Introduction

The linear optical properties of coordination compounds of transition-metal ions have been widely investigated.[1,2,3] These properties are strongly influenced by the electronic nature of the metal, the ligands and the remaining organic moiety that together constitute the chromophore. In coordination compounds, transition metal ions can serve as electron-releasing or -withdrawing agents, or as a constituent of the molecular bridge that provides electron coupling between other electron donating or accepting groups. As such, coordination compounds are often highly colored, featuring optical absorptions in the UV-Vis-NIR spectral regions that are either intra-valent (IVCT), metal-to-metal (MMCT), ligand-to-metal (LMCT), or metal-to-ligand (MLCT) charge transfer in origin. Further, if these transitions are of appropriate oscillator strength and the chromophores noncentrosymmetric, such species are also ideally suited for characterization using second-order NLO spectroscopic methods.

According to the two-level model[4,5] the static hyperpolarizability β_0 can be given by:

$$\beta_0 = \frac{3\,\Delta\mu_{12}\left(\mu_{12}\right)^2}{\left(E_{\max}\right)^2}$$

where μ_{12} is the transition dipole moment, $\Delta\mu_{12}$ is the dipole moment change, and E_{max} is the maximal energy of the CT absorption. β_0 is a measure of the intrinsic NLO response under non-resonant conditions. Therefore, such complexes can form the molecular basis of active photonic components, such as optical frequency-doublers or electro-optic modulators. Taking advantage of properties inherent to chromophoric coordination compounds has led to the design of supermolecules that exhibit extremely large first hyperpolarizabilities at telecommunication-relevant wavelengths, as well as systems that offer the capability for molecular-level switching of the first hyperpolarizability.

Experimental – Materials and Techniques

Materials. The syntheses of the compounds discussed in this review have been previously reported. For all the optical measurements – linear and nonlinear – the products were dissolved in the appropriate solvents. All solvents were used as obtained from ACROS. Linear absorption spectra were taken before and after the nonlinear optical measurements to ensure temporal and optical stability of the chromophores. For the (spectro)electrochemical experiments, an optically passive (transparent) electrolyte Bu_4NPF_6 was used to ensure conductivity. In this case the solvent (dichloromethane) was dried (P_2O_5) and distilled.

Techniques. The linear optical spectra were taken on a Perkin-Elmer Lambda 900 spectrometer. The first hyperpolarizabilities β (= second-order nonlinear molecular polarizability) for these compounds were determined by the hyper-Rayleigh scattering (HRS) method. This incoherent scattering technique is used because of the ionic nature of the compounds. The measurements are performed at 1064 nm.[6,7] The external reference method was used to determine the β-values of the compounds and p-nitroaniline (PNA) was used as standard. However, it was sometimes necessary to be able to discriminate between scattered (HRS) and emitted (by multi-photon fluorescence, MPF) photons. Therefore femtosecond HRS was used.[8,9] With such short femtosecond pulses, temporal resolution between immediate (nonlinear) scattering and the time-delayed (multi-photon) fluorescence is possible. Our instrument is the Fourier-transform implementation in the frequency-domain of this principle in the time-domain. These measurements are performed at 800 nm and/or 1300 nm. The standard was Crystal Violet (800 nm) and Disperse Red 1 (1300 nm). The (spectro)electrochemistry was performed using a Perkin-Elmer model 263A potentiostat. In-situ electrochemical switching of the molecular hyperpolarizability in combination with real-time hyper-Rayleigh scattering is performed in specially designed electrochemical optical cells.[10]

Results

Enhancing the value of the first hyperpolarizability. The first successful strategy for enhancing the first hyperpolarizability for this material type was to change from the historically used metallocenes with an out-of-plane charge-transfer to the compounds with an in-plane charge-transfer.[11,12]

Extension of the conjugated π-electron bridge between donor and acceptor moieties has been shown, both theoretically and experimentally[13,14] to lead to a maximum hyperpolarizability in metalorganic complexes for relatively short lengths. A study of *trans*-[$Ru^{II}(NH_3)_4(L^D)(L^A)$]$^{3+}$ (L^D = electron-rich ligand, L^A = pyridyl pyridinium ligand) (see Figure. 1) on the effects of the polyene chain extension on the linear absorption and NLO properties has revealed the phenomenon of blue-shifting the MLCT band with the addition of each CH=CH unit (Table 1). A decrease in the static hyperpolarizability is noticed with the extension beyond 2 double bonds (Figure 2). The *trans*-1,3-butadienyl bridge is optimal according to our measurements. This is in contrast to related purely organic dyes, for which β_0 increases up to at least 8 double bonds.[15]

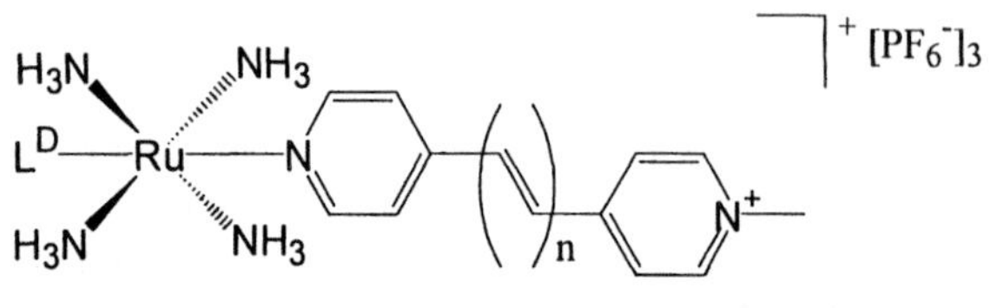

n = 0, 1, 2, 3; L^D = NH$_3$, 1-methylimidazole(mim)

Figure 1. Structures of trans-[Ru^{II}(NH$_3$)$_4$(L^D)(L^A)]$^{3+}$ *investigated*

Table 1. Linear Absorption Data for *trans*-[Ru^{II}(NH$_3$)$_4$(L^D)(L^A)]$^{3+}$

Compound L^D	n	λ_{max} / nm	ε / dm^3 mol^{-1} cm^{-1}
NH$_3$	0	590	15800
NH$_3$	1	595	16100
NH$_3$	2	584	18700
NH$_3$	3	568	17500
mim	0	602	16200
mim	1	604	16200
mim	2	592	21400
mim	3	570	21900

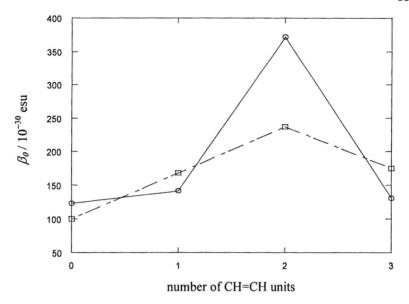

Figure 2. Plot of the Static Hyperpolarizability β_0 as a Function of the Number of CH=CH units Present in the Molecule

Enhancement of β_0 can also be achieved by changing the co-ligands, in order to increase the electron donating strength of the Ru(II) center.[16] For example, replacement of the neutral ammine group in the trans position by the thiocyanato anion results in a bathochromic shift of the MLCT absorption (see Table 2). This trend is obviously consistent with the expectation that the thiocyanate is a stronger electron donor when compared with NH_3 and therefore causes the largest destabilization of the Ru(II)-based HOMO. The HRS measurements were performed at 1064 nm and the estimated static hyperpolarizabilities β_0 were obtained by application of the two-state model. The data show larger β_0 values for the compounds with the thiocyanato ligands than for their counterparts with the ammine ligands. These increases can be traced to lower transition energies E_{max} (or a larger λ_{max} value) and higher transition dipoles μ_{12} (or a larger extinction coefficient ε).

Table 2. Example of the Influence of the Extension of the π-bridge and Donor Ligand in _trans_-[Ru^{II}(NH$_3$)$_4$(L^D)(L^A)]$^{n+}$ (n = 2 or 3) Complexes

Compound L^D	L^{A*}	λ_{max} / nm ε / dm^3 mol^{-1} cm^{-1}	β_{1064} / 10^{-30} esu β_0 / 10^{-30} esu
NH$_3$	MeQ$^+$	590	750
		15800	123
SCN$^-$	MeQ$^+$	628	963
		20000	247
NH$_3$	PhQ$^+$	628	858
		19300	220
SCN$^-$	PhQ$^+$	672	957
		21300	343

* MeQ$^+$ = N-methyl-4,4'-bipyridinium; PhQ$^+$ = N-phenyl-4,4'-bipyridinium

Coupled-oscillators for Unusual Frequency Dispersion Effects and Large-Magnitude Dynamic Hyperpolarizabilities. An alternative strategy shows that appropriate coupling of multiple charge transfer oscillators can give rise to supramolecular systems which feature substantial dynamic hyperpolarizabilities (β_λ).[17] Examples of such supermolecular NLO chromophores are shown in Figure 3. These systems exploit an ethyne-elaborated, highly polarizable porphyrinic component and a metal polypyridyl complex that serves as an integral donor (D) and acceptor (A) element. The rigid cylindrically π-symmetric connectivity between the [porphinato]zinc(II) and the metal(II)polypyridyl units in these systems align the CT transition dipoles of the chromophoric components in a head-to-tail arrangement. Thus, coupled oscillator photophysics and metal-mediated cross-coupling can be exploited to elaborate high β_λ supermolecules (Table 3). For example, the β_{1300} value determined for Ru-PZn is 2.5 times larger than that determined for any other chromophore at this wavelength, and highlights that this design strategy enables fabrication of new classes of chromophores that possess extraordinarily large β_λ values at telecommunication-relevant wavelengths. It is also important to appreciate that coupling of multiple oscillators does not necessarily give rise to supermolecules in which every CT transition possesses a $\Delta\mu_{12}$ value of identical sign. Such spectroscopic characteristics can give rise to chromophores that manifest an oscillatory dependence of the dynamic hyper-polarizability as the incident irradiation wavelength is varied. RuPZnOs (Table 3) features porphyrin B-and Q-state-derived transitions that have $\Delta\mu_{12}$ values of opposite sign; thus in addition to the frequency dependent sign of the resonance enhancement factor, appropriate engineering of the wavelength-specific relative contribution and sign of $\Delta\mu_{12}$ for a given CT transition constitutes an additional tool to tune the magnitude of the dynamic hyperpolarizability, which can potentially enable the development of novel materials with enhanced and more selective NLO properties.

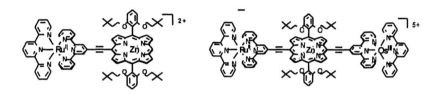

Ru-PZn   **Ru-PZn-Os**

Figure 3. Structures of Molecules Showing Coupled-Oscillator Photophysics

Table 3. Dynamic Hyperpolarizabilities (β_λ) of [porphinato]zinc(II) complexes

compound	β_{800} / 10^{-30} esu	β_{1064} / 10^{-30} esu	β_{1300} / 10^{-30} esu
Ru-PZn	<50*	2100	5100
Ru-PZn-Os	240	4500	860

* upper limit corresponds to the smallest HRS signal that can be measured accurately

Reversible switching of the first hyperpolarizability. Several switching schemes at the molecular level have been envisioned. Earlier schemes used isomerization and tautomerization, causing changes in the nature and/or degree of conjugation between electron donor and acceptor. Another approach is based on lowering the donating capacity of the electron donor, or the withdrawing capacity of the acceptor group. This has been achieved by protonating a dimethylaminophenyl donor group on an azafulleroid. The perturbation is created by adding a strong acid (trifluoroacetic acid) and can be reversed by adding a strong base (Hünig's base). A reversible molecular switch based on pH has therefore been demonstrated.[18]

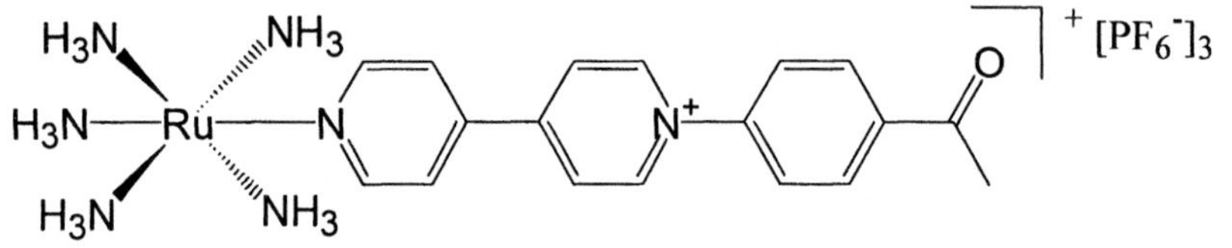

Figure 4. Structure of pentaammine ruthenium(II)

A more attractive approach is to use oxidation to lower the electron donating power of a metal that is involved in a MLCT process. This can be achieved by adding an appropriate oxidant, and the process can be reversed by adding a suitable reductant. This was first realized with a Ru(II)-pentaammine donor-based complex (for example, see Figure 4 in which the acceptor ligand has a N-acetylphenylpyridinium unit). The oxidation of the metal was achieved by adding hydrogen peroxide to the solution. This is reflected in the total bleaching of the MLCT band (see Figure 5(a)). At the same time the HRS signal completely vanishes (see Figure 5(b)). The re-reduction is performed by adding hydrazine. The original compound is re-gained which results in the re-appearance of the MLCT band and the HRS signal.[19] The error bars indicate 15% error.

A similar switching effect has been found for a nitrothiophene-substituted octamethylferrocene complex (Figure 6).[20] The oxidized octamethylferrocenium form was also isolated and chemically-induced redox-switching experiments were carried out with each starting material. However in this case the oxidation was done by addition of Bu_4NBr_3 and the reduction using hydrazine. The Fe(II) form has a strong green color due to absorptions at 644 nm and 447 nm, which are ascribed to two different D→A charge-transfer transitions. In the oxidized Fe(III) complex the CT transitions are replaced by a characteristic weak LMCT transition of the octamethylferrocenium unit at 851 nm, resulting in a yellow color (Figure 7). The switching effect is illustrated in Figure 7b which shows the alternation in the HRS signal by alternately oxidizing and re-reducing the Fe(II) form (solid lines). The dotted lines are the results obtained starting with the Fe(III) form and alternately reducing and oxidizing it.

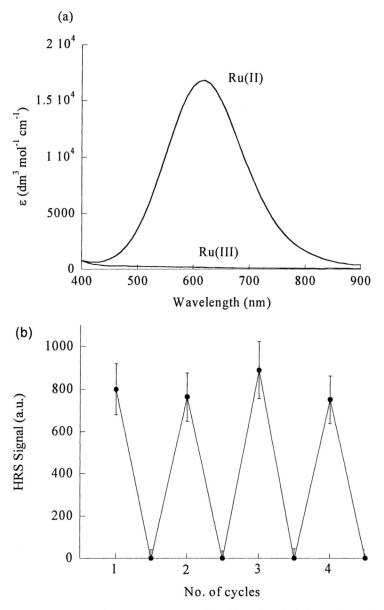

Figure 5. (a) Linear absorption spectra of Ru(II) and Ru(III) form; (b) redox-switching between Ru(II) and Ru(III) measured by HRS

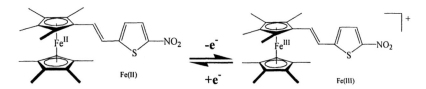

Figure 6. Structures of Molecules used for Combined Electrochemistry and Hyper-Rayleigh Scattering Experiments.

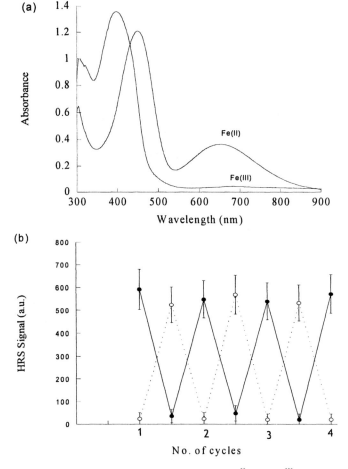

Figure 7. (a) Absorption Spectra of the Fe^{II} and Fe^{III} Compound (conc. $4.00 \times 10^{-5}M$, optical path 6 cm) (b) redox switching between Fe^{II} and Fe^{III} (conc. $4.00 \times 10^{-5}M$).

Due to their readily accessible redox potentials and the stability of the compound in both reduced and oxidized form, this system is very attractive for performing in-situ electrochemistry and hyper-Rayleigh scattering.[21] Some restraints are imposed on the design of a combined electrochemistry and HRS cell. The optical path should remain free (x-axis) as to be able to focus a high power beam into the cell. At the same time, the optical path should be free also in the perpendicular direction of the incoming beam (y-axis). Evenso, it is important that the necessary electrodes (rotating Pt-working electrode, a Pt-counter electrode and a Ag/Ag$^+$ reference electrode) do not induce additional scattering. Therefore, a custom cell was designed as shown in Figure 8. Note that the large volume size (15 cm^3) is necessary to meet all the previously mentioned conditions. This large volume size is also responsible for the long operating times required for complete oxidation or reduction.

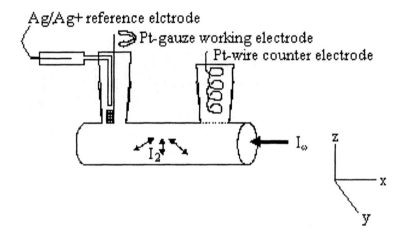

Figure 8. Schematic View of the Combined Electrochemistry/HRS cell

The oxidation was performed at a potential of 0.8 V for about 30 min. and the reduction at -0.7 V for 25 min. The switching of the second-order nonlinear response was in accordance with the variation of the linear optical data (Figure 9). The MLCT band disappears together with the hyper-Rayleigh signal and both signals re-appear upon reduction of the compound. Taking into account the difference in resonance enhancement, the dispersion-free value for the first hyperpolarizability can be switched between 100 and 10 × 10^{-30} esu.

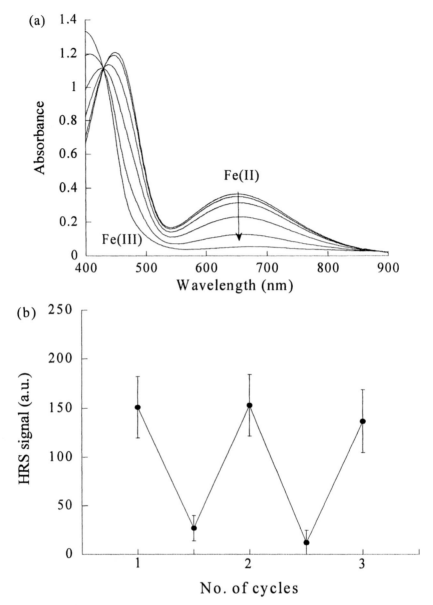

Figure 9. (a) Linear Absorption Data for Electrochemical Oxidation of Fe(II) to F(III) (conc. 4.00 × 10⁻⁵ M; Optical Path Length 6 cm) (b) Electrochemical Switching of the HRS response between Fe(II) and Fe(III).

Conclusion and Outlook

High-stability coordination compounds of transition metal ions offer an attractive alternative to conventional NLO chromophores based on traditional all-organic frameworks. Indeed, comparative studies have shown that Ru(II) pyridyl ammine complexes have larger β_0 responses than their 4-(dimethylamino)phenyl counterparts.[22,23] Additionally, appropriate coordination compounds can replace both the electron releasing dialkylamino or electron-withdrawing nitro groups of classic NLO chromophores to provide hyper-polarizable structures featuring augmented thermal stability. While both non-centrosymmetric organic and coordination compounds can manifest low-lying CT transitions that feature large transition dipole moments, coordination compound-derived supermolecular chromophores possess electronic structural features that make possible new opportunities to both enhance and modulate the magnitude of the molecular first hyperpolarizability. Such insights have enabled the design of new classes of chromophores that possess the largest hyper-polarizabilities yet determined at practical telecommunication wavelengths, and make possible new strategies to switch and gate NLO responses voltammetrically.

Acknowledgments

This research was supported by the National Fund for Scientific Research (FWO-V, G.0297.04), the Belgium government (GOA/2000/03) and the KULeuven (IUAP P51/03) (IA and KC). IA is a Postdoctoral fellow of the FWO-Vlaanderen. MJT thanks the Office of Naval Research for generous financial support. Studies carried out by the group of Coe were supported by the UK Engineering and Physical Sciences Research Council (grants GR/L56213, GR/M93864 and GR/R54293). The EPSRC (UK) and the COST organization, through Action D-14-WG0011, is also thanked for support (JAM).

References

1. Cheng L.T.; Tam, W.; Meredith, G.R.; Marder, S.R. *Mol. Cryst. Liq. Cryst.* **1990**, *189*, 137-153.
2. Calabrese, J.C.; Cheng, L.T.; Green, J.C.; Marder, S.R.; Tam, W. *J. Am. Chem. Soc.* **1991**, *113(19)*, 7227-7232.
3. Whittall, I.R.; McDonagh, A.M., Humphrey, M.G.; Samoc, M. *Adv. Organomet. Chem.* **1998**, *42*, 291-362.
4. Oudar, J.L.; Chemla, D.S. *J. Chem. Phys.* **1977** , *66(6)*, 2664-2668
5. Oudar, J.L. *Chem. Phys.* **1977**, *67*, 446-457.
6. Clays, K.; Persoons, A. *Rev. Sci. Instrum.* **1992**, *63(6)*, 3285-3289.

7. Hendrickx, E.; Clays, K.; Persoons, A. *Acc. Chem. Res.* **1998**, *31(10)*, 675-683.

8. Olbrechts, G.; Strobbe, R.; Clays, K.; Persoons, A. *Rev. Sci. Instrum.* **1998**, *69(6)*, 2233-2241.

9. Olbrechts, G.; Wostyn, K.; Clays, K.; Persoons, A. *Opt. Lett.* **1999**, *24(6)*, 403-405.

10. Asselberghs, I.; Clays, K.; Persoons, A.; McDonagh, A.M.; Ward, M.D.; McCleverty, J.M. *Chem. Phys. Lett.* **2003**, *368(3-4)*, 408-411.

11. Long, N.J. *Angew. Chem. Int. Ed. Engl.* **1995**, *34(7)*, 826-826.

12. Malaun, M.; Kowallick, R.; McDonagh, A.M.; Marcaccio, M.; Paul, R. L.; Asselberghs, I.; Clays, K.; Persoons, A.; Bildstein, B.; Fiorino, C.; Nunzi, J.-M.; Ward, M.D.; McCleverty, J.A. *J. Chem. Soc., Dalton Trans.* **2001**, *2001(20)*, 3025-3038.

13. Coe, B.J.; Jones, L.A.; Harris, J.A.; Brunschwig, B.S.; Asselberghs, I.; Clays, K.; Persoons, A.; Garín, J.; Orduna, J. *J. Am. Chem. Soc.* **2003**, *125(4)*, 862-863.

14. Coe, B.J.; Jones, L.A.; Harris, J.A.;Brunschwig, B.S.; Asselberghs, I.; Clays, K.; Persoons, A.; Garín, J.; Orduna, J. *J. Am. Chem. Soc*; **2004**, *126(12)*, 3880-3891.

15. Barzoukas, M.; Blanchard-Desce, M.; Josse, D.; Lehn, J.-M. ;Zyss, J. *Chem. Phys.* **1989**, *133(2)*, 323-329.

16. Coe, B.J.; Jones, L.A.; Harris, J.A.; Sanderson, E.E.; Brunschwig, B.S.; Asselberghs, I.; Clays, K.; Persoons, A. *J. Chem. Soc., Dalton Trans.* **2003**, *2003(11)*, 2335-2341.

17. Uyeda, H.T.; Zhao, Y.; Wostyn, K.; Asselberghs, I.; Clays, K.; Persoons, A.; Therien, M.J. *J. Am. Chem. Soc.* **2002**, *124(46)*, 13806-13813.

18. Asselberghs, I.; Zhao, Y.; Clays, K.; Persoons, A.; Comito, A.; Rubin, Y. *Chem. Phys. Lett.* **2002**, *364(3-4)*, 279-283.

19. Coe, B.J.; Houbrechts, S.; Asselberghs, I.; Persoons, A. *Angew. Chem., Int. Ed.* **1999**, *38(3)*, 366-369.

20. Malaun, M.; Reeves, Z. R.; Paul, R.L.; Jeffery, J.C.; McCleverty, J. A.; Ward, M.D.; Asselberghs, I.; Clays, K.; Persoons, A. *Chem. Commun.* **2001**, *(01)*, 49-50.

21. Asselberghs, I.; Clays, K.;Persoons, A.; McDonagh, A.M.; Ward, M.D.; McCleverty, J.A. *Chem. Phys. Lett.* **2003**, *368(3-4)*, 408-411.

22. Coe, B.J.; Harris, J.A.; Clays, K.; Persoons, A.; Wostyn, K.; Brunschwig, B.S. *Chem. Comm.* **2001**, *(17)*, 1548-1549.

23. Coe, B.J.; Harris, J.A.; Brunschwig, B.S.; Garín, J.; Orduna, J.; Coles, S.J.; Hursthouse, M.B. *J. Am. Chem. Soc.*, **2004**, *126(33)*, 10418-10427.

Author Index

Subject Index

568